Emily Sparkes

Understanding Mammalian Locomotion

UNDERSTANDING MAMMALIAN LOCOMOTION

Concepts and Applications

Edited by
JOHN E.A. BERTRAM

WILEY Blackwell

For general information on our other products and services or for technical support, please contact our Customer Care Department within the United States at (800) 762-2974, outside the United States at (317) 572-3993 or fax (317) 572-4002.

Wiley also publishes its books in a variety of electronic formats. Some content that appears in print may not be available in electronic formats. For more information about Wiley products, visit our web site at www.wiley.com.

Library of Congress Cataloging-in-Publication data applied for

ISBN: 9780470454640 (Hardback)
9781119113737 (EPDF)
9781119113720 (epub)
9781119113713 (oBook)

Cover image: Runner Skeleton
Credit: Eraxion

Caption: 3D rendered illustration – runner anatomy
Collection: iStock / Getty Image Plus

Set in 10/12pt ITC Garamond by SPi Global, Pondicherry, India
Printed in Singapore by C.O.S. Printers Pte Ltd

1   2016

To my mentors, colleagues and students – all of whom have been responsible for guiding me to see a wonderful world I had previously been unaware of.

Contents

List of Contributors

John E. A. Bertram
Department of Cell Biology and Anatomy, Cumming School of Medicine, University of Calgary, CA

Audrone R. Biknevicius
Heritage College of Osteopathic Medicine, Ohio University, USA

Yvonne Blum
Royal Veterinary College, London, UK

Sharon Bullimore
Cheltenham, UK

David R. Carrier
Department of Biology, University of Utah, USA

Monica A. Daley
Department of Comparative Biomedical Sciences, Royal Veterinary College, University of London, UK

Hartmut Geyer
Carnegie Mellon University, Pittsburgh, PA, USA

Anne K. Gutmann
New Balance Sports Research Lab, Lawrence, MA, USA

John W. Hermanson
Department of Biomedical Sciences, College of Veterinary Medicine, Cornell University, NY, USA

Elvedin Kljuno
Russ College of Engineering and Technology, Ohio University, USA

Susanne Lipfert
Human Motion Engineering, Lausanne, Switzerland

M. Maus
Imperial College London, London, UK

D. Maykranz
Friedrich Schiller University, Jena, Germany

Stephen M. Reilly
Department of Biological Sciences, Ohio University, USA

Daniel K. Riskin
Department of Biology, University of Toronto-Mississauga, CA

J. Rummel
Friedrich Schiller University, Jena, Germany

Andre Seyfarth
Institute for Sport Science, Technical University of Darmstadt, Germany

James R. Usherwood
Structure and Motion Lab, Royal Veterinary College, University of London, UK

Preface

Locomotion is one of the three key functional capacities of mammals, along with feeding and reproduction. An understanding of locomotion is imperative to understanding how adaptive evolution has allowed an organism and its ancestors to exploit the environmental opportunities and deal with the constraints and obstacles encountered in daily activity. Comparative analysis of terrestrial locomotion is an area of widespread interest, but the diversity of forms, as well as physiological and behavioral differences, make a comprehensive analysis of all animal forms a challenge to either produce or assimilate.

In this volume, the main focus is on a single group of specific interest – in this case, my colleagues and I have (largely) chosen the terrestrial mammals (I say largely, because some of the concepts are mentioned in the context of other clades, where these species serve as particularly good examples of the ideas being discussed, or where a discussion of the generality of the concept(s) is of value). The book presents a collection of chapters that will give the reader insight into some important factors that influence how mammals move using legs to travel in the terrestrial environment, and how we can use this understanding to interpret form and performance. Although each chapter should stand on its own, the collected pieces are intended to sum to a good introduction to the field of terrestrial locomotion and the concepts that are useful (imperative?) when delving into these issues.

As the title indicates, there are two main objectives of this volume. One is to identify some important concepts that should be understood by those with an interest in investigating terrestrial locomotion. These are meant to serve as a starting-point for discussions of the issues involved in investigating the "how and why" of terrestrial locomotion. Even though detailed analysis is important in order to verify precisely how a system works, it is conceptual understanding that allows the perception of patterns and the interpretation of meaning that will not be evident based on observation alone, regardless of how detailed or precise that observation (measurement) is. Concepts allow the interpretation of the patterns that exist in nature, and provide the insight into why the patterns exist as they do. As Aristotle stated; *"We have to examine the reasons for all these facts, and others cognate to them; that the facts are such is clear from our Natural History, we have now to ask the reasons for the facts."* (Aristotle, *On the gaits of Animals*, ASL Farquharson, trans.).

Sometimes, the most important concepts are evident only in highly obscured behaviors, hidden as apparently smaller features of more impressive aspects of mammalian movement. These "little things" are sometimes very difficult to observe, indeed, but can actually be the determining features of an important functional behavior, such as locomotion. I learned this early in my career, when I was investigating arboreal locomotion in gibbons. Following many others, I was able to describe how gibbons moved across a series of handholds but, no matter how closely I observed and analyzed these motions, discovering the reasons why these animals chose the movement paths they did was elusive. As often happens when delving into the natural world, some aspects of the motions were trivial to interpret, while other features did not fit into any obvious category.

However, once I had established a collaboration with a colleague who knew very little about gibbons, but knew a great deal about the dynamics of objects physically interacting with each other, it was possible to realize that our current perspective on gibbon locomotion lacked a critical feature – the recognition that energetic loss occurs when the path of the mass of the organism changes abruptly (see Chapters 5 and 6 for a specific discussion of this). Ironically, avoiding this energetic loss is so important to successful brachiation that it is, for all intents and purposes, eliminated from the locomotion of gibbons (at least when brachiating). The implication of this, of course, is that the most important dynamic feature of this animal's locomotion is not directly observable!

It is only through an analysis of alternative movement strategies (that are not used by the organism, because they are so disadvantageous) that the key determinant can be identified. Sometimes, the most obvious aspects of movement are not the most important, and some factors that appear diminishingly small can be far more influential than expected. Hopefully, this volume will provide a number of such "insights" that will build toward a sophisticated perspective on mammalian locomotion.

The second objective of this volume is to introduce and discuss how these concepts can be applied. Thus, many of the chapters provide specific examples of how fundamental concepts are applied to locomotion research to interpret the meaning of the form and function of locomotion. However, it is fully acknowledged that the interpretation of observations through the formulation and application of concepts continues to generate controversy now, just as it has done in the past. When dealing with systems as complex as those occurring in the natural world, working to a final and unambiguous conclusion on a system is a hard-forged and iterative process. As with our personal progress in learning, however, the objective of our exploration of the natural world is not so much to rid ourselves of all questions, but to replace simple questions with ones that are more and more sophisticated.

George Bernard Shaw proposed that *"All great truths begin as blasphemies"*. Following from that, I might contend that all lesser truths begin as controversies. And, as much as we might eventually be interested in great truths, most of us spend our day-to-day research lives working on lesser truths (where the practical potential to document progress is much more achievable). The scientific method has many complexities, but one view of scientific progress is the systematic identification and resolution of controversies. With regard to the locomotion of terrestrial mammals, this book is intended to assist with that process.

There is no intention here to produce an authoritative compendium of answers but, instead, the volume should be viewed as a guide to some important questions currently present in locomotion studies – an attempt to lead the reader to the edge of the "envelope", as it were. The exploration of concepts and their application, with the inevitable revelation (and not necessarily immediate resolution) of controversies is the scientific process – a much broader and intellectually rich endeavor than is the simple application of the scientific method.

Rather than try to provide proof to resolve controversies, an important objective of this volume is to expose these controversies within the context of the concepts on which they are based. Identifying and discussing these controversies serves the purpose of stimulating the next generation of researchers to embrace the discussion and to work to expand our understanding of this important aspect of mammalian biology, to solve those particular controversies and to replace them with the next set, as thought radiates into currently unoccupied "logical space". It is really for the next generation of researchers in locomotion biomechanics that this volume was assembled.

This is a particularly interesting time to be studying locomotion. We are the beneficiaries of a suite of new technical tools that allow the measurement and analysis of a wide range of animal movements and novel concepts. These promise to provide new insights into both the "how" and "why" of the movements we observe, and the morphological specializations they are dependent on.

Of even greater importance, however, is my contention that the field of locomotion mechanics is currently undergoing a revolution of perspective of such substance that it borders on a complete paradigm shift. Mammalian terrestrial locomotion has been of academic interest since academic interest originated (see the Aristotle reference above, for instance, or Chapter 1), yet substantial questions remain, in spite of sophisticated investigative tools currently at our disposal. For instance, the fastest quadrupedal gait – the gallop – demonstrates a number of apparently ambiguous mechanical features that have defied the understanding of the mechanical function of this gait (although it remains extremely well-described). That such a roadblock to understanding the meaning of the gait exists, in the face of highly technical advances in analysis techniques, indicates that it is not a technical limitation that prevents our full understanding of this important locomotion strategy, but a conceptual one.

The general approach of the volume is to explore the dynamics of locomotion in terrestrial legged mammals, with some consideration of the mechanics behind these motions. In the process, we are particularly searching for some general principles that can be used to understand and explain how morphology (form) and behavior (action) influence locomotory performance, broadly defined.

Often, locomotion is considered in terms of the neural control responsible for the active motions that can be observed. It will be noticed that there is not much mention of neural control in this volume. This is not because neural control is not acknowledged to be of critical importance in locomotion, but because the focus of the discussions move toward understanding how the biological entity, the organism, interacts and is influenced by its physical environment. The brain is, indeed, the driver of the locomotory system and, just like the driver of an automobile, it is ultimately in charge and directs the motion of the vehicle. However, like any good driver, an individual operating a vehicle must find his/her way between the boundaries of the specific roadway and obey the traffic regulations.

In one sense, this book can be considered an exploration of the "road map" that locomotion navigates, and the "highway regulations" governing how the brain must operate the system in order to be a successful "driver" of locomotion. No matter how much the brain would like it otherwise, in locomotion "F" (force) always equals "ma" (the product of mass and acceleration), and the influence of gravity cannot be avoided (no matter how fast an animal runs, it still weighs the same). Often, control of locomotion is approached from the perspective that "anything is possible", and the specific control patterns selected are analyzed with respect to the internal patterns and relationships between signals. Activity of the neural system, of course, results in activity of the limbs, trunk and other components of the organism. It is those patterns of motion that interact with the external world, and that interaction has consequences, both positive and negative, that we need to be aware of if we are to interpret locomotion effectively (even the central nervous system's role in that process).

One crucial aspect of observing and deciphering the interaction of the organism with its physical environment during the dynamic movements of locomotion is distinguishing the "phenomenon" of locomotion from the "mechanisms" of locomotion. Not much consideration is usually given to this, because the focus is more often on meticulously measuring (often under adverse circumstances, working with uncooperative animals) the dizzying array of movements that occur at all levels of system organization. It is the contention here, however, that the "dizzying array" of motions can only really be interpreted if what they accomplish for the organism can be identified. What is accomplished for the organism is the phenomenon of locomotion, while most of the readily observable motions constitute the mechanism(s) through which the phenomenon is achieved. The chapters in this work try to identify features of the phenomenon of locomotion, and the structural form of mammals organized to achieve locomotion, and to describe its relationship to the mechanisms responsible for making it happen. In so doing, the discussions direct attention to a new and, hopefully, insightful perspective on mammal locomotion.

A range of authors appear in this volume. They were selected on two criteria: that they had something interesting to contribute, and had demonstrated expertise. It will be noted that many of these individuals are not at the end of their careers, unlike we see in many summary texts designed to exploit the perspectives of those with extensive experience in the field. Rather, in keeping with my prediction that the field is moving through an important change of perspective, I have invited individuals to participate with me in this volume who are contributing to those changes.

Some of the chapters were invited because they provide a description of the important concepts, but most were invited because they described the application of those concepts, so they could relay some of the important controversies. All chapters were evaluated by one, or more, independent reviewers. Some of these reviewers came from within the ranks of the contributors to this volume, while others were solicited externally, because of their particular expertise. For their contributions in this regard, I would like to particularly thank R. McNeill Alexander, University of Leeds, J. Max Donelan, Simon Fraser University, Brock Fenton, University of Western Ontario, and Colin Pennycuick, University of Bristol. As always, however, the views portrayed in the chapters of this book are those of the authors, with whom the responsibility for producing and approving each contribution lies.

JOHN E. A. BERTRAM

CHAPTER ONE

Concepts Through Time: Historical Perspectives on Mammalian Locomotion

John E. A. Bertram

Department of Cell Biology and Anatomy, Cumming School of Medicine, University of Calgary, CA

1.1 INTRODUCTION

This chapter is meant only as a subjective description of some important landmarks along the route that has brought the field of terrestrial (primarily) mammalian locomotion to its current position. I have not been particularly interested in documenting the sequence of specific events, but have instead tried to follow the often circuitous path of ideas as they originate, are modified and are passed along to influence others.

In evaluating the current status of the field of terrestrial locomotion, I find that those of us working in this area are part of a long, and often very illustrious, community of individuals who seek out novel and creative methods of divining the constraints and opportunities exploited by animals in attaining movement within the physical world. The motivation for the development of this volume is to focus attention on the role of mechanics in understanding animal locomotion and, particularly, that of terrestrial mammals. The field is currently undergoing a substantive change in perspective, as new

Understanding Mammalian Locomotion: Concepts and Applications, First Edition.
Edited by John E. A. Bertram.
© 2016 John Wiley & Sons, Inc. Published 2016 by John Wiley & Sons, Inc.

technologies allow the critical evaluation of dynamic features of gait, and novel conceptual approaches are being assessed (although as we will see, few ideas in a field with such a long history can be considered truly novel). Overall, this volume is designed to stimulate the discussion of these newly arising opportunities, and this chapter is designed to "set the stage" for such a discussion.

Historical evaluations relevant to the field of locomotion science appear with some regularity, either as a contribution designed to put the field in historical context (Cappozzo *et al.*, 1992; Medved, 2001; Ashley-Ross and Gillis, 2002) or integrated within a discussion of the state of the field at the time of writing a more comprehensive assemblage of knowledge on the topic (Howell, 1944; Gambaryan, 1974; Walker, 1972). My interest is in emphasizing the origin of unifying (or dividing?) concepts, rather than exhaustively mapping the history of the field. To this end, I will trace my personal impression of key conceptual breakthroughs, whether in the fields of animal or human biomechanics and physiology or, indeed, even if it strays into robotics or the fundamental mechanics that ultimately underlies our field. I will liberally add my "interpretation" of events and the stimulus that led to them. When doing so, I will provide the evidence as I see it, but in few cases will this be conclusive, so all such descriptions should be evaluated as only the personal opinion of an interested observer.

1.2 THE ANCIENTS AND THE CONTEMPLATION OF MOTION

Undoubtedly, prehistoric man observed and wondered at the remarkable abilities of animals to move, likely largely motivated by the elusiveness of the prey they pursued, but also by wonderment at the fleetness of some terrestrial runners. The cave paintings from Lascaux, France (approximately 15,000 BC) demonstrate that our distant ancestors were, indeed, keen observers of the form and movement of mammals, indicating a fundamental interest in understanding and interpreting animal motion (Figure 1.1).

The movement of animals and their relationship to the locomotion of humans also holds an inherent interest for those trying to understand the world they observe. Such was the case with Aristotle (384–322 BC), an individual who has had a great influence on philosophical thought and the foundations of modern science. Among Aristotle's influential writings was *De motu Animalia* (On the Gait of Animals). Aristotle's interest in locomotion derived from the fundamental question of the difference between passive matter and the activity of living things, placing questions of locomotion at the foundation of major philosophical issues regarding the makeup and function of the physical world.

Among other surprising observations listed in *De motu Animalia*, Aristotle noted vertical motion of the human when walking, by observing the "zig-zag" movement of the shadow against a flat wall (as translated by Farquharson, 2007). This was particularly in reference to how the swing limb must flex in order to pass under the individual, but the observation was astute and important in recognizing the complexity of motions required to produce effective locomotion, and that locomotion involved multidimensional movement of the body as well as the limbs.

FIGURE 1.1

The cave painting from Lascaux, France, known as the "Third Chinese Horse". This painting depicts a galloping horse with arrows approaching its back, taken from the series referred to as the "Chinese horses". The cave was discovered by adventurous boys in 1940. The painting likely dates from 15,000 BC.

Aristotle also described the limb motions of quadrupeds as they walked, but these "observations" seem to have come more from theoretical analyses than from direct observation. As will be described below, some two millennia later, Borelli revised some of these descriptions to reflect more accurately the actual movements of walking quadrupeds as a means to explain some of the reasons that motions occur as they do.

1.3   THE EUROPEAN RENAISSANCE AND FOUNDATIONS OF THE AGE OF DISCOVERY

Although it is well known that Leonardo da Vinci (1452–1519) studied fish swimming and bird flight, it is not so well known that he also studied human and animal terrestrial locomotion. His approach to the investigation of structure, including meticulous anatomical studies, paved the way for considering the mechanisms of movement. Indeed, his work, in many ways, heralded the beginning of the European Renaissance. He performed many artistic studies on the form of animals and people in motion. For many of these studies, it is difficult to tell whether the objective of the work was to understand the function of the system, or to observe the natural motions in order to improve his art – in many cases, for da Vinci, these seem to have been one and the same.

Of particular note are his drawings of individuals walking up and down stairs, standing from a seated position and stopping from a run, all of which represent the mechanical realities of each of these circumstances. His pen and ink studies of individuals and animals laboring, demonstrating a variety of positions, are reminiscent of Muybridge's photographic studies that followed nearly 500 years later. Da Vinci's notebooks indicate that he was collecting material for a never-completed work on mechanics, a work that would have given us a complete view of his understanding of motion, including that of animals and people.

As keen an observer as he was, however, he did make some errors. He clearly believed, as all of us do as children, that heavier objects fall more rapidly than lighter ones. Our modern perspective of the world benefits from the luxury of the accumulated understanding that we call knowledge, served to us in the form of education.

The extensive notebooks of da Vinci hold numerous examples of his understanding of locomotion mechanics, some of which have proven correct, while others were not so successful. Leonardo (more or less) correctly observed that, "*Man and every animal undergoes more fatigue in going upwards than downwards, for as he ascends he bears his weight with him and as he descends he simply lets it go.*" However, he profoundly misunderstood the role of impulse, and suggested, "*A man, in running, throws less of his weight on his legs than when he is standing still. In like manner the horse, when running, is less conscious of the weight of the man whom it is carrying; consequently many consider it marvelous that a horse in a race can support itself on one foot only. Therefore we may say regarding weight in transverse movement that the swifter the movement, the less the weight towards the centre of the earth.*" (p. 150, Richter and Wells, 2008).

It is now eminently clear that, provided we remain in the Newtonian realm, we weigh the same regardless of how fast we run. It is the subtle strategies employed to deal with this fact that make the understanding of running gaits, in people and other mammals, a challenge to understand.

A substantial influence on the origins of biomechanics were the works of Galileo Galilei (1564–1642), particularly *Discourses on Two New Sciences*. One of these "New Sciences" was the formal western origin of mechanics. The principles outlined formulated the basis of understanding the mechanics of biological movement. Indeed, many examples described in the *Discourses* are biological, indicating that, in Galileo's mind, there was no definitive difference between the mechanics of the organic world and that constructed by humans. TA McMahon (1975) referred to Galileo's analysis of scaling issues, noting ironically that a woodcut from *Discourses* (Drake, 1989, p. 127 and 128) differed somewhat from Galileo's own discussion and analysis of scaling mechanics, and appeared to represent a scaling relationship intermediate between constant shape geometric "isometry" and the changing proportions of "static stress similarity", which Galileo suggested was necessary to preserve mechanical support function over large size changes. From more thorough analyses, McMahon described the consequences of this alternative scaling model, which he termed "Elastic Similarity" (1973, 1975; see also Chapter 8).

Giovanni Borelli (1608–1679) is noted for applying Galileo's principles of mechanics to the action of the musculoskeletal system – "*Borelli then shows what the forces of various muscles must be: for example, the force exerted by the biceps when a weight of 28 lbs is being held by the hand with the arm extended horizontally is 560 lbs*"

(Des Chene, 2005). Borelli's major contribution, *De Motu Animalium* (On the movement of animals, published posthumously in two parts in 1680 and 1681), is full of examples that would not be out of place in a modern introductory university course on musculoskeletal mechanics. Borelli worked on numerous problems in parallel with Newton, whose career and influence was rising toward the end of Borelli's life.

Although he studied planetary movement and pendular motion, Borelli never arrived at the conceptual formulation of gravity or the laws of motion that were some of Newton's greatest contributions. One of his most profound errors was to assert that, in some cases, forces were not "equal and opposite" but that, within the body, they could be unequal. However, he was extremely influential in describing the function of muscles, and also in his description of locomotion, particularly quadrupedal walking. With meticulous illustrations and well-founded arguments, Borelli was able to dispense with Aristotle's contention of diagonal-only foot contacts in quadrupedal walking, and was able to describe the features of balance associated with a stability tripod, as later championed by Sir James Gray (1944, 1968). The arguments were apparently convincing enough that the depiction of animal walking in sculpture and painting began to change following the publication of Borelli's book.

Another 17th century luminary was Claude Perrault (1613–1688). Like Borelli, he was instrumental in applying mechanical concepts to the action and organization of the musculoskeletal system. He introduced some important concepts, such as the spring-like capacity of muscles and joint position, representing the equilibrium between protagonist and antagonist muscles. Perrault also astutely observed that muscles function through "introduction of the spirituous substance brought by the nerves from the brain" (Des Chene, 2005).

As da Vinci had influenced thought and helped to initiate the Renaissance, Isaac Newton's (1643–1727) formulation of the laws of mechanics and the explicit role of gravity initiated a new era of enlightened modern science. Newton provided the theoretical framework with which to evaluate the role of the organic components of animal and human function in the physical world, within which they operate and from which their motions are influenced. Newton's Laws of Mechanics form the foundation of all modern biomechanics and find their way, at least in implied assumptions, into basically all modern work in the field.

1.4   THE ERA OF TECHNOLOGICAL OBSERVATION

With the beginning of the modern era and the application of novel technological advancements came the possibility of observing and quantitatively measuring aspects of human and animal movement that had previously not been available. Goiffon and Vincent (1779) are widely acknowledged as an important early example of applying technical evaluation to gait studies. A bell system was attached to the front hooves of horses, and the difference in ring pattern for different gaits was evaluated. As da Vinci's studies of motion and Borelli's descriptions of quadrupedal walking influenced the interpretation of how motion is shown in art, one stated purpose of the Goiffon and Vincent evaluation of the equine gait was to inform artists with regard to how the gaits of horses should be depicted in order to properly represent their actual motion.

In a conceptually parallel but more sophisticated approach, Marey (1884) used pneumatic bulbs attached to the feet of horses that operated an armature recording pressure changes on a smoked drum carried by a rider. From this apparatus, fairly accurate determinations of the foot contacts were documented for all of the standard equine gaits. Marey also applied a similar approach to the study of human locomotion. His work overlapped with that of Muybridge, and in his later career, he adopted photographic approaches originated by Muybridge and added to these by developing several novel technological innovations of his own – for instance, the photographic "rifle" (1892) and the development of modern-style multiple-frame "movies" (1899).

The ability to technically observe locomotion achieved a watershed point with the work of Muybridge (1887). Through the use of a series of shuttered still cameras operated by trip wires, Muybridge was able to produce high-quality serial still image photographs through the range of motions of a wide variety of human and animal activities. The motivation for this innovative approach apparently derives from his involvement in a bet regarding whether a trotting horse has any portion of its stride without contact with the ground. Settling this bet to the satisfaction of all involved required clear demonstration of the gait cycle, with adequate time resolution to indicate how the motions were produced.

Muybridge recognized that his novel technique provided a totally new perception of (and perspective on) movement. He produced a remarkable data set of images from as wide a variety of animal and human motion as he could manage. Included in these were figures of humans doing everyday tasks that, though composed of evenly timed separate images, resemble the "snap-shot" views drawn by da Vinci centuries previously. It is likely that da Vinci would have liked to produce the series as Muybridge had, if the technology had been available to him. Muybridge's original compilation (Muybridge, 1887), ten volumes in total, has been reproduced in a variety of abridged editions that remain remarkably useful through to the current day (Hildebrand, 1962, 1989; Bertram and Gutmann, 2008).

Imaging motion yields a range of information about the process of locomotion, but the analysis of much of the mechanics is facilitated by also measuring the forces involved. Amar (1916, cited in Jarrett *et al.*, 1980) developed a mechanical force-reactive platform that had a great deal of early influence on the analysis of human locomotion, particularly in adapting technology to the needs of amputees returning from World War I. Elftmann (1934) later developed a reliable mechanical force platform system for human locomotion. Manter (1938), working with Elftmann, devised a multiple mechanical force plate system (two platforms in series) appropriate for analysis of quadrupedal locomotion, and used these to analyze the locomotion of the cat.

Gray (1944) recognized that Manter's results indicated that horizontal forces generated by the limbs of quadrupeds create turning moments around the animal's center of mass, which can ultimately influence the measured vertical forces at each foot. Due to the complications arising from redistributed forces in quadrupeds moving at non-steady speeds, in his highly influential 1968 compendium of animal locomotion mechanics, Gray advised stringent control of horizontal accelerations for quadrupedal gait analyses. Unfortunately, the difficulty in controlling animal motion meant that this advice was not always followed appropriately, and the "stringent" limits on speed variation and acceleration became relaxed, as force platform analysis became more common with the commercial availability of strain gauges and piezoelectric sensors that formed the basis of electronic force plates (Cavagna *et al.*, 1963; Heglund, 1981; Biewener *et al.*, 1988). Just as Muybridge set the groundwork for the eventual application of high-speed imaging and cinematography in the study of locomotion, Manter's force plates played a similar role for the electronic analysis of force in quadrupedal locomotion (Roberts, 2005).

For human locomotion studies, Elftmann's force plate analyses (1934) set the stage for all who followed – and there were many, once electronic force plates and, even later, microcomputers, made digital management of data and analysis possible. Elftmann's innovations in understanding gait continued through the late 1960s, when he created novel investigative approaches. In 1966 he was the first to consider the metabolic cost of locomotion as a "surface" within gait parametric space – a perspective on the interaction between the physical processes of locomotion and the metabolic consequences that was ahead of its time.

1.5   PHYSIOLOGY AND MECHANICS OF TERRESTRIAL LOCOMOTION – COST AND CONSEQUENCES

Ancient man surely recognized that locomotion requires the investment of metabolic energy, and that this was the case even for constant speed movement on level ground. The reason that such motion is metabolically expensive is not obvious, though. Like a wheel moving across an even substrate, the motion of any mass perpendicular to the action of gravity should require very little energy to maintain motion once it has been established. Identifying the source of the cost of locomotion, however, is key in understanding what must be accomplished in order to produce locomotion. As Borelli famously observed, *"A perpetual law of nature consists of acting with the smallest work..."*. This statement implies that, if walking costs energy, the cost must be unavoidable. The question of "what about walking determines that it must cost?" formed the focus of much work in the early part of the 20th century.

A.V. Hill is renowned for his work on muscle mechanics and the formulation of a robust model describing the mechanical behavior of muscle (see Chapter 3). Hill (1927) originally assumed that the cost of locomotion came largely from the viscosity of muscular movement (this was known as the "viscoelastic" model of muscle; see Gasser and Hill, 1924). Hill's research group attempted to apply the results of their groundbreaking muscle experiments to whole-body behavior. Hill's group observed that walking and running on level ground appeared to require little mechanical work (although it did require substantial metabolic energy), "save that of air resistance" (Furusawa *et al.*, 1927). As a result, they suggested that, *"The whole of the mechanical energy liberated is used in overcoming the frictional resistance of the body itself, particularly the "viscosity" of the muscles"* (p. 32–33).

These ideas were very influential at the time and, in spite of much contrary evidence, they remained influential for decades following. For instance, Rashevsky mentions muscular viscosity in his analyses in 1948, *"Hence the inertial forces are not the limiting factor in determining ϕ* (where ϕ is the angle between limb axis and the substrate at initial contact – editor's comment), *and we must conclude, in agreement with A.V. Hill, that the viscous forces are the main source of resistance in continuous running."* (Rashevsky, 1948, p. 20).

Early on, however, Fenn contended that Hill's results did not entail a complete accounting of the mechanical work accomplished during walking and running. He explains the purpose of an important pair of papers on running: *"Hill, 1927, who has discussed in many valuable papers the problems connected with rapid movements of muscles has regarded frictional loss as the limiting factor in a fast run and as in fact the only item of considerable importance."* (Fenn, 1930a, p. 584). Fenn's conclusion was that a complete accounting of mechanical cost was necessary in order to determine how much metabolic "cost" remained to be explained by muscle "viscosity".

Fenn and colleagues performed some meticulous studies of kinematics and kinetics that rival modern attempts, albeit with substantially less sophisticated (but nevertheless cleverly utilized) equipment. Working with CA Morrison of the Eastman Kodak Company, they performed kinematic analysis on human runners using images taken from the newly emerging technology of cine film. At that time, such cameras were driven by hand crank, which resulted in variable framing rates. Frame rate was precisely calculated from measurements of a falling croquet ball (consequently a known acceleration) included in the view. Kinematic measurements were taken from images at intervals of approximately 0.016 sec, or just over 60 Hz. Kinetic measurements were derived from a mechanical force platform styled after that developed by Amar (1916, 1919, 1920).

From these studies, Fenn (1930a) organized the mechanical work of locomotion into two measurable categories: *"The external work of running may be divided into two parts – 1, the movements of the limbs in relation to the center of gravity of the whole body and 2, movements of the whole body as represented by the movements of its center of gravity."* (p. 433, Fenn, 1930b). These "movements of its center of gravity" are, of course, the subtle variations in height and forward speed that occur over the course of a normal stride – the "zig-zag" pattern recognized by Aristotle. Fenn also calculated that work against the drag of the air was not necessarily negligible, particularly for running. He concluded, however, that the production of what we now refer to as "internal" and "external" work (Cavagna and Kaneko, 1977) were responsible for the vast majority of mechanical "cost" incurred in locomotion.

Fenn's early work was technically sound, but appears to have been ahead of its time, and the approach was really only explored fully beginning in the late 1950s. Some other early workers took a similar technical approach to make some valuable breakthroughs in understanding. Much like Hill and Fenn, Margaria was at the forefront of investigations into the function of muscle (Margaria *et al.*, 1933; Margaria and Edwards, 1934) that led to an interest in how those muscles functioned in locomotion. His monograph on investigations of locomotion cost (1938) stands as an example of considerable conceptual insight, but it was largely neglected by the English-speaking scientific world until the work was reviewed by Margaria himself in book form some decades later (Margaria, 1976).

With the advent of electronic, rather than mechanical, force plates (Cavagna *et al.*, 1963) and, following this, the development of computer-based analysis techniques, the types of analyses originated by Fenn could be done effectively for a wide variety of species (Jayes and Alexander, 1978; Heglund *et al.*, 1982; Biewener, 1983). Ultimately, this led to the proposal that two fundamental mechanisms were responsible for limiting the cost of locomotion in animals that use legs to move across a substrate – namely, pendulum-like exchange of gravitational potential energy and kinetic energy of motion, and spring-like storage and return of kinetic energy through elastic strain energy in muscles and tendons (Cavagna *et al.*, 1977).

The approach formulated by Cavagna and colleagues has grown to be remarkably influential over the last four and a half decades, and forms the basis of much of the approach to the mechanics of animal locomotion currently in use. Due to this perspective, however, the field was turned toward an emphasis on the mechanisms available for energy *recovery*, and this came at the expense of clearly identifying the source of energy loss. It was, after all, the identification of the source of loss that motivated Fenn's original locomotion studies.

The determination of the mechanical work of locomotion for most of these studies is based on the assumption that mechanical work can be estimated as a function of the positive increases in total energy over an integral number of strides (complete gait cycles) – basically the approach utilized by Fenn. Blickhan and Full (1992) discuss some of the complexities involved with the assumptions on which such analyses are based. Donelan *et al.* (2002a) demonstrate that the standard use of the force plate to monitor center of mass motion underestimates the mechanical work done in human walking, because substantial work is done by one limb on the other during the transition between foot contacts. It is certain that a similar situation occurs for quadrupedal walking, albeit in an even more complicated manner, due to the potential for multiple interactions between limbs.

As influential as the internal and external work approach has been, there are some nagging inconsistencies generated from it and apparent paradoxes exposed. Human walking emerges with one of the greatest recovery levels measured (bested only by penguins – Griffin and Kram, 2000), and yet only reaches approximately 60–70% of the energy available to recover. This seems remarkably high to those who assume all kinetic energy must be lost, if not recovered, while others wonder why not higher? Quadrupeds universally appear rather poor at recovery, in the order of 35–50% (Cavagna *et al.*, 1977).

This type of analysis also appears to grossly underestimate the cost for larger mammals. For instance, "*The mechanical work rate of the horse exceeds the rate at which its muscles consume energy over its entire range of speeds, indicating that springs must supply the difference between energetic input and output.*" (Taylor, 1994). Producing more apparent mechanical work than metabolic energy consumed should raise concerns about breaching the Laws of Thermodynamics. As Taylor implies, however, it is routine to invoke elastic energy recovery to replace the unexplained mechanical cost with minimal metabolic investment. Elastically acting structures had been assumed for an extended period, and work such as the Camp and Smith (1942) treatise on the digital ligaments of the horse specifically characterized the properties of these structures in the context of the animal's locomotory capabilities.

Although elastic-like structures exist and have been shown to act as energy recovery systems (Thys *et al.*, 1972), it does not appear to be easy to identify structures in which the required quantity of energy storage and return is accomplished during locomotion. For instance, although substantial elastic energy storage has been anticipated in horse galloping, it has been remarkably difficult to identify elastic components operating in such a way in either the back (Alexander, 1988) or the distal limbs (Pfau *et al.*, 2006).

More recently, the concept of compliance as an integral feature of stability has also drawn substantial attention (McMahon, 1985; Geyer *et al.*, 2006; and see Chapter 7), where combined elastic-like and pendular features operate together to produce the mechanical features of even highly "pendular" gaits, such as that of the human walk.

1.6 COMPARATIVE STUDIES OF GAIT

Much of Gray's analyses of mammalian locomotion utilized a variety of species as examples for fundamental mechanical concepts he meant to illustrate. His work, however, did not involve an attempt to classify mammalian locomotion comprehensively, but worked toward distinguishing the mechanisms that affected how general categories of animal types function. Gray had a great talent for cutting past the details of complex morphology to identify the mechanical factors that underlay the detailed design of an animal but, in order to accomplish this, it was necessary for him to neglect much of the anatomical detail of any given species.

The alternative approach is to organize the details of animal motion and, from the result, to identify the factors responsible for affecting the opportunities and limitations of each form. Howell (1944) produced one of the most complete early treatises on specializations of animals for speed. Within this work, Howell systematically described the gaits of mammals, with particular interest in what makes them move fast, or fast for long distances. An ultimate goal of this work was to determine the phylogenetic relationship of locomotion types and, to this end, the work was ahead of its time. Unfortunately, key analytic techniques were not available at that time, so much of the classification depended on subjective determinations. The wealth of information in Howell's work, however, has meant that it receives continued reference (Biewener, 1983; Gatesy and Middleton, 1997; Griffin *et al.*, 2004).

Hildebrand (1965, 1977) revolutionized gait studies by developing a systematic classification of gaits that quantified not only foot placement, as had been done previously (Marey, 1873; Howell, 1944), but also phase relationships and contact timing features between limbs. His most useful formulation of this defined the gait of mammals on the basis of two independent features – the percentage of the complete stride that a selected foot is in contact with the ground, and the interval that a forelimb follows the hind limb placement. In this way, fundamental gaits emerged from the gait formula (Figure 1.2), and it was possible to easily compare the gaits of many different quadrupedal species (see Chapter 2). Although this classification has been instrumental in comparing gaits of different forms, it does not involve any functional or mechanical explanations for the patterns observed.

Gambaryan (1974) collected a substantial amount of material available from comparative biomechanical and anatomical analyses, and organized it according to phylogenetic relationships. This monograph represented a compilation of both understanding that had been generated in the English-speaking literature, of which Gambaryan was aware, and a large and independent Russian-speaking literature, of which the English speaking scientific world appeared largely unaware. Although many of the gait classifications have been regarded as somewhat arcane (does anyone refer to a dorsostable dilocomotory gait?), the work remains one of the most comprehensive collections of detailed morphology related to mammalian locomotion.

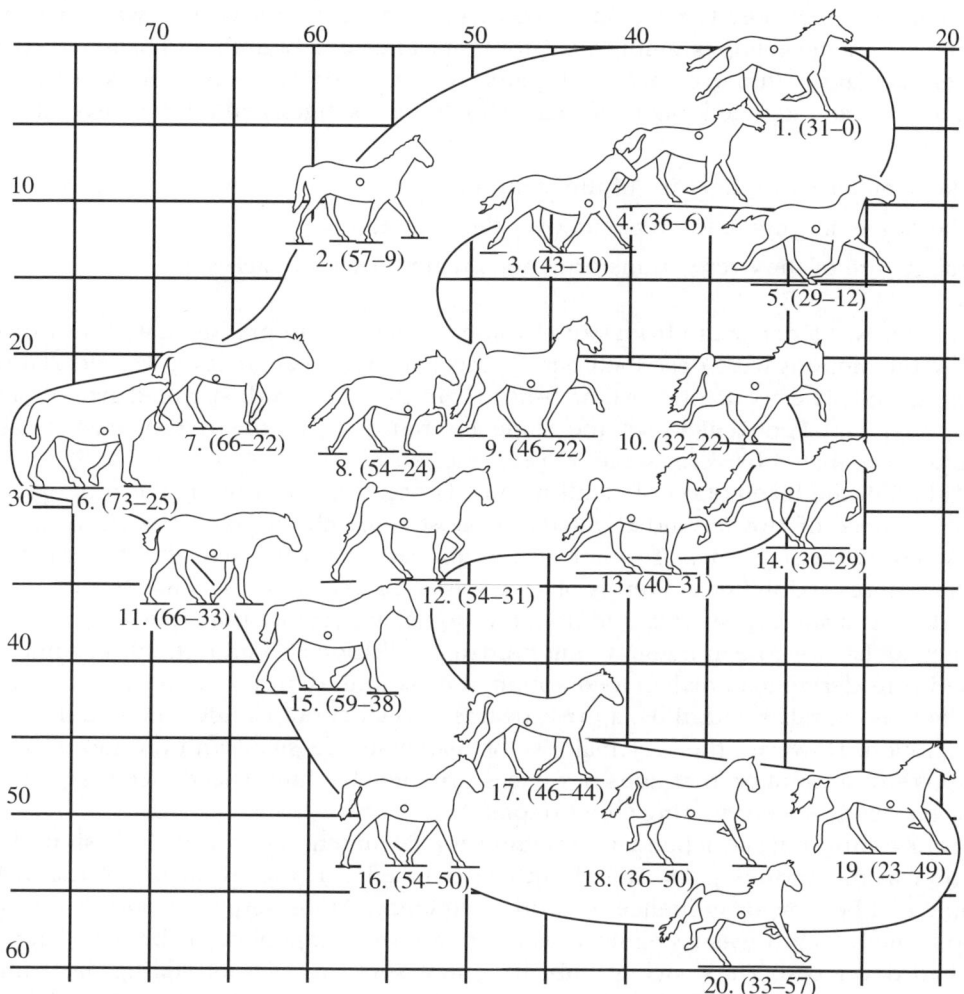

FIGURE 1.2

A "Hildebrand plot" of the symmetrical gaits utilized by horses (Hildebrand, 1965). By comparing the proportion of the stride that each hind foot was on the ground (x axis) with the proportion of the stride interval that the forefoot contact followed hind foot contact on that side (y axis) a distinct pattern of gait forms emerged. In this plot, the gait formulae, as determined by these two proportions, are listed for each of the small circles. The silhouette of the animal representing each gait form at the time of rear foot contact is also shown.

Although CJ Pennycuick has been better known for his work with bird flight, he also made important contributions to the documenting and analysis of ungulate gait. In the inimitable manner of those leaders in the field, such as Gray, Hildebrand and Alexander, Pennycuick (1975) was able to simplify the analysis of complex gait patterns to focus on functional features that had key importance in the production of effective locomotion.

He suggested that only three fundamental gaits should be recognized: walk, trot, and canter (where the gallop is simply a fast form of canter, and bounds and half-bounds are canter-like gaits with one or both limb pairs contacting together, rather than distributed in time). It was suggested that these gaits should be distinguished on the basis of:

a. the symmetry or otherwise of the stepping pattern;
b. the range of average foot contact over the gait; and
c. the nature of the energy transformations which occur at each step.

This latter was a departure from gait classifications that had predominated previously, where the motions were described but the mechanics, and energetics, were left largely unexplained. The work of Cavagna and others that followed, such as that of Gray, Alexander and Pennycuick, set the stage for returning to the rigorous mechanical analyses originated by Fenn some 50 years earlier.

Although the locomotion of small mammals was treated in the major monographs such as those of Howell and Gambaryan, most research attention has been paid to gaits of the larger mammals. To some extent, this can be attributed to attention garnered by the impressive abilities in terms of speed or endurance of these animals, or because key domestic species, such as the dog and horse, are readily trainable. The horse, in particular, has long been integrated in human civilization and important to commerce. When considering mammalian locomotion, it is natural to turn to models that are convenient and familiar and, thus, a great deal is known about canine, feline and equine locomotion. However, the vast majority of mammals are small and are more or less adapted to locomotion that does not necessarily involve high speed (see Chapter 10), or even steady speed running (see Chapter 11).

After establishing a dependable method for documenting and distinguishing limb motion in the various gaits, Hildebrand recognized that the dynamics of the swing limb could be a major influence on stride frequency. Stride frequency and its relationship to limb length was recognized as an important component of the gait diagrams he had developed. Thus, swing limb dynamics were one way of adding mechanical explanations to the gait descriptions (quantified via the diagrams) that he had developed. He also recognized that to drive the limb at frequencies that differed substantially from their natural swing period could require large and undesirable energetic expense. From there, he hypothesized that this may be influential in morphological differences in distal limb form (Hildebrand, 1985).

Although these theories are basically sound, the predictions met with some problems when tested specifically. Taylor *et al.* (1974) compared cheetahs, gazelles and goats and found that the metabolic cost of locomotion did not match the expectations of reduced moment of inertia in the distal limb. This is most likely because swing limb mechanics are only one portion of the energetic determinants in locomotion. However, the idea of integrating motion with the passive mechanical swing of the limb did point to a perspective of optimizing swing limb dynamics. This, ultimately, led to the ballistic model of human leg swing (Mochon and McMahon, 1980a, 1980b (again challenged experimentally; Mena *et al.*, 1981; Selles *et al.*, 2001) and has found its way into aspects of numerous mechanical models of bipedal (McGeer, 1990a; Garcia *et al.*, 1998; Collins *et al.*, 2001) and quadrupedal locomotion (Herr and McMahon, 2000, 2001).

1.6 RE-INTERPRETING THE MECHANICS: A FORK IN THE ROAD, OR SIMPLY SEEING THE OTHER SIDE OF THE COIN?

I pointed out above that much of the focus of comparative locomotion mechanics and energetics has been drawn to the consideration of mechanisms for passive and near-passive *energy recovery*. This is accomplished either through the exchange of gravitational potential energy with kinetic energy, emulating pendulum-like exchange, or through strain energy exchange with kinetic energy, as in the bouncing of a spring-mass system. In contrast, the investigation of bird flight energetics progressed in a substantially different conceptual direction, and one that might provide an interesting alternative for terrestrial locomotion.

Although comparable fluctuations in the vertical position of the center of mass (CoM) and multidirectional velocities occur during flapping flight, the consideration of energy recovery was not emphasized in the analysis of flight (although some exchange occurs). Instead, the focus remained, as it had started in legged locomotion with Hill and Fenn, on identifying the source of the *energetic costs* involved, as determined by the dissipative mechanisms responsible for energy loss to the environment in which the organism functions (Raspet, 1960). The analysis of flight then progressed to determining what features of the morphology and activity replaced the associated losses (Tucker, 1973). In the case of flight, the basic dissipative mechanisms can be readily identified as the modes of drag that result from the dynamic interaction of the organism with its aerial environment, and those actively generated by the production of lift. One possible reason that the analysis of terrestrial locomotion did not progress along the same lines as that of flight might be that identifying the source of cost is not necessarily as intuitive for legged locomotion (Bertram *et al.*, 2006; Kuo, 2007).

As mentioned above, Hill's original viscous muscular loss was replaced by mechanical considerations related to inertial motions of the body and its components. However, these, in turn, led to conclusions that required remarkably effective energy exchange mechanisms, particularly for larger running mammals. Is effective recovery the key to understanding the mechanical consequences of adaptive form and behavior in locomotion? Has the source of cost, the fundamental dissipative mechanism, been identified for terrestrial locomotion? There are currently two main factors that are viewed as the "source" of cost in legged locomotion. That is, two mechanisms are identified as the cause of the main dissipation when moving on limbs – one biological, and the other physical. As will be seen, both have long histories in locomotion analysis, and ultimately they may well be intimately related.

1.7 THE BIOLOGICAL SOURCE OF COST

It has long been recognized that positive work had to be produced by the musculature to raise the body against gravity and to accelerate the limbs, and also that the momentum of the body and its components could do work "on" the muscles through stretching them while they actively attempted to contract. This latter reversed, or negative, work could be demonstrated, but was essentially lost because, except for stretching elastic elements, it could not be utilized further in locomotion. That is, it could only be converted to heat and lost from the system. Thus, the biological cost of locomotion, the route through which metabolic resources were converted to active locomotion, had two main components: the generation of positive work within the muscles in the process of accelerating the body and its parts; and the negative work involved with decelerating those components.

Although positive and negative work, and their relative metabolic cost, had been recognized at the level of the muscle fiber (Fenn, 1924), Margaria was one of the first to demonstrate that such cellular mechanisms had a direct influence on whole-body metabolic cost in locomotion, by showing that the apparent efficiency of uphill and downhill walking and running asymptote to the same efficiency limits as single muscle fibers (0.25 for positive and −1.2 for negative work; Margaria, 1938, 1963). Margaria (1976) recognized the value of passive energy exchange, whether the gravitational E_P–E_K exchange in walking or the strain E_P–E_K exchange in running.

However, he also recognized that these models did not fully account for the losses involved in locomotion – what Margaria termed "wasted" mechanical work (p. 103): "*The resistance to progression can therefore be considered as substantially due to the negative work that is performed at each step in walking and running: to maintain a constant speed of progression this must be compensated by an equal amount of positive work.*" (p. 101); "*… the "resistance" met by the subject when walking or running on the level appears to be essentially met by the negative work that is performed at each step which must be compensated by an equal amount of positive work in the first phase of the following step: practically all the energy spent in walking and running on the level is utilized to meet this resistance.*" (p. 105). He also concluded that, at specific slopes, little negative work is actually performed (p. 104).

However, if one important driving force in the "design" of natural systems is limiting metabolic energy expenditure, as the energy recovery models suggest and as Borelli anticipated nearly 400 years ago, we have to wonder at the required "cost" of negative work. Is this a limitation (or flaw in the design) of the muscular system, or something physically required for legged locomotion? That muscles should use metabolic energy to absorb mechanical work was taken as a given until very recently.

1.8 THE PHYSICAL SOURCE OF COST (WITH BIOLOGICAL CONSEQUENCES) – THE ROAD LESS TRAVELED

It appears to have come as some surprise to early investigators, utilizing force plates to study the physical interaction of the limb with the ground during locomotion, that contact involved a horizontal deceleration in the first half of contact, followed by a re-acceleration in the following portion of the contact. However, we now know that this is a necessary consequence of transferring contact between limbs for steps of functional length. In other words, steps of a reasonable length will require that the falling mass of the individual be redirected by the next contact limb, and this requires an impulse with a component in the rearward direction that allows the body mass to "vault" over the limb. The deceleration that occurs in the first half of contact is where negative work is almost exclusively performed. If such forces are "required", then what are the consequences and, if there are consequences, how can such a contact be "optimized"?

Nicolas Rashevsky is well known as a key figure in developing mathematical biology – the application of mathematical modeling to a diverse range of questions in biology. Although his interests ventured into most areas of biology, he appeared intrigued by the mechanics of locomotion, publishing a number of papers on the topic (Rashevsky, 1944, 1946, 1948). Rashevsky was a physicist with an interest in applications to biology, and his work is renowned for being both insightful and almost entirely theoretical. This may explain why his locomotion analyses were met with some resistance (or neglect) by experimental biologists. He remained convinced that Hill's viscous muscle cost was an important consideration and, following Fenn's approach, he attempted to calculate what viscous damping "must" be by estimating the other important mechanical costs associated with locomotion. He concluded that only the hind limbs were really necessary for locomotion (possibly as a precursor to arguments he intended regarding the processes through which humans developed bipedalism). He concluded that quadrupedalism must exist for alternative reasons (what these might be he did not specify, but he implied that they would not be for generating thrust).

Rashevsky's analysis of limbed locomotion is best known from discussions published in the first part of his two-volume treatise on mathematical biology (Rashevsky, 1960, 1961). Although this work has been highly influential in many areas of biology, the implications of his ideas on limb function appear to have had marginal influence on thinking regarding animal locomotion. His discussions of running and jumping evolved over time. His 1948 analysis, largely reiterated in the 1960 volume, included a novel cost that did not appear in his very early models. This newly considered "cost" was the loss of momentum and, consequently, energy that results from the deflection of the animal's center of mass trajectory when the limb makes contact with the substrate (p. 266, 1960).

The first statement of this approach to the problem appears as, *"Running is essentially a series of consecutive jumps. During each jump the animal is for a while completely without contact with the ground, its center of gravity describing a parabola during the 'flight' phase. If at the end of each jump the total kinetic energy of the animal were completely lost, then the mechanism of running would be identical with the mechanism of jumping. Actually only a fraction of the total kinetic energy is lost at the end of each 'flight' phase. Therefore the theory of continuous running is somewhat different from the theory of jumping."* (p. 12, 1948).

This conclusion followed from the dynamics of interacting bodies, and indicates that Rashevsky treated the contact of the limbs and their influence on the center of mass trajectory as a rigid body collision (since this particular analysis did not include spring-like strain energy storage and return). Rashevsky simplified his model by having the limb make contact directly under the center of mass, thus limiting the collision loss to the vertical direction: *"During the 'flight' stage either the extremity which caused the propulsion or another one swings forward, hitting the ground when the body falls downward at the end of the parabolic flight phase. For simplicity let us consider the case in which the extremity hits the ground in such a position that* (a line between the foot contact and the animal's center of mass – editor's comment) *is vertical at the moment of impact. In this case only the vertical component of the velocity is lost or, at any rate, affected."* (for a more complete explanation of collision dynamics in the context of legged locomotion see Chapter 5).

It was possibly this simplifying assumption restricting the effect to vertical motion, and the unrealistic contact configuration, that caused this feature of the mechanics of limb locomotion to be neglected by the biological field – or possibly it was that Rashevsky did not explicitly describe this feature of the model, or discuss its importance for locomotion energetics (instead, he went on to discuss its implication to the form of limbs and the calculation of ultimate running speed). In any case, the vertical loss Rashevsky alluded to results from momentum and energy loss associated with deflecting the center of mass trajectory as a result of limb contact. It is determined by the physical interactions of the body with its substrate, and does not depend on functional properties of the musculoskeletal system, such as the metabolic cost of negative work (although negative work is one mechanism through which this loss can occur).

MG Bekker's 1956 book *The Theory of Land Locomotion* has had a great deal of influence on vehicular design, particularly for battlefield military transport, but also for vehicles as far afield and specialized as the "lunar rover" moon transport vehicle (Bekker, 1985). The first chapter of his 1956 book, however, is dedicated to discussing general features of legged locomotion, with particular reference to the principles affecting animal locomotion. Bekker followed Rashevsky's general approach, albeit in a more realistic formulation, and utilized a momentum-balance, collision-based approach to determine where the physical loss in the system originated. Although his work has been influential in vehicular design, this approach has largely gone neglected in the biological consideration of the mechanics of locomotion until quite recently.

VA Tucker, more renowned for contributions to flight physiology than terrestrial locomotion, recognized that walking and running were apparently highly inefficient modes of transportation, compared to swimming or flying. Having dealt effectively with the physiology of flying (Tucker, 1973), he recognized that substantially different constraints accounted for the "costs" associated with legged locomotion. Identifying logically that the metabolic cost of negative work in the muscles accounts for much of this apparent inefficiency, Tucker (1975) proposed that a system that eliminated the motivation to have the limb deflect the CoM at contact would avoid this important source of locomotion cost.

Tucker proposed that, to avoid such loss, the system, "...*applies force to the center of mass at right angles to its direction of motion. No work is done to change the velocity, for work is the product of a force and a displacement that are parallel to one another. When the force is at right angles to the displacement, the muscles that supply the force can neither do work nor have work done on them. The result is that the body is accelerated – that is, its velocity is changed to a new direction – at no expense for muscular work.*" (Tucker, 1975 p. 418). This perspective led him to suggest some interesting devices that would not have this loss, and to propose that bouncing was quite possible without the assistance of strain energy storage and return.

Just as Gray had begun with an analysis of fish form and locomotion, R McNeill Alexander acquired his initial interest in comparative morphology from working on a variety of aquatic forms. This changed during the 1960s to a more and more terrestrial perspective, and his work eventually became remarkably influential in modern human, animal and even legged robot research. This can be attributed largely to his approach which, again much like Gray, sought to find the fundamental mechanical factors that influenced form and performance. Alexander's work is characterized by relatively simple experiments that validate propositions based on stringent theoretical foundations, and is directed more to opening entire new directions of investigation, rather than meticulously documenting all aspects of each such system.

Although Alexander worked on a myriad of questions and structures in biology, his contributions to locomotion are defined by two basic directions of inquiry. The first is the analysis of form in (largely) musculoskeletal structures, including skeletal proportions (Alexander *et al.*, 1979), and the role of elastically behaving structures in locomotion (Alexander and Bennet-Clark, 1977). He also developed models of walking and running, working from basic mechanical principles. To both of these areas, a key feature of his approach was to develop optimization models through the identification of the key competing factors affecting the performance of the system.

Of particular importance in the consideration of the mechanics governing legged locomotion is a set of models of bipedal walking and running that were able to predict many aspects of human locomotion performance (Alexander, 1976). One main feature of these models was the consideration of kinetic energy loss associated with diverting the center of mass velocity vector at foot contact: "*The part $1/2mu_b^2$ of the kinetic energy* (referring to the horizontal component of the kinetic energy – editor's comment) *is carried over to the next step but the part $1/2mv_b^2$* (referring to the vertical kinetic energy – editor's comment) *is absorbed by the muscles as the descent is halted, and has to be replaced by work done by the muscles which give the center of mass the vertical component of velocity $+ v_b$ needed to start the next step.*" (Alexander, 1976, p. 494).

In this analysis, Alexander assumed that the angles with which the limb moved while on the ground were conveniently small, so the characterization of the loss as basically vertical, and the remaining kinetic energy as horizontal, are reasonably close. However, the component of the kinetic energy lost would actually be that in line with the axis of the new stance limb, having both horizontal and vertical components, and what remains would be the component perpendicular to the new stance limb (Garcia *et al.*, 1998; Donelan *et al.*, 2002b; Ruina *et al.*, 2005). In the model, Alexander associated the energy loss with the momentum of the body doing negative work on the muscles of the limb. This was in spite of the fact that the simple model in which this concept was introduced did not have a jointed limb in which negative work could be absorbed. Rather, the model identified this loss as a consequence of the geometry involved in diverting the travel of the center of mass. Although muscular absorption of work is a realistic mechanism, the loss dictated by the model originates with the dynamics of the supporting limb's deflection of the center of mass of the organism (see also Chapters 5 and 6).

As discussed above, Margaria (1976) attributed the cyclical losses involved in locomotion to negative work within the muscles. However, he also noted that force applied down the axis of the limb is the ultimate source of the work that must be actively absorbed by the muscles, and that such forces come from the geometry of the limb contact and the center of mass (as also described by Rashevsky, 1948, 1960; Bekker, 1956; Alexander, 1976). Margaria recognized that the horizontal component of the contact is involved with generating the horizontal deceleration, and that this would decrease the horizontal kinetic energy of the moving body. Likewise, the vertical component of the contact indicated loss in the vertical direction.

However, he did not expressly state that this implied that the component perpendicular to the contact strut (at right angles to the limb, as described by Tucker, 1975) was unaffected by the contact and would remain (leaving some portion of both vertically and horizontally oriented momentum to assist in the re-lifting the body in the next step): *"The negative work performed at each step is substantially that caused by the deceleration of the body when the forward foot strikes the ground."* (p. 105). Margaria, as Rashevsky had done previously, discussed the consequences of quadrupedal gait, suggesting that quadrupedalism had substantial disadvantages (p. 108). However, this conclusion depends on assumptions about the distribution of the body mass between limbs and the costs of other factors, such as stability.

Although Cavagna and colleagues (1976, 1977) influenced the direction of the field toward a focus on the mechanisms of energy recovery, they were aware of the mechanical energy loss that results from contact of the new stance limb and its redirection of the center of mass path. Citing Alexander's 1976 bipedal model, this group noted, *"In the simplest inverted pendulum system which has been used to characterize walking – the stiff-legged walk of Alexander (1)-E_{kv}* (where E_{kv} is the vertical kinetic energy – editor's comment) *would be lost from the system in each step when the front foot hits the ground and the transfer of kinetic energy into gravitational potential energy could only take place between E_{kf}* (referring to kinetic energy of forward motion – editor's comment) *and E_p* (referring to potential energy – editor's comment).*"* (Cavagna *et al.*, 1977, pp. R245 and 246).

Thus, Cavagna *et al.* (1977) recognized that it was likely that energetic complexities existed during the transition between supporting limbs. However, they elected to focus on the novel and physically compelling recovery mechanisms that they had identified. Due to this logically reasoned and extremely influential paper, the entire field was drawn away from the concept of energy loss at the transition, and drawn to the concept of identifying the proportion of mechanical energy recovered within the stride cycle. From this, the focus turned toward analyzing the morphological and behavioral adaptations available to allow for optimum recovery. Implicit in this approach, however, is the assumption that all available energy would be lost if specific strategies for recovery were not implemented – contrary to the implications of Alexander's analysis (or Bekker or Rashevsky).

Cavagna *et al.* (1977) was remarkably influential in the area of comparative mechanics of legged locomotion. Originating largely from the influence of this one paper, the concept of physically determined dissipative loss seems to have been neglected following its publication. Many factors combine to influence the focus of a field, but it is interesting to note that the key acknowledgement of the geometry associated with determining this aspect of loss (the relationship between ground reaction force generated by limb contact and the path of the CoM), as described by Alexander's model (and quoted above), comes in a sentence that is split between pages, with a large, important and complex figure intervening between the two halves of the sentence.

Is it possible that such a distracting presentation of a key sentence in an influential paper, due only to the serendipity of typesetting limitations, could have contributed to downplaying the importance of the concept and led to its neglect by almost the entire field? Certainly, it did not help to direct attention to this potentially important, but admittedly unintuitive, alternate perspective. On the other hand, it may have been inevitable that the field was drawn in different directions by the complexity of the issue of loss and recovery. The potential confusion is apparent if one considers a wheel: is the wheel a mechanically effective system because it has near perfect recovery or inconsequential loss?

McGeer (1990a, 1990b) recognized that inverted-pendular motion of a mass on a supporting limb could be added to McMahon's ballistic swing limb model (Mochon and McMahon, 1980a, 1980b). By further recognizing that the main source of loss in this system comes from the inevitable mechanical energy losses that occur when the center of mass path is diverted from one support limb to the next, McGeer was able to optimize the system to minimize this loss. Thus, he created a passive dynamic "pseudo-bipedal" walking machine with knees and feet. Note that this was by no means the first passive dynamic walking machine, as there had been such toys available for over 100 years (Fallis, 1888). However, none of these had knees, and none depended on dynamic stability (i.e., the McGeer walker was only stable while moving, and it was essentially unable to stand when not moving).

The original McGeer walker was dynamically stable only in the forward-rearward direction. Lateral stabilization was artificially engineered by having laterally paired legs that provided a wide base of support to prevent sideways pitching. Although it only had two "feet", it was a functional "biped" in only two dimensions. However, this did demonstrate that forward-rearward passive stabilization was not particularly difficult in motion.

Ruina and colleagues refined the McGeer walker to become a truly passive dynamic biped (Figure 1.3; Collins *et al.*, 2001). They then added simple actuators to demonstrate that powered walking with a minimal system was possible (Collins *et al.*, 2005).

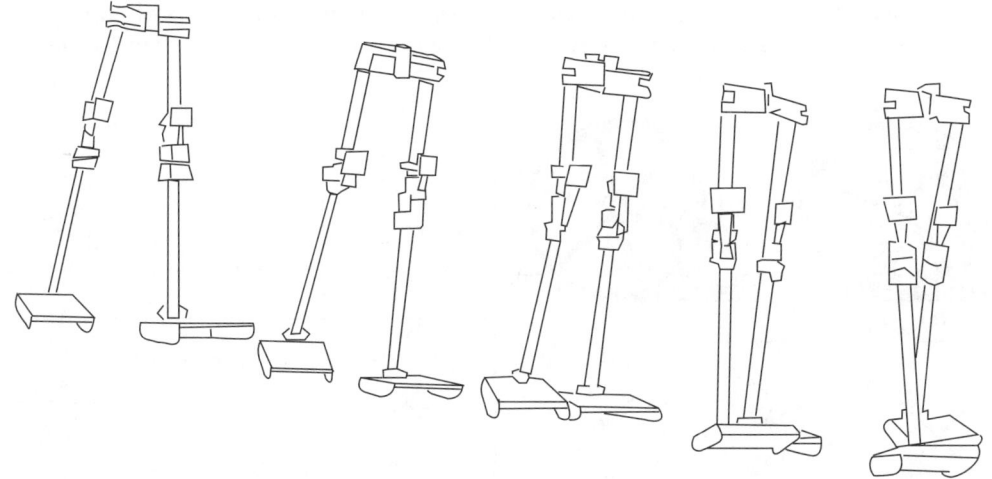

FIGURE 1.3

Tracing of a passive dynamic walking robot. This machine is composed of two legs, each with knees and a shaped foot. It has no motors or control, but "walks" down a shallow slope powered by gravity. Many characteristics of human walking are spontaneously generated by this machine, including pelvic tilt and rotation and the turning motion of the stance foot during mid to late stance (from a video supplied by A. Ruina).

Interestingly, the proportions that make the system work well are eerily similar to those of a bipedal human (McGeer, 1993). This observation raises the question of whether human proportions might be fundamentally based on exploiting subtle mechanical opportunities for energy conservation and control that had not previously been appreciated.

Smith and Berkemeier (1997) applied a similar passive dynamic approach to quadrupedal locomotion, using a very simple model, with some interesting and unexpected results (see discussion of this model in Chapter 6). For instance, their analysis suggests that many aspects of bipedal and quadrupedal walking are similar, though quadrupedal walking should be *more* efficient than bipedal (contrary to Margaria's expectation). However, the quadrupedal system appears naturally unstable with respect to the phase relationship between limbs. If verified, we might anticipate that such phase instability would have to be actively determined – for instance, through a neural cycling system like the central pattern generator. At this time, quadrupedal passive dynamics has not been thoroughly or explicitly analyzed, largely because the addition of a second pair of limbs substantially increases the complexity of dynamic interactions of the components of the system (trunk, limbs, etc.) and the substrate. However, this remains one of the more compelling current questions in mammalian locomotion.

The successive contacts of the limbs are analogous to the contact of spokes of a rimless wheel (McGeer, 1993, p. 280). Even though passive dynamic walking machines are fairly recent in origin, the analogy of the rimless wheel has quite a long history. Although not explicitly described, the "zig-zag" motion of Aristotle derives from the limbs acting as spokes of a wheel. This analogy was explicitly indicated by Gray (1959), Margaria (1976) and McGeer (1993) (Figure 1.4). The rimless wheel is the simplest depiction of the substrate interactions of the contacting support limb, while neglecting any aspects of swing limb mechanics. Though simple, it has formed the

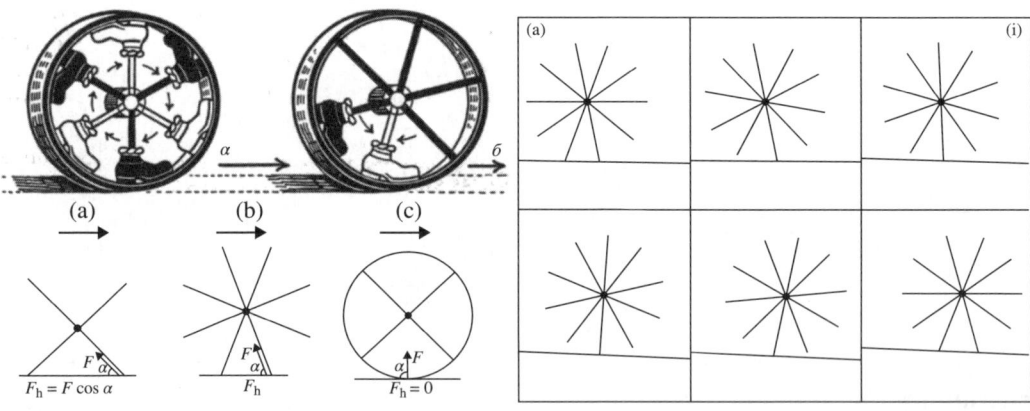

FIGURE 1.4

Some examples of using the rimless wheel as a starting point for analyzing the dynamics of bipedal locomotion. Above left: illustration from Gray (1959, p. 19). Bottom left: illustration from Margaria (1976, p. 106). Right: a portion of an illustration from McGeer (1993, p. 280).

starting point of a new set of analytic models of limbed locomotion (Alexander, 1976; Garcia *et al.*, 1998; Smith and Berkemeier, 1997). Implicit in these models is the assertion that the contact dynamics makes up an important (critical?) component of the mechanical interaction of legged locomotion. Unfortunately, this aspect of the mechanics appears to have been largely neglected until recently, and we currently await explicit application of these ideas to the analysis of limb function and the consequences of their form on determining animal performance and energetics.

1.9   CONCLUSIONS

It is likely that transfer between energy types – be it pendular, elastic, or the relatively under-investigated rotational kinetic energy – plays a substantial role in the economical production of locomotion and the provision of stability. However, it is also likely that dissipative effects of contact dynamics also play a role in determining how mammalian systems best utilize their opportunities for moving in the terrestrial environment. Yet to be determined are the relative contributions of dissipative and recovery components, and how these opportunities are provided for by specific behavioral and morphological "strategies". Currently, the implications of each are not well understood, largely because most analyses have not recognized that loss and recovery might be integrated in non-intuitive ways.

   In briefly reviewing the history of animal locomotion studies, I have neglected a great deal of very important and insightful work that has been done on specific features of the musculoskeletal system, including such issues as muscle chains of the limbs or trunk, connective tissues and their properties, joint function and the relationships between all of these. However, such detailed information only leads to general understanding of the organism, its functional abilities and constraints, when it is appropriately placed within the context of an understanding of the system of which it is a part. At this juncture, it is not certain that the system as a whole, even modeled at a simplistic level, is properly understood. It is hoped that the chapters of this volume will help to stimulate the discussion that will put such information in the context it deserves.

▌ REFERENCES

Amar J (1916). Trottoir dynamographique. *Comptes Rendus de l'Académie des Sciences – Series I* **163**, 130–132.

Amar J (1919). *Revue de Chirurgie* LVII, 539.

Alexander RMcN (1976). Mechanics of bipedal locomotion. In: Spencer PS (ed). *Perspectives in experimental biology, Volume 1, Zoology*. Pergamon Press, Oxford, pp. 493–504.

Alexander RMcN (1988). Why animals gallop. *American Zoologist* **28**, 237–245.

Alexander RMcN, Bennet-Clark HC (1977). Storage of elastic strain energy in muscle and other tissues. *Nature* **265**, 114–117.

Alexander RMcN, Jayes AS, Maloiy GMO, Wathuta EM (1979). The allometry of the limb bones of mammals from shrews (*Sorex*) to elephant (*Loxodonta*). *Journal of Zoology, London* **189**, 305–314.

Amar J (1920). *The Human Motor*. New York: EP Dutton, pp. 282.

Ashley-Ross MA, Gillis GB (2002). A brief history of vertebrate functional morphology. *Integrative and Comparative Biology* **42**, 183–189.

Bekker MG (1956). *Theory of land locomotion*. University of Michigan Press, Ann Arbor.

Bekker MG (1985). The development of a moon rover. *Journal of the British Interplanetary Society* **38**, 537.

Bertram JEA, Gutmann A (2008). Motions of the running horse and cheetah revisited: fundamental mechanics of the transverse and rotary gallop. *Journal of the Royal Society Interface* **6**, 549–559.

Bertram JEA, Ruina A, Srinivasan M (2006). *Collision costs in terrestrial gaits*. V World Congress of Biomechanics, Munich, July 2006.

Biewener AA (1983). Locomotory stresses in the limb bones of two small mammals: the ground squirrel and chipmunk. *Journal of Experimental Biology* **103**, 131–154.

Biewener AA, Blickhan R, Perry AK, Heglund NC, Taylor CR (1988). Muscle forces during locomotion in kangaroo rats: force platform and tendon buckle measurements compared. *Journal of Experimental Biology* **137**, 191–205.

Blickhan R, Full RJ (1992). Mechanical work in terrestrial locomotion. In: Biewener A. (ed). *Biomechanics: Structures and Systems, A practical approach*. IRL at Oxford Press, New York, pp. 75–96.

Borelli G.A. (1680–1). *De Motu Animalium*. Bernabo, Rome. Translation (1989) as *On the Movement of Animals*. Springer, Berlin.

Braun M (1992). *Picturing time: The work of Etienne-Jules Marey*. The University of Chicago Press.

Camp LC, Smith N (1942). Phylogeny and function of digital ligaments of the horse. *Memoirs of the University of California* **18**, 69–123.

Cappozzo A, Marchetti M, Tosi V (1992). *Biolocomotion: a century of research using moving pictures*. Promograph, Rome.

Cavagna GA, Kanenko M (1977). Mechanical work and efficiency in level walking and running. *Journal of Physiology* **268**, 467–481.

Cavagna GA, Saibene FP, Margaria R (1963). External work in walking. *Journal of Applied Physiology* **18**, 1–9.

Cavagna GA, Heglund NC, Taylor CR (1977). Mechanical work in terrestrial locomotion: two basic mechanisms for minimizing energy expenditure. *American Journal of Physiology - Regulatory, Integrative and Comparative Physiology* **233**, 243–261.

Collins S, Wisse M, Ruina A (2001). A three-dimensional passive-dynamic walking robot with two legs and knees. *International Journal of Robotics Research* **20**, 607–615.

Collins S, Ruina A, Tedrake R, Wisse M (2005). Efficient bipedal robots based on passive-dynamic walkers. *Science* **307**, 1082–1085.

Coleman MJ, Chatterjee A, Ruina A (1997). Motions of a rimless spoked wheel: a simple three-dimensional system with impacts. *Dynamical Systems* **12**, 139–159.

Des Chene DD (2005). Mechanisms of life in the seventeenth century: Borelli, Perrault, Regis. *Studies in History and Philosophy of Biological and Biomedical Sciences* **36** (2), 245–260.

Donelan JM, Kram R, Kuo AD (2002a). Simultaneous positive and negative external mechanical work in human walking. *Journal of Biomechanics* **35**, 117–124.

Donelan JM, Kram R, Kuo AD (2002b). Mechanical work for step-to-step transitions is a major determinant of the metabolic cost of human walking. *Journal of Experimental Biology* **205**, 3717–3727.

Drake S (1989). *Galileo Galilei, Discourses on two new sciences* (2nd edition). Wall & Emerson, Inc. Toronto.

Elftmann, HO (1934). A cinematic study of the distribution of pressure in the human foot. *Anatomical Record* **59**, 481–491.

Fallis GT (1888). *Walking toy*. US Patent No. 376,588.

Farquharson ASL (2007). *On the gait of animals by Aristotle.* (http://etext.library.adelaide.edu. au/a/aristotle/gait/complete.html) eBooks@Adelaide, University of Adelaide Library.

Fenn WO (1924). The relation between the work performed and the energy liberated in muscular contraction. *Journal of Physiology* **58**, 373–395.

Fenn WO (1930a). Frictional and kinetic factors in the work of sprint running. *American Journal of Physiology* **92**, 583–611.

Fenn WO (1930b). Work against gravity and work due to velocity changes in running. *American Journal of Physiology* **93**, 433–462.

Furusawa K, Hill AV, Parkinson JT (1927). The dynamics of "sprint" running. *Proceedings of the Royal Society of London. Series B* **102** (713), 29–42.

Gambaryan PP (1974). *How animals run.* Chapter 2: History of the study of locomotion, pp. 14–19. John Wiley & Sons, New York.

Garcia M, Chatterjee A, Ruina A (1998). The simplest walking model: stability, complexity and scaling. *Journal of Biomechanical Engineering* **120**, 281–288.

Gasser HS, Hill AV (1924). The dynamics of muscular contraction. *Proceedings of the Royal Society B: Biological Sciences* **96**, 398–437.

Gatesy SM, Middleton KM (1997). Bipedalism, flight, and the evolution of theropod locomotor diversity. *Journal of Vertebrate Paleontology* **17**, 308–325.

Geyer H, Seyfarth A, Blickhan R (2006). Compliant leg behaviour explains basic dynamics of walking and running. *Proceedings of the Royal Society B* **273**, 2861–2867.

Goiffon and Vincent (1779). *Memoire artificielle des principes relatifs a la fidele representation des animaux tant en peinture, qu'en scultpture. I partie concernant le cheval.* (Memoire on the artificial principles of faithful representation of animals both in painting and sculpture. Part I concerning the horse). Alfort. (reference from Gambaryan, 1974).

Gray J (1944). Studies in the mechanics of the tetrapod skeleton. *Journal of Experimental Biology* **20**, 88–116.

Gray J (1959). *How animals move.* Pelican/Penguin Books, Harmondsworth, UK.

Gray J (1968). *Animal Locomotion.* Weidenfield & Nicolson, London.

Griffin TM, Kram R (2000). Penguin waddling is not wasteful. *Nature* **408**, 929.

Griffin TM, Kram R, Wickler SJ, Hoyt DF (2004). Biomechanical and energetic determinants of the walk-trot transition in horses. *Journal of Experimental Biology* **207**, 4215–4223.

Heglund NC (1981). A simple design for a force-plate to measure ground reaction forces. *Journal of Experimental Biology* **93**, 333–338.

Heglund NC, Fedak MA, Taylor CA, Cavagna GA (1982). Energetics and mechanics of terrestrial locomotion. IV. Total mechanical energy changes as a function of speed and body size in birds and mammals. *Journal of Experimental Biology* **97**, 57–66.

Herr HM, McMahon TA (2000). A trotting horse model. *International Journal of Robotics Research* **19**, 566–581.

Herr HM, McMahon TA (2001). A galloping horse model. *International Journal of Robotics Research* **20**, 26–37.

Hildebrand M (1962). Walking, running and jumping. *American Zoologist* **2**, 151–155.

Hildebrand M (1965). Symmetrical gaits of horses. *Science* **150**, 701–708.

Hildebrand M (1977). Analysis of asymmetrical gaits. *Journal of Mammalogy* **58**, 131–156.

Hildebrand M (1989). The quadrupedal gaits of vertebrates. *Bioscience* **39**(11), 766–776.

Hildebrand M, Hurley JP (1985). Energy of oscillating legs of fast-moving cheetah, pronghorn, jackrabbit and elephant. *Journal of Morphology* **184**, 23–31.

Hill AV (1927). *Muscular movements in man: the factors governing speed and recovery from fatigue.* McGraw-Hill, New York.

Howell AB (1944). *Speed in mammals.* Chicago University Press, Chicago.

Jarrett MO, Moore PR, Swanson AJG (1980). Assessment of gait using components of the ground reaction force vector. *Medical & Biological Engineering & Computing* **18**, 685–688.

Jayes AS, Alexander RMcN (1978). Mechanics of locomotion of dogs (*Canis familiaris*) and sheep (*Ovis aries*). *Journal of Zoology* **185**, 289–308.

Kuo AD (2007). The six determinants of gait and the inverted pendulum analogy: A dynamic walking perspective. *Human Movement Science* **26**, 617–656.

Margaria R (1938). Sulla fisiologia e specialmente sul consumo energetico della Marcia e della corsa a varie velocità ed inclinazioni del terreno. *Atti della Accademia Nazionale dei Lincei Memorie, serie VI* **7**, 299–368.

Margaria R (1976). *Biomechanics and energetics of muscular exercise*. Clarendon Press, Oxford, 146 pp.

Margaria R, Edwards HT (1934). The source of energy in muscular work performed in aerobic conditions. *American Journal of Physiology* **108**, 341–348.

Margaria R, Edwards HT, Dill DB (1933). The possible mechanism of contracting and paying the oxygen debt and the role of lactic acid in muscular contraction. *American Journal of Physiology* **106**, 689–715.

Margaria R, Cerretelli P, Aghemo P, Sassi G (1963). Energy cost of running. *Journal of Applied Physiology* **18**, 367–370.

Marey E-J (1873). *La machine animale: Locomotion terrestre et aeriemme*. Paris, Balliere. Translated as: *Animal Mechanism: A treatise on terrestrial and aerial locomotion*. 3rd ed. New York, Appleton, 1884.

Marey E-J (1892). Le fusil photographique. *La Nature* **10**(1), 320–330.

Marey E-J (1899). *La chronophotographie*. Guathier-Villars (ed). Paris.

McGeer T (1990a). Passive dynamic walking. *International Journal of Robotics Research* **9**, 68–82.

McGeer T (1990b). Passive dynamic running. *Proceedings of the Royal Society of London. Series B* **240**, 107–134.

McGeer T (1993). Dynamics and control of bipedal locomotion. *Journal of Theoretical Biology* **163**, 277–314.

McMahon TA (1973). Size and shape in biology. *Science* **179**, 349–351.

McMahon TA (1975). Allometry and biomechanics: limb bones in adult ungulates. *American Naturalist* **109**, 547–563.

McMahon TA (1985). The role of compliance in mammalian running gaits. *Journal of Experimental Biology* **115**, 263–282.

Medved V (2001). *Measurement of Human Locomotion*. Chapter 1.2: The history of locomotion measurement, p. 5–14. CRC Press LLC, Boca Raton.

Mena D, Mansour JM, Simon SR (1981). Analysis and synthesis of human swing leg motion during gait and its clinical applications. *Journal of Biomechanics* **14**, 823–832.

Mochon S, McMahon T (1980a). Ballistic walking. *Journal of Biomechanics* **13**(1), 49–57.

Mochon S, McMahon T (1980b). Ballistic walking: an improved model. *Mathematical Biosciences* **52**, 241–260.

Muybridge E (1887). Animals in motion. Brown LS (ed). (1957). New York, NY: Dover Publications, Inc. (Dover Publications edition).

Pennycuick CJ (1975). On the running of the gnu (*Connechaetes taurinus*) and other animals. *Journal of Experimental Biology* **63**, 775–799.

Pfau T, Witte TH, Wilson AM (2006). Centre of mass movement and mechanical energy fluctuation during gallop locomotion in the Thoroughbred racehorse. *Journal of Experimental Biology* **209**, 3742–3757.

Rashevsky N (1944). General theory of quadrupedal locomotion. *Bulletin of Mathematical Biology* **6**, 17–32.

Rashevsky N (1946). Some considerations on the structure of quadrupedal extremeties. *Bulletin of Mathematical Biology* **8**, 83–93.

Rashevsky N (1948). On the locomotion of mammals. *Bulletin of Mathematical Biology* **10**, 11–23.

Rashevsky N (1960). *Mathematical biophysics: Physico-mathematical foundations of biology. Form and locomotion in some quadrupeds*, pp. 262–269. Vol. 2, 3rd revised edition. (first edition published 1938). Dover Pub., New York.

Rashevsky N (1961). *Mathematical principles in biology and their applications.* Pub. 114, American Lecture Series. Charles C. Thomas, Springfield.

Raspet A (1960). Biophysics of bird flight. *Science* **132**, 191–200.

Richter IA, Wells T (2008). *Leonardo Da Vinci Notebooks.* Oxford University Press, New York (original publication 1952).

Roberts TJ (2005). A step forward for locomotor mechanics. *Journal of Experimental Biology* **208**, 4191–4192.

Ruina A, Bertram JEA, Srinivasan M (2005). A collisional model of the energetic cost of support work qualitatively explains leg sequencing in walking and galloping, pseudo-elastic leg behavior in running and the walk-to-run transition. *Journal of Theoretical Biology* **237**, 170–192.

Selles RW, Bussmann JB, Wagenaar RC, Stam HJ (2001). Comparing predictive validity of four ballistic swing phase models of human walking. *Journal of Biomechanics* **34**, 1171–1177.

Smith AC, Berkemeier MD (1997). *Passive dynamic quadrupedal walking.* Proceedings of IEEE International Conference on Robotics and Automation, pp. 34–39.

Taylor CR (1994). Relating mechanics and energetics during exercise. In: Jones JH (ed). *Comparative Vertebrate Exercise Physiology: Unifying physiological principles*, Volume 38A, Advances in Veterinary Science and Comparative Medicine (CE Cornelius, RR Marshak, EC Melby, eds), Academic Press, pp. 181–215.

Taylor CR, Shkolnik A, Dmi'el R, Baharav D, Borut A (1974). Running in cheetahs, gazelles, and goats: energy cost and limb configuration. *American Journal of Physiology* **227**, 848–850.

Thys H, Faraggiana T, Margaria R (1972). Utilization of muscle elasticity in exercise. *Journal of Applied Physiology* **32**, 491–494.

Tucker VA (1973). Bird metabolism during flight: Evaluation of a theory. *Journal of Experimental Biology* **58**, 689–709.

Tucker VA (1975). The energetic cost of moving about. *American Scientist* **63**, 413–419.

Walker WF (1972). Body form and gait in terrestrial vertebrates. *Ohio Journal of Science* **72**(4), 177–183.

Wells T (1952). *Leonardo Da Vinci Notebooks.* Oxford World Classics, Oxford Univ. Press, Oxford.

CHAPTER TWO

Considering Gaits: Descriptive Approaches

John E. A. Bertram

Department of Cell Biology and Anatomy, Cumming School of Medicine, University of Calgary, CA

2.1 INTRODUCTION

A gait is a strategy for using the body to generate movement through the physical environment. The term is most commonly applied to the footfall patterns of legged gaits in terrestrial animals – particularly that of mammals – but the concept applies to identifiable movement patterns of any locomotion. For example, specific gaits and changes between gaits have been documented in crabs (Blickhan and Full, 1987), fish (Webb, 1994), swimming frogs (Nauwelaerts *et al.*, 2001) and flying birds (Hedrick *et al.*, 2002).

If a gait represents a strategy for moving, why does an organism choose that particular strategy, or why does an organism change strategies for different conditions or speeds? In several of the chapters in this volume specific factors will be discussed that attempt to answer these and related questions (see, for instance, Chapter 6). An important initial step in approaching such questions in the context of legged terrestrial

Understanding Mammalian Locomotion: Concepts and Applications, First Edition.
Edited by John E. A. Bertram.
© 2016 John Wiley & Sons, Inc. Published 2016 by John Wiley & Sons, Inc.

mammals is describing the differences in the manner of movement of each identifiable gait. This descriptive step is required, so that the details of the movement pattern can be organized and assessed with regard to any mechanical or physiological explanations that may emerge from a more functional analysis. Description of the gait, however detailed and quantified, does not in itself "explain" the gait – that requires an understanding of the dynamic consequences of the movement pattern (where the term "dynamic" is used in its mechanical context, referring to physical interactions involving accelerations and acting with respect to time).

Common gait patterns in terrestrial mammals have been identified and classified, based on shared features of commonly expressed movement patterns. The most easily recognized and fundamental quadrupedal terrestrial gaits are the walk, trot and gallop. Often, the canter is also considered, but the canter does not differ qualitatively from the gallop, and it can be considered to be simply a slow-speed version of the gallop. Likewise, the mechanics of other bounding gaits commonly used particularly by smaller mammals, such as the bound and half-bound, will be discussed at more length later (see Chapter 5) and will be described as having substantial fundamental features in common with galloping gaits. This chapter will focus only on how the motions in such gaits are currently categorized.

Hildebrand formally identified the symmetric and asymmetric gaits as both fundamental and fundamentally different. In symmetric gaits, both feet of a pair (fore or hind) move in a manner equally spaced in time (Hildebrand, 1966). The normal bipedal walk and run are symmetric gaits, as are the quadrupedal walk and trot. The quadrupedal walk appears to have much in common with the bipedal walk (Alexander, 1984), and the quadrupedal trot appears functionally similar to the bipedal run, except that a fore-hind pair of limbs contacts alternately, rather than the alternating individual left-right pelvic limbs of the biped (Cavagna *et al.*, 1977).

In asymmetric gaits, the movement and timing of each limb of a pair is not equal. The gallop and canter of the quadrupeds are asymmetric, where limbs on either side of the trunk function somewhat differently than their contralateral partner. The bipedal skipping gait, most notably used by the Sifaka (*Propithecus* sp., Figure 2.1), is an asymmetric bipedal gait.

2.2 DEFINING THE FUNDAMENTAL GAITS

Mammalian gaits have been discussed by a great many individuals, each coming from a range of perspectives and interests with regard to their purpose for exploring the motions observed. Not surprisingly, much of the terminology used in the discussion of gait derives from horsemanship, itself a diverse field with nomenclature that has a variety of origins. The discussion of gait can easily become mired in subtle details that are not of great functional value, especially when the objective is a concise understanding of the movement patterns used and their value in the context of the motions, species and environment involved. A valuable description of the fundamental mammalian gaits, generated with the purpose of application in a diverse zoological context, is provided in Dagg (1973). These definitions are reproduced here, in the hope that the simple descriptions will promote a more consistent use of the terms.

Lands on
left rear limb

Launches from
right rear limb

FIGURE 2.1

Tracings of the distinctive two contact bound, or bipedal skip, of the Sifaka (*Propithecus* sp.). Sifakas are primarily arboreal climbers and leapers, but they use the skip to cross open spaces between trees. These animals bound between sequentially placed hind foot contacts, while the body is maintained in an upright position without the forelimbs making contact with the substrate.

From Dagg (1973):

- "**Walk:** The slowest gait, in which two, three or four legs support the body at any one time. It is symmetrical, with the left legs repeating the movements of the right legs, half a stride later. It can be subdivided into various components (Hildebrand, 1966).

- "**Running walk:** A quick walk sometimes called an *amble*, with the footfalls the same as those in the walk. At least one hind and one front leg usually support the animal. This gait cannot be sustained for long periods.

- "**Trot:** The symmetric gait of intermediate speed, in which two diagonal legs usually support the body when it is in contact with the ground. It is not synonymous with the slower "walking trot" defined by Hildebrand (1967).

- "**Pace:** The symmetric gait of intermediate speed, in which two lateral legs usually support the body when it is in contact with the ground. Pace, not rack, is the accepted word for this gait in harness racing of horses around the world.

- "**Gallop:** The fastest gait, in which the body is often unsupported following a push-off with the hind legs and sometimes with the forelegs. It is asymmetrical, with the right and left legs doing different movements in a stride. This gait includes the canter and the lope. These movements may intergrade with those of the bound in many species.

- "**Bound:** A fast gait in which the front legs and back legs move together in pairs. The *half-bound* is included in this category. The push-off is always from the hind feet but, in the bound, the animal lands on both forefeet together. Since this cuts down the velocity of forward motion considerably, many animals put one front leg down and then the other, which is the half-bound.

- "**Stott:** This gait, also called the "pronk" or "spronk", is performed with all four legs taking off and landing together. During the period of suspension, the legs hang down vertically from the body (*editor's note*: the term stott is of Scottish derivation and means "bounce". The term pronk has Afrikaans origin. Numerous African gazelle species utilize this gait. The word has an origin related to the term "prance").

Although a useful organization of the terminology, some of Dagg's original "editorial" comments cannot be depended upon completely. The running walk, also known as the tölt, that is used by the Icelandic horse breed (or the Paso Fino South American breed), for instance, is a fast-moving gait that can be sustained as long as any other fast moving gait (Biknevicius *et al.*, 2006). I might also suggest that, at least for bipedal humans, it is possible to run in place, while it is not possible to use a true "walk" in place. The walk, therefore, is not necessarily the "slowest" gait, at least under some circumstances.

It would also be possible to define a large number of subtle variations to any of these gaits, as organisms are capable of a substantial range of coordination patterns, as suits their particular movement challenges. Two other general gaits should not be neglected: the ricochetal bounding hop, commonly referred to as a salutatory gait; and the asymmetric bipedal skip.

The ricochetal bounding hop, or saltatory gait, uses the paired hind limbs in a bouncing gait characteristic of kangaroos and several of the smaller rodent species (spring hares, genus *Pedetes*; kangaroo rats, genus *Dipodomys*; jerboas, subfamily Cardiocraniinae; gerbils, subfamily Gerbillinae and hopping mice, genus *Notomys*). Although the sportive lemurs (primates, family Lepilemuridae) are primarily arboreal, they use saltatory locomotion when on the ground.

The asymmetric bipedal skip is uncommon. It is most notably used by the Sifaka when on the ground (Figure 2.1). Like the sportive lemurs, the Sifaka is a primarily arboreal lemur. The bipedal skip is also employed by some birds and the jerboas under some circumstances (Alexander, 2004).

2.3 CLASSIFYING AND COMPARING THE FUNDAMENTAL GAITS

In order to compare the movement patterns of different gaits (whether these are gaits as utilized in different circumstances by the same organism, or gaits employed by different species), it is necessary to have a system that allows quantification of the distinguishing features characteristic of each. Although this might seem an obvious starting point for understanding terrestrial locomotion, it is not at all obvious how movement should be quantified.

Each animal utilizes complex motions as it extends the limbs, contacts the substrate, bears load, moves over the limb during support, and then folds, swings the limb forward and prepares for the next contact. Throughout the gait cycle motions of joints, the position of limb and body components and the force applied to the ground will dynamically change. Beyond the individual limb context, each limb of a pair functions relative to its contralateral partner, and, for quadrupedal forms, the cranial pair of limbs function relative to the caudal pair. At the whole-animal level, gait patterns function with variables that depend on amplitude, frequency and phase, while all of these have effects on the motion of the body, as well as inter-connected elements such as the neck, head, back and tail (not to mention internal organs that can also slosh around inside the body cavity). What, then, should be quantified to characterize each gait?

In observing the action of the limbs during terrestrial locomotion, it is apparent that legged locomotion progresses through a series of steps mediated by the contact of the limbs with the substrate, and that these individual steps repeat as a cycle of footfalls. The repetitive sequence of footfalls that includes a contact for all limbs involved in locomotion is referred to as a *stride*. Thus, a stride is the most basic repeating unit of the gait, and includes the contact of all limbs from the initial contact, through each of the limbs involved in ground contact and support in sequence, until the initial limb makes its subsequent contact and initiates the next stride.

Gaits are most often quantified using the methods developed by Hildebrand (1965, 1966). In developing these methods, Hildebrand observed that previous classification systems did not reflect the duration of different phases of the moving quadruped. Prior to Hildebrand's work, gaits were traditionally depicted as contact sequences. Vincent and Goiffon (1779) are usually acknowledged as the first to try to objectively assess the patterns involved in equine gait. They used small bells attached to the hooves of horses, and noted the auditory sequence of footfalls distinctive to each gait. Marey (1873) later recorded pressure changes from pneumatic sensors attached to the hooves. In these analyses, the succession and duration of support (limb contact with the ground) and swing (no contact between the limb and the ground) phases, for all four limbs, are depicted along a progressive time axis (Figure 2.2).

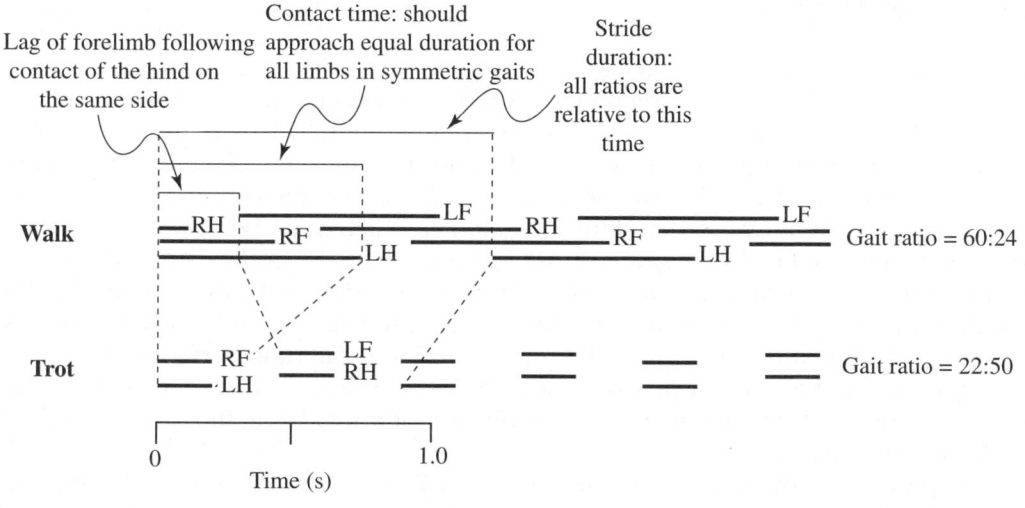

FIGURE 2.2

Comparison of the walk (above) and trot (below) gaits, using the graphical method originally devised by Marey (1873). The sequence of contacts reads from left to right, where the bold horizontal line indicates contact. Duration of contact or non-contact of each limb is indicated relative to a horizontal time axis. Hildebrand (1965) employed Marey's gait diagram to define three fundamental variables that characterize symmetric gaits. The measurements used to calculate the gait ratio (Hildebrand, 1965) are indicated and corresponding points of each gait are depicted with dashed lines. LH = left hind contact; LF = left fore contact; RF = right fore contact; RH = right hind contact.

Following Marey's measurements, Raabe (1883) developed a circular plot of limb contact and non-contact to depict the patterns that represent the cyclic nature of equine gaits. Although such graphs provide a visual indication of the gait that allows a subjective comparison (between different gaits such as the walk and trot, for example), they do not provide easily quantifiable or meaningful comparisons, especially when comparing between species (where size differences can generate differences that may or may not be functionally meaningful – see Chapter 9). Based on the time sequence of footfall contacts, Hildebrand calculated a number of variables that allow quantitative comparison between different gaits, speed, individuals or species (Figures 2.2 and 2.3).

The objective of a good gait metric is to identify the smallest number of variables that distinguish effectively between the broadest range of gaits. Hildebrand was unable to identify an analysis comprehensive enough to include both the symmetric and asymmetric gaits which was also able to distinguish subtle differences between species within each of these two fundamental modes of limb use. As a result, he developed analyses specific for, and limited to, either the symmetric (Hildebrand, 1965, 1966) or the asymmetric gaits (Hildebrand, 1977). Different variables are calculated depending on whether the study involves a symmetric or an asymmetric gait so, for these analyses, the type of gait has to be assessed prior to its evaluation in order for the correct measurements to be made.

2.4 SYMMETRIC GAITS

Symmetric gaits can be characterized by three measurable features: stride duration, duty factor, and phase lag between limbs. Valid comparisons between the locomotion of animals that may be of very different size, limb length, or operating at different speeds, will depend critically on a method of normalizing the absolute values measured. This is in order to bring the broad range of possible absolute values within the same comparative range, so that differences in gait pattern can be distinguished from differences due to the size or speed of particular species. The cyclic nature of limb use suggests that a fundamental feature of any gait is the stride cycle, as defined from the touch down of one limb to its next touch down. At faster speeds, or for smaller species, the stride cycle will occur over a brief time period, but the time period of the events occurring within the cycle will also be limited. Comparing specific events to the stride cycle of a species, then, provides a reasonable way of normalizing the actions involved with the gait pattern.

The proportion of support provided by each limb – the duration of the support relative to the stride time – is referred to as the *duty factor* of each limb (a term derived from the concept of duty cycle in engines). The duty factor is a key functional determinant of the gait, as it indicates the potential a limb has for contributing to the support and propulsion of the animal (Alexander *et al.*, 1980). It is only during ground contact that meaningful forces can be generated against the substrate in order to support the animal's body weight against gravity. It should be noted that, just because the limb is in contact with the substrate, it does not necessarily indicate that appreciable force is applied. However, under normal circumstances, this is the only opportunity for the limb to contribute appropriately in order to provide for effective locomotion, and

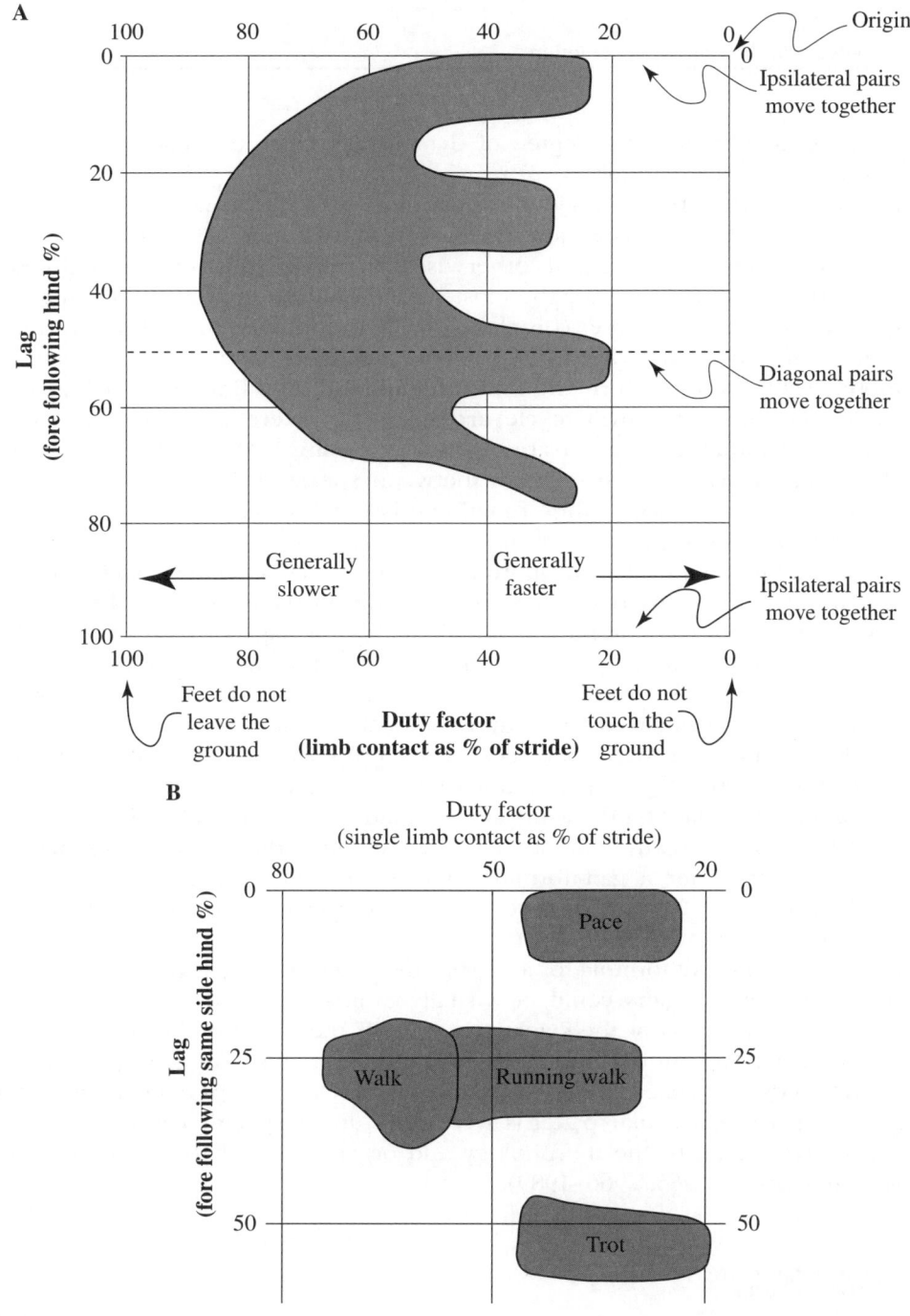

FIGURE 2.3

(**A**) The 'Hildebrand plot' of gait formulae for symmetric gaits. The shaded region indicates the gaits used by a wide range of terrestrial mammals (Hildebrand, 1965). (**B**) The gait formula region occupied by common symmetric equine gaits. The trot and pace are both running gaits – hence, smaller duty factors. In the trot, a diagonal pair of front and rear limbs contact simultaneously while in the pace, ipsilateral limbs move together. In the walk, the limb movements are generally equivalent (lag is approximately 25% between any limb), and the duty factor is relatively high (slow progression). The running walk (grounded walk – see Chapter 10, variously the paso or tölt, depending on horse breed) also has evenly distributed contacts, but the speed is greater (with a consequently lower duty factor).

it can be expected that some degree of dynamically changing load-bearing occurs during the contact of each limb.

Hildebrand (1965, 1966) based his quantitative characterization of symmetric gaits on a few important observations. In a symmetric gait, the stride time of the fore and hind limbs will be equal, otherwise their phase relationship will change over time. He also noted that, for the most part, the duration of ground contact will be equal for fore and hind feet (on close inspection, this is not absolutely correct (Bertram *et al.*, 2000), but the differences are not great enough to invalidate this assumption). As a result, information on forelimb and hind limb use, such as swing and stance portions of the stride cycle, are essentially interchangeable. Normalizing to stride cycle duration compensates for many timing features that change with speed between gaits, or as size differs between species (although see Chapter 9 for a discussion of the more subtle functional issues involved with the size scaling of locomotor function).

Even though near equivalent proportional contact occurs between limbs in a symmetric gait, there are a variety of phase relationships possible between the actions of fore and hind limbs. Hildebrand quantified this as the lag between contact of the forelimb that follows contact of the hind limb on the same side. Again as an absolute duration, this measure is sensitive to speed and size of animals compared, as well as gait used, so it was normalized to the duration of the stride cycle.

Hildebrand found that these two measures – proportion of stride cycle that each foot contacts the substrate, and proportion of the stride cycle that forelimb contact lags behind hind limb contact on the same side – could characterize all variations of symmetric gaits used by quadrupedal animals (Hildebrand, 1966). These two measures could be used as either a gait formula, where the number pair characterizes the symmetric gait (Figure 2.2), or could serve as the coordinates on a plot (often referred to as a "Hildebrand plot" (Figure 2.3)).

By measuring the gait formula for a variety of species, the organization and relationship of the symmetric gaits could be visually evaluated. The best example of this derives from the analysis of the symmetric gaits of the domestic horse (Hildebrand, 1965), where well-recognized gaits occupy specific regions of the gait plot (see Figure 1.2). Hildebrand used the identification of gait relationships to begin the process of discovering why particular patterns were used under specific circumstances, and what this might mean to the morphology and organization of the limbs of specific species (Hildebrand, 1965, 1966, 1989).

2.5 ASYMMETRIC GAITS

For asymmetric gaits, the two limbs of a pair are not equally spaced in time. Conventional terminology emphasizes spatial position over temporal sequence in characterizing the role of the limb (which may appear confusing). The limb referred to as "trailing" is actually the first to make contact with the ground, while the "lead" limb is the second to make contact in the cycle; however, it remains the limb in contact at the cranial-most position in the sequence (therefore positionally in front of the other limb of the pair, consequently in the "lead"). The asymmetry of contact within and between limb pairs results in an increase in the number of variables needed to characterize and compare asymmetric gaits. Hildebrand (1977, 1989) developed six variables to fully account for the variety of asymmetric gait strategies in use by mammals:

1. The duration of hind limb support (time during which one or both hind limbs are in contact with the ground).

2. The duration of forelimb support (time during which one or both forelimbs are in contact with the ground).

3. The time lag between hind limb contacts.

4. The time lag between forelimb contacts.

5. The relationship between limb pair distribution, where the last limb to contact (lead) is either on the same side in fore and hind pairs, or on the opposite side.

6. The phase relationship between pairs of limbs, as measured by the time interval between hind limb midstance and forelimb midstance as a proportion of stride duration (midstance is used, because fore and hind contact duration can be asymmetric as well as the phase relationship between the pairs).

Using Hildebrand's techniques for asymmetric gaits it is not possible to include all relevant analysis variables on a single graph. Consequently, it is necessary to use a number of key graphs to characterize important aspects of the movement pattern in order to compare between speeds, species or individuals. For instance, a plot of hind limb support against the phase difference between fore and hind pairs is quite inform-ative in evaluating the relationship between different types of gallop style (Figure 2.4), as it provides information on the presence or absence of a suspension phase (a ballistic portion of the stride, during which no limb is in contact with the ground).

The asymmetric gaits of quadrupeds can be complex to understand. Part of the confu-sion originates with the fact that many aspects of these gaits blend continuously across the generally defined patterns that characterize each gait. As Hildebrand himself stated, *"There is a continuum of gaits, but some possible gaits are avoided."* (1989, p. 766). Very subtle differences in the overlap between classified gaits are likely not functionally impor-tant; these gait patterns blend into each precisely, because the physical consequences of small relative timing shifts at the boundaries between specifically defined "gaits" are not great. For instance, the difference between a transverse gallop with a small suspension phase, and one without, is not fundamental. In both cases, the center of mass (CoM) will be moving in much the same manner (upward and forward as the fore lead limb comes off the substrate), but the suspension phase will be lost if the magnitude of the CoM velocity is not great enough to lift the trunk and the previous limb before the next hind limb makes contact. Nevertheless, it is useful to consider the fundamental asymmetric patterns: the transverse and rotary gallops; the half-bound; and its functional extension, the bound (though a true bound is actually a left-right symmetric gait – indicating how tricky it can be to classify quadrupedal gait concisely and unambiguously).

The asymmetric gaits are relatively fast-moving, and are often characterized by important vertical motions of the body, as they all tend to have a "bounding" aspect. This means that the limbs are used over relatively brief periods to reverse the direction of the falling mass of the animal, and they act to "transition" the trajectory of the animal, from forward and downward to forward and upward. In between these limb-mediated transitions, the mass tends to follow a ballistic trajectory, as the action

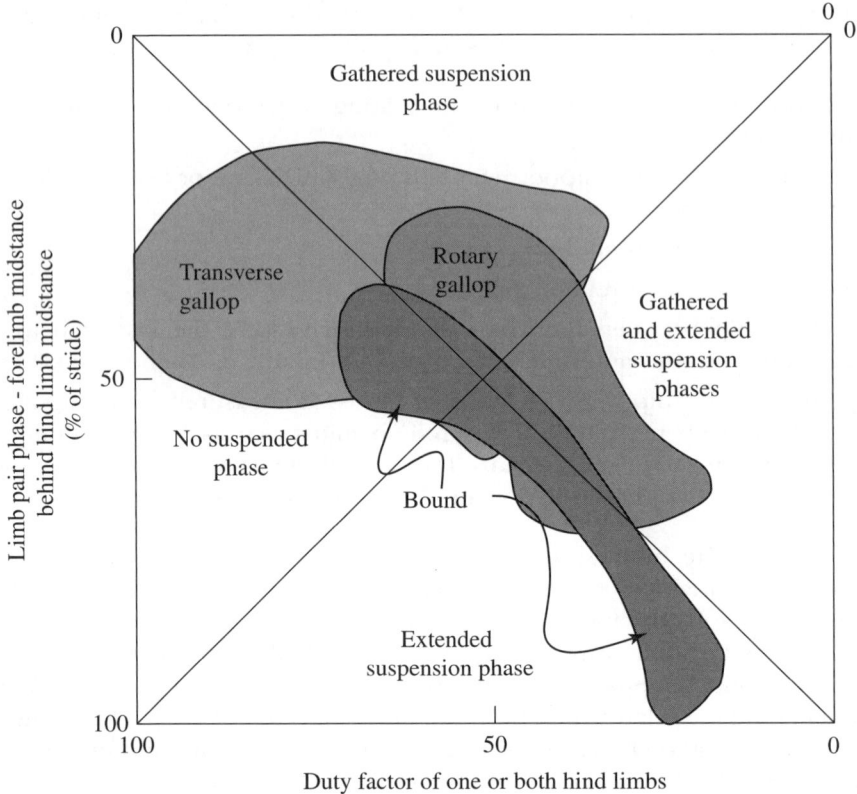

FIGURE 2.4

Plot of asymmetric gaits using Hildebrand's method. The horizontal axis represents duration of support as a proportion of the stride, while the *y*-axis represents the phase shift between pairs of limbs. This plot largely distinguishes the duration and type of suspension employed in gallop-type asymmetric gaits (adapted from Hildebrand, 1977).

of gravity passively reverses the direction of travel from forward and upward to forward and downward. Gallops and bounds are generally characterized by having some sort of ballistic flight phase in the stride cycle, where no feet are in contact with the substrate. This non-contact portion of the stride is also commonly referred to as a "suspension" phase, where the animal appears suspended over the substrate – at least when considering support by limb contacts.

The duration and proportion of the gait cycle involved with the non-contact (suspension) portion can vary considerably, depending on running speed and the species (with this latter largely influenced by length of back and limbs). This can even reach the extreme where no actual "flight" phase occurs – that is, where one or more feet are in contact with the substrate at all times during the stride, even though the gait is obviously a gallop. The lack of a suspension phase usually occurs in relatively slow-moving gallops, such as the slow gallops of a number of artiodactyls (quite common in the gallop of goats, for instance). A fully suspended phase need not be a characteristic of even the bound (a term implying an arcing flight phase) – for example, one might also consider the progression of an inchworm as an extremely slow "bound".

Even in those cases where no suspension phase exists, and at least one of the limbs is in contact with the substrate throughout the stride cycle, the path of the CoM of the animal still undergoes changes in direction over the stride – one actively mediated by force applied through the limbs (from downward to upward), and the other passively determined by the action of gravity (from upward to downward). Of interest in characterizing footfall patterns in the asymmetric gaits is how the limbs are used to influence the active transition and to exploit the opportunities presented by the passive transition.

Consider the ways in which the four limbs could be used to alter the direction of the animal's mass as it falls from the suspension phase in a run. One simple option would be to use all the limbs together at one time. This would result in a violent reversal of direction, as all four limbs apply force. With simultaneous extension of all limbs together, total ground reaction force can be high, so the animal will be launched for a substantial non-contact suspension "flight". Although such a gait may seem overly erratic, it is a viable option and is, in fact, used by several species. As defined by Dagg above, it is commonly termed the "pronk" (or stott, sprange or prange), and is used by a number of cervids and bovids as a display. Dagg (1973) lists three cervid and twelve African ungulate species that use the pronk, including the springbok (*Antidorcos marsupialis*), the impala (*Aepyceros melampus*) and the wildebeest (*Connochaetes taurinus*).

Hildebrand (1977) plots the pronk near the bottom right corner of Figure 2.4, just beyond the lower right limits of the bound. This position indicates the very brief period when the limbs are in contact with the ground – with a diminishingly small duty factor (possible because of the combined strength of all four limbs acting simultaneously). The abrupt transition from downward to upward in the pronk appears quite violent, and intuitively suggests that the motion requires substantial energetic investment. Although no direct measurements of this gait have been made, theoretical analyses (Ruina *et al.*, 2005) suggest that it is a mechanically, and presumably also metabolically, expensive gait to maintain. This is in fitting with its use as a display gait – potentially showing predators that the animal employing it has substantial athletic ability and is, consequently, not a good target to pursue.

Gaits in which contact by fore and hind limb pairs are spread in time can be categorized on a continuum from the bound to the gallop. In the true bound, fore and hind limb pairs are used together, and there is little or no lag between contacts within each pair. Smaller mammals, such as squirrels, often use this gait. Alternatively, limb contact within a pair can be spread in time, providing a trailing (initial contact) and lead (final contact) limb. In the gallop, both fore and hind limb pair contacts are distributed in time. Intermediate between the gallop and bound is the half-bound, where the contact of only one pair of limbs (usually the forelimbs) is spread, while the other pair functions in unison.

Although the gallops, half-bound and bound can be categorized distinctively, based on specifics of the footfall pattern employed, they have the common functional characteristic of using the fore and hind limbs in some sort of sequence to apply force to the substrate. These limb forces reverse the vertical direction of the CoM, and provide the momentum that allows the trunk to "bound", at least to some extent, between grouped limb contacts.

For a quadruped using limb pairs, where contact is distributed in time, there are two possible relationships between the fore and hind limbs – the inter-limb relationship (lead) can be the same front and rear, or opposite. If the lead is on the same side for fore and hind limb pairs, then the bounding gait is considered a transverse gallop. This is because limb contacts move in a transverse pattern across the animal's limbs (e.g. left hind, right hind, left fore, right fore). If the lead is on opposite sides for each pair, then the gait is considered a rotary gallop, because the contacts rotate around the limbs (e.g. left hind, right hind, right fore, left fore).

The localization of the pronk at the lower right corner of Figure 2.4 is a correct characterization of the gait, but it is not exclusive to this location. The use of all limbs in unison creates an ambiguity with the timing variables indicated on the y-axis of this particular plot. The pronk could also legitimately be plotted in the upper right corner of Figure 2.4, since neither forelimbs nor hind limbs trail each other. This reiterates the continuum nature of the patterns indicated on such plots, where some features that appear widely dispersed on the plot can actually be quite similar in function.

In the continuum from the bounds through gallops, the most distinguishing feature is that ground contact is sequenced, both along the front-rear axis and with regard to the trailing-leading contacts of a limb pair. These are usually higher speed gaits, and substantial distance can be traversed by the body between contacts. Most of these gaits involve a non-contact suspension phase. Given the dominant front-rear distribution of limb pairs, two possible non-contact conformations are common for the quadrupedal bound/gallop – collected suspension and extended suspension.

The distinction between a "collected" and an "extended" suspension is determined by which limb pair – fore or hind – that the animals launches from into the non-contact portion of the stride cycle. In the collected suspension, the body vaults from the front limbs. This means that the forelimbs are swinging rearward following their contact, while the hind limbs are swinging forward in preparation for their next contact. This places both fore and hind limbs under the trunk, in a "collected" pattern, while the animal is off the ground. In contrast, extended suspension is initiated from hind limb thrust. In extended suspension, the hind limbs are swinging rearward, while the forelimbs are extending forward in preparation for their next contact, and the animal naturally assumes a stretched-out, or extended, form as it launches into the next set of limb contacts.

Cursorial mammals are, by nature, fast-moving. High running speed increases the likelihood of non-contact suspension periods within the gait, whether it is a symmetric run such as the trot and pace, an asymmetric bipedal gait such as the skip, or quadrupedal bounds and gallops. Although either collected or extended suspension is possible in any of the quadrupedal bounds and gallops, certain types of suspension are often associated with particular gaits. The half-bound of a typical rabbit is an example (Figure 2.5). At modest speeds, the rabbit will use a half-bound with an extended suspension. Following the aerial bound initiated by the hind limb pair, the front limbs are extended into the contact and the hind limbs make contact before front limb contact is lost (even though, in position, the hind limbs will likely be placed ahead of the forelimbs due to flexion of the back). As speed increases, two features of contact change. The time of foot contact becomes shorter and the duration of non-contact becomes longer. At high enough speeds, a brief collected suspension emerges, where the forelimbs lose contact prior to hind limb touch down.

The transverse gallop is usually associated with a collected suspension (Figure 2.5). However, this can also be variable. Some forms that use the transverse gallop have no appreciable suspension phase. Those that do, such as the domestic horse, show a

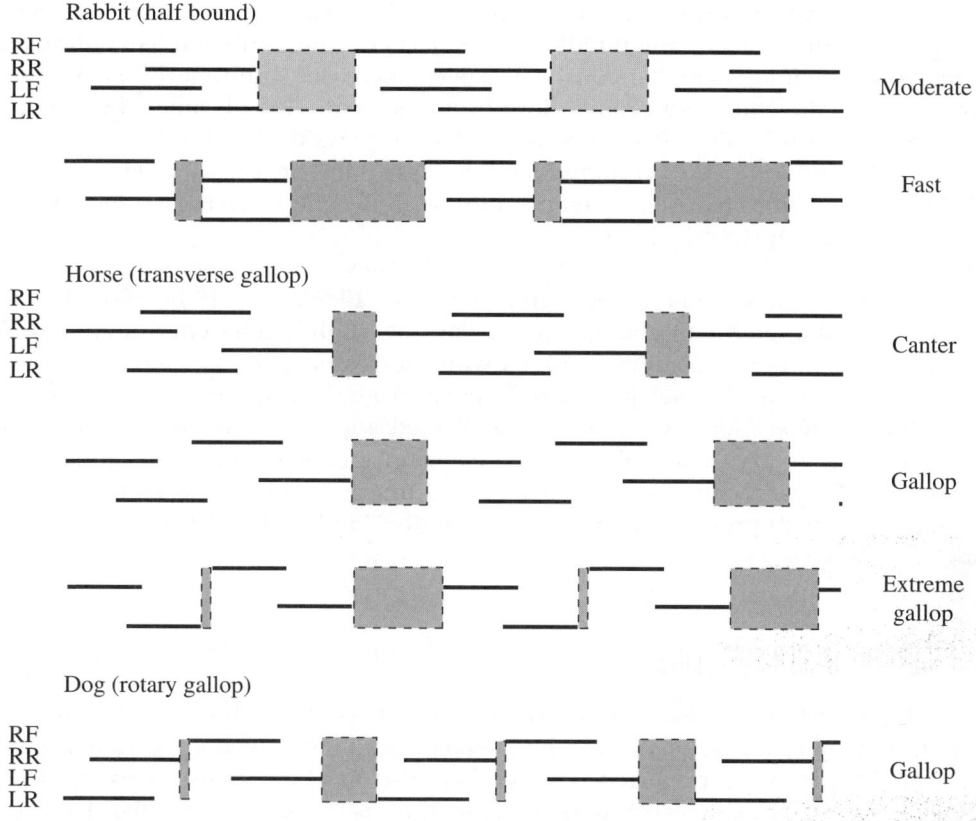

FIGURE 2.5

Comparison of foot contacts (horizontal bars) and non-contact periods (shaded rectangles) for bounds and gallops in common species. In the half-bound, the hind limbs operate nearly simultaneously. A non-contact, or suspension, phase of the gait occurs, following the powerful thrust of the hind limb pair. As speed increases, the forelimbs can lose contact with the substrate before the hind limbs make contact, producing a second, collected suspension. In the transverse gallop of the horse, contact is initiated by a rear limb following a collected suspension, and the limbs contact in sequence before another collected suspension is initiated by the lead forelimb. In the slower running canter, the footfall sequence is equivalent to the gallop, but the overlap of the lead hind and trailing forelimbs is extensive. The transverse gallop is usually characterized as having only a collected suspension phase (and this does not occur in many species) but, at extremely high speeds on a hard surface, it is possible for a brief extended suspension to occur as well. In the rotary gallop of the dog, which is commonly used for higher speed running in this species, both collected and extended suspension is evident. Timing of contacts and suspension periods for different species is not drawn to scale.

range of limb overlap as speed changes. The slow, three-beat gait known as the canter is simply a gallop in which the middle two contacts (a hind limb and its contralateral forelimb) occur almost simultaneously. As speed increases, the time of each contact decreases, and the contacts become more evenly spread. This becomes the four-beat sequence of the true gallop. The suspension phase follows the last forelimb contact, so is of the collected form. At even higher speeds, especially on a hard surface, it is possible for the hind limbs to lose contact prior to forelimb contact, giving a brief extended flight (Brown, 1955; Howell, 1944).

The rotary gallop is characteristic of dogs when they run at higher speeds, and it is the gait used by the cheetah at its highest speeds (Hildebrand, 1959, 1961; Hudson *et al.*, 2012). Although the footfall pattern seems only slightly different from that of the transverse gallop, the mechanics involved with the rotary gallop, as utilized by most highly cursorial forms, has substantial differences from the transverse gallop (Bertram and Gutmann, 2008). Most often, both a collected and an extended suspension are observed in the rotary gallop, though the collected phase is often substantially less than that of the extended phase. In this regard, the fast half-bound of the rabbit is similar to that of the rotary gallop, except that the hind limbs of the rabbit tend to work in unison.

2.6 BEYOND "HILDEBRAND PLOTS"

Hildebrand's techniques for characterizing gaits have been highly influential for comparing mammalian gaits, even if the characterization of asymmetric gaits requires more variables than are preferred. The method also has limitations when applied to irregular gait cycles, such as occur as the animal transitions between gaits. This latter problem arises because the analysis is dependent on the definition of a stride cycle that reflects the action of one limb within the sequence. However, in irregular gaits, one or more limbs may not contact in registry with respect to this defined cycle. As a consequence, the Hildebrand methods are difficult to apply to the analysis of gait transitions, or to other circumstances where a consistent cycle becomes obscured, such as when moving over complex or irregular substrates. It may well be that such irregular circumstances actually determine the value of the morphology and coordination pattern of a given species (see Chapter 11 and Lee *et al.*, 2007). The quantification of major shifts in coordination pattern may hold important insight into how, and why, many species move as they do. Proper investigation of this cannot be initiated, however, unless a method is available to quantify and classify how gait is influenced by these transient circumstances.

If the temporal boundaries of the gait cycle definition cause a problem in irregular locomotion, it is necessary to define an alternative approach that does not have this limitation. Abourachid (2003) developed an alternative approach that applies equally well to symmetric, asymmetric and irregular gaits, including gait changes. The technique involves quantification of limb contact patterns, as Hildebrand's approach does, but the analysis strategy is based on two observations.

First, all terrestrial vertebrates possess basically similar morphology and interactions between the two limbs of a pair, whether fore or hind. This observation holds true for terrestrial locomotion on limbs, regardless of the mode adopted, even in a case where the cranial pair of limbs differs greatly from the caudal pair (as in terrestrially moving bats, see Chapter 12). The fundamental relationship between the two limbs of a pair suggests that their actions are likely directly coordinated, and it should not be necessary to separate their analysis to characterize the gait.

Second, the forelimbs are the first to encounter environmental conditions, such as obstacles or changes in substrate compliance. Consequently, functional adjustments of gait likely originate with the anterior contacts. This point applies well to many common circumstances, but there are many where it does not. Consider, for instance, the rabbit discussed above. At high speed, the hind limbs follow the forelimbs in the cycle, but they contact the substrate ahead of the forelimb contact position.

Abourachid proposed that gait analysis should take these two general observations into account. Rather than analyzing gait as a succession of locomotor cycles compared in registry with contact of a hind limb, she proposed that gait should be considered as a succession of antero-posterior sequences (APS), starting with the cranial-most contact, and moving through to caudal contacts. Thus, the APS is defined as the combination of consecutive cycles of forelimbs, followed by consecutive cycles of the hind limbs. The limb chosen as reference to begin each AP sequence does not affect the analysis in the case of a symmetric gait since, by definition, the two limbs perform the same role. In an asymmetric gait, however, it will naturally be the trailing limb, because it makes contact first (Figure 2.6).

Three types of coordination of the limbs are considered in the APS approach: coordination between the forelimbs; coordination between the hindlimbs; and coordination between the two limb pairs – front and rear (Abourachid, 2003).

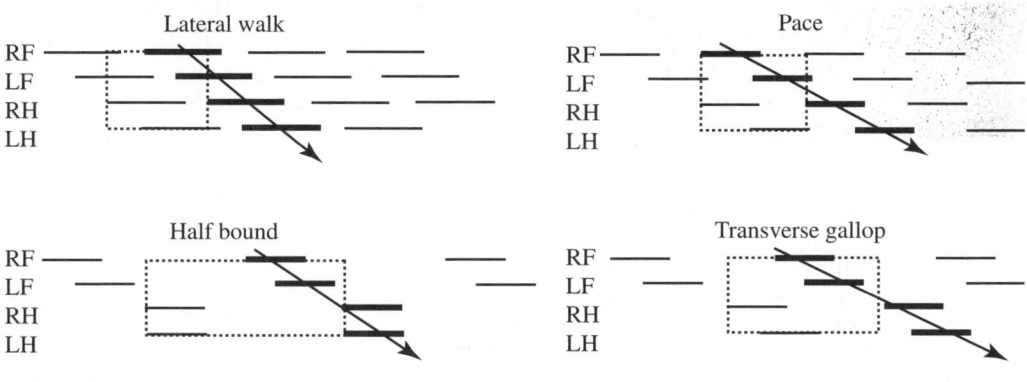

FIGURE 2.6

Limb contact patterns for several gaits. The stride cycle defined via Hildebrand technique, beginning with the right hind limb (RH) contact, is bounded by the dashed rectangular outline. The Anterio-posterior sequence (APS) stride cycle is shown with bold contact bars. The APS pattern is indicated by the arrow, which shows the logic antero-posterior limb movement by combining the cranial pair consecutive cycles, followed by consecutive cycles of the caudal pair (after Abourachid, 2003). Although each gait can be defined and characterized using the Hildebrand method, no consistent contact patterns emerge. In most quadrupedal gaits, a consistent pattern does emerge for the APS definition.

The coordination of these three functional units are measured by three variables, respectively (Figure 2.7):

1. Lag between contacts of the forelimb pair (fore lag, FL).
2. Lag between contacts of the hind limb pair (hind lag, HL).
3. Lag between the initial contact of the forelimb and hind limb pairs (pair lag, PL).

These variables are calculated as a percentage of the cycle of the reference limb, which is simply the time between contacts of the first limb (trailing forelimb) of the APS considered.

The APS method allows the calculation of these three variables regardless of the gait considered, whether symmetric or asymmetric. These variables are also sufficient to distinguish all gaits, thus providing an alternative method for describing the manner in which an individual or species utilizes their limbs in locomotion. The small number of variables required to characterize all gaits, and the adaptability of the analysis regardless of the regularity of gait, are distinct advantages for the analysis of irregular locomotion and gait transition strategies (Abourachid *et al.*, 2007; Maes *et al.*, 2008).

A newly developing approach to quantifying the differences between gaits, and particularly the patterns involved with gait change and irregular gaits, involves analysis of the residual phase relationship between limb movement patterns. Working with legged locomotion in insects, Rezven *et al.* (2009) refers to the analysis as a "kinematic phase" method, where the fine structure of the frequency and phase characteristics of limb motion are measured from video (with appropriately fast-framing rate). The position of individual feet relative to the body are generally periodic, especially for steady state (constant speed) level locomotion. The waveform describing the motion of each limb in steady state locomotion contains frequency and phase characteristics that can be quantified and compared to the waveforms generated during a perturbation, and following the accommodation strategy employed to re-establish stability.

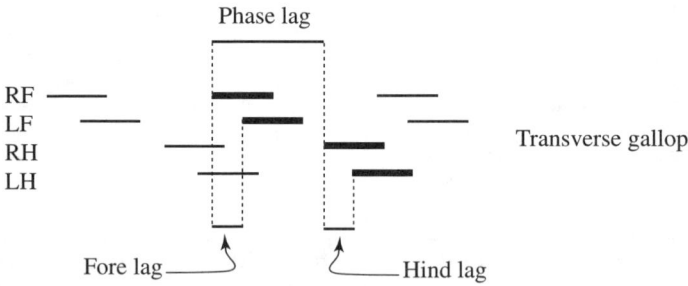

FIGURE 2.7

Characterization of gait using the Anterior-Posterior Sequence approach. The three variables used to quantify the gait can be measured regardless of the gait pattern. In this asymmetric example (transverse gallop), the timing variables that determine the APS value are indicated by bars associated with the dashed lines. Each dashed line indicates the specific contact involved with the measure. The lag between each pair of limb contacts is shown below, and the phase lag between the limb pairs is shown above (modified from Abourachid, 2003).

One method of quantifying and evaluating these differences is through the phase residual (relative phase difference) of movement patterns of individual limbs or coordinated sets of limbs. From this, it is possible to discriminate features of frequency and phase as they are altered through the perturbation and subsequent accommodation (Rezven *et al.*, 2009). Wilshin *et al.* (2012) successfully applied this technique to gait change in dogs as a method of following the walk-trot transition characteristics (Figure 2.8).

2.7 STATISTICAL CLASSIFICATION

One approach to the classification of gaits is through the determination of discriminating characters by statistically analyzing relationships between events in the gait cycle. The basic logic of the approach does not differ substantially from classic, manual characterization criteria of limb coordination (Gambaryan, 1974;, Hildebrand 1966, 1977). The availability of digital video and electronic analysis techniques, which allow convenient digitization of gait features, provides an opportunity for statistically based,

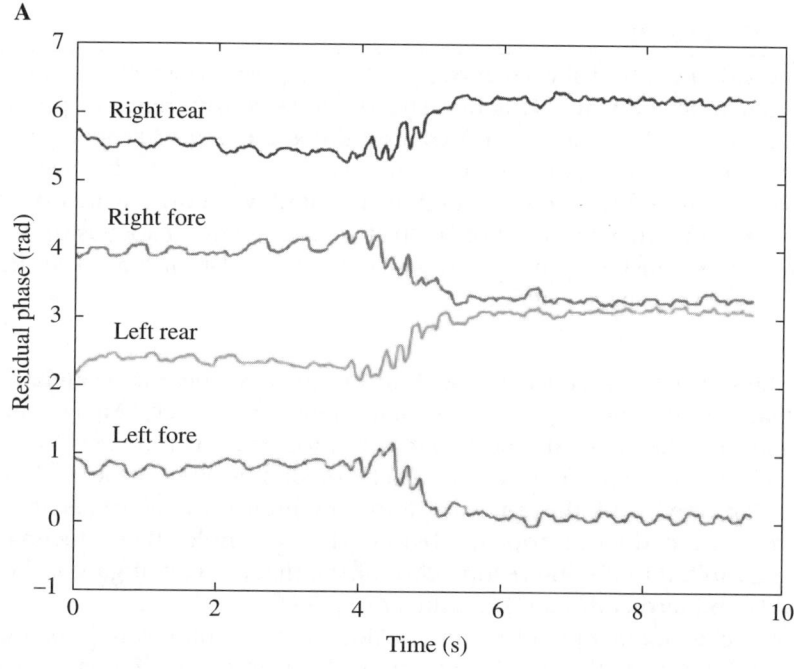

FIGURE 2.8

Residual phase plots of the walk-trot transition of dog. (**A**) Residual phase is plotted (radians) against time for individual limbs. Blue – right rear; red – right fore; cyan – left rear; green – left fore. The dog is walking at steady state when the trace begins, then transitions to a trot between the 4th and 5th second. Following the gait transition, the phase relationships settle to a consistent level and the animal maintains a steady state trot. Note that 0 and 2π (6.14) radians are equivalent values, indicating that, in the trot, the right hind and left fore work in synchrony and out of phase with the right fore and left hind.

B

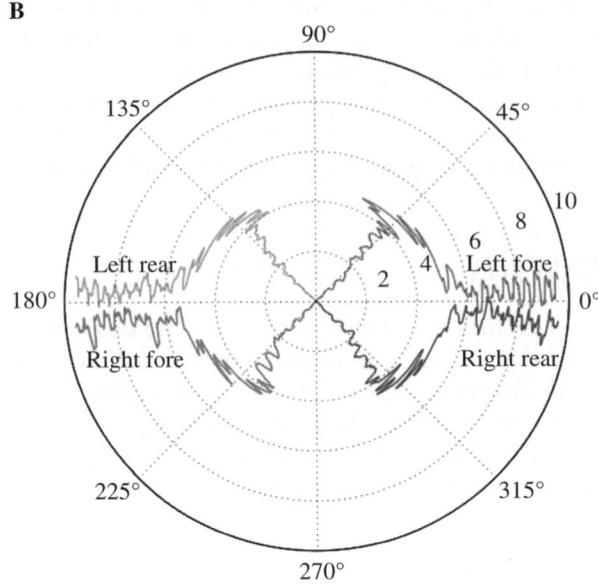

Figure 2.8 *(Continued)*

(B) The same gait transition shown above plotted on polar coordinates. Time increases from the center point outward. This plot shows the large difference in phase between the walk (interior of plot) and the trot (paired limbs moving in synchrony) following transition. Specific features of the transition strategy can also be seen. Note, for instance, that the phase change is initiated in each limb via a drift away from the phase relationship used in either the walk or the trot, as the treadmill speeds up to stimulate the transition. Unpublished plots courtesy of Simon Wilshin, Structure and Motion Laboratory, Royal Veterinary College.

automated classification of gaits using such techniques as Linear Discriminant Analysis, LDA (Robilliard *et al.*, 2007) and Multivariate Gaussian Density, MGD (Starke *et al.*, 2009). Such approaches have the potential to reduce analysis time and to allow larger scale analyses of locomotion under a variety of conditions. These techniques also increase the objectivity of the analysis, and can indicate similarities in movement patterns when these do not appear obvious. For example, these techniques have recently been applied to the interesting case of the distinctive tölt gait of the Icelandic horse breed (Robilliard *et al.*, 2007, Starke *et al.*, 2009).

The tölt is a classic example of a gait considered a "running walk", as illustrated in Figure 2.3. As the designation implies, the classification of this gait is ambiguous – is it a walk or is it a run? The question is important, because the mechanisms involved with generating a functional walk or run appear substantially different (Cavangna *et al.*, 1977, Lee *et al.*, 2011).

The tölt is a relatively fast-moving gait, but it uses the footfall pattern characteristic of a walk. Following Robilliard *et al.* (2007), Starke *et al.* (2009) used both MGD and LDA methods to determine that the tölt should fall within the classification of a run. This confirmed the results of an extensive mechanical analysis which showed the tölt has mechanical features that are more aligned with the bouncing of a run than the vaulting strut-like limb function characteristic of the walk (Biknevicius *et al.*, 2004, 2006). This also supports the contention of Bertram and Gutmann (2008) that the gait could be considered a "symmetric gallop" (where it is recognized that, by definition, a "gallop" should be an asymmetric gait, but such ambiguities can arise when different approaches to classifying gaits intersect).

2.8 NEURAL REGULATION AND EMERGENT CRITERIA

As distinct patterns of motion are ultimately generated by neural signals, it is common to assume that gait patterns are "determined" by neural control patterns. That is, it is commonly assumed that specific gait patterns only exist because of regulatory patterns inherent in the function or control of the neural system (whether determined centrally, peripherally or a combination thereof). Neural control of locomotion is, indeed, complex. Its inherent properties certainly influence possible strategies available in locomotion, just as the properties of muscular function limit the available possibilities of active motion, but neural control strategies must integrate with the functional opportunities presented by the interaction of the organism with its physical environment (see Chapters 5, 6 and 7).

A comprehensive discussion of the numerous models that have been proposed regarding neurally determined gait patterns is well beyond the scope of this chapter. However, one rich field of discussion regarding gait and its control derives from modeling central pattern generators (CPGs) of the central nervous system as a complex of coupled nonlinear oscillators (Pavlidis and Pinsker, 1977; Bay and Hemami, 1987; Rand *et al.*, 1988; Collins and Stewart, 1993). Recently, these concepts have been successfully applied to the control of legged robots (Ijspeert, 2008).

Although much of the evident central neural patterns, lesion effects and gait differences of the inter-segmental coordinating system can be successfully modeled using coupled oscillator mathematics, this does not mean that the gaits are *determined* by such nonlinear properties. Coupled oscillators may well model, or even be involved in, the function of coordinating the limbs and body to produce the gait but, ultimately, the gait exists because it is a mechanically effective way for the body of the organism to interact with its substrate to displace itself (see Chapters 5, 6 and 7). One question that needs to be addressed is "what are the limits of influence of the neural control process, the consequences of mechanical function and the role of integration between these two?"

The CPGs are a critical feature of the process of gait production, but they do not explain why quadrupedal gaits are restricted to a subset of the vast array of inter-limb coordination patterns possible. Such restriction to a limited subset of possibilities may well indicate that some criterion must be met in order for the gait to be functional. In other chapters, it is argued that this criterion should be based on the dynamics of the interaction of the animal's mass with the physical characteristics of the medium it travels on (or under, or through; see Chapter 5). However, if the criteria are dependent on the interplay of subtle mechanical influences, it may be difficult to discover the complex patterns of interconnecting influences that eventually determine the set of "reasonable" gait patterns. In such a circumstance, it would be valuable to evaluate patterns that emerge from the mathematical relationships of the gait. One such approach to defining the limits of gait pattern availability that appears quite interesting is the concept of "regular realizability" (McGhee, 1968; McGhee and Jain, 1972).

Regular realizability uses a gait matrix to characterize the phase relationships between limbs, where contact and noncontact of individual limbs are coded in a binary fashion (1 = contact, 0 = noncontact). Added to this is a contact duration matrix vector that is used to distinguish contact periods. Up to this point, the procedure has much in common with Hildebrand's gait formula, except that a different operational approach is used (matrix algebra). Functional gaits are objectively defined from this matrix by applying an optimality criterion – that of regular realizability, which requires a matrix duration vector, where the state of each limb within a complete cycle is equivalent. This criterion produces a subset of the possible limb contact patterns that have some interesting parallels to the gaits employed by mammals.

It was found that six bipedal gaits fit the regular realizable criterion (McGhee and Jain, 1972). These were walking, running, two forms of skipping (alternating lead and constant lead), and two gaits that, at the time, were not naturally found. One of these appears much like the newly discovered pendular run (Srinivasan and Ruina, 2006), a gait that has been found to occur naturally in some birds (Usherwood, 2010 – see Chapter 6). The sixth regularly realizable bipedal gait appears much like the "hop" portion of the triple jump event in athletics.

McGee and Jain (1972) calculated that a total of 5040 quadrupedal gaits can theoretically be distinguished, at least in terms of possible sequence relationships. Of these, 45 qualify as regularly realizable, but many do not appear to be used naturally. Of the gaits that are used by quadrupeds, however, only two that are known do not meet the regular realizable criterion, and both of these can be considered context-specific or marginal gaits. One is the amble, as used by the elephant. The elephant amble has recently been characterized as a running walk (Ren and Hutchinson, 2008), and the gait appears to be influenced by special limitations demanded by the large size of these animals. The other quadrupedal gait that does not match regular realizability is the equine fast canter that some horses are trained to produce. This is an artificially induced gait that would not likely occur naturally. If analyzed closely, it is likely that many of the artificially induced equine performance gaits would not match many natural functional criteria.

Why should regular realizability be a criterion that influences whether a gait is useful to an organism or not? Although analyzed using the matrix vector representation of contact and duration, ultimately the analysis depends on physical characteristics, such as stride length, duty factor, inter-limb phase, and the morphological features of the organism that determine how the movement of the limbs can occur. All of these are functionally important, as well as quantifiable features of gait. As such, the mathematical criterion on which regular realizability is based will be related, directly or indirectly, to functional criteria and aspects of performance.

However, unless the results of a criterion such as regular realizability can be translated into functional features of the gait or the animal that produces it, the analysis itself does not add anything to our understanding of gait. Gaits produce a footfall pattern, and those patterns can be analyzed (described) in a number of ways. Many of the analyses discussed above are designed to organize those patterns. This is valuable as a first step in developing an understanding of the gait, but organization of the footfall patterns in itself, regardless of how discriminating or sophisticated, does very little to provide an understanding of why the gaits exist as they do, and which of the possible patterns provide functionality while others are functionally prohibited.

2.9 MECHANICAL MEASURES AS DESCRIPTIONS OF GAITS

Ultimately, locomotion is a mechanical interaction between the animal and the substrate it moves on. The strategy used to manage this mechanical interaction must be functionally relevant (and cost-effective, since energy is a limiting resource) within the bounds of the properties of the biomaterials from which the organism is constructed. Even though gaits originate from the coordination of the neuromuscular system, they do not just happen, but represent at least an approach to an optimization of the interaction of the animal with the physical properties of its supporting environment. To understand this optimization, it is necessary to consider the factors important to the functioning of the animal that must be reconciled to produce effective locomotion.

The mechanics involved with gait is a main focus of most of the chapters of this book, and many chapters discuss mechanical aspects of gait at some length and detail. Within the context of evaluating gaits as cyclical patterns of movement, it is important to recognize how such patterns can translate into mechanical features of the interaction of the animal with its substrate. As always, however, it is necessary to discriminate "descriptions" of gait from "comprehension and understanding". Even though mechanical characterization of gait may be directly related to the function of locomotion, measurements of the forces, moments, momentum, impulse or energy involved are only different forms of description, unless interpreted in the context of the functions required to produce gait.

Often neglected in this process is the consideration of what alternatives exist. It is common to accept the gaits we observe as the only alternatives possible. It is more likely, however, that the gaits we observe exist because they have functional characteristics that have advantages over the alternative patterns that could exist. It is impossible to evaluate these advantages – the real determinants of gait – without considering the gait in the context of how else the organism might move.

2.10 CONCLUSION

The gait analysis methods discussed in this chapter are quantitative, and can often distinguish between quite subtle features of the way an animal utilizes its limbs for locomotion. Although useful in documenting similarities and differences in gait, it should be recognized that all of these methods constitute only descriptions of the gaits utilized, and are not explanatory in themselves. That is, defining that a particular species uses a specific gait under certain circumstances, or even that most species utilize the same gait under the same circumstances, or that a given species does not utilize the gait under certain circumstances, does not explain why these gait patterns are utilized (*or not*).

Quantitative definition of the motions involved derived from these analysis techniques can be the initial step in addressing why the motions occur as they do. Although driven by neural activity, locomotion is ultimately a mechanical phenomenon. Neither a description of the external motions, however quantitative, nor an analysis of the neural activity will be adequate to explain the function of the motions. Locomotion is a physical interaction between the organism and its environment that will be influenced, to greater or lesser degrees, by both internal and external functional consequences. To comprehend the consequences of either the motions, or the control strategies that influence those motions, it is necessary to understand the mechanics within the global context of the strategies available. The gaits expressed to generate legged locomotion are the solution to a complex problem, but we are only just beginning to define the problem. Describing gaits defines the "answer" that these organisms have arrived at, but gives us little insight into the "question" being addressed.

▍REFERENCES

Abourachid A (2003). A new way of analysing symmetrical and asymmetrical gaits in quadrupeds. *Comptes Rendus Biologies* **367**, 625–630.

Abourachid A, Herbin M, Hackert R, Maes L, Martin V (2007). Experimental study of coordination patterns during unsteady locomotion in mammals. *Journal of Experimental Biology* **210**, 366–372.

Alexander RMcN (1984). The gaits of bipedal and quadrupedal animals. *International Journal of Robotics Research* **3**, 49–59.

Alexander RMcN (2004). Bipedal animals, and their differences from humans. *Journal of Anatomy* **204**, 321–330.

Alexander RMcN, Jayes AS, Ker RF (1980). Estimates of energy cost for quadrupedal running gaits. *Journal of Zoology* **190**, 155–192.

Bay JS, Hemami H (1987). Modeling of a neural pattern generator with coupled nonlinear oscillators. *Biomedical Engineering* **34**, 297–306.

Bertram JEA (2004). New perspectives in brachiation mechanics. *Yearbook of Physical Anthropology* **47**, 100–117.

Bertram JEA, Chang Y-H (2001). Mechanical energy oscillations of two brachiation gaits: measurement and simulation. *American Journal of Physical Anthropology* **115**, 319–326.

Bertram JEA, Gutmann A (2008). Motions of the running horse and cheetah revisited: fundamental mechanics of the transverse and rotary gallop. *Journal of The Royal Society Interface* **6**, 549–559.

Bertram JEA, Lee DV, Case HN, Todhunter RJ (2000). Comparison of the trotting gaits of Labrador Retrievers and Greyhounds. *American Journal of Veterinary Research* **61**, 831–837.

Bertram JEA, D'Antonio P, Pardo J, Lee DV (2002). Pace length effects in human walking: "Groucho" gaits revisited. *Journal of Motor Behavior* **34**, 309–318.

Biknevicius AR, Mullmeaux DR, Clayton HM (2004). Ground reaction forces and limb function in tölting Icelandic horses. *Equine Veterinary Journal* **36**, 743–747.

Biknevicius AR, Mullmeaux DR, Clayton HM (2006). Locomotor mechanics of the tölt in Icelandic horses. *American Journal of Veterinary Research* **67**, 1505–1510.

Blickhan R, Full RJ (1987). Locomotion energetics of the ghost crab. II. Mechanics of the centre of mass during walking and running. *Journal of Experimental Biology* **130**, 155–174.

Brown LS (1955). Eadweard Muybridge and his work. In: *Animals in motion*, Eadweard Muybridge, 1957. Dover Publications, Mineola, NY, pp. 9–10.

Cavagna GA, Heglund NC, Taylor CR (1977). Mechanical work in terrestrial locomotion: two basic mechanisms for minimizing energy expenditure. *American Journal of Physiology – Regulatory, Integrative and Comparative Physiology* **233**, 243–261.

Collins JJ, Stewart IN (1993). Coupled nonlinear oscillators and the symmetries of animal gaits. *Journal of Nonlinear Science* **3**, 349–392.

Dagg AI (1973). Gaits in mammals. *Mammal Review* **3**, 135–154.

Gambaryan PP (1974). *How mammals run: anatomical adaptations*. Hardin, H (transl.). New York, Wiley, pp. 367.

Goiffon and Vincent (1779). *Memoire artificielle des principes relatifs a la fidele representation des animaux tant en peinture, qu'en scultpture. I partie concernant le cheval.* (Memoire on the artificial principles of faithful representation of animals both in painting and sculpture. Part I concerning the horse). Alfort. (reference from Gambaryan, 1974).

Griffin TM, Main RP, Farley CT (2004). Biomechanics of quadrupedal walking: how do four-legged animals achieve inverted pendulum-like movements? *Journal of Experimental Biology* **207**, 3545–3558.

Hedrick TL, Tobalske BW, Biewener AA (2002). Estimates of circulation and gait change based on a three-dimensional kinematic analysis of flight in cockatiels (*Nymphicus hollandicus*) and ringed turtle-doves (*Streptopelia risoria*). *Journal of Experimental Biology* **205**, 1389–1409.

Heglund NC, Cavagna GA, Taylor CR (1982). Energetics and mechanics of terrestrial locomotion. III. Energy changes of the centre of mass as a function of speed and body size in birds and mammals. *Journal of Experimental Biology* **97**, 41–56.

Hildebrand M (1959). Motions of the running cheetah and horse. *Journal of Mammalogy* **40**, 481–495.

Hildebrand M (1961). Further studies on locomotion of the cheetah. *Journal of Mammalogy* **42**, 84–91.

Hildebrand M (1965). Symmetrical gaits of horses. *Science* **150**, 701–708.

Hildebrand M (1966). Analysis of the symmetrical gaits of tetrapods. *Folia Biotheoretica* **6**, 9–22.

Hildebrand M (1967). Symmetrical gaits of primates. *American Journal of Physical Anthropology* **26**, 119–130.

Hildebrand M (1977). Analysis of asymmetrical gaits. *Journal of Mammalogy* **58**, 131–156.

Hildebrand M (1989). The quadrupedal gaits of vertebrates. Biosci. **39**, 766–775.

Howell AB (1944). *Speed in animals, their specialization for running and leaping*. University of Chicago Press, Chicago, IL, 270 pp.

Hudson PE, Corr SA, Wilson AM (2012). High speed galloping in the cheetah (*Acinonyx jubatus*) and the racing greyhound (*Canis familiaris*): spatio-temporal and kinetic characteristics. *Journal of Experimental Biology* **215**, 2425–2434.

Ijspeert AJ (2008). Central pattern generators for locomotion control in animals and robots. *Neural Networks* **21**, 642–653.

Jayne BC (1988). Muscular mechanisms of snake locomotion: an electromygraphic study of the sidewinding and concertina modes of *Crotalus cerastes*, *Nerodia fasciata* and *Elaphe obsoleta*. *Journal of Experimental Biology* **140**, 1–33.

Lee, DV, McGuigan, MP, Yoo, EH, Biewener, AA (2007) Compliance, actuation, and work characteristics of the goat foreleg and hindleg during level uphill and downhill running. *Journal of Applied Physiology* **104**, 130–141.

Lee DV, Bertram JEA, Anttonen JT, Ros IG, Harris SL, Biewener AA (2011). A collisional perspective on quadrupedal gait dynamics. *Journal of The Royal Society Interface* **8**, 1480–1486.

Maes LD, Herbin M, Hackert R, Bels VL, Abourachid A (2008). Steady locomotion in dogs: temporal and associated spatial coordination patterns and the effect of speed. *Journal of Experimental Biology* **211**, 138–149.

Marey E-J (1873). *La machine animale: Locomotion terrestre et aeriemme*. Paris, Balliere; translated as *Animal Mechanism: A treatise on terrestrial and aerial locomotion*. 3rd edition. New York, Appleton, 1884.

Manter JT (1938). The dynamics of quadrupedal walking. *Journal of Experimental Biology* **15**, 522–540.

McGhee RB (1968). Some finite state aspects of legged locomotion. *Mathematical Biosciences* **2**, 67–84.

McGhee RB, Jain AK (1972). Some properties of regularly realizable gait matrices. *Mathematical Biosciences* **13**, 179–193.

Nauwelaerts S, Aerts P, D'Août K (2001). Speed modulation in swimming frogs. *Journal of Motor Behavior* **33**, 265–272.

Pavlidis T, Pinsker HM (1977). Oscillator theory and neurophysiology. *Federation Proceedings* **36**, 2033–2035.

Raabe C (1883). *Locomotion du cheval, cadran hippique des allures marchées*. Paris.

Rand RH, Cohen AH, Holmes P (1988). Systems of coupled oscillators as models of central pattern generators. In: Cohen AH (ed). *Neural control of rhythmic movements*. John Wiley & Sons.

Revzen S, Koditschek DE, Full RJ (2009). Towards testable neuromechanical control architecture for running. In: Sternad D (ed). *Progress in Motor Control*. Springer Science + Business Media, LLC.

Robilliard JJ, Pfau T, Wilson AM (2007). Gait characteristics and classification in horses. *Journal of Experimental Biology* **210**, 187–197.

Ruina A, Bertram JEA, Srinivasan M (2005). A collisional model of the energetic cost of support work qualitatively explains leg sequencing in walking and galloping, pseudo-elastic leg behavior in running and the walk-to-run transition. *Journal of Theoretical Biology* **237**, 170–192.

Srinivasan M, Ruina A (2006). Optimization of a minimal biped model discovers walking and running. *Nature* **439**, 72–75.

Starke SD, Robilliard JJ, Weller R, Wilson AM, Pfau T (2009). Walk-run classification of symmetrical gaits in the horse: a multidimensional approach. *Journal of The Royal Society Interface* **6**, 335–342.

Usherwood, JR (2010). Inverted pendular running: a novel gait predicted by computer optimization is found between walk and run in birds. *Biology Letters* **23**(6), 418–421.

Vincent, A-F, Goiffon, GC (1779). *Mémoire artificielle des principes relatifs a la fidelle représentation des animaux tant en peinture qu'en sculpture*. Paris, Alfort.

Webb PW (1994). The biology of fish swimming. In: Maddock L, Bone Q, Rayner JMV (eds). *Mechanics and physiology of animal swimming*. Cambridge University Press, Cambridge, pp. 45–62.

Wilshin, SD, Hayes, GC, Porteus, J, Spence, AJ (2012) Describing gait transitions and the role of symmetry in control. *Integrative and Comparative Biology* **52** (Suppl. 1), e1–e201.

CHAPTER THREE

Muscles as Actuators

Anne K. Gutmann[1] and John E. A. Bertram[2]

[1] New Balance Sports Research Lab, Lawrence, MA, USA
[2] Department of Cell Biology and Anatomy, Cumming School of Medicine, University of Calgary, CA

3.1   INTRODUCTION

The purpose of this chapter is to consider what skeletal muscle is and how it operates, in order to gain a better understanding of how the functional properties of muscle facilitate effective locomotion. Although there are various types of muscle in the body, this chapter will focus exclusively on striated skeletal muscle, since only this type of muscle is directly used to power and control locomotion. We also emphasize a historical exploration of the concepts and their development, because many aspects of basic muscle function, including its physical basis and implications for locomotion, remain not entirely settled. Evaluating current concepts of muscle function in the context of how these concepts were arrived at provides a valuable perspective on their potential limitations.

Understanding Mammalian Locomotion: Concepts and Applications, First Edition.
Edited by John E. A. Bertram.
© 2016 John Wiley & Sons, Inc. Published 2016 by John Wiley & Sons, Inc.

3.2 BASIC MUSCLE OPERATION

Since its creation in the late 1950s, cross-bridge theory has been a widely (although not universally) accepted model of muscular contraction. Although cross-bridge theory does not provide a complete or perfect model of muscle function, many aspects of muscle function can be explained by cross-bridge theory. Therefore, we will approach muscle operation primarily from this well-regarded perspective.

3.2.1 Sliding filament theory – the basis for cross-bridge theory

In 1954, two groups – H. E. Huxley, working with R. Niedergerke, and A. F. Huxley, working with J. Hanson – independently reached the conclusion that change in muscle length was caused by two sets of inter-digitating protein filaments, actin and myosin, sliding past one another within the sarcomere. This sliding interaction defined the contractile unit of muscle. Prior to this proposal, the prevailing theory was that protein folding caused changes in muscle length.

Sliding filament theory was based primarily on the observation that the A-bands of striated muscle do not change length during passive shortening or lengthening of muscle. A-bands are structurally anisotropic regions ("A" for anisotropic) that contain myosin (Figure 3.1) and are characterized by high birefringence. Birefringence, or "double refraction", is the decomposition of a single ray of light into two rays. Materials with high birefringence produce a large angle between the two rays while materials with low birefringence produce a small angle between the two rays. Differences in birefringence translate into differences in brightness under a polarized light microscope. The observation that the A-bands remain at constant length while the sarcomere lengthens suggested that the actin filaments slide relative to the myosin filaments in the A-bands. Huxley and Hanson (1954) also showed that the actin filaments do not change length by measuring actin filament lengths after the myosin had been removed from muscle at various muscle lengths.

In intact muscle, it is difficult to measure the actin filament length, because the I-bands, (isotropic regions of the sarcomere characterized by low birefringence) correspond to the portion of the actin filaments that extend beyond the zone of actin-myosin overlap, so do not indicate the entire length of the actin filaments (Figure 3.1). Since Huxley and Neidergerke (1954) and Huxley and Hanson (1954) did not observe changes in protein filament length during passive shortening or stretching, they concluded that any changes in muscle length must be due to changes in the amount of overlap of the protein filaments, rather than to folding or contraction of the polymer. It should be noted that both Huxley and Neidergerke and Huxley and Hanson, as well as other researchers, did in fact observe changes in A-band length during active shortening and stretching. Both early groups attributed these length changes to optical artifacts. See Pollack (1983) for an overview of research reporting changes in A-band length.

3.2.2 Basic cross-bridge theory

Although H.E. Huxley and R. Niedergerke (1954) strongly hinted at a mechanism that could explain sliding filament theory when they wrote of a "*relative force between actin and myosin ... generated at a series of points in the region of overlap in each sarcomere*",

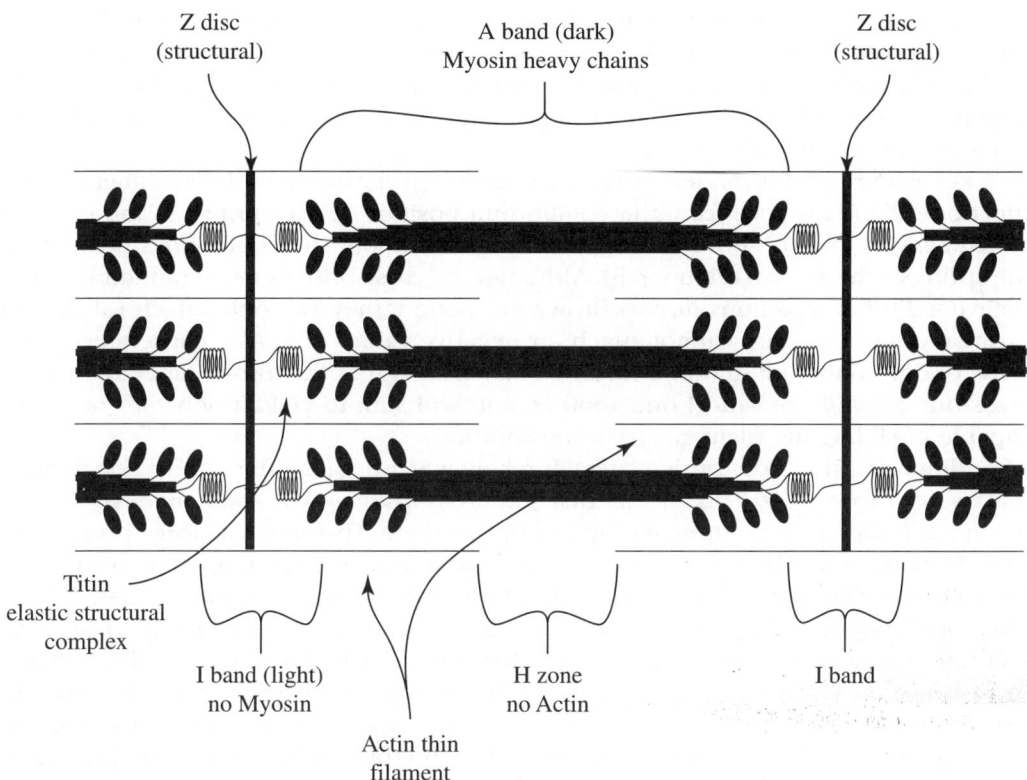

Z disc (structural)

A band (dark) Myosin heavy chains

Z disc (structural)

Titin elastic structural complex

I band (light) no Myosin

H zone no Actin

I band

Actin thin filament

FIGURE 3.1

The organization of the sarcomere, the fundamental contractile unit of muscle. Conglomerates of myosin molecule "tails" form the backbones of the myosin filaments. The "heads" of the myosin molecules project from the backbones forming the cross-bridges. During active concentric contraction, the cross-bridges attach to the surrounding actin filaments and undergo a conformational change that pulls the actin and myosin filaments past one another. This increases the overlap of the actin and myosin filaments, and decreases the distance between the Z-lines, shortening the sarcomere. Recent studies indicate that titin attaches myosin to the Z-lines, and may provide some of the passive elastic properties observed in the sarcomere. The A-bands correspond to the myosin filaments; the I-bands correspond to the space between myosin filaments; and the H-bands correspond to the space between actin filaments.

where *"the tension per filament should be proportional to the number of these points, and therefore to the width of this zone of overlap"*, a model that could quantitatively describe such a mechanism would not be formulated for several more years.

In 1957, A. F. Huxley proposed that the actin and myosin filaments within a muscle could be modeled as rigid rods that slide past one another by the action of cross-bridges (note that this model includes numerous quantifiable features, but it remains a model). He modeled cross-bridges as small sliding elements attached to the myosin filaments by linearly elastic springs. These sliding elements temporarily attach to certain sites on the actin filaments, allowing the cross-bridge springs to generate the force necessary to move the filaments past one other (Figure 3.2). Although cross-bridges can only move a small distance while attached, large changes in sarcomere length can be achieved by the cross-bridges going through many cycles of attachment, force generation, and detachment.

A key feature of this model is that the rates of attachment, f, and detachment, g, are functions of the distance from the equilibrium position of the springs to the point of attachment, x, and these functions are asymmetric about the equilibrium position of the springs, where $x = 0$ (Figure 3.3). Although the cross-bridges move randomly about their equilibrium positions due to Brownian motion, they can only attach for certain positive values of x, and cannot attach for negative values of x. As a consequence, the cross-bridges only pull in one direction. This ensures that the forces generated by the cross-bridges will not cancel one another, and will sum to yield a non-zero net force capable of sliding the filaments past one another.

Additionally, the cross-bridges detach when $x > h$ because the rate of attachment, $f(x)$, is zero for $x > h$. This means that the cross-bridges will spontaneously detach when $x > h$ for a muscle stretched by an applied force. This prevents force production from increasing indefinitely during eccentric contractions (stretching of active muscle by an externally applied load), according to $F = kx$, where k is the stiffness of the cross-bridge spring. The model also describes the average force expected from each cross-bridge. Although the force that any cross-bridge will generate for a given attachment-detachment cycle will vary with x, the average force over many attachment-detachment cycles will be the same for all cross-bridges. This is because the average distance from equilibrium will be the same for all cross-bridges, since any given x is

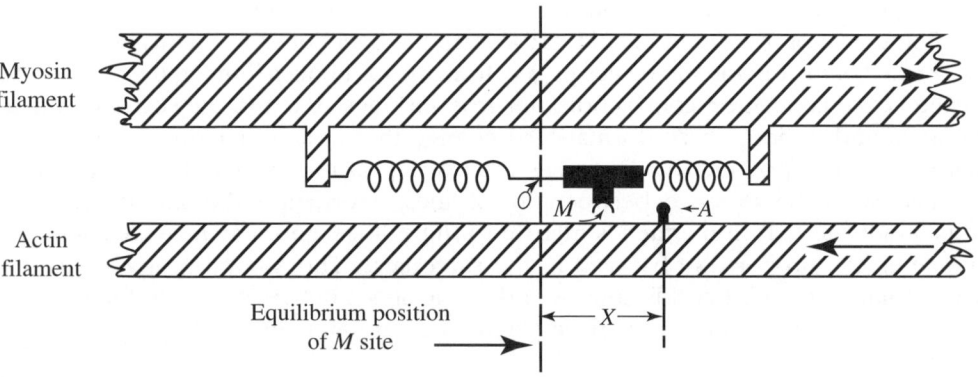

FIGURE 3.2

Diagram from Huxley (1957) illustrating the key features of the basic cross-bridge model. The actin and myosin filaments are modeled as rigid rods that slide past one another via action of cross-bridges, small sliding elements attached to the myosin backbone via springs, which can temporarily attach to sites on the actin filament to generate the force necessary to move the filaments past one another. The cross-bridge is labeled M for myosin attachment site, and the actin attachment site is labeled A. The distance from the equilibrium position of the springs to the point of attachment is x.

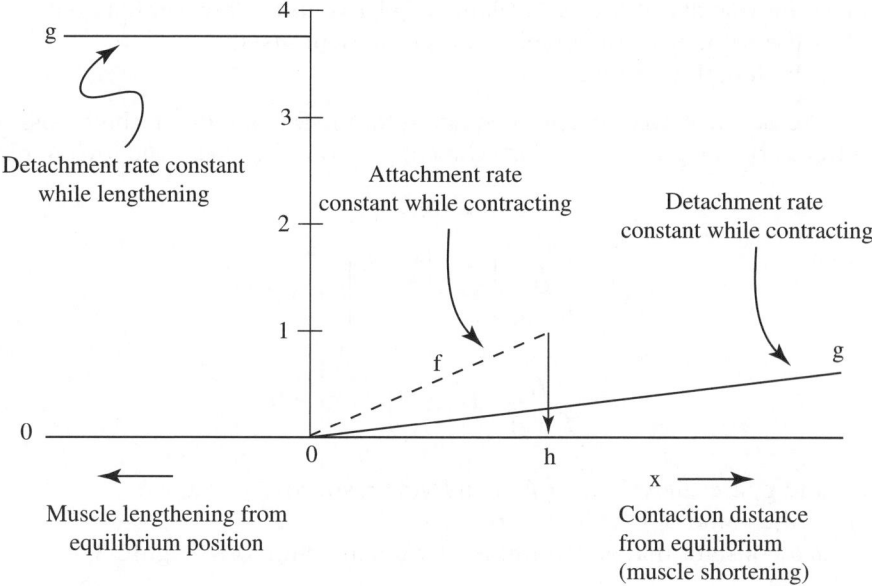

FIGURE 3.3

Graph based on Huxley (1957) showing how the rate constants for attachment, f, and detachment, g, vary with the distance from the equilibrium position of the cross-bridge springs, x. Cross-bridges can only attach for $0 < x < h$, where $f(x) > 0$. This ensures that the cross-bridges will only pull in one direction, and that they will eventually detach when stretched past h during an eccentric contraction.

determined by the random effects of Brownian motion, while the stiffness, k, should be the same for all cross-bridges.

Using Huxley's cross-bridge model, it is possible to write a differential equation (Equation 3.1) that describes the rate of change of the fraction of attached cross-bridges in terms of the rate constants for attachment and detachment, $f(x)$ and $g(x)$ respectively, and the fraction of attached cross-bridges, n, assuming that the behavior of a given cross-bridge does not depend on the states of the other cross-bridges or the history of the system:

$$\frac{dn}{dt} = (1 - n) f(x) - ng(x) \tag{3.1a}$$

This may also be written as:

$$-v\frac{dn}{dx} = f(x) - n\big(f(x) - g(x)\big) \tag{3.1b}$$

or

$$-\frac{sV}{2}\frac{dn}{dx} = f(x) - n\big(f(x) - g(x)\big) \tag{3.1c}$$

Where: v is the velocity of the actin filament relative to the myosin filament
V is the velocity of sarcomere shortening (lengths/sec)
s is the length of a sarcomere.

Equation 3.1c can be solved to get an equation for the fraction of attached cross-bridges for any given shortening velocity and attachment position of the actin and myosin filaments

$$n = 0 \quad x > h$$

$$n = \frac{f_1}{f_1 + g_1}\left(1 - e^{\left(\frac{x^2}{h} - 1\right)\frac{\varphi}{V}}\right) \quad 0 < x < h$$

$$n = \frac{f_1}{f_1 + g_1}\left(1 - e^{-\frac{\varphi}{V}}\right)e^{\frac{2xg_2}{sV}} \quad x < 0 \qquad (3.2)$$

Where: f_1 and g_1 are the values of $f(x)$ and $g(x)$ respectively at $x = h$
g_2 is the value of $g(x)$ for $x < 0$
φ is a constant that is determined by initial conditions (Figure 3.4).

Equation 3.1a can also be set equal to zero to derive the fraction of attached cross-bridges under steady-state (equilibrium) conditions:

$$n_{eq} \frac{f(x)}{f(x) + g(x)} \qquad (3.3)$$

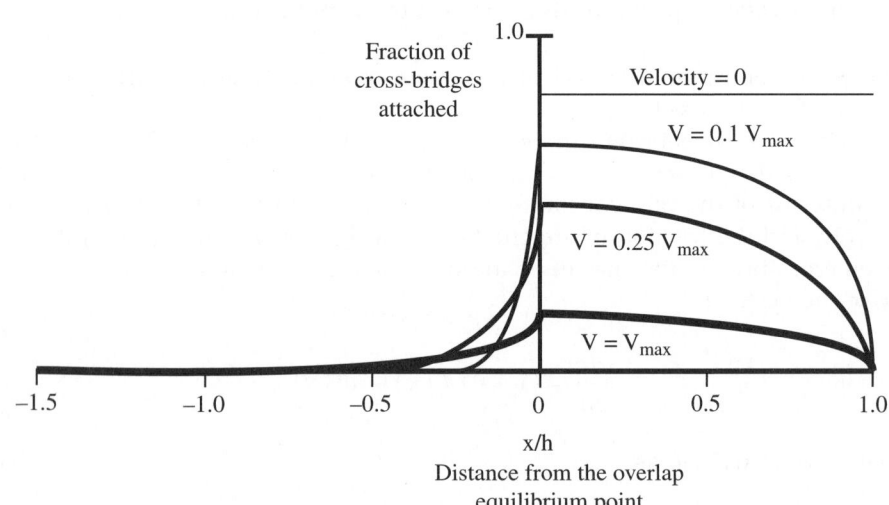

FIGURE 3.4

Plot based on Huxley (1957), showing the fraction of cross-bridges, n, as a function of the relative position of attachment of the cross-bridges, x/h, and sarcomere shortening velocity, V.

It is worth noting that the actual existence of cross-bridges had not yet been directly verified when Huxley proposed his model. Therefore, cross-bridge theory represented a purely theoretical explanation for the observed sliding behavior of the actin and myosin filaments within the sarcomere. However, the mathematical formulation of Huxley's model allowed the predictions of cross-bridge theory to be tested against actual experimental results.

3.2.3 Multi-state cross-bridge models

One problem with Huxley's 1957 cross-bridge model was that it could not account for variable filament spacing. Sarcomeres shorten more or less at constant volume. This implies that the sarcomeres must increase in diameter as they shorten (indeed, this increase in diameter can be readily observed in contracting muscle) and, therefore, the spacing between the actin and myosin filaments must change. The original Huxley model had no means of accommodating changes in filament spacing, since it had sliding cross-bridges. However, electron microscopy soon shed light on the structure of the cross-bridge, suggesting a mechanism.

Isolated heavy-meromyosin molecules (the larger digestion product derived from the complete myosin molecule), viewed under an electron microscope using a shadow-casting technique, showed a thin arm and a globular head (Figure 3.5A). This suggested that cross-bridges might be modeled as elastic arms, with heads at one end, that could interact with attachment sites (Figure 3.5). Also, biochemical studies found compliant regions in the myosin at the connections of the arm with the myosin backbone, and of the head with the arm. It was proposed that these compliant regions could act as hinges. Thus, the rotation of the arm could adjust for different filament spacings, at

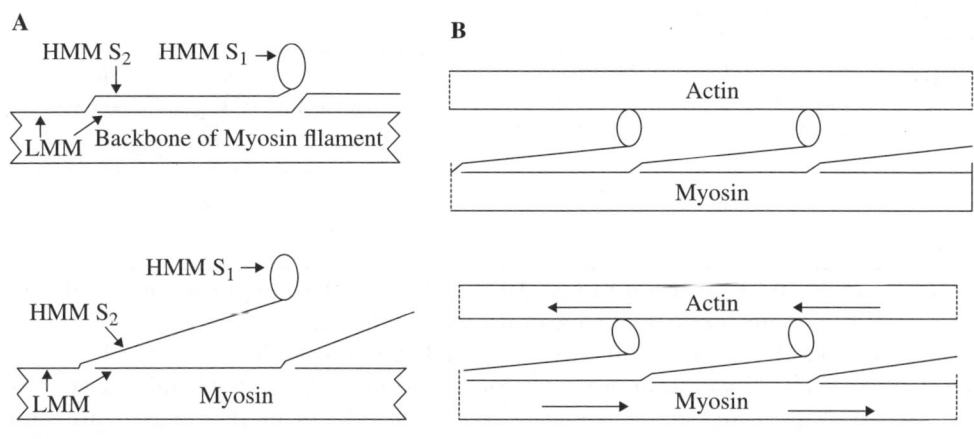

FIGURE 3.5

Diagram from Huxley (1969), illustrating the hinged arm model of the cross-bridge. **(A)** Heavy meromyosin (HMM) has a globular head (HMM S1) which attaches to the actin filament and a thin arm (HMM S2) which adjusts for variable filament spacing. **(B)** Force is generated by rotation of the cross-bridge head.

least to some degree, while the rotation of the head produced force (Figure 3.5B) (Huxley, 1969). This structural information set the stage for more complex, multi-state cross-bridge models.

The Huxley 1957 version of the cross-bridge model is often referred to as a two-state model, because the cross-bridges are in one of two states – either completely attached, or completely detached. The two-state model, however, could not explain some dynamic aspects of muscle behavior. For instance, it could not explain the time course of the tension observed when muscle undergoes a rapid, stepwise length change in certain *in vitro* experiments. During such a length change, there is initially a nearly instantaneous response, followed by a quick recovery response, and then a slow recovery response (Figure 3.6).

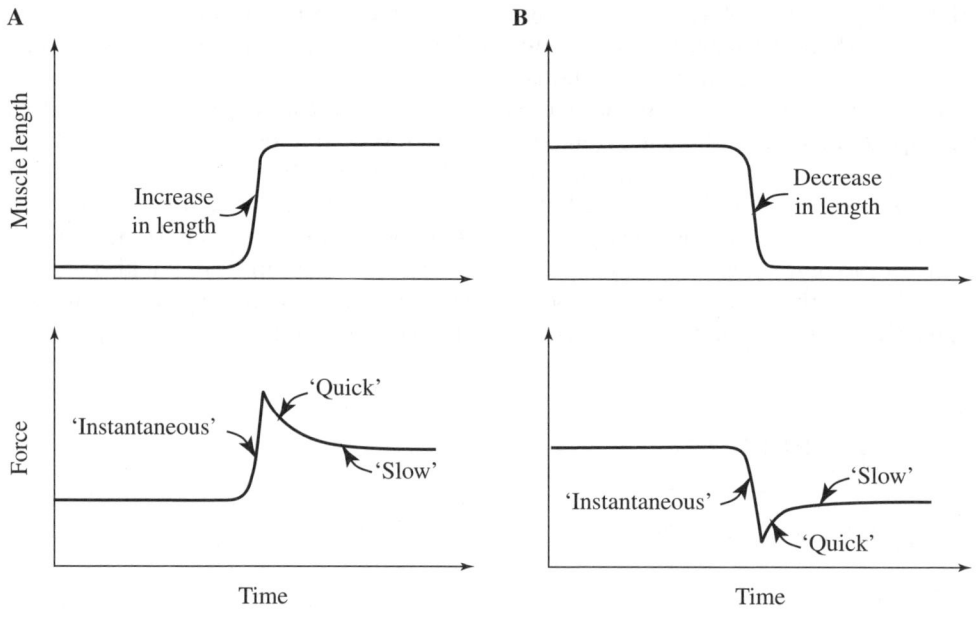

FIGURE 3.6

Plots based on Huxley and Simmons (1971), showing the experimentally observed response of active muscle fibers to a rapid stepwise length change. (**A**) An increase in fiber length results in force increase while (**B**) a decrease in fiber length results in force decrease. The "instantaneous" tension response occurs first, and is linear with a very steep slope; the "quick" tension response occurs next and is a non-linear curve; and the "slow" tension response occurs last and is nearly linear, with a shallow slope. This behavior implies that the "instantaneous" tension response is governed by the elastic properties of the cross-bridge arms; the "quick" tension response is governed by the rate constants associated with the rotation of the cross-bridge heads through various stable attachment states; and the "slow" tension response is governed by the rate constants associated with the attachment and detachment of the cross-bridges.

In the simple two-state model, there is only one rate constant governing attachment, and this can only account for a single type of response to length change. To account for the "instantaneous", "quick", and "slow" responses, Huxley and Simmons (1971) formulated a multi-state model. During a quick stepwise length change in this model, the arm of the cross-bridge is first stretched (or compressed) elastically, then the cross-bridge head rotates through various stable attachment states, releasing some of the elastic strain energy. Finally, the cross-bridge detaches and the cycle begins again. Thus, the "instantaneous" response is governed by the passive elastic properties of the cross-bridge arm, the "quick" recovery is governed by the non-linear rate constants associated with the rotation of the cross-bridge head through various stable attachment states, and the "slow" recovery is governed by the rate constants associated with attachment and detachment of the cross-bridges.

Biochemical studies performed after the multi-state model was first proposed provided corroborating evidence for the hinged-arm, multi-state behavior of cross-bridges. Rayment *et al.* (1993) combined high-resolution crystallographic information, obtained via x-ray diffraction, with electron density maps, obtained via electron microscopy, to create a structural model of the actin-myosin complex during cross-bridge binding. According to their model, as ATP hydrolyses into ADP + P^+, a cleft in the head of the cross-bridge begins to close, and this causes the cross-bridge to deform (storing elastic strain energy) and weakly bind to actin (weakly bound state). P^+ is then released, allowing the cleft to close completely and the cross-bridge to fully bind to actin (strongly bound state). Next, ADP is released, and the cross-bridge springs back into its original conformation while still attached to the actin (the power stroke). Finally, ATP binds to the cleft in the head of the cross-bridge, which causes the cleft to open and allows the cross-bridge to detach from the actin. These actions complete the attachment, force generation, and detachment cycle.

3.3 SOME ALTERNATIVES TO CROSS-BRIDGE THEORY

One interesting alternative to cross-bridge theory is the reptation model formulated by Pollack (1990). In this model, the cross-bridges do not actively contribute to force production but, rather, serve a passive, structural role, binding the myosin filaments to one another and creating a regular myosin network (Figure 3.7A). In the absence of ATP, the actin filaments bind to the cross-bridges in a state of rigor. If ATP becomes available, some of the bonds are broken, allowing portions of the actin filament to move via thermal agitation. If the bonds are more susceptible to being broken in one direction than the other, then directional motion could occur, and the filaments would slide across each other in a snake-like manner (Figure 3.7B).

Another possibility is that translation of the actin filaments across a static myosin network might be driven by active length changes occurring within the actin filament itself (Yanagida *et al.*, 1984). This hypothesis is supported by the fact that relatively large length changes have been observed in actin filaments under physiological conditions (Schutt and Lindberg, 1990) – evidence that length changes can occur without requiring "sliding" of filaments.

The main advantage of these alternative models is that they provide contractile mechanisms that do not rely on rotation of the cross-bridges. Cross-bridge rotation has

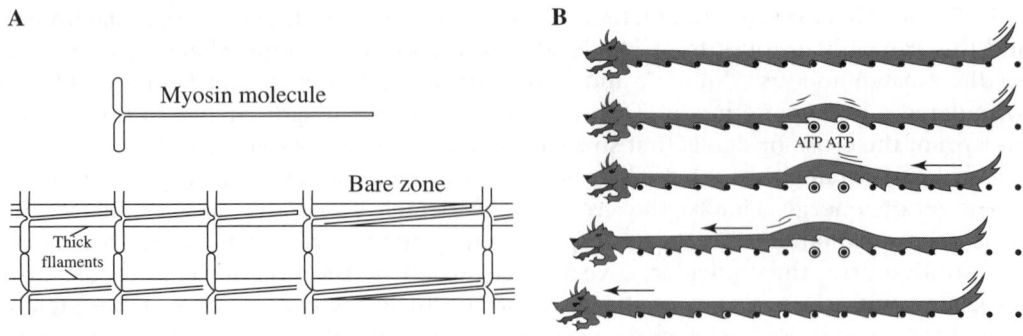

FIGURE 3.7

Illustrations from Pollack (1990), showing how cross-bridges might (**A**) join together to create a structural myosin network, which would (**B**) provide a substrate for actin reptation. ATP allows the actin filaments to detach from the myosin network, so that the filaments can crawl along the myosin, powered either by thermal agitation or active length changes of the actin filaments.

never been directly observed, so it is possible that cross-bridges do not actually undergo this conformational change. However, neither of these alternative models have been formulated in a mathematical manner, so they cannot be used to make testable, quantitative predictions about muscle behavior.

3.4   FORCE PRODUCTION

The amount of force that a muscle can produce is not constant under all operating conditions. In the following section, we will examine how force production changes with muscle length and muscle contractile velocity. These are the two key features of muscle function and contractile capacity within the musculoskeletal complex that define the muscle's capacity to act as the actuator of the locomotor system.

3.4.1   Isometric force production

The maximum amount of force each sarcomere (contractile unit) can produce depends on the overlap of the actin and myosin filaments. If extended to a macroscopic scale, this means that, at any contractile state, the amount of force a muscle can produce depends on its overall length, because muscle length is determined by the filament overlap length within the many sarcomeres arranged end-to-end in series. The relationship between force and filament overlap or, equivalently, force and muscle length, is generally assumed to hold for both isometric (constant length) and non-isometric (dynamic) contractions (although see the discussion of stretch-induced force enhancement and shortening-induced force depression in section 3.2.3).

The force-length relationship for skeletal muscle is obtained by measuring the force that a muscle (or some muscular subunit such as a muscle fiber, myofibril, or sarcomere) can produce when maximally activated under isometric conditions across a series of different muscle (or muscle subunit) lengths (Huxley, 1957; Gordon *et al.*, 1966).

The sarcomere force-length relationship is the quintessential version of this relationship, because it most clearly illustrates the connection between the force-length relationship and cross-bridge theory (Figure 3.8; Huxley, 1957).

For a range of optimal sarcomere lengths, maximum force production capacity is achieved (Figure 3.8A, ii–iii). This is referred to as the "plateau" of the force-length relationship. For sarcomere lengths longer than optimal, force production capacity decreases (Figure 3.8A, i–ii). This is referred to as the "descending" limb of the force-length relationship. For sarcomere lengths shorter than optimal, force production capacity also decreases (Figure 3.8A, iii–v). This is referred to as the "ascending" limb of the force-length relationship.

Cross-bridge theory does a good job of explaining the descending limb and plateau of the experimentally observed force-length relationship, but it does not fully explain the ascending limb. According to cross-bridge theory, the amount of force generated by a sarcomere should be proportional to the number of attached cross-bridges and, hence, the degree of filament overlap (Huxley, 1957). This is because cross-bridges are uniformly spaced across the lengths of myosin filaments (with the exception of a small gap in the center of the filament), the fraction of attached cross-bridges per unit length is constant for isometric contractions, and the force generated per cross-bridge is constant.

On the descending limb of the force-length relationship, force increases with increasing filament overlap as more cross-bridges are able to attach, until all the cross-bridges are able to attach and force is maximized (Figure 3.8B, 1–2). Across the plateau of the force-length relationship, force remains constant, since no new cross-bridges are added with increasing overlap, due to a gap in the cross-bridges at the center of the myosin filament (Figure 3.8B, 2–3). This is consistent with the predictions of cross-bridge theory.

On the ascending limb, the force generated decreases with increasing filament over-lap, even though all the cross-bridges should still be able to attach. There is speculation that force production may decrease for lengths shorter than optimal, due to interaction between overlapping actin filaments, or cross-bridges pulling actin filaments in the wrong direction as they attach to the opposite actin filament (Figure 3.8B, 3–6) (Gordon et al., 1966). For sarcomere lengths less than approximately three-quarters of optimal length, force production may further decrease, due to myosin filaments folding as they collide with the Z-lines, thin regions of connective material at the ends of each sarcomere (Figure 3.8B, 5–6) (Gordon et al., 1966). However, none of these mechanisms have been fully modeled or verified, so no quantitative predictions about the force-length relationship can be made for the ascending limb.

It is possible to shift the peak of the force-length relationship for a whole muscle or muscle group within an individual by specific training. The force-length relationship for leg extensor muscles peaks at shorter muscle lengths in cyclists, because cyclists train with their leg in a relatively flexed leg position. By contrast, the force-length relationship for leg extensor muscles peaks at longer lengths in runners, because runners train with their leg in a relatively extended position (Herzog et al., 1991). Such shifts could potentially be accomplished by decreasing (in the case of cyclists) or increasing (in the case of runners) the number of sarcomeres in series in the muscle fibers, in order to change the muscle length at which optimal overlap of the actin and myosin fibers within a sarcomere is achieved.

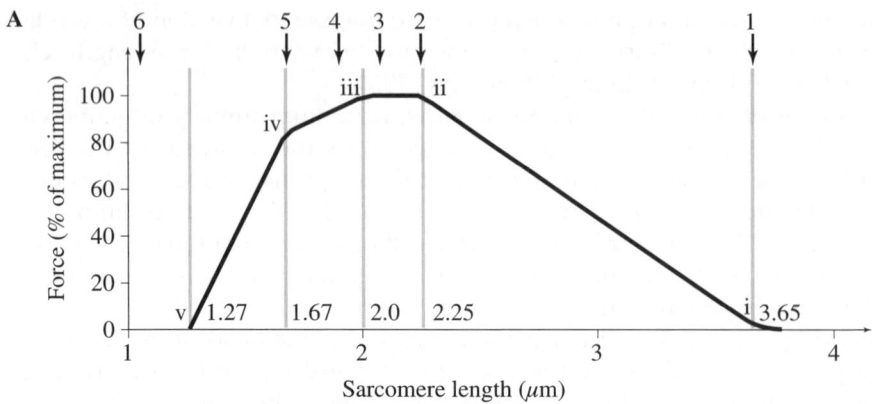

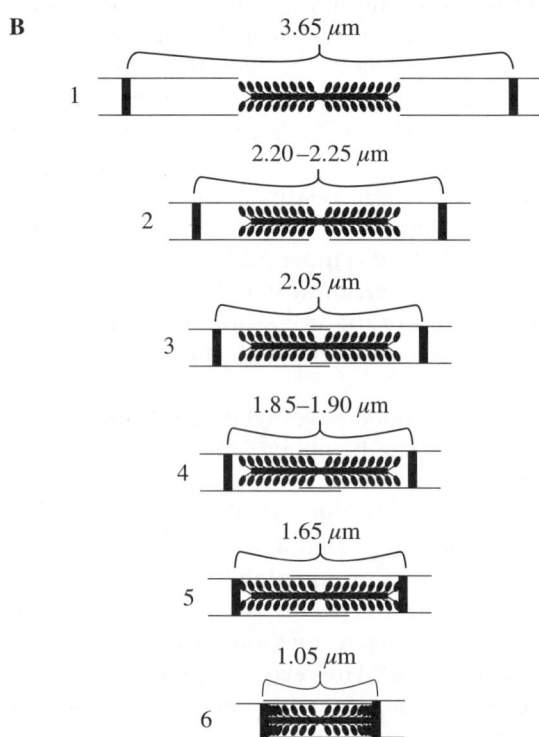

FIGURE 3.8

(A) Plot following Gordon *et al.* (1966), showing the sarcomere force-length relationship for frog muscle. The experimentally determined transition points are labeled i–v, and the transition points predicted using the cross-bridge model are labeled 1–6. There is reasonably good agreement between the two. (B) The relative positions of the actin and myosin filaments for the theoretical transition points are illustrated below. The mysosin filaments are represented as horizontal rods with hair-like projections (cross-bridges), and the actin filaments are represented as smooth horizontal rods. The vertical rods represent the Z-lines that mark the end of the sarcomere.

3.4.2 Non-isometric force production

The maximum amount of force that a muscle can produce depends not only on muscle length, but also on contractile velocity (how fast the muscle shortens or stretches while activated). The force-velocity relationship for skeletal muscle is obtained by measuring the maximum amount of force that a muscle (or muscle subunit) can produce when contracting at various velocities at a given muscle length. This relationship was observed many times (e.g., Fenn and Marsh, 1935; Gasser and Hill, 1924; Levin and Wyman, 1927) before anyone succeeded in formulating a theory that could explain it.

In 1938, A.V. Hill was finally able to solve the puzzle by taking a thermodynamic approach. In the first part of his iconic paper, *The heat of shortening and the dynamic constants of muscle*, he showed that the force-velocity relationship could be derived from data about the heat of muscular contraction. He noted that the energy a muscle consumed (beyond that needed to produce isometric force) when shortening a given distance, x, while producing a given constant force, F, depended on both the work done, Fx, and the distance shortened according to the relationship $E = Fx + ax = (F + a)x$, where E is the "extra" energy (above isometric) consumed during shortening, and a is a constant. This can be equivalently written as $dE/dt = (F + a)v$, where v is the velocity of shortening.

He also noticed that the rate of heat liberation is linearly related to muscle load according to $dE/dt = b(F_0 - F)$, where F_0 is the maximum force that can be sustained under isometric conditions $dE/dt = b(P_0 - P)$. Equating these two relationships gives $F = \dfrac{b(F_0 + a)}{v + b} - a$, the force-velocity relationship for concentric (shortening) contractions, where a and b are constants (Figure 3.9; Hill, 1938). There was no way to pre-determine what the constants a and b should be, so Hill adjusted the constants to fit his equation to data obtained from frog skeletal muscle actively shortening under various loads.

Later, A.F. Huxley (1957) showed that the force-velocity relationship for concentric contractions could also be derived from cross-bridge theory. Huxley obtained an equation for the force-velocity relationship by substituting the equation for the fraction of attached cross-bridges, Equation 3.3, into the equation for the average force per cross-sectional area of muscle (stress, P),

$$P = \frac{msk}{2l} \int_{-\infty}^{\infty} nx\, dx \tag{3.4}$$

Where: m is the number of cross-bridges (myosin attachment sites) per cm³ of muscle, l is the distance between the attachment sites on the actin filaments. This yields:

$$P = \frac{msw}{2l} \cdot \frac{f_1}{f_1 + g_1} \cdot \left\{ 1 - \frac{V}{\varphi}\left(1 - e^{-\frac{\varphi}{V}} \right)\left(1 + \frac{V}{2\varphi}\left(\frac{f_1 + g_1}{g_2} \right)^2 \right) \right\} \tag{3.5}$$

He then showed that this equation could be made to produce force-velocity results that agreed with those of Hill by adjusting the free parameters

$$\frac{w}{\varepsilon}\left(\frac{f_1 + g_1}{f_1} \right) \quad \text{and} \quad \frac{f_1 + g_1}{g_2}$$

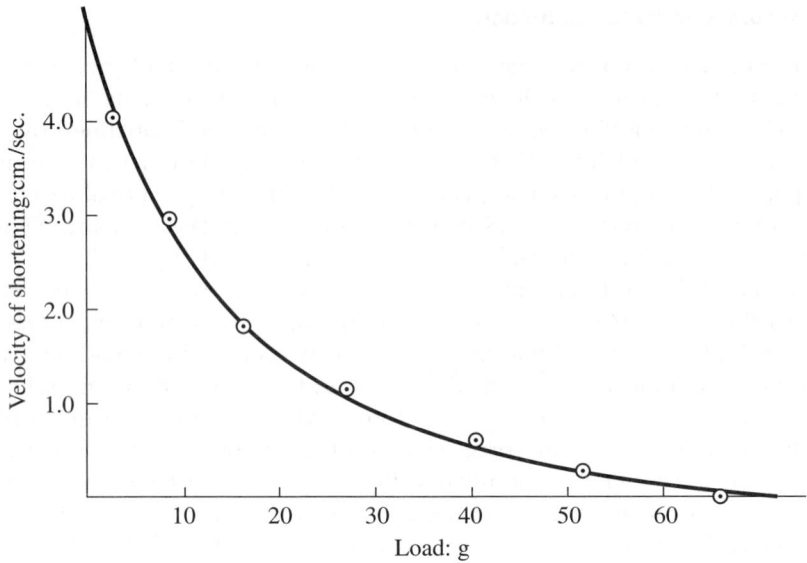

FIGURE 3.9

Plot from Hill (1938), comparing the force-velocity relationship derived from thermal measurements (line) to direct velocity measurements for concentric contractions under constant load (circles). The equation for the force-velocity relationship is $F = \dfrac{b(F_0 + a)}{v + b} - a$, where F is force, F_0 is the force developed under isometric conditions, v is velocity, and a and b are constants related to the rate of heat liberation. These constants are adjusted to fit the hyperbolic equation to the experimental data (Note: the force and velocity axes are opposite to those in the rest of this text).

Where: w is the maximum work done by any given cross-bridge, ε is the energy released by hydrolysis of one ATP to ADP + P$^+$ during each cross-bridge cycle, and by adjusting the ratio of g_1/f_1 so that the "maintenance" heat liberated during isometric contraction agreed with Hill's experimental findings.

Although Hill's thermal model and Huxley's cross-bridge model of the force-velocity relationship are both able accurately to predict the force-velocity relationship for concentric (shortening) contractions, neither does well in predicting the relationship for eccentric contractions (stretching muscle that is actively trying to contract) (Figure 3.10). Mathematically, the hyperbolic equation Hill derived for concentric contractions predicts negative (compressive) forces for eccentric contractions. No such discontinuity occurs in real life – real muscles produce resistive tensile force when stretched, particularly while activated.

Huxley's cross-bridge model does better and predicts positive (tensile) forces for eccentric contractions, but the predictions of this model are quantitatively inaccurate. Huxley's model predicts that the ratio of the slopes of the force-velocity relationship for slow stretching (eccentric) and slow shortening (concentric) should be 4.33, whereas it is experimentally shown to be nearly 40% greater – about 6 (Katz, 1939). It also predicts that the force-velocity curve should asymptotically approach a value of

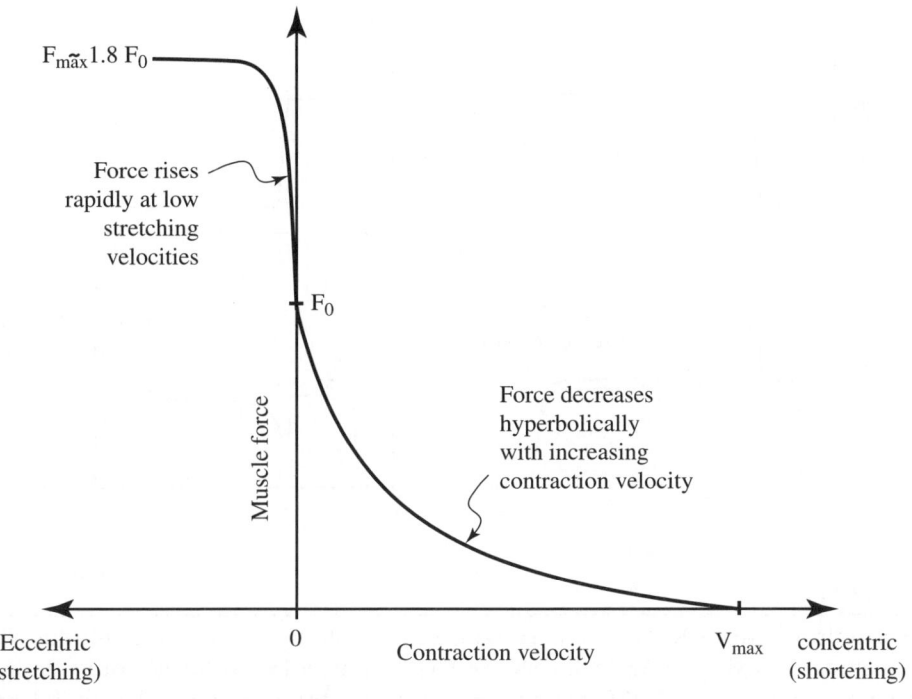

FIGURE 3.10

The complete force-velocity relationship for skeletal muscle, including both concentric (shortening) and eccentric (elongation) contractions. F_{max} is the maximal tension developed during active stretching, F_0 is the tension developed under isometric conditions, and v_{max} is the maximal shortening velocity.

5.33 F_0 for fast stretching, but the experimentally determined value of the asymptote is much lower – about 1.8 F_0 (Katz, 1939).

Finally, it should be mentioned that some dynamic aspects of muscle behavior (e.g., force enhancement and force depression) cannot be explained by either the force-length or force-velocity relationship. If a muscle is activated, and then stretched to a longer final length and held in isometric tension at the new length, the amount of force produced is greater than the amount of force that can be produced when the muscle is held in isometric tension at a constant length without being actively stretched (Figure 3.11A) (Abbott and Aubert, 1952,; Morgan et al., 2000).

Conversely, if a muscle is activated and then relaxed to a shorter final length and held in isometric tension at the new length, the amount of force produced is less than the amount of force produced when the muscle is held in isometric tension at a constant length without actively shortening (Figure 3.11B) (Abbott and Aubert, 1952; Morgan et al., 2000). Such stretch-induced force enhancement and shortening-induced force depression cannot be predicted by the force-length and force-velocity relationships (Rassier and Herzog, 2003), but would likely affect force production during the dynamic movements associated with locomotion.

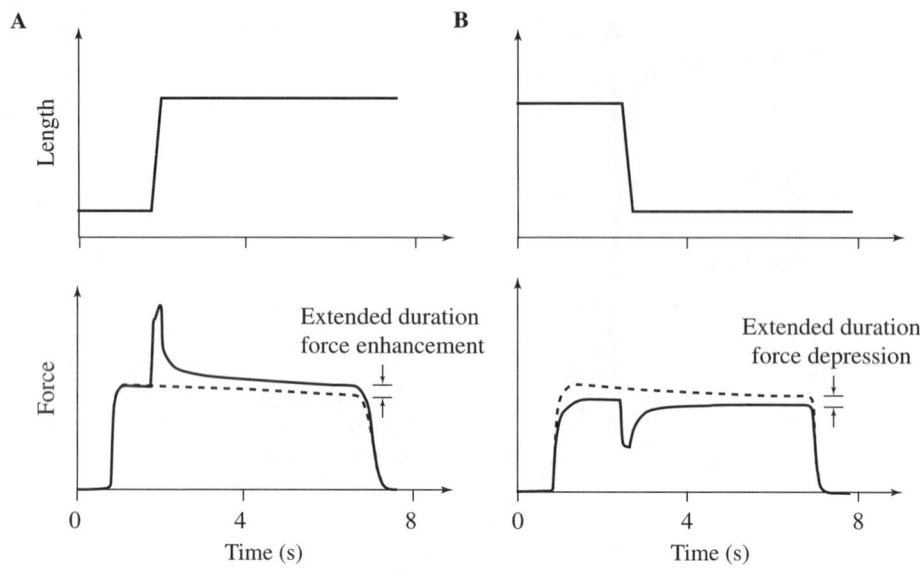

FIGURE 3.11

Graphs redrawn from Rassier and Herzog (2003), showing (**A**) stretch-induced force enhancement above isometric levels, and (**B**) shortening-induced force depression below isometric levels. Solid curves represent the force generated under isometric conditions by previously stretched (force enhancement) or shortened muscle (force depression). Dashed curves represent force generated during normal isometric contractions (constant fiber length), conducted at the same fiber length as that attained after stretching or shortening.

3.5 THE HILL-TYPE MODEL

In addition to deriving the force-velocity relationship of muscle from thermodynamic considerations, A.V. Hill also contributed a highly influential phenomenological model of muscle. In the third section of his 1938 paper, Hill modeled the contractile behavior of frog leg muscle as a two-component system, composed of an active contractile element in series with a passive elastic element (Figure 3.12A). The contractile element followed the hyperbolic force-velocity relationship, $F = \dfrac{b(F_0 + a)}{v + b} - a$, which he had derived earlier in the paper. The series elastic component followed the linear equation of a simple spring, $F = kx$, where k is spring stiffness and x is deformation. Hill recognized that some damping was biologically inevitable, but considered this damping to be of minor importance.

The Hill-type model reflects the mechanistic perspective of the industrial age. In the mid-to-late 1800s there was substantial interest in characterizing the non-linear properties of biological and non-biological materials. The properties of these materials were modeled as combinations of springs and dashpots (viscous damping components that added rate and time-dependent effects). The two most notable of these simple

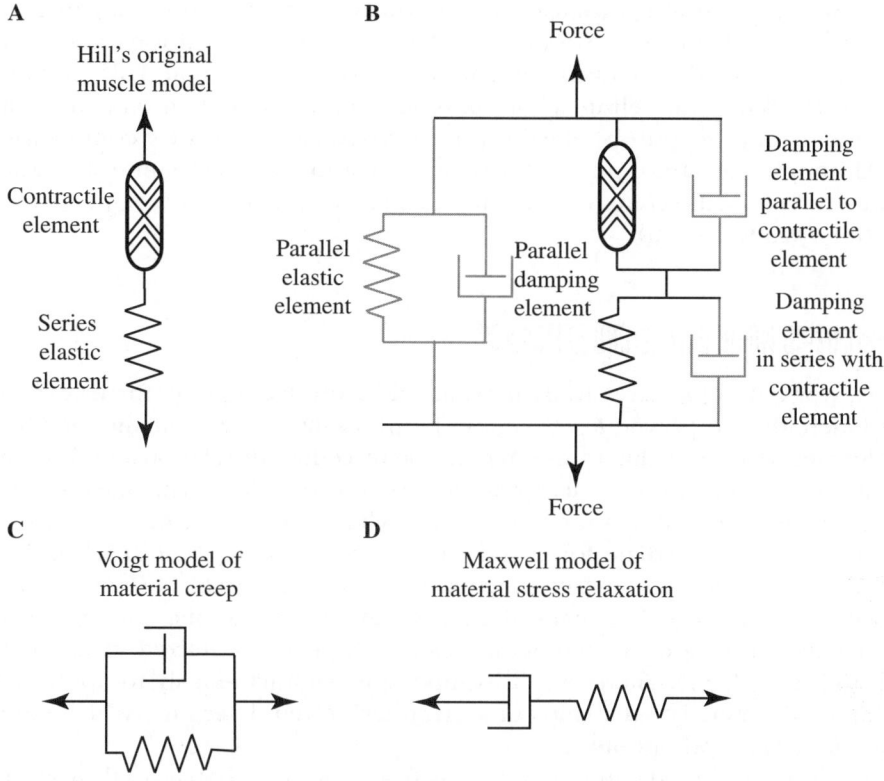

FIGURE 3.12

The "Hill-type" phenomenological model of muscle, as originally proposed by A.V. Hill in 1938, consisted of: (**A**) a contractile element in series with an elastic element. (**B**) A variety of additional elastic and damping components can be added to the original model. Each element can be represented with a relatively simple mathematical relation. Such phenomenological models were in common use prior to Hill's application to muscle. Examples include the (**C**) Voigt and (**D**) Maxwell models of viscoelastic behavior in materials.

mechanical models are the Voigt model (Woldemar Voigt, 1850–1919), which consisted of a spring and viscous damper in parallel (Figure 3.12C), and the Maxwell model (introduced in 1867), composed of a spring and damper in series (Figure 3.12D). The Voigt model described creep under load, and the Maxwell model described stress relaxation over time. Another, slightly more complex, model is the standard linear solid or Kelvin model, composed of a spring and dashpot in series, with both in parallel with a second spring. These simple models are capable of elegantly modeling specific behaviors of materials with relatively complex viscoelastic properties, such as biomaterials.

The Hill-type model derives its usefulness in part from its simplicity and, in part, from the fact that it can readily be adapted to meet the requirements of the circumstances under which it is used. It was not devised to represent accurately the biological process of muscular contraction (i.e., it was not a mechanistic model).

Rather, it was a phenomenological model constructed of simple machines whose behavior replicated the bulk properties of muscle. Many additions and modifications have been made to (and interpreted into) the original Hill model (e.g., Figure 3.12B). When modeling whole muscle, a parallel element is usually added. This element may be purely elastic, purely viscous, or various combinations of these. Although the Hill-type model of muscle function has some limitations, it continues to be used extensively to model the bulk properties of muscle during locomotion (Winter, 1990).

3.6 OPTIMIZING WORK, POWER, AND EFFICIENCY

In theory, work, $W = \int \vec{F} \cdot d\vec{x}$, can be maximized by producing high forces over a large change in length, and power, $P = \vec{F} \cdot \vec{v}$, can be maximized by producing high forces at high velocities. However, due to the hyperbolic force-length relationship, high muscle forces can only be achieved at near-zero contraction velocities. This means that work is maximized at near-zero contraction velocity, where force is greatest, while power is maximized at an intermediate force and contraction velocity, about $0.3F_0$ and $0.3v_{max}$, where the product of force and contraction velocity is the greatest (Figure 3.13). The efficiency of doing work (mechanical energy output / metabolic energy consumed) also varies as a function of contractile velocity, and is maximized at about $0.2v_{max}$ (Figure 3.13; Hill, 1939). Therefore, it is impossible simultaneously to optimize work, power, and efficiency. The demands of a given task often determine which parameter animals emphasize and optimize.

The leg design of bipedal jumpers such as frogs and bush babies (galagos) suggests that these animals use a strategy that increases the ability of muscles to do work while decreasing the power demands placed on the muscles. These jumpers have a highly flexed leg posture and relatively long rear legs. The flexed leg posture results in a high ground reaction force to muscle lever arm ratio. This increases the muscles' ability to do work by allowing the leg muscles to produce high forces by operating at modest contraction speeds (Figure 3.13).

Note that the work generated by the muscle must equal the work done by the limb segment, although the force to distance ratio of each can be altered by changing the ratio of the lever arms (lever advantage of a joint can alter relative force required across the joint, but not the amount of work accomplished). The long rear legs allow the feet to maintain contact with the ground longer while the leg extends. This decreases average power requirements, $P_{ave} = \Delta W / \Delta t$, by increasing the amount of time available for the muscles to accomplish the task, Δt. This interpretation of leg design agrees well with the results of Roberts and Marsh (2003), which showed that the model of frog jumping that best replicated the kinematics of real frogs maximized muscle work, but did not maximize average power.

Why might jumping animals such as frogs and bush babies optimize work rather than power? Operating muscle at the contraction velocities needed to generate peak power reduces the ability of a given volume of muscle to generate force (Figure 3.13). Therefore, a greater volume of muscle would be required to meet the force requirements of jumping. However, increasing muscle volume would increase total body weight and, hence, increase the amount of work that the muscles would need to do to

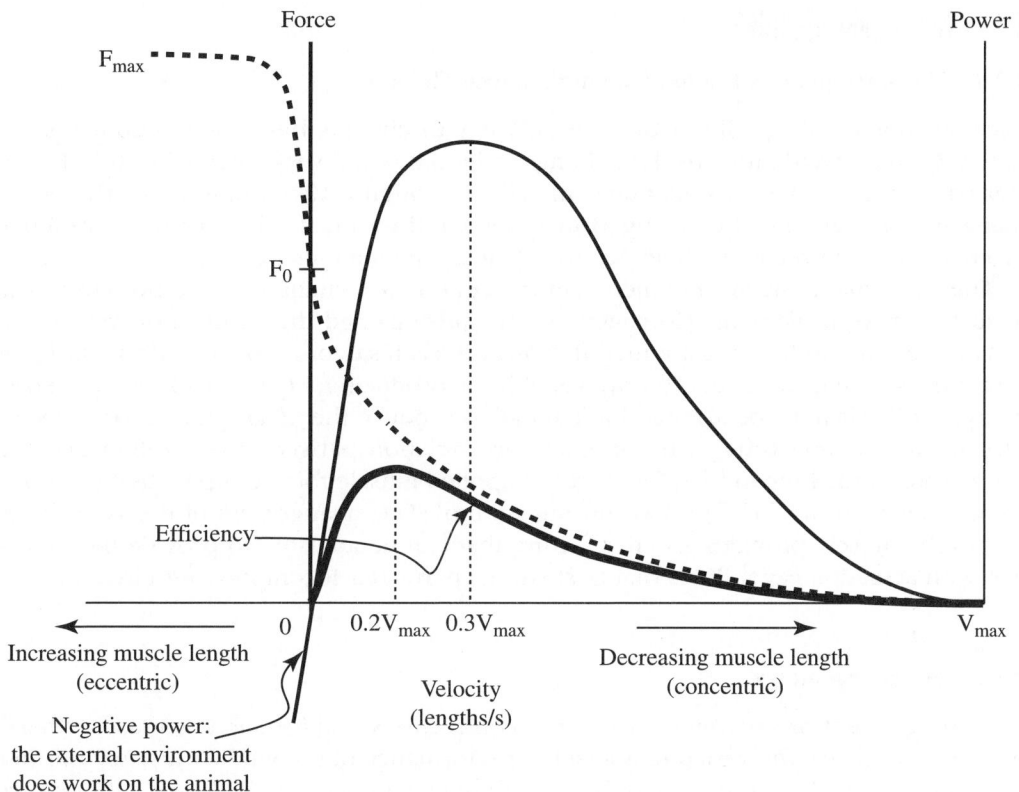

FIGURE 3.13

Plot showing how muscle force, power and efficiency (mechanical energy output / metabolic energy consumed) vary as a function of contractile velocity. Maximum force occurs at zero velocity (isometric contraction), while maximal power occurs at approximately $0.3v_{max}$, and maximal efficiency occurs at approximately $0.2v_{max}$.

accomplish the jump. Additionally, muscle volume would ultimately be constrained by the geometry of the limb. A leg with a bulky muscle would have difficulty flexing. This would reduce the amount of time the feet are in contact with the ground which, in turn, would increase the power the muscles must generate to produce a jump with a given height. Thus, these animals likely find that optimizing work, rather than power, allows them to achieve higher jump heights, given the properties of skeletal muscle and the mechanical requirements of jumping.

The objective of jumping in species that depend on a one-off jump are relatively well defined; survival of the individual will depend on its ability to jump to escape predation or successfully reach the next tree. However, even in this case, there are trade-offs that must be considered when optimizing performance (as discussed above). This example highlights the importance of studying muscle function in the context of the physiology of the organism and the physical properties of the environment.

3.7 MUSCLE ARCHITECTURE

3.7.1 The sarcomere as the fundamental contractile unit

Each sarcomere is capable of up to a 70% length change, but, since sarcomeres are only 1-5 μm in length, the absolute change in length is still very small – just 0.7–3.5 μm (for context, note that most mammalian cells are roughly 10 μm in length). Therefore, many sarcomeres must be arranged in series within a myofibril in order to produce macroscopic length change between the skeletal attachments points.

The force that a given sarcomere can generate is determined by the overlap of the actin and myosin filaments (force-length relationship) and the contraction velocity of sarcomere (force-velocity relationship). However, each sarcomere or myofibril (string of sarcomeres acting in series) is only capable of producing ≈1 μN of force. Therefore, many myofibrils must be arranged in parallel to produce the magnitude of force necessary for locomotion (due to the moment arm relation between the forefoot and the calcaneous of the human foot, for instance, applied muscle force of up to 18 times body weight can occur for brief periods during a jump). The arrangement of the sarcomeres within the muscle provides another feature that can be adjusted to provide the organism with actuation capabilities that best suit its particular locomotion requirements.

3.7.2 Muscle geometry

The arrangement of sarcomeres (or other contractile subunits, such as muscle fibers) within a muscle can be varied to adjust the performance of the whole muscle. Consider the simple case where there are only two subunits within a muscle. If two subunits are arranged in series, then the muscle can shorten twice as far in a given amount of time and, consequently, it can reach twice the maximum shortening velocity that it would reach if the two subunits were arranged in parallel. However, the maximum force the muscle can generate when the subunits are in series will be only half as great as if the same subunits were arranged in parallel (subunits arranged in series pull on each other, so the total force is determined by the maximum force production capacity of a single subunit).

Conversely, if two subunits are arranged in parallel, then the muscle can produce twice the amount of force that it could if the two subunits were arranged in series (subunits arranged in parallel each pull on the load directly, so their forces sum). However, the muscle will only be able to shorten half the distance in a given amount of time and, consequently, will reach half the velocity. The parallel arrangement also allows the muscle to generate a submaximal force at half the cost of the series arrangement, since only half the muscle mass (volume) needs to be activated to produce a given amount of force.

It should be noted that a given volume of muscle will be capable of the same amount of work (and power) per contraction cycle, regardless of whether or not the subunits within the muscle are arranged in series or in parallel. Different muscle arrangements might emphasize force production, F, or length change, dx, but they cannot alter the amount of work, $W = \int \bar{F} \cdot d\bar{x}$, that can be done by a given volume of muscle. See Table 3.1 (Edgerton *et al.*, 1986) for a comparison of the functional implications of series and parallel contractile unit arrangement.

TABLE 3.1■ A comparison of the maximal mechanical capabilities of muscle for muscle subunits arranged in series (A) and in parallel (B). The small graphs at the bottom show force vs. time plots (left) and force vs. velocity plots (right) for the two muscle geometries (from Edgerton *et al.*, 1986)

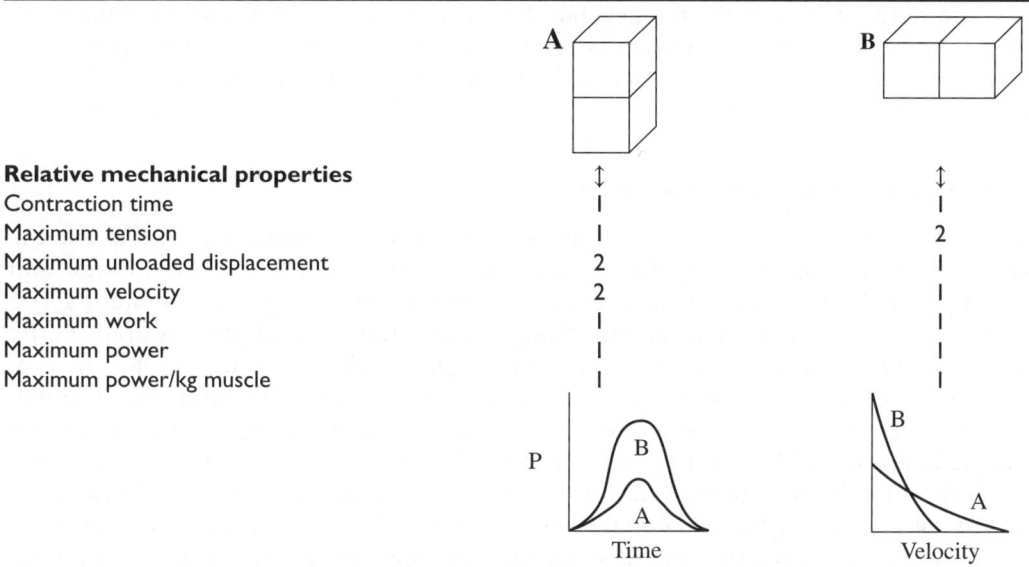

Relative mechanical properties	A	B
Contraction time	1	1
Maximum tension	1	2
Maximum unloaded displacement	2	1
Maximum velocity	2	1
Maximum work	1	1
Maximum power	1	1
Maximum power/kg muscle	1	1

Muscle fibers are rarely arranged strictly in series or purely in parallel *in vivo*. Long, narrow muscles do exist that are composed of a relatively small number of long fibers, which correspond more or less to the series arrangement, and there are complex pennate muscles that are composed of a relatively large number of short fibers, which correspond more or less to the parallel arrangement.

Muscles that must contract over a long functional range are long and strap-like, with muscle fibers arranged in parallel. This geometry increases the contraction range but reduces the amount of force that can be produced per volume of muscle tissue. Large length changes are produced by summing the incremental length changes produced by a great many sarcomeres in series. However, the force that can be produced for a given muscle volume is relatively low, because the parallel configuration of the muscle fibers reduces the number of muscle fibers that can act on a given tendon. These muscles often reposition limb segments when segments are not in contact with the ground, and reposition the head, neck and components of the head. The extraocular muscles, for instance, move the eye within its orbit. Their parallel fiber arrangement allows them to rapidly adjust the direction of gaze, but the task does not involve bearing substantial load.

Muscles that must produce high forces, on the other hand, are relatively short, with muscle fibers arranged in a pennate configuration. This geometry increases the amount of force that can be produced, but decreases the contraction range. Large forces can be generated for a given muscle volume, because the pennate configuration of the muscle fibers increases the number of fibers that can act on the associated tendon.

However, the contraction range is relatively short, because the incremental length changes are summed over fewer sarcomeres in series within the shorter muscle fibers. These muscles generally have a remarkably small lever arm with respect to the joint. This takes advantage of the increased force production capacity of this muscle geometry, and reduces the length changes that the muscles must undergo when doing work on the limbs. This type of muscle is often used to generate force to support body weight while limbs are in contact with the ground, and perform other tasks that involve bearing high loads.

3.7.3 Elastic energy storage and return

The amount of mechanical work that muscles must do in locomotion, and hence the metabolic cost of generating this work, can be reduced through the storage and return of elastic strain energy. Energy that might otherwise be lost can sometimes be stored as elastic strain energy by stretching tendons and other elastic materials in the body. This energy can be returned later in the gait cycle, when the tendon recoils. However, storing elastic strain energy does incur a metabolic cost. In order to stretch a tendon, a muscle must generate force, and generating muscle force consumes metabolic energy, although the cost will be relatively low if force is generated isometrically. Hill (1938) showed that, for concentric contraction (muscle shortening), the cost of generating force is equal to the cost of producing isometric force plus the cost of doing work, and the cost of stretching the muscle a given distance. Therefore, the cost of generating a given amount of force isometrically is necessarily lower than the cost of generating the same amount of force while shortening the muscle to do positive work.

There is good evidence that such elastic mechanisms are used by animals as diverse as humans (Ker *et al.*, 1987), horses (Minetti *et al.*, 1999), camels (Alexander *et al.*, 1982), turkeys (Roberts *et al.*, 1997), and wallabies (Biewener and Baudinette, 1995), in bouncing gaits such as trotting, running, bounding and the saltatory hopping of the macropods. In these gaits, a portion of the kinetic energy of the center of mass is stored during landing, and returned during take-off (Cavagna *et al.*, 1977). However, the specific role of elastic energy storage and return in many gaits, and in the context of the mammalian limb and trunk anatomy, is currently coming under renewed scrutiny. Although it appears extremely likely that elastic storage and return is important to the mechanics and energetics of bouncing gaits in many mammals, recent analyses suggest that it may not be the defining feature of these gaits, even though bouncing motions occur (Litchwark and Wilson, 2007, 2008; Pfau *et al.*, 2006; Ruina *et al.*, 2005; and see Chapters 5 and 6, plus Chapter 7 for an alternate view).

3.7.4 Damping/energy dissipation

Any energy consumed by a muscle that does not do work to increase the potential, kinetic, or elastic strain energy of the body is ultimately dissipated. For example, energy is dissipated as heat when a muscle contracts isometrically, and also when a muscle changes length while either shortening or being stretched. In the case of isometric contraction, no external work is done, so all the energy consumed by a muscle is lost in the form of heat. Even when a muscle does work during concentric contractions, the efficiency (proportion of metabolic energy translated to mechanical energy) is not large – approximately 25% (McMahon, 1984), which implies that about 75% of the energy consumed by muscles is ultimately dissipated as heat.

In the case of eccentric contractions (stretching active muscle by applying an external load to the muscle), when a muscle does "negative work" on the body, it not only consumes energy to perform the contraction, but it also degrades mechanical energy to heat, reducing the total energy of the body. Consequently, it is not possible to retain such negative work within the muscle for later use. However, negative work done by the muscle-tendon complex can, in some circumstances, be retained as elastic strain energy and used later in the gait cycle.

Although Hill (1922) originally concluded that energy dissipation was primarily due to viscous damping of the form $F_d \propto v$ (where F_d is the damping force and v is the contractile velocity of the muscle), he eventually disproved his own theory by showing that the energy dissipated as heat during constant velocity concentric contractions (beyond that produced during isometric contractions) was proportional to the distance the muscle contracted, regardless of contractile velocity. In truly viscous damping, heat release would be proportional to both contraction distance and velocity, according to $E_d = F_d x \propto vx$, where E_d is the energy dissipated (Hill, 1938). The heat released during eccentric contractions is less than the heat released during isometric contractions, indicating that the mechanism of force production is more passive when the contractile element is stretched while actively trying to contract, than it is when contracting against a resistive load.

Additionally, the way muscles are attached across joints can affect the amount of energy dissipated during a given movement. Single-joint muscles can produce any of the movements necessary for locomotion. However, they sometimes must do work against one another (i.e., one single-joint muscle must contract and do positive work (add energy), while another single-joint muscle needs to stretch and do negative work (dissipate energy)). Although it is theoretically possible to eliminate some of the wasted energy by using two-joint muscles, it appears that the configuration of the thigh muscles (important locomotion muscles) in most animals (except kangaroos) is such that the animals cannot take advantage of this strategy (Alexander, 2003). It is unlikely that a means of decreasing metabolic cost of locomotion has been left unexploited by nature. This enigma suggests we do not fully understand the implications of the muscle-joint configurations possible, or the consequences each provides to performance.

3.8 OTHER FACTORS THAT INFLUENCE MUSCLE PERFORMANCE

3.8.1 Fiber type

Skeletal muscles are composed of a variety of muscle fiber types, each with slightly different contractile properties. The most extreme examples are fast-twitch (type II) and slow-twitch (type I) fibers. Both fast- and slow-twitch muscle can produce approximately the same maximum isometric force per cross-sectional area but, as the name implies, fast-twitch fibers can contract more quickly and, consequently, produce more power than slow-twitch fibers (Figure 3.14). Fast-twitch fibers fatigue quickly, because

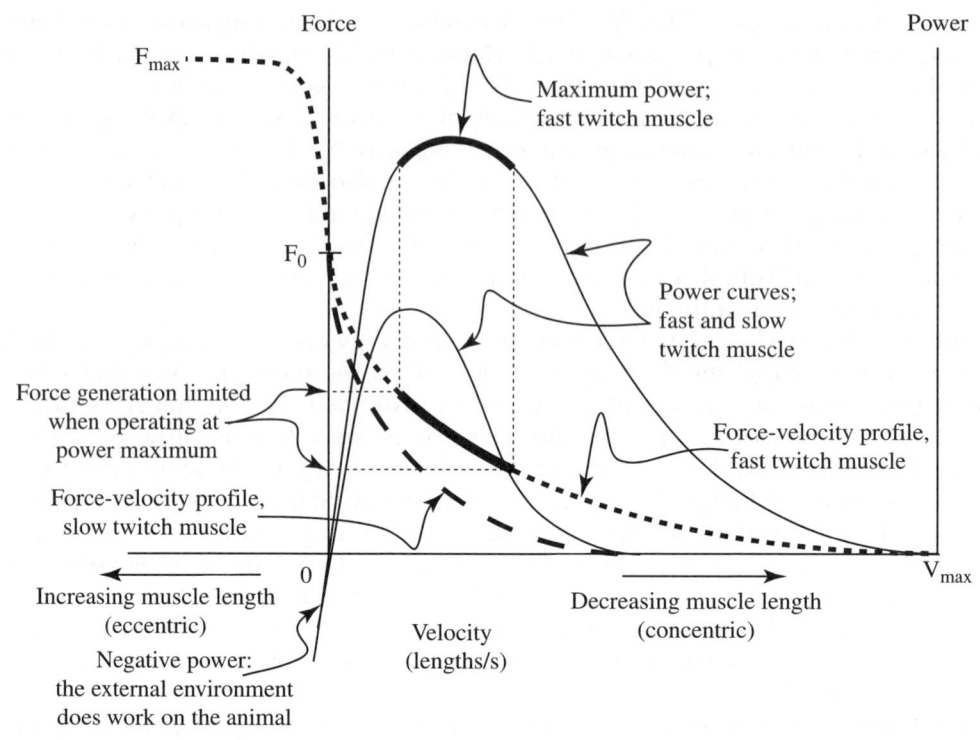

FIGURE 3.14

Illustration of the force-velocity and power-velocity relationships for fast- and slow-twitch muscle. Both fast- and slow-twitch muscle can produce about the same amount of isometric force per unit cross-sectional area, but fast-twitch muscle can contract much more quickly $((v_{max})_{fast} > (v_{max})_{slow})$ and, thus, produce more power $((P_{max})_{fast} > (P_{max})_{slow})$ than slow-twitch muscle. However, slow-twitch muscle has better endurance.

aerobic metabolic pathways are unable to provide enough metabolites to meet the high power demands of these muscles. Most of the power used by fast-twitch fibers is supplied by anaerobic glycolysis.

By contrast, slow-twitch fibers contract relatively slowly, but have better endurance, because aerobic metabolic pathways are able to provide most of the power needed to meet the lower power demands of these muscles. The oxygen used in oxidative metabolism in slow-twitch fibers is supplied by a dense network of capillaries, generally giving muscle groups with these fibers a darker appearance. Intermediate muscle types have properties that are between those of the two extreme fiber types. The contractile properties of these muscle fibers are largely determined by variations in the myosin heavy chain composition, and other features of their cellular physiology, that can be evaluated through histochemical analysis.

The ratio of fiber types within a muscle or a functional muscle group impacts the types of tasks for which a muscle or muscle group is best suited. A muscle that is predominantly composed of fast-twitch fibers is better suited for tasks that require high power production but are relatively short in duration. A muscle that has more slow-twitch fibers is better suited for tasks that have relatively low power requirements but last for relatively long durations. Similarly, individuals with different proportions of fiber types within the same muscle may be better suited for different activities. Individuals with a relatively high fraction of fast twitch fibers are better suited for activities that require high power production and less endurance, such as sprinting or weight lifting. Individuals with a relatively high fraction of slow twitch fibers are better suited for activities that have low power requirements but require more endurance, such as cross-country or marathon running.

Although it may not be possible to convert a fast-twitch fiber into a slow-twitch fiber, it is definitely possible to convert one intermediate fiber type to a different type of intermediate fiber and to increase the volume of one type of fiber through specific training for power (in the case of fast-twitch fibers) or endurance (in the case of slow-twitch fibers) – at least within genetically predetermined limits for a given individual or species (Fitz and Widdrick, 1996).

3.9 ACTIVATION AND RECRUITMENT

Although the maximum amount of force a muscle can produce at a given length and contractile velocity is determined by the muscle's force-length and force-velocity properties, most muscles usually operate well below their maximum capacity. Muscle activation can be modulated to control muscle contractions with relative precision.

Muscle activation can be modulated by selective activation of a fraction of the total muscle volume (recruitment). Muscle is organized into bundles of muscle fibers (motor units), where each motor unit is innervated by a single neuron. This unique feature of the somatic system adds important flexibility to the coordination and control of muscle activity. It is possible to control the amount of force that the muscle, as a whole, produces by activating some motor units but not others. If more motor units are activated, force production is increased. Motor unit size varies – some motor units contain just two or three muscle fibers (e.g., those in the eye, where fine control is critical) while others contain thousands of fibers (e.g., those in large leg muscles, where rapid recruitment and powerful response are critical).

Usually, *in vivo* motor units are recruited in order of small to large, and slow- to fast-twitch fibers, to produce a smooth increase in muscular tension. Often, during sustained, sub-maximal muscular contractions, motor units are rotated into and out of activation in order to allow some motor units a chance to recover, while others take their place to avoid whole muscle fatigue and catastrophic decline in functional capacity. Thus, the level of muscular fatigue may vary from motor unit to motor unit within a muscle, and changes in muscle function may only arise when fatigue progresses to the extent that inappropriate motor units are recruited. This strategy makes it difficult to assess fatigue at the muscular level, since the level of fatigue can vary substantially, even within a single muscle.

Muscle activation can also be modulated by rate coding, or varying the frequency of the activation signal. As the signal increases in frequency, the force peaks induced by each action potential start to overlap and sum to produce higher levels of force. Maximum force is achieved when the frequency of stimulus is high enough that the individual force peaks are no longer distinguishable. This state is called fused tetanus. However, muscles rarely (if ever) reach tetanus during normal locomotion. This is most likely because tetanic contractions would rapidly cause muscular fatigue, and could damage muscle tissue. Muscles usually receive much lower, sub-tetanic stimulation frequencies, and smooth force production is achieved by simultaneously activating multiple motor units.

3.10   WHAT DOES MUSCLE DO BEST?

It is common to think of muscle simply as contractile tissue, which implies an ability to change length, and it certainly has this capacity. However, from a number of perspectives, muscle appears to be better at producing force than it is at changing length.

First, when active, the cross-bridges are, by default, engaged and ready to produce force; ATP must be added to disengage the cross-bridges and allow the actin and myosin filaments to slide past one another. Second, muscles use less metabolic energy when producing force isometrically than when shortening to do work. Third, the musculoskeletal system tends to use lever configurations that demand high muscle force but limit muscle length change. These pieces of biochemical, energetic, and anatomical evidence suggest that muscles might be better at producing force than changing length.

Still, in real life, there are many instances where muscles must change length to do positive work and generate power. Animals rarely move in straight, horizontal lines at constant speed, so energy must be added whenever an animal increases speed, turns, jumps, or climbs. Energy must also be added to replace energy that the animal loses to the environment via dissipative mechanisms such as heat, vibration, etc. Terrestrial mammals must add the energy needed for locomotion by shortening their skeletal muscles to do work (Alexander, 2003). Additionally, animals must perform many nonlocomotory tasks that require work in order to survive (operating the jaws is an obvious example).

Therefore, although muscles may be better at producing force, they must be adapted to be reasonably good at changing length as well. When interpreting the adaptations involved in allowing for these functional capabilities, it may be productive to consider the fundamental and inherent capabilities and constraints determined by the properties of the tissues involved.

▌ REFERENCES

Abbott BC, Aubert XM. (1952). The force exerted by active striated muscle during and after change of length. *Journal of Physiology* **117**, 77–86.

Alexander RMcN (2003). *Principles of Animal Locomotion*. Princeton University Press, Princeton, NJ.

Alexander RMcN, Maloiy GMO, Ker RF, Jayes AS, Warui CN (1982). The role of tendon elasticity in the locomotion of the camel. *Journal of Zoology, London* **198**, 293–313.

Biewener AA, Baudinette RV (1995). *In vivo* muscle force and elastic energy storage during steady speed hopping of tammar wallabies (Macropus eugenii). *Journal of Experimental Biology* **198**, 1829–1841.

Cavagna GA, Heglund NC, Taylor CR (1977). Mechanical work in terrestrial locomotion: two basic mechanisms for minimizing energy expenditure. *American Journal of Physiology* **233**, R243–R261.

Edgerton VR, Roy RR, Gregor RJ, Rugg S (1986). Morphological basis of skeletal muscle output. In: Jones, NL, McCartney, N, McComas, AJ (eds). *Human Muscle Power*, pp. 43–47. Human Kinetics Publishers, Champaign, IL.

Fenn WO, Marsh BS (1935). Muscular force at different speeds of shortening. *Journal of Physiology* **85**, 277–297.

Fitz RH, Widdrick JJ (1996). Muscle mechanics: adaptations with exercise training. *Exercise and Sport Sciences Reviews* **23**, 427–473.

Gasser HS, Hill AV (1924). The dynamics of muscular contraction. *Proceedings of the Royal Society B* **96**, 398–437.

Gordon AM, Huxley AF, Julian FJ (1966). The variation in isometric tension with sarcomere length in vertebrate muscle fibres. *Journal of Physiology* **184**, 170–192.

Herzog W, Guimarães AC, Anton MG, Carter-Erdman KA (1991). Moment-length relations of rectus femoris muscles of speed skaters/cyclists and runners. *Medicine & Science in Sports & Exercise* **23**, 1289–1296.

Hill AV (1922). The maximum work and mechanical efficiency of human muscles, and their most economical speed. *Journal of Physiology* **56**, 19–41.

Hill AV (1938). The heat of shortening and the dynamic constants of muscle. *Proceedings of the Royal Society London* **126**, 136–195.

Hill AV (1939). The mechanical efficiency of frog's muscle. *Proceedings of the Royal Society London B* **127**, 434–451.

Huxley AF (1957). Muscle structure and theories of contraction. *Progress in Biophysics and Biophysical Chemistry* **7**, 255–318.

Huxley AF, Niedergerke R (1954). Structural changes in muscle during contraction. *Nature* **173**, 971–973.

Huxley AF, Simmons RM (1971). Proposed mechanism of force generation in striated muscle. *Nature* **233**, 533–538.

Huxley HE (1969). The mechanism of muscular contraction. *Science* **164**, 1356–1366.

Huxley HE, Hanson J (1954). Changes in the cross-striations of muscle during contraction and stretch and their structural interpretation. *Nature* **173**, 973–976.

Katz B (1939). The relation between force and speed in muscular contraction. *Journal of Physiology* **96**, 45–64.

Ker RF, Bennett MB, Bibby SR, Kester RC, Alexander RMcN (1987). The spring in the arch of the human foot. *Nature* **325**, 147–149.

Levin A, Wyman J (1927). The viscous elastic properties of muscle. *Proceedings of the Royal Society London B* **101**, 218–243.

Lichtwark GA, Wilson AM (2007). Muscle fascicle and series elastic element length changes along the length of the human gastrocnemius during walking and running. *Journal of Biomechanics* **40**, 157–164.

Lichtwark GA, Wilson AM (2008). Optimal muscle fascicle length and tendon stiffness for maximizing gastrocnemius efficiency during human walking and running. *Journal of Theoretical Biology* **252**, 662–673.

Lutz GJ, Rome LC (1994). Built for jumping: the design of the frog muscular system. *Science* **263**, 370–372.

Martini FH (1995). *Fundamentals of Anatomy and Physiology*, 3rd Edition, pp. 310–314, Prentice Hall, Englewood Cliffs, NJ.

McMahon TA (1984). *Muscles, relexes, and locomotion*. Princeton University Press, Princeton, NJ.

Minetti AE, Ardigò LP, Reinach E, Saibene F (1999). The relationship between work and energy expenditure of locomotion in horses. *Journal of Experimental Biology* **202**, 2329–2338.

Morgan DL, Whitehead NP, Wise AK, Gregory JE, Proske U (2000). Tension changes in the cat soleus muscle following slow stretch or shortening of the contracting muscle. *Journal of Physiology* **522**, 503–513.

Pfau T, Witte TH, Wilson AM (2006). Centre of mass movement and mechanical energy fluctuation during gallop locomotion in the Thoroughbred racehorse. *Journal of Experimental Biology* **209**, 3742–3757.

Pollack GH (1983). The cross-bridge theory. *Physiological Reviews* **63**, 1049–1113.

Pollack GH (1990). *Muscles and molecules: uncovering the principles of biological motion*. Ebner & Sons, Seattle. pp. 1–38.

Rassier DE, Herzog W (2003). Considerations on the history dependence of muscle contraction. *Journal of Applied Physiology* **96**, 419–427.

Rayment I, Holden HM, Whittaker M, Yohn CB, Lorenz M, Holmes KC, Milligan RA (1993). Structure of the actin-myosin complex and its implications for muscle contraction. *Science* **261**, 58–65.

Roberts TJ, Marsh RL (2003). Probing the limits to muscle-powered accelerations: lessons from jumping bullfrogs. *Journal of Experimental Biology* **206**, 2567–2580.

Roberts TJ, Marsh RL, Weyand PG, Taylor CR (1997). Muscular force in running turkeys: the economy of minimizing work. *Science* **275**, 1113–1115.

Ruina A, Bertram JEA, Srinivasan M (2005). A collisional model of the energetic cost of support work qualitatively explains leg sequencing in walking and galloping, pseudo-elastic leg behavior in running and the walk-to-run transition. *Journal of Theoretical Biology* **237**, 170–192.

Schutt CE, Lindberg U (1990). Actin as the generator of tension during muscle contraction. *Proceedings of the National Academy of Sciences* **89**, 319–323.

Winter, DA (1990). *Biomechanics and motor control of human movement*. New York, Wiley.

Yanagida T, Nakase M, Nishiyama K, Oosawa F (1984). Direct observation of motion of single F-actin filaments in the presence of myosin. *Nature* **307**, 58–60.

CHAPTER FOUR

Concepts in Locomotion: Levers, Struts, Pendula and Springs

John E. A. Bertram

Department of Cell Biology and Anatomy, Cumming School of Medicine, University of Calgary, CA

4.1 INTRODUCTION

The analysis of locomotion, and the assessment of the function of the organ systems involved and responsible for it, should begin with the question, "What exactly is accomplished in locomotion?" Certainly, the organism moves from one location to another and, in technical terms, we can specify that the mass of the animal (as represented by the center of mass of the system) is displaced relative to its reference frame (under natural circumstances, its physical environment). For active locomotion, we generally assume that this is accomplished by activity of the musculoskeletal system in an ordered, cyclic and specifically directed manner – the motions that constitute the gait. Although this seems obvious, verging on trivial, understanding what is accomplished in locomotion is the source of substantial confusion in the analysis of locomotion. The confusion arises from not distinguishing the *phenomenon* of locomotion – the dynamics involved with displacing the animal's mass within the physical

Understanding Mammalian Locomotion: Concepts and Applications, First Edition.
Edited by John E. A. Bertram.
© 2016 John Wiley & Sons, Inc. Published 2016 by John Wiley & Sons, Inc.

environment – from the *mechanism(s)* that accomplish this movement. However, neither can be properly understood without recognizing the extent of influence that each has, and the role that each plays. The purpose of this chapter, and the following one, is to try to distinguish the phenomenon of legged locomotion from the mechanisms responsible, while also identifying the main mechanisms utilized.

Legged locomotion is almost universally accomplished using limbs in a repeating cycle, referred to as a gait (derived from Middle English indicating "way"). The gaits commonly utilized by terrestrial mammals are by no means the only solutions available for terrestrial locomotion. Rather, a myriad of other movement patterns are possible that will (eventually, and with limited economy) result in displacement of the center of mass relative to its environment. Some of these alternative and energetically ineffective movement strategies can appear quite absurd (as my colleague Jan Kowalczewski succinctly observed, "Silly walks are inefficient, that's what makes them silly.").

Our observations of highly functional animal locomotion bias us toward thinking that locomotion "needs" to involve well-coordinated and predominantly cyclic motions. We do, in fact, observe well-coordinated motions in the natural world, but the high degree of coordination is determined not by a fundamental requirement for locomotion *per se* to result. Rather, it is instead required to provide effective locomotion, where "effective" refers to a stable and controllable movement, in which the "cost" is limited. In this case, cost is in terms of the energy consumed to accomplish the motion.

Ultimately, metabolic energy is of concern to the organism, but it is likely that, in the case of locomotion, much of the metabolic cost derives from producing forms of mechanical energy and doing work within the physical environment. Interpretation of the function, form, organization and motions of the locomotory apparatus should involve consideration of the energetic consequences within the context of the advantages that the system offers over other potential options. In effect, understanding locomotion will require considering and understanding the cost(s) association with moving the organism from one location to another, as well as interpreting those costs within the context of what other options might be available.

Access to energy and its management is crucial to organisms, so that much of evolutionary adaptation (and learning survival strategies) involves optimizing energy utilization (Alexander, 2001). The way the animal moves is likely the best compromise available under the circumstances, although it may well be a challenge to understand all the factors that are included in the "compromise". Figuring out the role of the factors involved in the compromise selected, in terms of morphology, movement pattern, physiological strategy and so on, will provide substantial insight into the adaptive strategies that influence this aspect of the organism's survival.

The motions observed in the various forms of mammalian locomotion can be viewed as distinct solutions to the problems encountered when trying to move the animal's mass across its environment, utilizing the biological machinery (limbs, muscle actuators, control strategies) available to the organism. The solution selected will be influenced by the conditions of the terrestrial environment being confronted – whether that is the constant and ubiquitous factor presented by the magnitude of gravitational acceleration, the highly variable features of the surface properties on which the animal moves, or the objective of the movement itself (which can itself be as varied as fleet pursuit of prey, negotiating obstacles to access assets of the environment not available to competitors with less agility, or sustained endurance). In the physical circumstances of most terrestrial environments, the motions involved with locomotion represent reasonable solutions under the conditions present. However, these very satisfactory solutions may not be appropriate in another type of physical environment or circumstance.

There are two basic approaches to investigating function and mechanics of legged locomotion, neither of which, purely on its own, has much hope of providing a complete understanding of the trade-offs involved with arriving at the particular solution demonstrated by each species. These alternatives are:

1. perform a detailed analysis of the system and the action of all of its sub-parts, eventually working to put the myriad of components together to generate an understanding and appreciation of the integrated whole (observational analysis); or

2. model the system working from the simplest model that captures the key features of the function and add less critical features, in order of their impact on the function of the system, until the complete system has been reproduced (reductionist modeling).

The first approach is inviting because the parts, as they are used, are readily available to study. However, it is doomed largely because of the complexity of biological systems and the subtlety of their actions. If we understood the system completely to begin with, we might be able to reconstruct and interpret the contributions of all the components and their actions. In reality, though, we are largely investigating blindly, with little (verified) sense of why the complex components are assembled as they are. Without an overall perspective on the system and factors influencing performance, it is difficult to appreciate properly how each of the component actions contribute within the context of the overall function.

On the other hand, simple models may do well to inform us of key aspects of system function but, as simple (simplistic?) models, they fall very short of demonstrating the remarkable coordination and integration of the biological action that they attempt to emulate. Also, with simple models, it is possible that the apparent behavior of the model parallels that of the modeled system without capturing the determinant features of the original system. For such cases, we would misinterpret the source/purpose of the system behavior, because the model is not complete. To paraphrase Einstein: "Make your model as simple as possible, but no simpler" – indicating that it is, indeed, possible to simplify a model beyond its capacity to provide insight into the functioning of the system it models. Possibly an indication of Einstein's genus was being able to recognize the boundary between "as simple as possible" and too simple.

The solution, of course, is to apply both approaches in order to work towards a meaningful (and useable) understanding of the system, its function and the factors that allow its function within the constraints and opportunities provided by the interaction between the organism and its external environment. Observational analysis is necessary in the process, but should not serve as an end in itself – it is easy to formulate "just so" stories that explain observations, but assuming that a conception of an answer which matches observations under a given set of circumstances is the actual explanation can easily lead the focus away from the factors that truly underlie the behavior. Recognizing "how" a natural system operates is a necessary step in exploring the system and formulating an understanding of "why" it works the way it does. Such an understanding can then be explicitly tested (verified) in a variety of ways – including modeling (see Chapters 6, 7 and 15).

Simple (reductionist) models can be proposed as hypotheses of the functional implications of aspects of the system, then tested against the relevant features of the biological system as verification (see Chapter 6). As an example, consider the flight of a bat. The hand and digits of the bat wing form the structural support of the airfoil for the bat, but this particular arrangement does not necessarily define functionality in flight (i.e. there are a number of alternative methods of providing the structure of an airfoil – the feathers of birds, for instance). Understanding the function of the elements within the wing depends on understanding how the wing must function within the physical environment it encounters, in this case providing lift and overcoming drag. Certainly it would not matter if we documented every detail of the numerous muscle or nerve actions associated with the wing, if the basic lift-and-thrust function of the airfoil was not first properly identified. The wing, regardless of the details of its structure, serves first and foremost as a dynamic interface between the organism and its environment (Riskin *et al.*, 2008).

In many cases, the function of the components operating the system cannot be construed without the context of how the organism integrates within the physical environment in which it operates. On the other hand, an analysis of the organism's integration with its physical environment (the airfoil properties, in the case above) can lead to an understanding of what needs to be accomplished by the wing. From that, it can predict, at least at the general level, what is required of the actuators controlling the appendage. Eventually, the investigations become a process of deciphering more and more of the functional integration of the system, but this can be aided substantially by (correctly) identifying the fundamental features (concepts) on which function ultimately depends.

In the example above, the interaction between the organism and its physical environment is mediated by the aerodynamic effect of the wings, but the approach holds for the interaction of the limbs with the substrate in terrestrial locomotion as well. Of course, such an analysis will not provide information on the details of how this interaction is accomplished by the biological components of the limb, but it can help to identify the role of specific actions and activities by defining what "must" be accomplished to produce energetically cost-effective locomotion.

The purpose of the current chapter (and the next) is to introduce the mechanical principles that underlie the fundamentals of limb function in terrestrial mammals, emphasizing that the elemental role of the limb is to mediate the interaction of the terrestrial organism with the most influential portion of its physical environment – the surface on which it supports and propels itself. Footfalls and limb contacts within the gaits of terrestrial mammals (and the forces/impulses/work involved) mediate between the organism and its physical environment in a process that integrates the motion of one relative to the other. In this chapter, the mechanisms available for the limb to accomplish that role will be discussed. The following chapter will discuss what this integration accomplishes for the animal. Together, the concepts introduced are meant to provide a basis for interpreting how the details of the morphology and motion accomplish the required task of providing effective locomotion for the organism.

4.2   THE LIMB: HOW DETAILS CAN OBSCURE FUNCTIONAL ROLE

The following section provides a brief illustration of how focusing too much on details of morphology without proper context can make it difficult to interpret function based on form alone. When considering limb function, it is natural to look to the elements that make up the limb and to compare similarities and differences between species. The objective would be to identify consistent functional and morphological patterns that indicate important relationships between the form and function of the system. Although all mammalian – and, indeed, tetrapod – limbs are constructed based on the same general pattern (Romer and Parsons, 1986; Hildebrand and Goslow, 1995), substantial differences in form and proportion exist, even when considering only those mammals functioning in a terrestrial environment.

It is a consistent and long-held tenant of zoology (and other fields in the life sciences) that form and function are intimately related. Undoubtedly, this is the case (with some sensitivity to the fact that adaptive precision may not be absolute and that phylogenetic inertia – which may preserve structures beyond their functional usefulness – is always present). However, when looking at the complexity of a mammal moving swiftly across an unpredictable environment we might well ask, "What aspect of form; what specifics of function?" It may seem reasonable to begin by considering each element of the system individually, working towards eventually putting the pieces of the puzzle together. Unfortunately, without a functional context, the varieties of form represented by terrestrial mammals may prove more of an impediment than an opportunity. Take for instance the clavicle …

Humans, as primates, posses a direct bony connection between the pectoral and the axial skeleton, in the form of a jointed clavicular brace between the scapula and the sternum. The sternum represents the ultimate skeletal connection of the pectoral girdle to the thorax, as it is integrated into the musculoskeletal components that define both the thoracic and the appendicular components of the skeleton. As humans, it is natural that we consider the human condition the norm. However, a complete bony clavicular connection between the pectoral limb and the thorax is not particularly common.

Jenkins (1974) noted the variability of the mammalian clavicle, pointing out that it is universally present in only four orders of mammals, Primates being the most noteworthy. The others are the Chiroptera, the bats – a widespread group that has substantially different demands on its pectoral girdle than most quadrupeds – the Tubulidentata, currently represented only by the Aardvark, and the Dermoptera, the flying lemurs – again with substantially different pectoral functional demands than most terrestrial mammals. Primates, Chiroptera and Dermoptera reside within the same phylogenetic "neighborhood" (Shoshani and McKenna, 1998), so the presence of the clavicle within these particular groups may have a substantial phylogenetic influence.

The clavicle is present in most Insectivora and marsupials, but varies from rudimentary to well developed in different families among the Lagomorphs, rodents and edentates. In the carnivorans, the clavicle is either rudimentary (with many species bordering on vestigial) or absent. It is completely lacking as an ossified structure in the remaining aquatic (Cetacea, Pinnepedia, Sirenia) and terrestrial (Artiodactyla, Perissodactyla, Hyracoidea, Proboscidia and Philodota) orders. Although as a structural "brace", it may seem reasonable that this element is missing in the aquatic forms, some of the most cursorial of mammals are represented within the Artiodactyla and Perissodactyla, while these, combined with the Proboscidia, represent the largest of the extant terrestrial mammals (hippopotamus, rhinoceros and elephants), where it might be expected that including a bony "brace" for attachment of the forelimbs to the trunk would be a reasonable strategy.

Undoubtedly, the clavicular connection, whether bony or membranous, has a function in the attachment and integration of the limbs with the body, and many species may depend on this element as an important constituent of the locomotor apparatus. However, our bias as primates may provide us with a distorted impression of the role of the clavicle which, on the whole, does not appear to be particularly well represented or, presumably, useful in an array of mammals that express a vast range of locomotory function.

How, then, is the variable morphology of the clavicle to be interpreted? I argue that it is first necessary to define the function of the limb for each species, then identify the role of the clavicle in providing whatever component of that particular function it is responsible for. Within this context, then, it will be possible to interpret the specific morphology present – keeping in mind that it is well within possibility that a particular function can be provided equally well by a variety of system organizations and element forms, and that numerous non-locomotory influences may be involved with the expression of the form and organization observed.

So within what context does the clavicle reside? With limited direct bony attachment, stabilization and control of the pectoral girdle, the supporting foundation of the limb and its attachment to the thorax is predominantly provided by a muscular sling with connections to the axial skeleton dorsally and ventrally (Figure 4.1). Although much is often made of the individual action of these muscles in anatomical texts, the true function of the muscles is as components of the supporting sling, and they work together to transfer loads between the limbs and the trunk. In this context, it is most common for these muscles to actively resist loads applied through the limbs to maintain the position of the scapula and pectoral girdle on the trunk (Davis, 1949; Dyce *et al.*, 2002; Carrier *et al.*, 2006).

The point here is that the interaction between the center of mass (CoM) and the substrate is complex enough that a variety of functional strategies may well be able to provide quite similar opportunities for effective locomotion, with some exploiting the properties of the clavicle, while others avoid its constraints. In fact, the optimization of a particular element may not be very specific, with a broad range of options satisfying the requirements of the system. Therefore, it is very difficult to infer function from form, except in the most general or the most specific sense (both extremes involve limitations on the complexity considered).

In order to interpret the form and activity of locomotory structures, it is necessary to define their functional role(s) within the complete system. Following this, it is possible to interpret the contribution of specific features of the system to the function of the entire complex. Because we are discussing biological systems rather than engineered ones, it is also necessary to keep in mind that the form-function relationship may be obscured by competing functions – say, for instance, respiration in the case of the pectoral attachment to the ribs of the thoracic cage (Bramble and Carrier, 1983), or phylogenetic constraints where the functional demands applied to ancestors within the lineage remain in descendent lineages that may not employ the components in the same way. Presumably, such "anachronisms" would not remain if they were overly

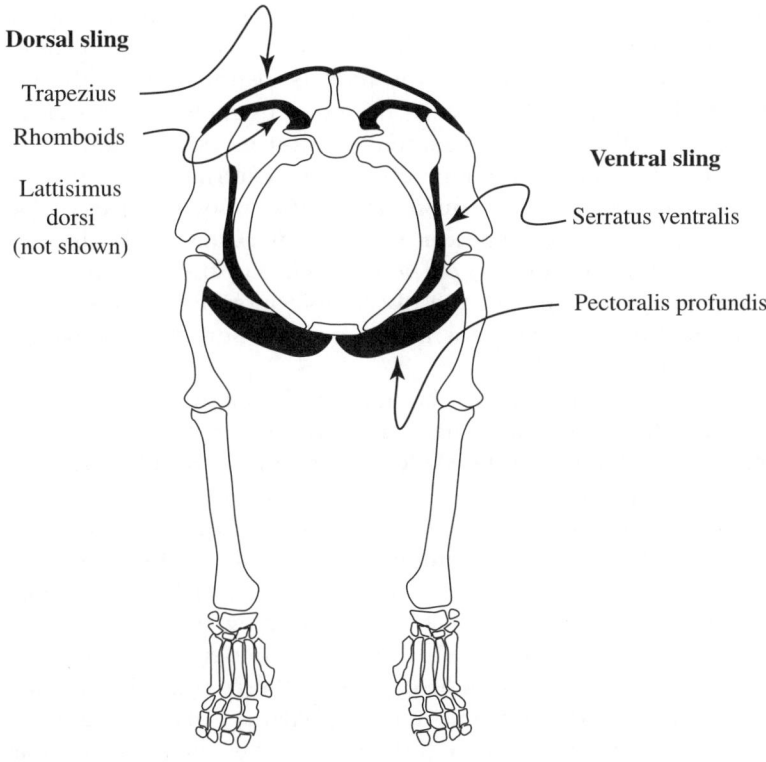

Dorsal sling

Trapezius

Rhomboids

Lattisimus
dorsi
(not shown)

Ventral sling

Serratus ventralis

Pectoralis profundis

FIGURE 4.1

Diagram of the muscle sling that forms the functional attachment of the pectoral girdle to the trunk. This sling is the main support of the trunk, even in those forms that possess a clavicle.

deleterious to the animal (even though any persistent remnant related to a previous function can be problematic to our interpretation of the morphology).

4.3   LIMB FUNCTION IN STABILITY AND THE CONCEPT OF THE "EFFECTIVE LIMB"

4.3.1   Considering the mechanisms of stability

The specific actions of muscles within the limb are complex, indeed. The dynamic nature of locomotion means that, essentially, all features of limb function change in some way over the course of the stride, so the role and activity of all the sub-components (from nerve impulse and muscle mechanics to joint contact forces and torques, and all other components in between) vary over each cycle. Later, I will show that changes in the action of one limb can also affect the role of other limbs, so the complexity becomes multi-dimensional. If we are interested in understanding how the limb functions to provide for locomotion, and how to interpret the differences in function and form between different mammals that utilize their limbs for locomotion across solid surfaces, then how should such dynamic complexity be addressed?

One approach that holds some promise is to not focus on the myriad processes involved with the limb fulfilling its role in locomotion (the mechanism of its action), but instead focus on what the role of the limb is in transport of the organism (the phenomenon of locomotion). Ultimately, the consequence of the total action of the limb (muscle-tendon complex and ligament tension working against bone and joint resistance) is to mediate the interaction between the external physical world (substrate and gravity) and the mass of the organism. The limbs apply impulses (impulse $\mathbf{I} = \int \mathbf{ma}\, dt$, where m is mass, $\mathbf{a}$ is acceleration and t is time) that result in changes in the momentum (momentum $\mathbf{p} = m\boldsymbol{v}$, where $\boldsymbol{v}$ is velocity) of the organism and its component parts. Together, these sum to the observable motion of the organism, and our challenge is to identify what happens in this process and, more importantly, explain why it occurs as it does.

Although a great many specifics of the limbs and their action vary between species, the limbs in general function as articulated linkages, where the joints form the articulation. The action of the joint, and the skeletal components it links, will result from the integrated effects of the active control, the dynamic (inertial) properties of components (body and limb masses at rest and in motion) and the interaction of all with the substrate (and its properties). It is common to think that active control dominates – where the brain signals the actions that *make* locomotion happen. Control-response motions are important for many active tasks operating in a complex and changing environment. Very slow locomotion, which might be better termed "stepping", is much like stationary standing and does, indeed, require direct control to elicit the required responses – control of body and limb motion, balance, maneuverability, and so on. Effective locomotion at any speed greater than "very slow", however, appears somewhat different. This results from two important factors.

First is that a substantial degree of the stability that an organism has during locomotion occurs primarily in the form of "dynamic stability", where the system is actually statically unstable at any point in the cycle, but the continuum of motions over the gait cycle result in a stable, repeating cycle (a statically unstable structure will fall over or collapse if left alone). Consider a tall lamp stand with a heavy base standing on a flat surface. The lamp is statically stable, in that its stability can be perturbed, for instance by a small bump, and it will then rock back and forth but eventually return to its stable stance. Rigid objects standing still on a flat horizontal surface contacting the ground at not less than three points have a simple criterion for static stability – the downward projected position of the center of mass of the object must be within the region enclosed by the surface contact points (or, in a technically precise definition, the horizontal position of the center of mass must be enclosed by the convex hull of the contact points).

Sometimes, this notion is used to characterize the stability of animals, either standing still or walking slowly (Gray, 1944, 1968). For instance, a quadrupedal animal standing still with extended limbs is much like a table – quite stable when all its feet are on the ground. Having four widely separated legs makes for greater control authority over the body orientation – greater changes in body orientations may be generated with the same effort (given the same body). However, during locomotion, both the position of the CoM, placement of the feet on the ground and flexion-extension of the limbs change over the course of a stride. It has been shown that the CoM does not stay within a tripod of contact points for most gaits commonly used (Alexander, 2002). The human form, with a high position of the center of mass (generally residing in the lower abdomen just below the navel) and operating over much of the stride with a single limb in contact with the ground, is an example of a system that is not a statically stable system.

So how do humans stabilize their top-heavy body with their small feet? An object like a human body, that is statically unstable when uncontrolled, can be stable while moving by cycling through a movement pattern that returns to pass through its original position. Such stability is referred to as "dynamic stability".

Dynamical systems theory gives us tools and the language to meaningfully character- ize the stability of dynamic systems, linear or nonlinear, including the locomotor system of bipeds. Here, a periodic pattern such as that of a walking human or a robot (Figure 4.2) can be characterized in terms of a "limit cycle". Such a periodic pattern can be stable or unstable; a periodic pattern is said to be stable (or "attractive") if small deviations from the periodic pattern gradually decrease, so that the original periodic pattern is eventually recovered after a small perturbation. The size of perturbations that can be successfully rejected (overcome), and how fast the original periodic pattern is recovered may be used to quantify how effective a particular periodic pattern is as a stabilizing control strategy.

It is often assmed that human walking requires substantial dynamic feedback to main- tain stability, while standing stationary, in contrast, depends much less on active response to sensory feedback. The degree to which the human system "depends" on its sensory feedback in each case is not well determined, but it is certain that standing can *only* be accomplished by active maintenance of balance (due to the high CoM and the small base of support). Maintaining stationary balance will depend on responding to sensory indica- tions of the balance status at any given moment (Park *et al.*, 2004; Alexandrov *et al.*, 2005).

The tacit assumption in dynamic movements, such as walking and running, is that stability is mainly due to active neuromuscular control based on feedback from periph- eral and central balance sensors (Duysens *et al.*, 2002). Undoubtedly feedback plays a

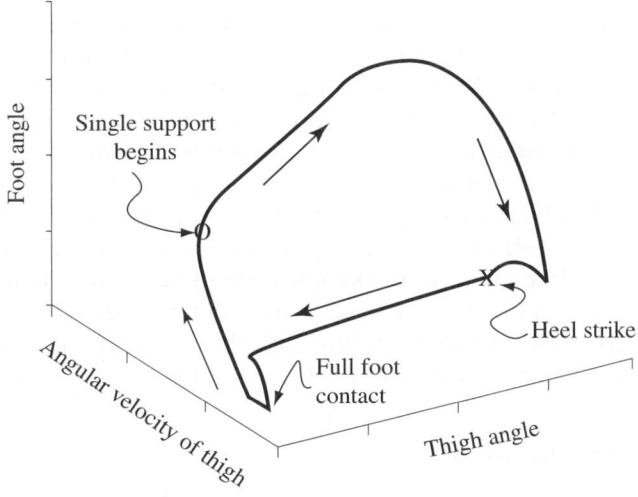

FIGURE 4.2

A limit cycle diagrammed in state space, consisting of thigh angle, angular velocity of thigh and foot angle for a computer-generated anthropomorphic bipedal walking model (redrawn from Srinivasan *et al.*, 2008).

substantial role, especially in dealing with unpredictable environmental perturbations. However, the extent to which such feedback control is important can only be determined by first understanding what can be accomplished without sensory feedback. Evidence suggests that an unexpected degree of stability can originate with the fundamental dynamics of the system itself.

In a series of papers that opened a new approach to walking dynamics, McGeer (1990, 1993) showed that robots with no explicit motors or sensors could walk stably – albeit downhill – using gravitational potential energy to replace inevitable dissipation losses. Such passive-dynamic walking machines are remarkably simple – and they have no actuators (motors) and no control system – but they are able to walk effectively. However, they are limited in what else they can do. Passive-dynamic walking machines are able to walk well down a slight incline, but are unable to maneuver or even stop (they are unstable without forward progression – see Figure 1.3). Further, they can resist only relatively small perturbations – especially if these perturbations are not in alignment with the spontaneous motions involved with the gait.

In spite of these limitations, the ability of these machines to "walk" likely holds substantial insight into the mechanics of walking – both human and, by extension, quadrupedal. Although it may seem that getting everything coordinated to generate a stable passive dynamic walking cycle would be a daunting task, it turns out that such completely passive walkers work remarkably well for a fairly wide range of parameter choices, indicating that it may actually be easier to be stable in walking than in static standing, where any system with a jointed limb requires active balancing (McGeer, 1993).

One can, perhaps, reconcile the surprising nature of these machines by noting that even most bicycles are passively stable (self-stabilizing without active balancing and steering), even when carrying a rider, for speeds between 13–30 km h^{-1} (Meijaard *et al.*, 2007). At slightly lower speeds, such as 10 km h^{-1}, the system is no longer inherently stable, and neuromuscular control by the rider is needed to actively balance the bicycle. This is generally not apparent to the rider, however, until extremely slow speeds are reached.

The inherent stability of walking dynamics likely also integrates with active control where a greater dependence is placed on passive stability when available, and this is augmented by neuromuscular control when circumstances require. Since static standing (which overtly appears quiet) actually requires substantial continuous active balance control, due to our top-heavy construction, as we slow to a stop, balance maintenance shifts from (at least partially) passively dynamic to completely actively mediated. However, we feel no great change in control as the transition occurs, probably because stability is such an important feature of daily life. Neural control strategies below our overt level of consciousness automatically intercede. Recently, McGeer's ideas have been extended to construct bipedal robots that are able to walk stably on level ground powered by minimal motor action (Collins *et al.*, 2005).

4.3.2 The role of the effective limb

Much of locomotion "control" is necessary as a means for limb action to integrate with (in the sense of "work with") the properties of the system as a whole, including interaction with the substrate on which the animal moves. In mammalian limbs, the integrated control functions through control of the jointed limb elements, their motions and their combined physical properties (stiffness, compliance, elasticity, damping, extension/rotation, etc.).

The basic organization of a joint involves an articulation, partially stabilized through the joint surface interaction, and ligamentous thickenings of the fibrous joint capsule, around which flexor and extensor muscles apply torque. If the joint rotates during this activity, then power is produced at the joint (the dot product of torque and angular velocity in three dimensions, or the product of torque and angular velocity in planar motion). The instantaneous powers of each joint in the system can be summed (along with inertial characteristics of the system) to determine the mechanical effect of the limb interaction with the substrate on the path of the body mass of the animal (usually characterized as the CoM, the point around which the distributed mass is evenly distributed).

The procedures required to rigorously determine how each component of the limb functions, and influences the way the organism interacts with its supporting substrate, are quite involved. This level of analysis is primarily useful for evaluating how the interaction is accomplished – the "how" of the movement (currently most conveniently accomplished with some form of inverse dynamic analysis – Winter, 2009). This is in contrast to determining what that interaction means to the locomotion of the organism – the "why" of movement that determines whether the particular motions are the most appropriate under the given circumstances.

At the most fundamental level, the limb, regardless of its conformation, acts as the intermediary for forces applied between the mass of the animal and the substrate. It is this role of the limb that must be understood first, before details of the manner that the limb accomplishes this role can be interpreted. The context of the function must be provided prior to interpreting details of the features of the limb and the rest of the relevant components that contribute to the function of the system as a whole. From this perspective, we can consider the concept of the "effective limb", where the mechanical function is largely independent of the morphology, though the morphology and physiology will, in the end, determine its functional capacity.

Not to confuse the issue too badly, but it is possible to utilize the concept of an effective limb in different, useful ways. In the following, the effective limb will be described in terms of the effect the limb has on the body of the organism, where each limb acts specifically as part of the interaction between the mass of the animal and the substrate on which the organism moves. This is a useful conceptualization when evaluating the role of a particular limb in support or motion. Alternatively, it is possible to consider the net (or combined) influence of all limbs in contact with the substrate as acting as an "effective limb". This net effect of contact is what influences the motion of the organism's complete mass. Such a net conceptualization is particularly useful when considering the effects of different strategies of locomotion, as in a comparison between different gaits utilized by an organism or similar gaits in different species (Lee *et al.*, 2011).

4.4 LEVERS AND STRUTS

Given that the mammalian limb operates through levered articulations (joints in which muscle-tendon groups apply moments to alter the angular position of the joint), it is natural to assume that whole limbs act as levers themselves. Interestingly, a lever-like function of the limb on the body is not as common in terrestrial locomotion as we might assume.

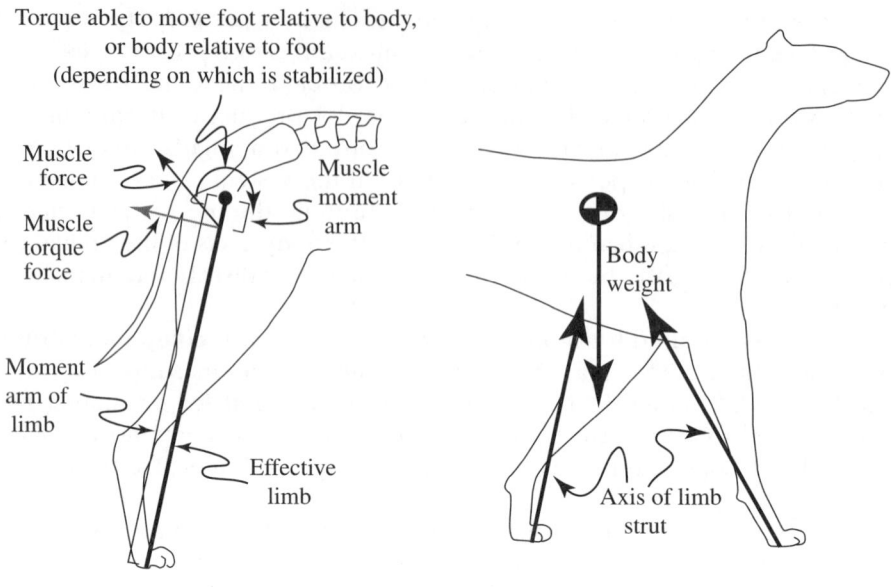

Torque able to move foot relative to body,
or body relative to foot
(depending on which is stabilized)

Muscle
force

Muscle
moment
arm

Muscle
torque
force

Moment
arm of
limb

Effective
limb

Body
weight

Axis of limb
strut

FIGURE 4.3

Comparison of lever (left) and strut (right) function of the limb (adapted from Gray, 1944). The lever requires the action of muscles extrinsic to the limb (at its attachment to the trunk), whereas the force balance can be accomplished with intrinsic muscles when the limb is used as a strut (even if each limb applies a horizontal force to the trunk). Note that the action of the limb is considered in terms of its effect on the system (the effective limb), rather than as a complex of elements.

Although the loads themselves are transmitted through the tissues of the limb, the effect of loads (forces) applied between ground contact and the attachment of the limb to the body does not follow the path of the anatomical limb (which may be complex – as in, "crooked as a dog's hind leg"). In terms of the effect of the load on the stability or motion of the body, the physical route of the connection *within* the limb between contact and the body is irrelevant to its effect on the whole animal (although it will determine the within-limb loads and torques that are required to apply the contact forces). The mechanically "effective limb" is the line of force acting between the substrate contact and the limb's connection to the trunk. The effective limb defines the transfer of load and/or torque between these two points. The structure must be capable of sustaining the force and torque, but how it accomplishes this is irrelevant to the effect that those forces and torques have on the body (and the consequent motion of the animal).

The effective limb functions as a mechanical lever if force is produced perpendicular to the axis of the limb, this being accomplished by the application of a torque (turning force) at the point of attachment of the limb to the body (Figure 4.3). In this case, the effective limb must resist loads both along its length and in bending between the two end points. In theory, a lever could be subject to torques applied at either end but, in the case of the mammalian foot on a terrestrial substrate, any torque applied at the ground must be either generated or resisted by a torque at the limb's attachment.

The situation can be more complex where gripping forces and free movement of body mass can allow for other conditions, such as in an arboreal environment.

The alternative to lever function is the extensible strut, where load is transmitted down the mechanical (support) axis without a turning moment applied (i.e., limb force passes through the center of rotation of the proximal attachment). In the articulated limb, strut-like function is provided by the structural support acting between the points of contact of the limb with the body and the limb with the substrate. The straight-line distance between the limb-substrate and limb-body contacts is altered by flexion and extension of the articulated joints of the limb but, if no turning moment is applied at the body connection, such a limb would act as an extensible strut rather than a lever (Figure 4.3). As Gray (1944) succinctly described: "*In so far as it (the limb) functions as an extensible strut, the limb is extended by its own intrinsic musculature and the mechanism of locomotion is essentially that of a punt when propelled by a pole. In so far as it acts as a lever, it is operated by its extrinsic musculature and the mechanism is essentially similar to that of a canoe propelled by a two-handed paddle, the distal end of which remains fixed.*" (p. 116).

Lever and strut function is equally possible for the articulated mammalian limb, and interpretation of the function of the limb in different species – and, particularly, different gaits – requires distinguishing how the limb mediates the interaction between the CoM and the substrate. Of course, the distinction between these two options is not absolute and, under most circumstances, there will be measurable effects of both strut-like and lever-like action. The conditions defined as pure strut or lever function are mechanically explicit, and are rarely realized precisely in a natural system (especially over any appreciable time during the complex dynamic changes that occur during locomotion).

For the condition of balanced forces – namely, when the force of the body is exerted at the center of the limb's articulation with the body, and when tension in the muscles (and ligaments) connecting the elements of the limb (including its connection to the body) and the reaction of the ground balances that of the body on the limb, then the limb is in equilibrium. If the resultant forces of all the extrinsic muscle and ligament tensions are zero, the limb acts as a strut. Alternatively, if the extrinsic limb muscle forces – those that attach the limb to the body – are not zero, then the limb exerts forces on the ground at right angles to the mechanical axis of the limb, and it acts as a lever. The differences between these two functional strategies are meaning-ful, because each type of function has its own set of capabilities, limitations and energetic consequences.

Utilization of the limb by any particular species in either a predominantly strut-like or lever-like manner is likely determined by the functional requirements involved with the locomotory task required, and the relative cost of creating locomotion using each mechanism under the conditions present. Following Gray's boating analogy above, shallow water makes effective paddling difficult, while deep water makes effective punting impossible. Similar trade-offs between the advantages of lever-like and strut-like function of the limbs in different circumstances influence the manner in which the limb is most effectively utilized (even though the limb may look similar, and has the same functional components, in each mode of use).

For models of human bipedal walking moving steadily on a level surface, for instance, it has been shown that a strut-like limb use can involve substantially lower total mechanical energy investment than powering the system via lever-based torque (Kuo, 2002). Some functions of the human limb require lever-like function, and this will also be true for quadrupeds. For example, balancing the off-axis position of the CoM and the stance limb in human walking requires thigh adductor muscle action in order to stabilize inertial-based moments applied around the hip. Without these stabilizing moments (torques applied to the limb acting as a lever), the individual would collapse.

Evidence of the consequences of using the limb in a strut-like manner versus using it as a lever comes also from some interesting studies of running around a relatively tight curve, as is commonly done in running races on indoor tracks. It has been shown that the centripetal force generated by running on a bend has a substantial negative effect on running speed in humans (Usherwood and Wilson, 2006; and see Chapter 6), indicting that force applied down the axis of the human leg is a limiting feature of running speed. This, in turn, suggests that the human pelvic limb functions largely as a strut in running (see also Weyand *et al.*, 2000, 2010).

The greyhound, however, shows no comparable effect during bend running, and can negotiate curved tracks with no apparent negative effect. This would indicate that, in the greyhound gallop, the limbs are likely dominated by a lever-like function (Williams *et al.*, 2009). Of course, the gallop is a substantially different gait than the bipedal human run (Bertram and Gutmann, 2009; Minetti *et al.*, 1999), which is mechanically comparable to the quadrupedal trot and pace (see Chapter 2). Given the similarities between the orientation of the limbs in the quadrupedal trot and bipedal run, it is likely that limb function in the trot is also dominated by a strut-like function, just as in the vaulting action of the walk. However, all gallops are not necessarily equivalent (Hildebrand, 1959; Bertram and Gutmann, 2009; Biancardi and Minetti, 2012), leaving the opportunity for both strut-dominated and lever-dominated options even within a given gait.

4.5 GROUND REACTION FORCE IN GAITS

It was noted above that lever-like function requires the limb to generate force perpendicular to the effective limb at the ground (to resist torque applied at the limb attachment to the body). Horizontally applied forces at foot contact occur transiently under almost all locomotion circumstances, but this does not mean that limbs always act as levers. Horizontal forces are generated in a strut-like limb when its long axis is inclined from the vertical at contact. Loads passing through the long axis of the limb can apply an oblique force to the ground, which has a horizontal component that must be resisted by a horizontal force at the foot contact. For a limb that acts as a stiff strut, the force on the limb deflects the path of the CoM. For example, strut-like limb contact in walking causes a braking force in the early part of contact as the CoM vaults over the stance limb. In the second part of contact, the CoM is re-accelerated as the body falls forward after mid-stance.

It is common to resolve the force applied to the ground by limb contact into vertical and horizontal components. This can be a useful operation in locomotion analysis, just as it is in elementary ballistics, because gravitational acceleration only affects the vertical component. However, this dissociation of vertical and horizontal forces should be done judiciously. The total ground reaction force represents the combined

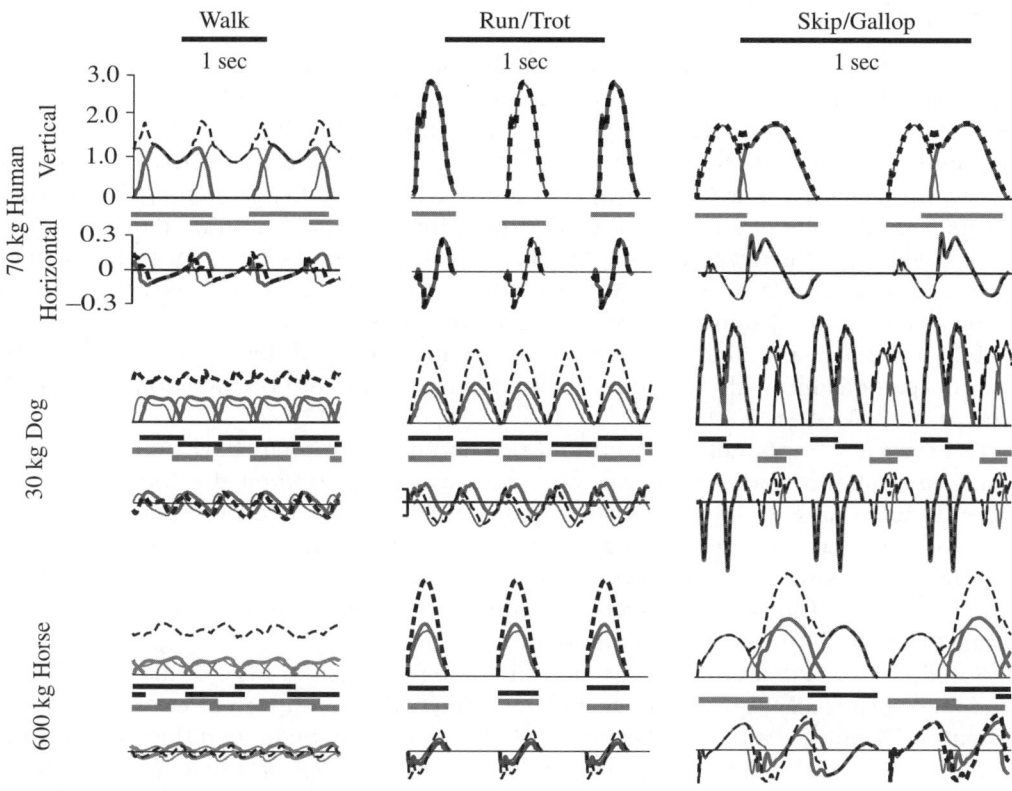

FIGURE 4.4

Comparison of ground reaction forces for the human, dog and horse at three gaits. Force given normalized to body weight. Note that the time axis is identical for all species at each gait (time interval bar at top), but it changes between gaits to aid in visualizing features of each contact as speed increases. Individual fore and hind limb loads are indicated in gray, with forelimb represented by a thick line and hind by a thin line (in the human, these lines indicate left and right). The net effect of all simultaneous applied force on the CoM is given by the black dashed line. The representation of complete ground reaction over a contact sequence can differ substantially from the consideration of a single limb alone. The footfall pattern of each gait is given between the vertical and horizontal force profile, where each bar represents the contact of a given limb; black indicates a forelimb, and gray a hind limb.

effect of the interaction between the animal's mass and the substrate that supports it. This has a net effect on the CoM of the animal, and results from the combined action of all limbs simultaneously in contact with the substrate (Figure 4.4). Isolation of vertical and horizontal components of the ground reaction force, or separation of one limb contact from the other, may allow insight into some aspects of limb function, but it may well obscure other features of what is actually being accomplished to the whole body in locomotion.

4.5.1 Trot

In the quadrupedal trot, diagonal front and hind limbs contact approximately at the same time, with each diagonal pair alternating contact, and a brief non-contact phase between (see Chapter 2). The net ground reaction force profile of the trot is remarkably similar to that observed in the bipedal run, indicating that the net effect of the alternating diagonal pair of quadrupedal limbs acts on the CoM much as the alternating individual limbs of a running biped (where the diagonal fore-hind pair can act on the whole body effectively as a single limb, and the system behaves much like a mass bouncing on a springy support). Although the net ground reaction force acting on the body results from combined fore and hind contacts, the magnitude, duration and timing of the fore and hind limb forces are usually slightly different (Lee *et al.*, 1999), and this can change with the slope the animal moves on and the position of the CoM (Lee *et al.*, 2004; Lee, 2011).

The vertical component of the ground reaction in the trot may have an impact spike resulting from deceleration of the foot (and lower limb) when the foot strikes the ground (note that $F = ma$, so it is often quite valuable when interpreting force profiles to identify the specific mass being accelerated, noting that the acceleration involved can be either positive or negative in some circumstances). The main portion of the ground reaction force (GRF) profile is unimodal (a single hump), with the forelimb contact usually of greater magnitude than the hind.

Forelimb forces that are greater than hind limb forces is the case for many terrestrial mammals, especially those of intermediate size (with the important exception of the primates – Kimura *et al.*, 1979; Demes *et al.*, 1994), but can be quite different in smaller mammals. The difference in load distribution between front and rear limbs is due to the position of the foot-ground contact and the fore-aft position of the CoM (Gray, 1968; Raichlin *et al.*, 2009). There are a number of functional reasons and constraints determining the position of the CoM, and this varies between species (somewhat) and even between individuals (if measured carefully enough).

Net horizontal force in the direction of travel for trot (and bipedal run) indicates initial braking, followed by reacceleration during each contact. Initial braking force may seem counter-productive, where it might appear that an advantage would be gained by not having the braking component. The braking portion of the contact is a natural consequence, however, of using the limb as a strut, as placement of contact of a strut-like limb in front of the CoM necessarily results in a breaking effect. Note, for instance, even though the first step of a human sprint involves only acceleration, the second and subsequent contacts involve some degree of braking within each stride (Jacobs and van Ingen Schenau, 1992; Hunter *et al.*, 2005).

In bipedal acceleration, the limb contacts the substrate ahead of the CoM after about the third step, explaining much of the braking impulse (if the limb acts as a strut and contacts ahead of the CoM, a component of the force acts between the contact point and the CoM, causing a braking effect). The second step in bipedal acceleration, whether human (Mero, 1988) or avian (Roberts and Scales, 2002), involves a braking impulse, even though initial foot contact is behind the CoM. The limb appears to be used as a strut, but with the ground reaction vector passing behind the CoM. Without forward acceleration, this would cause the individual to pitch forward. However, it is likely that the turning moment generated around the

CoM counters the pitching backward that would be caused by the horizontally applied impulse at the ground contact (as it does for positive acceleration in the quadruped (Lee *et al.*, 1999; and see below)).

If the braking impulse (where impulse = $\int F\,dt$, the area under the horizontal force/time curve) is equal to the acceleration impulse, then velocity lost in the first half of the contact will be replaced in the second half. In this case, the CoM will have the same horizontal velocity at the end of contact (when it goes into the non-contact phase where horizontal speed changes only negligibly due to air resistance) as it did at the beginning of the contact. This is referred to as steady speed trotting, meaning that it is steady average speed, even though the actual horizontal velocity changes constantly over the contact.

Steady speed trotting involves slightly different force profiles in the horizontal direction for the fore and hind limbs, just as observed for the vertical force component. In both fore and hind limbs, horizontal force transitions from braking to accelerating over the course of the contact. In most mammals and trotting circumstances, however, the forelimb contacts are characterized by predominantly braking force, while the hind limbs show a predominance of acceleration (see Figure 4.4). This is largely explained by the position of the foot relative to the CoM over the course of the contact. The net horizontal force at any given instant will be the sum of the fore and hind horizontal forces.

In steady speed locomotion over a flat horizontal surface, there is some logic to considering vertical and horizontal forces independently, as action in the vertical direction does not affect the horizontal, and horizontal forces do not affect the vertical. This is decidedly *not* the case if even small net positive or negative accelerations are involved. Horizontal accelerations, whether positive (resulting in velocity increase in the direction of travel) or negative (resulting in velocity decrease in the direction of travel) can only be applied at the surface of the substrate.

The CoM of the animal, however, is some distance above the ground surface. Horizontal acceleration applied at the ground surface will result in a turning moment around the CoM (Figure 4.5). With an unbalanced torque applied, the CoM will act as the axis of rotation; for a positive net acceleration, the front of the animal will tend to tip upward, while a negative net acceleration will cause the head end to tip downward and the rear end to tip upward. For relatively small positive and negative accelerations, the axis of the animal's body will remain horizontal, which means that the vertical load born by the fore and hind limbs must change (Figure 4.5) – that is, the distribution of vertical loads will shift between fore and hind limbs.

It is disconcerting to find that what an animal does in the horizontal direction can alter measurements made in the vertical direction, but this is a direct consequence of the physics of the system. This was noted by Gray (1944, 1968), and is the basis of the requirement for quadrupedal gait studies that use only a single force sensing plate (where subtle accelerations and fore-aft redistributions of force between individual limbs cannot be measured) to be rigorously restricted only to steady speed locomotion. If multiple force plates are employed, so that independent footfalls of simultaneously contacting fore and hind feet can be assessed while the animal moves naturally with modest changes in speed, the turning moment can be used to provide information that is difficult to measure in other ways. (Lee *et al.*, 1999; Bertram *et al.*, 2000).

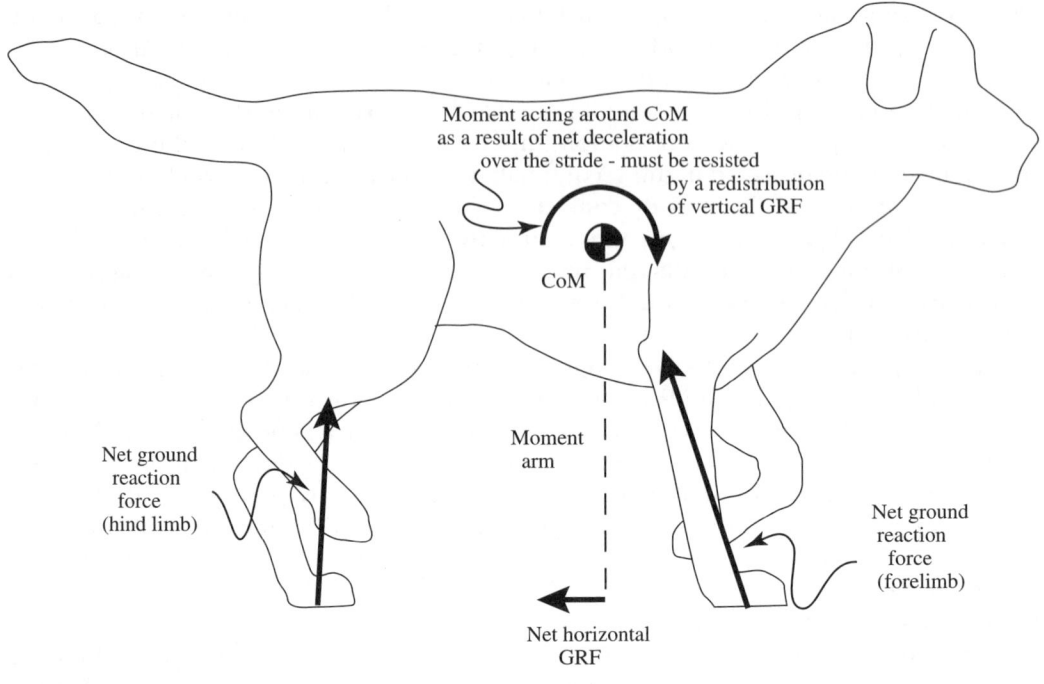

FIGURE 4.5

Diagram of the interaction between horizontal acceleration (negative acceleration depicted) and the moment applied to the center of mass (CoM). Net braking will tend to cause the animal to tip forward. To maintain balance, load increases in the forelimb contact and decreases in the hind limb. Unless the animal is in a perfect steady state (zero acceleration), horizontal forces directly influence vertical force distribution.

4.5.2 Walk

Although the ground reaction force of the trot has some complexities to consider, it is straightforward to interpret, in contrast to other quadrupedal gaits. The vertical impulse distribution between the fore and hind limbs in the quadrupedal walk parallels that observed in the trot, dependent as it is on proportion of body weight supported by each limb set (and results from the uneven cranial-caudal distribution of mass in many mammals). The profile of the vertical force is generally not as symmetric as that of the trot, however. Instead, a distinctive shoulder is observed for both the fore and hind limb contact, the specifics of which depend on which species is considered, its size and conformation, and the speed it is walking (see Figure 4.4).

The asymmetric vertical force profile in the quadrupedal walk arises from the same mechanism as the distinctive (but largely symmetric) double-hump vertical force profile seen in the human bipedal walk (Figure 4.4). In bipedal walking, the strut-like limb contacts the substrate and applies a braking impulse while changing the direction of the CoM as it falls forward off the previous stance limb. The CoM vaults over the new stance limb, with the mass rotating around the axis of the ankle. This produces a centripetal force that acts opposite to gravity as the mass passes over the vertically oriented stance limb like an inverted pendulum, decreasing the force applied by body weight at mid-stance (Alexander and Jayes, 1978a, 1978b). As the angle of the limb extends, in preparation for the next limb contact, the centripetal force is no longer in line with gravity, so its effect on the vertical component of force diminishes in the latter part of the contact (but gravitational force is no longer opposing the centripetal force of the limb rotation, so the end of stance is where the foot is most likely to lose contact with the ground (Usherwood, 2005; and see Chapter 6)).

Quadrupedal walking involves multiple simultaneous contacts, but something similar to the CoM motion management seen in the bipedal walk occurs in this case as well (Griffin *et al.*, 2004). However, the multiple contacts mean that the symmetry of the "inverted pendulum" motion is obscured by the effect of multiple limbs – some fore and some hind – all acting on and contributing to the overall CoM motion (see Chapter 5). Multiple limb interaction between the substrate and the CoM will have an influence on the GRF of each individual limb contacting the substrate, which is somewhat surprising, since we generally think that the limbs influence the motion of the CoM, not the other way around. Of course, the issue here is that the role of the limbs is to move the CoM via force applied at the substrate. Any load applied to the CoM (as from another limb) will affect what needs to be accomplished by the other limbs to produce the optimal CoM trajectory under the specific conditions that the animal is contending with.

4.5.3 Gallop

The limbs transmit force between the ground and the body to affect the motion of the mass of the animal in asymmetric gaits such as the canter (a slow, three-beat gallop) and the full gallop (a four-beat gait), just as they do in the walk and trot. However, the high-speed nature of the galloping gait and the fore-hind sequencing of contacts (in contrast with the more laterally sequenced symmetric gaits) has made the mechanical purpose of the specific leg sequencing of the gallop a challenge to interpret.

Two distinctive types of gallop have been identified, based on the sequence that individual limbs bear load during the stride. These two forms are the transverse gallop (typified by the equine gallop) and the rotary gallop (typified by the high-speed gallops of the cheetah and greyhound) (Hildebrand, 1959; and see Chapter 2). A number of features of these two gallop forms differ, and may have important meaning for the mechanics of each (Bertram and Gutmann, 2009; Biancardi and Minetti, 2012). The ground reaction force profile of each limb is reasonably consistent for the gallop in general. In Figure 4.4, the ground reaction forces (normalized to body weight and stride duration) are plotted for the horse and dog (Merkens *et al.*, 1993; Walter and Carrier, 2007).

In the gallop and walk, the overlap and sequence of limb contacts combine to influence the trajectory of the CoM of the animal. Although ground reaction forces are routinely measured for individual limb contacts, the meaning of each on the locomotion of the animal can only be interpreted in the context of their combined effects on the animal as a whole. Interpreting the specific loads applied by each limb requires evaluation of the consequences of the gait on the motion of the mass of the animal. This may well require that individual limbs be evaluated within the sequence of contacts that make up the complete stride cycle, with subsequent determination of the net effect on the dynamics of the interaction of the whole animal with its substrate.

4.6   THE CONSEQUENCE OF APPLIED FORCE: CoM MOTION, PENDULA AND SPRINGS

As described in Chapter 1, the mechanical approach to the analysis of comparative locomotion experienced a resurgence in the 1960s and 1970s, possibly spurred by Gray's influential volume (Gray, 1968), but also due to the emergence of electronic analytical tools that made complex dynamic analyses more tractable. Among these works, one paper stands as a watershed that summarized a perspective on locomotion that had been building since the 1930s (Margaria, 1976). This was the contribution by Cavagna, Heglund and Taylor, entitled *Mechanical work in terrestrial locomotion: two basic mechanisms for minimizing energy expenditure* (Cavagna *et al.*, 1977). This contribution marked a collaboration between Cavagna, who had been working on the mechanics of whole body motion in humans (Cavagna *et al.*, 1963, 1964, 1971, 1976), with the comparative perspective of Taylor (Taylor *et al.*, 1970, 1974; Taylor and Roundtree, 1973). The group recognized that fundamental mechanisms available to reduce the mechanical cost of locomotion appeared to be utilized by an extremely wide range of terrestrial animals, including bipedal mammals and birds, as well as quadrupedal mammals.

The two basic mechanisms referred to are:

1. a pendulum-like exchange between kinetic and potential energy that characterizes walking gaits; and

2. a bounce-like exchange between kinetic energy of the CoM and strain energy of elastic components of muscle and connective tissues in running gaits.

Cavagna *et al.* (1977) demonstrated that in walking gaits, oscillations of potential and kinetic energy of the CoM are relatively out of phase, allowing for exchange between the two energy forms. These energy fluctuations are in phase for bipedal running and quadrupedal trotting gaits. Since energy cannot be exchanged between two types that increase and decrease in synchrony through the stride cycle, running appeared to be remarkably wasteful. However, it was recognized that, in the run, the CoM was simultaneously at its lowest and slowest point at mid-stance, requiring that the supporting limb deflect under the applied load. This offered the opportunity for strain energy to be stored in elastic structures (tendons, muscles and ligaments) and returned to re-accelerate the system.

Much has been made of the pendular exchange of potential and kinetic energy in animal locomotion. Following substantial work on the subject, beginning with Fenn (1930) and Margaria (1938), Cavagna *et al.* (1963) formulated and described the concept of the exchange (transduction) between potential and kinetic energy in human walking as the CoM of the individual moves over the stance limb. In human bipedal walking, it was observed that the CoM passes over a relatively stiff (strut-like) stance limb and, in doing so, rises until it reaches its highest point when directly over the supporting limb. As the highest point, the body has its greatest potential energy (E_p) at mid-stance in the bipedal walk ($E_p = mgh$, where m is mass, $\mathbf{g}$ is the acceleration due to gravity and h is height above the substrate).

At mid-stance, however, the CoM is also traveling at its slowest speed, and consequently has the lowest kinetic energy ($E_k = 1/2mv^2$). After mid-stance, gravity accelerates the CoM forward and downward and E_p is exchanged with E_k in the same manner as a pendulum, except that, in this case, the pendulum is inverted, with the mass rotating around the pivot of the foot and ankle. With the next footfall contact, the kinetic energy derived from the fall from the previous stance can be converted, at least partially, back to E_p, when the body mass rises as it vaults over the extended stance limb. In this way, the energy generated to initially elevate and accelerate the body mass can be conserved, as it is exchanged between E_p and E_k over the course of successive stride cycles.

The E_p–E_k exchange has been observed in the walking gaits of a wide range of animals, and has been used as a definition of the mechanics that distinguishes walking and running gaits (Cavagna et al., 1977; Biknevicius and Reilly, 2006). Many animals do, indeed, show some degree of pendulum-like exchange, where changes in E_k are accompanied by changes of opposite phase in E_p (Griffin and Kram, 2000; Heglund et al., 1982; Minetti et al., 1999). However, most quadrupeds do not appear particularly good at this exchange – measured exchange for walking in quadrupeds appears to range from about 35–50% (Cavagna et al., 1977). If the exchange of these energy forms is important to economical locomotion, some compromise appears to be made by quadrupeds.

The evaluation of pendulum-like and bounce-like exchange is predicated on the very reasonable assumption that increases in CoM E_p require work (against gravity). However, as this approach is currently used, the corollary, where the decrease in E_p indicates inevitable loss unless converted to another energy form, may not be so straightforward to measure effectively. The recovery approach depends on following energy exchanges *within* the system (the organism) without directly measuring any losses (or identifying the mechanism of loss (Bertram and Hasaneini, 2013; and see Chapter 5)).

The work acting to move the mass of the organism relative to the external world is referred to as the external work, W_{ext}. In an organism moving at a constant average speed, the work performed on the CoM can be decomposed into two components – that in the vertical direction working against gravity, and that in the horizontal direction, performing work that alters the horizontal velocity of the system slightly as the limbs cycle through the contact pattern of the stride.

The sum of the positive increments of vertically directed energy ($E_p + E_{kv}$) over the stride yields the positive work performed in the vertical direction to lift the CoM against gravity (W_v). The sum of positive increments of kinetic energy in the forward direction, E_{kf}, over the stride, yields the positive work performed in the horizontal (forward) direction (W_f); this is required to replace decreases in horizontal velocity resulting from braking effects of limb contacts.

The work performed over the stride can be considered in terms of mechanical power by dividing each component by the stride period, resulting in a measure of mechanical power in the vertical and forward directions (P_v and P_f). Cavagna et al. (1977) found the magnitude of these two power measures to be very similar in a range of walking birds and mammals, suggesting the possibility that they may be effectively exchanged. The combined power invested in the vertical and horizontal directions can be summed to describe the external mechanical power (P_{ext}), the power applied between the CoM and the supporting substrate that results in the motion of the CoM over the stride cycle (see McMahon, 1984, Figure 8.3, pp. 192–193). Power in the lateral direction should also be included, but it is generally of vanishingly small magnitude in mammals with predominantly parasagittal motion.

In walking, if E_p increases without a decrease in E_k, then work must be performed by the leg (leg work). If E_p decreases without an increase in E_k, then work must be performed by the CoM on either the muscle and its tendon (negative work) or passive tissues (energy loss). This energy is lost as heat unless it is stored as strain energy in elastic structures.

Cavagna *et al.* (1977) evaluated the effectiveness of the exchange between E_k and E_p by comparing the magnitude of vertical and forward power ($|P_v| + |P_f|$) to the apparent work expended, P_{ext}. P_{ext} is calculated from the positive increments of the total energy, where $E_{tot} = (E_p + E_{kv} + E_{kf})$. Recovery is then calculated as a percentage, comparing the positive mechanical power ($|P_v| + |P_f|$) minus the external power (P_{ext}, which represents the power required if no exchange occurs) to the measured positive mechanical power ($|P_v| + |P_f|$), so that:

$$\text{Recovery}\,(\%) = \left(\left(\left(|P_v|+|P_f|\right)-\left(P_{ext}\right)\right)\Big/\left(|P_v|+|P_f|\right)\right)\times 100 \qquad (4.1)$$

The calculation of energetic recovery compares the changes of E_p and E_k of the complex distributed mass organism to the exchange expected for a perfect, point mass pendulum operating in a vacuum (no loss to air resistance). The utility of such a comparison depends largely on the assumption that this method of tracking energy exchange reveals a fundamental feature of mechanically economical locomotion, and that the point mass pendulum is one example of a perfect exchange system that sets a limit that legged locomotion systems could aspire to. Implicit in this approach to the analysis of mammalian locomotion is the assumption that energy recovery is a feature of locomotion that can be modified in the function of the system and adapted for in the morphology or behavior of the organism, although this is rarely discussed specifically.

The general similarity of magnitude and opposite phase of E_p and E_k in walking naturally suggests a direct exchange. In running, these two energy forms are in phase, and E_{kf} can be very much larger than E_p, so a similar exchange is not possible. Instead, Cavagna *et al.* (1977) suggested that E_k could be exchanged with strain energy in structures with elastic properties, such as tendons, ligaments and components of muscle. It is technically quite difficult to make these measurements directly in order to confirm this exchange, because invasive measurement of force and deformation of internal structures would be required while the animal ran. Instead, a clever indirect method that implied elastic-like energy exchange was used to validate this hypothesis.

Cavagna *et al.* (1977) calculated the mass specific power (rate of work performed per kg of body mass) used for raising the CoM against gravity (from the vertical ground reaction force), and the mass-specific power required for reaccelerating the CoM in the forward direction after the impact of the limb(s) with the ground caused a horizontal deceleration (from the horizontal ground reaction force). These values were summed over the stride to calculate the mass-specific power for the total external work performed in the stride, P_{ext}. It was argued that muscle efficiency (γ) was the proportion of mechanical power performed relative to the metabolic power (P_{metab}) invested, and they proposed the relationship:

$$\gamma = \left(\left(P_{ext} + P_{int}\right) - P_{elastic}\right)/P_{metab} \tag{4.2}$$

Where: P_{int} is the "internal mechanical power" – the mechanical power required to move components of the body (primarily the limbs) relative to the CoM of the organism, $P_{elastic}$ is the mechanical power stored and released elastically.

At the time this work was performed (1977), there were no measures of P_{int} for quadrupedal mammals, so a minimum efficiency ($\gamma' = P_{ext}/P_{metab}$) was calculated, which evaluated the proportion of external mechanical power that could be accounted for through metabolic power (where metabolic power was determined from indirect calorimetry assuming maximum efficiency of muscle at all points in the stride – itself a large assumption). For the dog, at a slow (1.4 m/s) to a fast trot (4.2 m/s) γ' increased from 0.25 to 0.27 and, in the bouncing saltatory gait of the kangaroo, γ' went from 0.24 at slower speeds to an amazing 0.76 as these animals approached 8.3 m/s (30 km/hr).

Since it is known, from muscle physiology, that the limit of γ is about 0.25 (McMahon, 1984; He et al., 2000), any value of γ' over 0.25 must indicate a power source in addition to metabolically supplied power. Since the calculation of γ' left out a known component of the mechanical cost of motion – namely, P_{int} – any value of γ' in the vicinity of 0.25 would likely indicate a substantial contribution from a non-metabolically derived power source. In fact, subsequent measurements by this group (Fedak et al., 1982) suggest that P_{int} increases with speed to the point where it could account for all of the metabolic power of fast running in larger mammals (potentially any mammal larger than a mid-size dog).

So, if the power required to move the CoM relative to the ground in most mammals requires more efficient conversion of metabolic power to mechanical power than is possible in animals, and the power required to wiggle the limbs relative to the body can have substantial additional costs, how is the power of locomotion supplied? According to this perspective of locomotion dynamics, the answer is provided in Equation 4.2 – power available from elastic recoil ($P_{elastic}$) can provide power in a passive or near-passive manner (little metabolic investment is required) for continuing cycles of motion (steady state locomotion, where energy is exchanged throughout the cyclical motions). Power provided in this way will decrease the total mechanical power that must be accounted for by metabolic power supply. Thus, it was concluded that the capacity for animals to perform these running gaits was critically dependent on the reduction of power requirements via the storage and return of energy through elastic structures within the body and limbs.

Other gaits available to the quadruped, such as the gallop, do not fit either the inverted pendulum or spring-like models well, and it was surmised that these complex asymmetric gaits use a combination of the two mechanisms (Cavagna et al., 1977). However, there was no obvious rationalization for why the combination is utilized, what the advantages are or what constraints act on the motion in gaits like the gallop. Alexander (1993) hypothesized that certain forms of the gallop provide the opportunity to store strain energy in structures of the back. Horizontal inertia of the rear part of the body would flex the back and rear limbs at the hip as inertia swings the rear portion of the body forward during forelimb contact. This energy would be returned to the stride when the rear limbs made contact with the substrate and powered the next non-contact phase of the gait. However, this strategy would likely not work for those gallops with a non-contact phase between front and rear contacts (where the stored energy would likely be released prematurely), and it has been difficult to identify elastic structures capable of storing and returning adequate energy.

4.7 ENERGY EXCHANGE IN LOCOMOTION – VALUABLE OR INEVITABLE?

Substantial evidence indicates that exchange between E_p and E_k occurs in legged loco-motion, and the only exception so far appears to be the extremely slow locomotion of the tortoise (Zani *et al.*, 2005). At the slow speed of the tortoise walk, very little E_k exists at any stage of the gait cycle, so it is not available for exchange. One key factor to consider with regard to the exchange between E_p and E_k is whether features of organic adaptation can increase pendular exchange and, with it, economy of loco-motion. Due to the action of gravity, E_p will be spontaneously converted to E_k, unless prevented from doing so. Likewise, available E_k can be directed by a strut-like stance limb to the vertical direction, spontaneously increasing E_p.

In order for either of these *not* to occur, active stabilization of the mass or absorp-tion of energy needs to take place, which, for instance, may take the form of the vaulting stance limb shortening as load is applied, thus absorbing available E_k. If this available energy is not stored as strain energy, then lengthening of active compo-nents of the jointed limb will be required to control the motion of the CoM and absorb this energy. In such a case, the energy is absorbed as negative work per-formed by the limb.

Pendulum-like exchange can be readily demonstrated in the legged locomotion of most terrestrial vertebrates. However, at issue is whether this is of real importance in either the control or functional economy of mammal locomotion. Any mass distributed at a distance from a point of rotation, and possessing some momentum, will act as a pendulum. It is an important feature of the physical world in which the animal func-tions, and is likely exploited as such – but is this the critical feature that governs effective function in terrestrial locomotion?

To determine this, it is important to understand how the action of the limbs and their integration with motion of the body mass influence locomotion. However, it is not likely that adaptive features of mammalian morphology can have a substantial effect on the fundamental aspects of whole body pendular exchange (there may be value in considering the distribution of mass in the limb, however; see Hildebrand and Hurley, 1985; Kilbourne, 2013).

Take, for example, the case of a brachiating gibbon. If there was ever an animal that appeared to utilize pendular exchange, it is the arm-swinging locomotion of the gib-bon. Pendular exchange has been directly demonstrated in the gibbon and siamang, and behaviors exploiting the potential for pendular exchange have long been observed (Fleagle, 1974; Prueschoft and Demes, 1984). However, it has been noted that a brachiator will only be truly pendular at one speed (Swartz, 1989; Chang *et al.*, 2000; Bertram and Chang, 2001), and pendular dynamics do not explain much regarding how brachiation operates. The pendular perspective does not help to identify the major challenges that these animals face in locomotion, or what determines the energetic cost of this mode of travel (see Chapter 5).

Potential-kinetic energy exchange occurs in both brachiation and terrestrial legged locomotion (just as it undoubtedly does in the undulating flight of some bats and birds) but, in order to use this fact in interpreting the structure or function of locomoting systems, it must be determined that such exchange is manageable by the organism. As we will see in the following, exchange happens, and it is utilized when it does, but it is not likely the driving feature of function or design in legged locomotion of terrestrial mammals.

4.8 MOMENTUM AND ENERGY IN LOCOMOTION: DYNAMIC FUNDAMENTALS

Momentum is the product of mass and velocity, where a large mass moving at a relatively slow velocity can have the same momentum as a smaller mass moving at a rapid pace. Similarly, a larger mass moving at a relatively slow speed can have the same kinetic energy as a smaller mass moving more rapidly, where kinetic energy is commonly written $E_k = 1/2mv^2$ (though the v here should more formally be written $|v|$, the magnitude of velocity, which is speed). Even though these two physical quantities – momentum and kinetic energy – relate to aspects of locomotion, they have fundamental differences that affect the way they can be used to characterize the dynamics of locomotion.

Momentum is a vector quantity, where all dimensions are independent (and linear momentum is independent of angular momentum). Momentum represents a physical condition of the object. This is in contrast to kinetic energy (or any energy for that matter), which is a scalar quantity (has only magnitude) and represents the state of the system at that point in time. Even though the units of kinetic energy and work are equivalent, and kinetic energy can cause work to be done, kinetic energy is simply a measure of the potential for work to be done by the momentum of the system. Energy can reasonably be considered to be simply the "currency" of work.

Energy can be exchanged between forms, just as monetary currency can be exchanged. A classic example is the simple pendulum, where kinetic energy at the bottom of the swing is spontaneously converted to potential energy as the momentum of the weight is constrained by the supporting cord to perform work against gravity. The potential energy is again spontaneously converted back into kinetic energy as gravity does work, accelerating the object downward (as gravity applies an impulse $[I = \int F_g \, dt]$ to the mass, which increases its momentum). This is a good example of a closed system (neglecting loss to fluid resistance as the weight moves through the air, or hysteresis in the supporting cord as tensile forces change through the swing cycle); energy is neither lost nor generated, but spontaneously changes between forms within the system.

In the pendulum described above, the total energy of the system remains constant, so the total energy state of the system is the same regardless of where in the cycle it is measured. This is not the case for momentum. At the top of the swing, when the weight reverses directions, velocity is zero, so it follows that momentum is zero; conversely, at the bottom of the swing, momentum is maximum. A change in momentum requires the application of an impulse. In the case of a simple pendulum (neglecting air resistance and hysteresis), this impulse comes from the force of gravity acting on the mass of the object (and the impulse results from this force acting over a period of time).

When two objects interact, each applies an impulse to the other. The total momentum after an interaction between objects is equal to the total momentum of all objects involved just prior to the interaction, unless work is done on, or by, components of the system (as in limbs). This is the case even if the apparent momentum effects are undetectably small. For example, the momentum of the earth must change as a result of the pendulum described above applying an impulse to it through the supporting structure of the swing. Note that when considering the origin of the impulse, interaction occurs outside the basic pendulum that was originally considered the "system". It will become apparent (Chapter 5) that it may be more convenient for the understanding of the energetics of locomotion to follow the route mechanical energy flows into and out of the system (organism), than it is to track how energy moves between forms within the system (the pendulum and spring-mass approach).

Unlike energy, however, momentum is a vector quantity; the different dimensions of momentum (directions) are independent, and momentum in one direction cannot be converted to another direction. Similarly, an impulse in one direction cannot affect the momentum in another direction (it is a little more complex in the swing, where two impulses are applied – one from gravity, increasing momentum, and one from the supporting cord, redirecting that momentum through the swing arc). It is particularly important to consider these features in legged locomotion, where impulses are imparted between the supporting substrate and the organism's mass at each footfall.

4.9   ENERGY – LOST UNLESS RECOVERED, OR AVAILABLE UNLESS LOST?

It is self-evident that locomotion using limbs costs metabolic energy, even going downhill, where gravitational acceleration could potentially negate many costs associated with level or uphill movement (Margaria, 1938, 1968; figure reproduced in McMahon, 1984, pp. 212–213). Why does locomotion cost so much energy? Pendulum-like exchange in human walking can account for up to 65% of the mechanical energy required in each step of the bipedal human (Cavagna *et al.*, 1976), but it is important to ask, "where does the 35% that is lost go?"

Effective use of the pendulum means that only 35% of the energy of level, steady speed locomotion needs to be provided by the muscles – but why is 65% cost avoidance the limit (and recall that this appears much lower in quadrupedal forms). Implicit in the "two basic mechanisms" approach to the mechanics of legged locomotion is the assumption that all energy is lost unless it is recovered through some mechanism. However, is this a reasonable assumption? Would it not be just as valid to contend that all energy generated in locomotion should be available, unless lost through some mechanism? It is critical to identify how energy is actually "lost" in legged locomotion, in order to interpret the adaptations that can be responsible for limiting how much is lost, and through what mechanisms. These are the issues necessary to understand how "cost-effective" locomotion is produced in biological systems, and to properly interpret the morphological adaptations and functional strategies that contribute to it.

Muscles are used as mechanical actuators that convert metabolic (chemical) energy to mechanical force, length change and, consequently, work (see Chapter 3). This is somewhat analogous to how an automobile converts chemical energy in combustion to power the rotation of the drive wheels. The rate of fuel consumption in any internal combustion automobile will increase as it climbs a grade. In decent, it will also consume fuel as long as the engine is running but, at a steep enough decline, the momentum produced by the impulse applied through gravitational force on the vehicle will mean that no extra energy needs to be added to maintain speed (and beyond this slope brakes will be needed to absorb excess momentum). In the automobile, brakes are applied at relatively little cost – the engine need not contribute directly, and any "cost" associated with brake action is deferred to maintenance required to repair wear.

The passive brake strategy is not much available in legged locomotion, and deceleration must be actively generated through muscular effort (requiring an immediate energetic investment). An automobile can act in just the same way as an animal if an engine retarder brake is used. Switching to a lower gear, for instance, where momentum of the vehicle is absorbed, speeds up the rotational velocity of the engine, all the while burning extra fuel in the cylinders. When legs are used to decrease speed, the muscles actively have to resist the tendency for the joints to fold. If the force generated by the braking action of the limb stretches the muscles while they are actively trying to contract, work is done *on* the muscles. This is referred to as negative work (see Chapter 3). Negative work done by the muscles in decelerating the mass of the animal and limbs, plus the requisite positive work needed to accelerate the system back to its original speed, is usually identified as the major source of "cost" in legged locomotion.

One could ask, "why has the biological system been set up so that energy is lost through active muscle work?" It is possible that this is a limitation of the system, where muscles of adequate force generating capacity to resist lengthening under imposed external loads would be too bulky to be feasible (or would expend more metabolic energy preventing negative work than is used in the negative work). Analysis of human walking identifies negative work as a cost, but suggests that it does not seem to dominate the cost (Soo and Donelan, 2010). There is no reason to think that the limbs of quadrupedal mammals operate in a substantially different manner. If not explicitly negative work performed by the muscles (and the positive work required to "pay back" those losses), where does the cost of legged locomotion arise?

In the next chapter, the source of mechanical "cost" in legged locomotion is explored. It will be demonstrated that the mechanical cost of locomotion resides largely in the dynamics of the interaction between the CoM path and the substrate on which the animal moves. This interaction is mediated by the action of the limbs, and explicit costs, such as negative work or deformation of tissues or the substrate, are largely mechanisms through which features of physically required momentum and energy balance are maintained.

▌ REFERENCES

Alexander RMcN (1993). Why animals gallop. *American Zoologist* **28**, 237–245.

Alexander RMcN (2001). Design by numbers. *Nature* **412**, 591.

Alexander RMcN (2002). Stability and manoeuvrability of terrestrial vertebrates. *Integrative and Comparative Biology* **42**, 158–164.

Alexander RMcN, Jayes AS (1978a). Vertical movements in walking and running. *Journal of Zoology (London)* **185**, 27–40.

Alexander RMcN, Jayes AS (1978b). Optimum walking techniques for idealized animals. *Journal of Zoology (London)* **186**, 61–81.

Alexandrov AV, Frolov AA, Horak FB, Carlson-Kuhta P, Park S (2005). Feedback equilibrium control during human standing. *Biological Cybernetics* **93**, 309–322.

Bertram JEA, Chang YH (2001). Mechanical energy oscillations of two brachiation gaits: measurement and simulation. *American Journal of Physical Anthropology* **115**, 319–326.

Bertram JEA, Gutmann A (2009). The running horse and cheetah revisited: fundamental mechanics of two galloping gaits. *Journal of The Royal Society Interface* **6**, 549–559.

Bertram JEA, Hasaneini SJ (2013). Neglected losses and key costs: Tracking the energetics of walking and running. *Journal of Experimental Biology* **216**, 933–938.

Bertram JEA, Lee DV, Case HN, Todhunter RJ (2000). Comparison of the trotting gaits of Labrador Retrievers and Greyhounds. *American Journal of Veterinary Research* **61**, 831–837.

Biancardi CM, Minetti AE (2012). Biomechanical determinants of transverse and rotary gallop in cursorial mammals. *Journal of Experimental Biology* 4144–4156.

Biknevicius AR, Reilly SM (2006). Correlation of symmetrical gaits and whole body mechanics: debunking myths in locomotor biodynamics. *Journal of Experimental Zoology Part A: Comparative Experimental Biology* **305**, 923–934.

Bramble DM, Carrier DR (1983). Running and breathing in mammals. *Science* **219**, 251–256.

Carrier DR, Deban SM, Fischbein T (2006) Locomotor function of the pectoral girdle 'muscular sling' in trotting dogs. *Journal of Experimental Biology* **209**, 2224–2237.

Cavagna GA, Saibene FP, Margaria R (1963). External work in walking. *Journal of Applied Physiology* **18**, 1–9.

Cavagna GA, Saibene FP, Margaria R (1964). Mechanical work in running. *Journal of Applied Physiology* **19**, 249–256.

Cavagna GA, Komarek L, Muzzolini S (1971). The mechanics of sprint running. *Journal of Physiology, London* **217**, 709–721.

Cavagna GA, Thys H, Zamboni A (1976). The sources of external work in level walking and running. *Journal of Physiology, London* **262**, 639–657.

Cavagna GA, Heglund NC, Taylor CR (1977). Mechanical work in terrestrial locomotion: two basic mechanisms for minimizing energy expenditure. *American Journal of Physiology* **233**, R243–R261.

Chang YH, Bertram JEA, Lee DV (2000). The external force and torques generated by a brachiating White-handed gibbon (*Hylobates lar*). *American Journal of Physical Anthropology* **113**, 201–216.

Collins S, Ruina A, Tedrake R, Wisse M (2005). Efficient bipedal robots based on passive-dynamic walkers. *Science* **307**, 1082–1085.

Davis DD (1949). The shoulder architecture of bears and other carnivores. *Fieldiana Zoology* **81**, 285–305.

Demes B, Larson SG, Stern JT, Jungers WL, Biknevicius AR, Schmitt D (1994). The kinetics of primate quadrupedalism: "hindlimb drive' reconsidered. *Journal of Human Evolution* **26**, 353–374.

Dyce KM, Sack WO, Wensing CJG (2002). *Textbook of Veterinary Anatomy* (3rd edition). WB Saunders, Philadelphia, pp. 840.

Fedak MA, Heglund NC, Taylor CR (1982). Energetics and mechanics of terrestrial locomotion. II Kinetic energy changes of the limbs and body as a function of speed and body size in birds and mammals. *Journal of Experimental Biology* **97**, 23–40.

Fenn WO (1930). Work against gravity and work due to velocity changes in running. *American Journal of Physiology* **93**, 433–462.

Fleagle JG (1974). The dynamics of the brachiating siamang (*Hylobates [Symphalangus] syndactylus*). *Nature* **248**, 259–260.

Gray J (1944). Studies in the mechanics of the tetrapod skeleton. *Journal of Experimental Biology* **20**, 88–116.

Gray J (1968). *Animal Locomotion*. London: Weidenfield & Nicolson. 479 pp.

Griffin TM, Kram R (2000). Penguin waddling is not wasteful. *Nature* **408**, 929.

Griffin TM, Main RP, Farley CT (2004). Biomechanics of quadrupedal walking: how do four-legged animals achieve inverted pendulum-like movements? *Journal of Experimental Biology* **207**, 3545–3558.

He ZH, Bottinelli R, Pellegrino MA, Reggiani C (2000). ATP consumption and efficiency of human single muscle fibers with different myosin isoform composition. *Biophysical Journal* **79**, 945–961.

Heglund NC, Cavagna GA, Taylor CR (1982). Energetics and mechanics of terrestrial locomotion. III Energy changes of the centre of mass as a function of speed and body size in birds and mammals. *Journal of Experimental Biology* **97**, 41–56.

Hildebrand M (1959). Motions of the running cheetah and horse. *Journal of Mammalogy* **40**, 481–495.

Hildebrand M, Goslow GE (1995). *Analysis of vertebrate structure*, 4th edition. J. Wiley, New York, pp. 657.

Hildebrand M, Hurley JP (1985). Energy of the oscillating legs of a fast-moving cheetah, pronghorn, jackrabbit, and elephant. *Journal of Morphology* **184**, 23–31.

Hunter JP, Marshall RN, McNair PJ (2005). Relationships between ground reaction force impulse and kinematics of sprint-running acceleration. *Journal of Applied Biomechanics* **21**, 31–43.

Jacobs R, van Ingen Schenau GJ (1992). Intermuscular coordination in a sprint push-off. *Journal of Biomechanics* **25**, 953–965.

Jenkins FA Jr (1974). The movement of the shoulder in claviculate and aclaviculate mammals. *Journal of Morphology* **144**, 71–83.

Kilbourne BM (2013). Scale effects and rotational inertia in the limbs of quadrupedal mammals. *Integrative and Comparative Biology* **53** Suppl. 1.

Kimura T, Okada M, Ishida H (1979). Kinesiological characteristics of primate walking: its significance in human walking. In: Morbeck ME, Preuschoft H, Gomberg N (eds). *Environment, behavior and morphology: Dynamic interactions in Primates*. Fischer, New York, pp. 297–311.

Kuo AD (2002). Energetics of actively powered locomotion using the simplest walking model. *Journal of Biomechanical Engineering* **124**, 113–120.

Lee DV (2011). Effects of grade and mass distribution on the mechanics of trotting dogs. *Journal of Experimental Biology* **214**, 402–411.

Lee DV, Bertram JEA, Todhunter RJ (1999). Acceleration and balance in trotting dogs. *Journal of Experimental Biology* **202**, 3565–3573.

Lee DV, Stakebake EF, Walter RM, Carrier DR (2004). Effects of mass distribution on the mechanics of level trotting in dogs. *Journal of Experimental Biology* **207**, 1715–1728.

Lee DV, Bertram JEA, Anttonen JT, Ros IG, Harris SL, Biewener AA (2011). A collisional perspective on quadrupedal gait dynamics. *Journal of The Royal Society Interface* **8**, 1480–1486.

Margaria R (1938). Sull fisiologia e specialmente sul consumo energetico della marcia e della corsa a varie velocità ed inclinizationi del terreno. *Atti della Accademia Nazionale dei Lincei. Memorie, serie VI* **7**, 299–368.

Margaria R (1968). Positive and negative work performances and their efficiencies in human loco-motion. *Internationale Zeitschrift für angewandte Physiologie einschließlich Arbeitsphysiologie* **25**, 339–351.

Margaria R (1976). *Biomechanics and energetics of muscular exercise*. Clarendon Press, Oxford, 146 pp.

McGeer T (1990). Passive bipedal running. *Proceedings of the Royal Society of London B* **240**, 107–134.

McGeer T (1993). Dynamics and control of bipedal locomotion. *Journal of Theoretical Biology* **163**, 277–314.

McMahon TA (1984). *Muscles, reflexes, and locomotion.* Princeton University Press, Princeton, 331 pp.

Meijaard JP, Papadopoulos JM, Ruina A, Schwab AL (2007). Linearized dynamics equations for the balance and steer of a bicycle: A benchmark and review. *Proceedings of the Royal Society of London A* **463**, 1955–1982.

Merkens HW, Schamhardt HC, van Osch GJ, Hatman W (1993). Ground reaction force patterns of Dutch Warmbloods at the canter. *American Journal of Veterinary Research* **54**, 670–674.

Mero A (1988). Force-time characteristics and running velocity of male sprinters during the acceleration phase of sprinting. *Research Quarterly for Exercise & Sport* **59**, 94–98.

Minetti AE, Ardigo LP, Reinach E, Saibene F (1999). The relationship between mechanical work and energy expenditure of locomotion in horses. *Journal of Experimental Biology* **202**, 2329–2338.

Park S, Horak FB, Kuo AD (2004). Postural feedback responses scale with biomechanical constraints in human standing. *Experimental Brain Research* **154**, 417–427.

Preuschoft H, Demes B (1984). Biomechanics of brachiation. In: Preuschoft, H, Chivers, DJ, Brockelman, WY, Creel, N (eds). *The lesser apes: evolutionary and behavioral biology.* Edinburgh University Press, Edinburgh, pp. 96–118.

Raichlin DA, Pontzer H, Shapiro LJ, Sockol MD (2009). Understanding hind limb weight support in chimpanzees with implications for the evolution of primate locomotion. *American Journal of Physical Anthropology* **138**, 395–402.

Riskin DK, Willis DJ, Iriate-Diaz J, Hedrick TL, Kostandov M, Chen J, Laidlaw DH, Breuer KS, Swartz SM (2008). Quantifying the complexity of bat wing kinematics. *Journal of Theoretical Biology* **254**, 604–615.

Roberts TJ, Scales JA (2002). Mechanical power output during running accelerations in wild turkeys. *Journal of Experimental Biology* **205**, 1485–1494.

Romer AS, Parsons TS (1986). *The vertebrate body*, 6th edition. Holt Rinehart and Winston, Orlando, pp. 679.

Shoshani J, McKenna MC (1998). Higher taxonomic relationships among extant mammals based on morphology, with selected comparisons of results from molecular data. *Molecular Phylogenetics and Evolution* **9**, 572–584.

Soo CH, Donelan JM (2010). Mechanics and energetics of step-to-step transitions isolated from human walking. *Journal of Experimental Biology* **213**, 4265–4271.

Srinivasan M, Raptis IA, Westervelt ER (2008). Low-dimensional sagittal plane model of normal human walking. *Journal of Biomechanical Engineering* **130**, 051017.1–051017.11.

Swartz SM (1989). Pendular mechanics and the kinematics and energetics of brachiating locomotion. *International Journal of Primatology* **10**, 387–418.

Taylor CR, Roundtree VJ (1973). Running on two and four legs: Which consumes more energy? *Science* **179**, 186–187.

Taylor CR, Schmidt-Nielsen K, Raab JL (1970). Scaling of energetic cost of running to body size in mammals. *American Journal of Physiology* **219**, 1104–1107.

Taylor CR, Shkolnik A, Dmi'el R, Baharav D, Borut A (1974). Running in cheetahs, gazelles and goats: Energy cost and limb configuration. *American Journal of Physiology* **227**, 848–850.

Usherwood JR (2005). Why not walk faster? *Biology Letters* **1**, 338–341.

Usherwood JR, Wilson AM (2006). Accounting for elite indoor 200m sprint results. *Biology Letters* **2**, 47–50.

Walter RM, Carrier DR (2007). Ground reaction forces applied by galloping dogs. *Journal of Experimental Biology* **210**, 208–216.

Weyand PG, Sternlight DB, Bellizi MJ, Wright S (2000). Faster top running speeds are achieved with greater ground forces not more rapid leg movements. *Journal of Applied Physiology* **89**, 1991–1999.

Weyand PG, Sandell RF, Prime DN, Bundle MW (2010). The biological limits to running speed are imposed from the ground up. *Journal of Applied Physiology* **108**, 950–961.

Williams SB, Usherwood JR, Jaspers K, Channon AJ, Wilson AM (2009). Exploring the mechanical basis for acceleration: pelvic limb locomotor function during accelerations in racing greyhounds (*Canis familiaris*). *Journal of Experimental Biology* **212**, 550–565.

Winter DA (2009). *Biomechanics and motor control of human movement*, 4th edition. Wiley, New York, pp. 370.

Zani PA, Gottschall JS, Kram R (2005). Giant Galápagos tortoises walk without inverted pendulum mechanical energy exchange. *Journal of Experimental Biology* **208**, 1489–1494.

CHAPTER FIVE

Concepts in Locomotion: Wheels, Spokes, Collisions and Insight from the Center of Mass

John E. A. Bertram

Department of Cell Biology and Anatomy, Cumming School of Medicine, University of Calgary, CA

5.1 INTRODUCTION

Complex and highly coordinated actions occur when a quadruped moves effectively within its environment. Even at a walk, numerous muscles respond with complex coordination, driven by the activity of the motor nerves. Forces are generated within the limb and axial skeleton, and these interact with the properties of the substrate in the gait cycle to generate locomotion. Meanwhile, sensory feedback from a number of sources inform the neural control centers about the organism's relation to its environment and the relationship between its limb components, to each other and to the rest of the body. All of these features can be monitored and described, but describing and knowing them does not necessarily lead directly to understanding what they accomplish, or why they occur as they do. In this chapter, the concepts that form the basis of an understanding of locomotion mechanics are described by focusing not on the complexities of the details observable, but on the consequences of the motions of the organism within its environment.

5.2 UNDERSTANDING BRACHIATION: AN ANALOGY FOR TERRESTRIAL LOCOMOTION

The Hylobatid apes (Gibbons and Siamangs) utilize a unique form of arboreal locomotion known as brachiation, in which the animal swings beneath overhead branch supports, using its pectoral limbs. Even casual observation will indicate that substantial pendulum-like motions occur in brachiation (Preuschoft and Demes, 1984). Inverted, pendulum-like energy exchange has been proposed as a crucial basis for over-ground walking gaits (Cavagna et al., 1977; see also Chapter 4), so it might be useful to evaluate the role of pendular energy exchange in animals whose environment offers the opportunity to readily exploit true pendular energy exchange.

Brachiating animals appear to utilize pendulum-based mechanical opportunities, both in simple swinging and to actively alter their center of mass (CoM) velocity using body repositioning (a form of parametric amplification), much like the "pump" maneuver used in playground swings (Fleagle, 1974). However, other aspects of their dynamic motion does not match that expected of a true, energy exchange-driven pendulum (Preuschoft and Demes, 1984; Swartz, 1989; Chang et al., 2000; Bertram and Chang, 2001). For a system that so obviously has the opportunity to utilize pendular exchange, it is perplexing that these animals do not appear to have adapted to exploit the advantages of spontaneous pendular energy exchange more effectively throughout their complete locomotion repertoire. Is there an alternative approach to the mechanics of this form of locomotion that can explain both why the system does and does not exhibit pendulum-like motions under different locomotion circumstances?

The pendulum-based form of "locomotion" (where locomotion is rather broadly defined) that is likely most familiar to us is the playground swing. Although the details of the momentum changes are complex (Wirkus et al., 1998), almost everyone learns at an early age to use muscle work to influence momentum in a precisely timed "pumping" motion (where the "pump" adds energy to the system). It is also common for the more ambitious to imagine they might be able to swing higher and higher, eventually swinging completely over the support, to carry on in an arc around the swing pivot for the full 360°. Everyone who tries, however, comes to realize that, on a normal playground swing, this is impossible. Although continual pumping increases the swing arc, so that each successive swing gets higher and higher (adding momentum and total energy – expressed as kinetic and potential energy at different portions of the swing), once the swing nears the horizontal and/or begins along the top of the arc around the pivot, gravity no longer maintains tension in the supporting cords of the swing. At this point, the cords supporting the swing buckle. Rather than continuing around the swing arc on the way back down, the mass of the individual will begin a vertically oriented ballistic fall, inside the arc described by the fully extended support cords (Figure 5.1). Once the individual's falling mass reaches the extended swing arc on the downside of the swing (when the support cords become fully extended once again), a small jolt is felt. Following this event, substantially less energy is left in the system, and the ensuing swing will be far lower than the previous one (swing height,

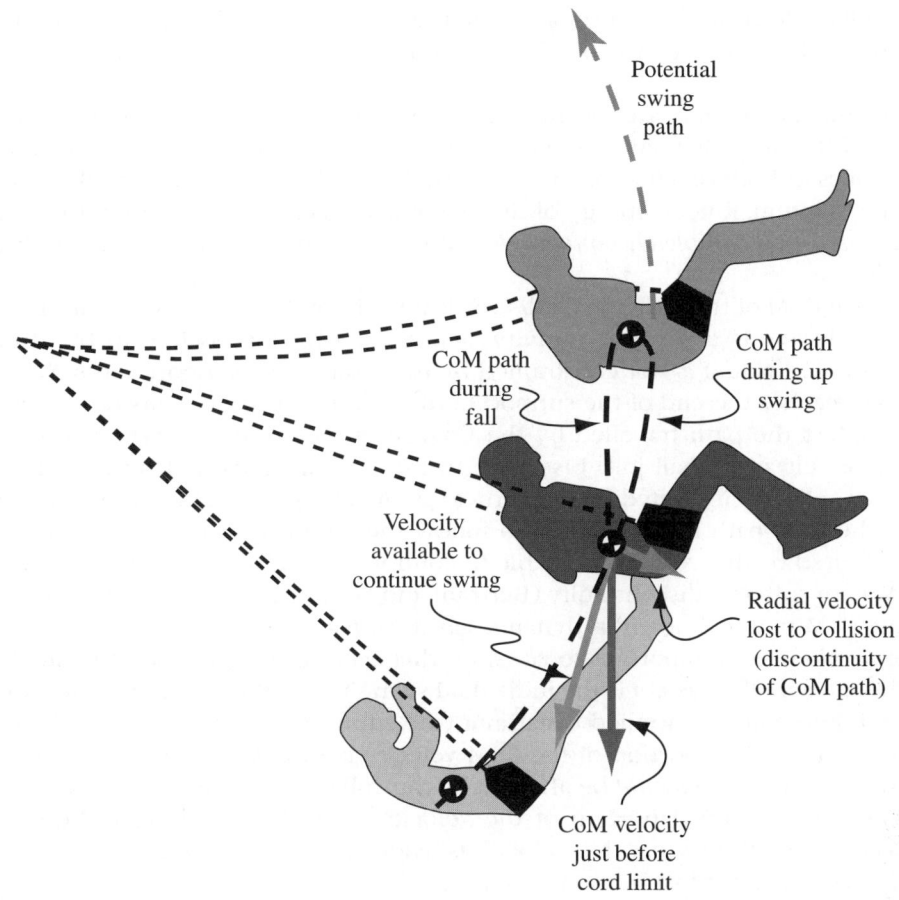

FIGURE 5.1

Tracings of a young person on a playground swing – sequence is indicated by shading, with lighter shading being earlier. When the swing extends above horizontal, the support cords allow the CoM to fall within the swing arc. When the CoM reaches the limit of the support cords, the CoM path is deflected. The radial velocity (in line with the support cord) is lost, and the speed is decreased (velocity available after cord limit is less than before) as the next swing progresses.

demonstrating maximum potential energy, will be a direct indication of the energy level of the system). The higher the swing, the greater the jolt and the greater the loss of energy, so adding energy to "go over the top" in such a swing becomes a futile effort. What mechanism is responsible for extracting energy from this system?

Swings completing 360° are commonplace in circuses and in videos available on the Web. All of these systems replace the flexible cords supporting the swing with rigid struts, and this prevents the mass of the individual from falling inside the swing arc. Instead, the rigidity of the support guides the mass of the individual smoothly around the swing arc, maintaining a constant distance from the axis of rotation. There is no opportunity for the jolt that loses so much momentum (and, consequently, kinetic energy). Even though substantial energy needs to be added to the system for it to travel a full circle around the support point, this can be gradually accumulated through the body manipulations that make up the pumping action – but only if that energy is not lost.

Momentum and energy are lost through the small jolt when the support cords divert the path of the individual, and understanding this process is critical to understanding the dynamics of both brachiation and terrestrial legged locomotion. Note that with the loss of momentum, kinetic energy of the swing also decreases, in spite of the fact that *potential energy is completely converted to kinetic energy* as the mass of the individual descends.

When the CoM of the individual falls vertically within the extended arc of the swing, it will accelerate as a result of gravity, just as in a normal swing. In this free-fall portion, the individual is not constrained by the limit of the support cords. When the individual reaches the end of the support cord, however, a severe constraint is applied and this alters the path travelled by the CoM. It is the abrupt alteration of the CoM path that is felt as a small jolt. Just prior to the jolt, the CoM path has a substantial downward velocity (indicated by the vertical arrow in Figure 5.1) while, just following the jolt, the CoM path is constrained to follow the arc of the circle at the limit of the cord (indicated by the 'velocity available to continue swing' arrow in Figure 5.1). Thus, the CoM path suffers a discontinuity (Bertram and Hasaneini, 2013). Such a discontinuity of the CoM path is known in dynamics as a *collision*.

Here is where it is important to consider this event in terms of what is "lost" from the system, where the system is the individual swinging. That portion of the individual's velocity in line with the applied constraint (the supporting cord) is involved with the collision, so the radial portion of the CoM velocity is lost. Any velocity perpendicular to the supporting cord *cannot* be affected by the collision (because velocity is a vector quantity). The non-radial portion of the velocity is available following the collision event to continue the swing; this velocity is available not because it was "recovered", but because it was not lost in the first place.

Due to the geometric relationship of the falling mass and swing arc constraint for a "too high" swing, a substantial portion of the available momentum applies a futile impulse through the swing cords to the support (ultimately, the mass of the earth, so it is determined by the interaction of two masses – the individual and the earth). Energy of the system (the swinging individual) is lost through the collision, and so the following swing height is reduced. The energetic loss can be substantial, even for what feels to the individual as a relatively small "collision". Understanding the role of collision-like losses explains much of the dynamics observed in brachiation (Bertram, 2004), and it also applies, in an inverted form, in legged locomotion (Bertram and Hasaneini, 2013).

For an ape using slow brachiation, when the next overhead support is within reach from the previous one, it is possible for the animal to swing on its outstretched arm and convert all kinetic energy to potential energy. This motion is similar to a standard pendulum, except the pivot point moves with each swing. Because the vertical motion of the CoM spontaneously stops its ascent at the highest point of the swing (when all kinetic energy is converted to potential energy), velocity and momentum are zero, so no collision is possible; thus, no loss will be incurred.

Under these circumstances, the inevitable exchange between kinetic and potential energy can be exploited, and the brachiator will behave much like a passive pendulum, except that alternating contacts will allow the animal to move through its environment. This motion is used, not so much to optimize the energy exchange (that will occur spontaneously in any case), but to avoid momentum loss that would occur if a different swing strategy was chosen. Pendulum-like recovery of energy, and the avoidance of collision-like loss in the exchange between handholds, require identical motions in slow brachiation so, consequently, they are complimentary and indistinguishable.

Of course, apes (particularly the lesser apes), as remarkably athletic animals, are also capable of contending with a variety of complexities of handhold opportunities, particularly that of spacing between support opportunities. In circumstances where the next handhold can be reached from the previous one, the objective will be to make the transition as smooth as possible. Siamangs, for instance, can use a variety of different strategies to accomplish this (Michilsens *et al.*, 2011, 2012), each strategy involving a relatively smooth transition and a minimization of the energy-absorbing jolt (collision).

Hylobatids are capable of very rapid brachiation, where the animals swing between supports that are further apart than their arm span. In this situation, the animals swing rapidly below each handhold, and release for a ballistic flight phase to reach the next available handhold. This gait is known as ricochetal brachiation. Distances of 2–3 meters between handholds are readily negotiated, and some observers report that these animals can clear gaps of as much as 9 m (Cannon and Leighton, 1994). This is quite remarkable dynamic behavior – consider taking a step of 2–3 m, let alone 9 m! In order to travel such horizontal distances, the animal must carry substantial vertical and horizontal speed into the take-off swing, so the threat of a large energetic loss resulting from a collision-like jolt when landing is important. Note that these collision effects need not involve "colliding" with another object, like the ground or the trunk of a tree, but can be the dynamic collisions involved with smaller discontinuities in the CoM path, just like the small jolt that can occur on a playground swing.

In order to maintain maximum momentum through the swing, a collision-like jolt must be avoided when contacting a new handhold. If the transition as the animal makes contact following the parabolic flight is tangential with the circular path of the CoM at the limit of its extended arm as it begins to swing beneath the support, then the transition is smooth, and no collision occurs (Figure 5.2). This turns out to be a critical feature of successful brachiation, and avoiding this collision-like loss determines the optimal strategies available in higher speed brachiation (Bertram *et al.*, 1999).

Although brachiation appears to be the ideal pendulum-based gait, and pendular exchanges between E_k and E_p are evident and valuable in the swing of these animals, effective brachiation depends ultimately on managing CoM velocity and avoiding circumstances in which energy is unnecessarily lost (as in small collision-like discontinuities in the CoM path – Gomes and Ruina, 2005). It is a simple dynamic concept, but some of the implications of this for the investigation of locomotion are disconcerting. For instance, collisions appear so important to brachiation that they are largely eliminated from the gait (a possibility in brachiation that results from the geometry of the interaction with the overhead superstrate, which is not as readily available to terrestrial forms moving on top of a substrate – but see Gomes and Ruina, 2011). This is disconcerting because the most important determinant of the gait dynamics in brachiation appears to be *not directly observable*. It is so important to the energetic effectiveness of brachiation that its negative effects are likely to be eliminated or

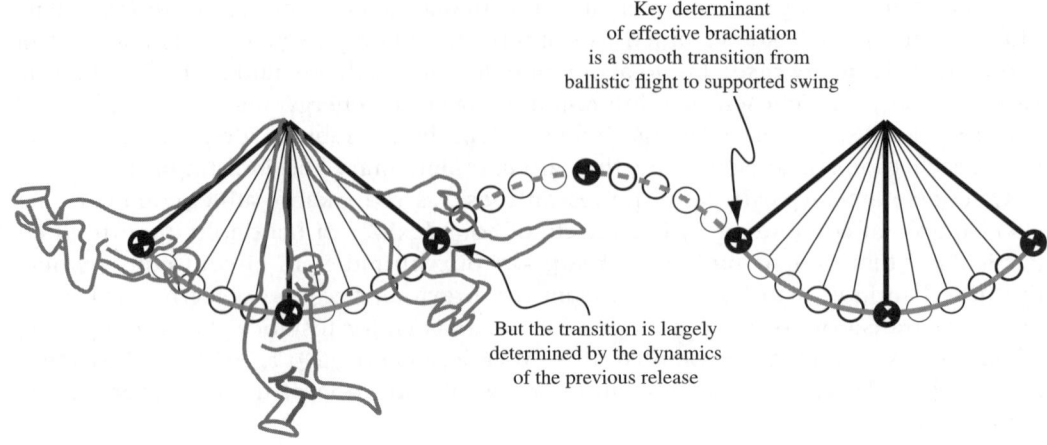

Key determinant
of effective brachiation
is a smooth transition from
ballistic flight to supported swing

But the transition is largely
determined by the dynamics
of the previous release

FIGURE 5.2

Higher speed ricochetal brachiation has a non-contact "flight" phase. Although the animal swings below the handhold much like a pendulum when in contact, the key to effective brachiation is maintaining momentum at the transition to the next handhold. The transition from the end of the ballistic flight to the beginning of the next swing needs to be smooth (tangent), or a "jolt" similar to that of the playground swing will remove momentum and cause a loss of energy.

minimized to the point of negligibility (as may be the case for a number of key factors involved in locomotion dynamics – Srinivasan, 2007).

The most natural approach to the study of brachiation (and potentially several other ways of moving on legs), starting with simply making observations of the locomotion mode as an investigative tool, yields negligible results expressly because the crucial factor determining how the motions should operate has been eradicated so that the locomotion can best occur. In order to determine the value of the movement strategy used in brachiation, it is necessary to properly model the system and evaluate the alternative ways available to move – alternatives that are not actually utilized because they are so ineffective. Substantial potential lies with this approach for the investigation of locomotion, but it requires both modeling sophistication and insight into the biological possibilities available.

As logical as the collision-based view of brachiation appears, there are two important deficiencies that the model, as stated, does not account for. One is that brachiation involves complex muscular activations and control that suggest that the coordinated motions are actively guided (Jungers and Stern, 1980; Stern *et al.*, 1980; Stern and Larson, 2002). Such evidence is not surprising, however, if the circumstances in which brachiation naturally occurs are considered. Although, on a theoretical basis, the arboreal environment provides the potential for highly effective locomotion, it is, in reality, a three-dimensionally complex environment to negotiate. In most circumstances, it is the environment that dictates step length for a brachiator, much like a person walking on unevenly spaced railroad ties or stepping-stones. The organism is forced to conform to the support opportunities available, which can require substantial muscle mediated adjustment and compensation on the part of the animal.

A second issue that conflicts with the above description of brachiation as a gait with the potential for extreme energetic effectiveness are direct metabolic measurements, where the metabolic cost of brachiation appears no different than terrestrial locomotion (Parsons and Taylor, 1977). Unfortunately, the studies that indicate this equivalence were conducted without an understanding of the dynamics determining effective brachiation. The analyses were done on spider monkeys (*Ateles* sp.), which are adept brachiators, even though the use of their prehensile tail may conflict with the precise definition of brachiation (which implies exclusive use of the forelimbs).

In order to obtain metabolic measurements through indirect calorimetry (analysis of respiratory gases), it was necessary to have the animals brachiate while maintaining a constant position relative to the gas analysis equipment. To accomplish this, a clever analog to the treadmill was developed – a rope mill, in which a loop of rope ran between pulleys at either end (like a clothes line). It is not certain what the tension in the rope mill was, but the study protocol likely did not provide the animals an opportunity to utilize the dynamics that allow the advantages of this locomotion strategy, even though they did use their pectoral limbs to move under a supporting superstrate. The conditions may well have been much like measuring human locomotion energetics when running in sand – a circumstance that does not allow the effectiveness of the human running system to be properly observed.

Even if further, more sophisticated analyses of the metabolic cost of brachiation do not indicate a substantial energetic effectiveness to the gait, this does not obviate the mechanical descriptions of the key determinants presented above. Locomotion – especially steady-state locomotion – is just one feature of the dynamics of the organism's relationship with its physical environment. As will be demonstrated in the next section, it could conceivably be possible for humans to walk with substantially reduced energetic cost. To do this, however, it would probably be necessary to sacrifice the ability to perform a number of other important activities, such as negotiating obstacles, climbing, sitting, or even standing.

The complex environment that brachiators contend with requires that they carry with them the means (neuromuscular capacity) to deal effectively with heterogeneous features they encounter daily. This added capacity will not come without increases in cost, whether used within the observed motion, or in some other aspect of the animal's struggle for survival. The more complex the demands on the organism's functional morphology, the more difficult it will be to concisely define a cost-benefit relationship between form, behavior and functional capabilities.

5.3 BIPEDAL WALKING: INVERTED PENDULUM OR INVERTED "COLLISION-LIMITING BRACHIATOR ANALOG"?

Human walking appears to be a relatively smooth progression as the individual swings the foot into place and transfers body weight to the next support limb. Relatively minor changes in forward velocity and vertical position of the CoM accompany this progression, with these motions mediated by muscular activity that stabilizes the limb and powers the movement (Rab, 1994; Waters and Morris, 1972). As mentioned above, however, it is necessary to identify what precisely is accomplished in walking before the observed details of the motions and activities can be properly interpreted. As in the analysis of brachiation, the essence of the dynamics may be captured with remarkably simple models (even if expanding those to include the complexities of the complete system is intimidating). And, also like brachiation, the management of momentum and energy in the transition between support limbs is a key feature of mechanically and metabolically cost-effective motion.

As one might imagine, quadrupedal gaits are dynamically more complex than the bipedal walking of humans. We also have a great deal of detailed information on the mechanics of human locomotion. As legged systems, however, quadrupedal and bipedal gaits share many fundamental features, and it is sometimes convenient to begin an exploration of terrestrial legged locomotion, and the gaits used to accomplish this, by understanding the less complex bipedal form.

In its simplest manifestation, bipedal walking has much in common with the sequential contacts of a rimless, spoked wheel, and it has been described as such numerous times (see Chapter 1, Figure 1.4). This extremely simple model of walking has stiff limbs, and knees are not required because the next spoke comes into position without the threat of the foot scuffing the ground on the return swing. It is easy to visualize that, if the wheel had enough spokes, the distance between contacts would be negligible, and the device would roll almost as easily as a wheel with a rim. The fewer the spokes, however, the more the wheel will bump at the contact of each spoke. Forward velocity will also go through larger changes with fewer spokes, as it accelerates forward as it passes the vertical and falls off one spoke, then decelerates with the jolt of the next spoke's contact. Also, the CoM of the system will move up and down more with fewer spokes, as it falls farther before the next contact occurs.

Together, the increased "bumpiness" of the motion sums to make forward progress more difficult. The distance between spoke contacts causes discontinuities in the path of the CoM (where the discontinuities make up the "bumpiness"). Each of these discontinuities is a collision, just as in the playground swing falling from a too-high position but, in this instance, it is a compressive collision caused by the strut-like spoke, instead of the tension collision occurring when the swing's support cords go taught (Figure 5.3). Although only a rough simulation of bipedal walking, even this simple model captures many features crucial to determining the fundamental dynamics of walking. Most importantly, it indicates the source of momentum and energy loss that inevitably occurs with each step when walking, even when the substrate is level and presents no obstacles that require active work to negotiate.

The bumpy progress of a rimless wheel moving along a smooth surface is basically equivalent to a wheel with a rim moving on a roughened surface. Consider a relatively small diameter wheel, such as that of a rollerblade or skateboard; motion on a smooth surface is easy, due to the low turning friction constructed into the bearings of the wheel. However, roughness of the surface (such as rolling over cobbled paving stones) increases the resistance to travel. Ostensibly, it would appear that friction increases, since the system will not roll so far for a given speed or push. Actually, it is the loss of momentum and energy due to the numerous small collisions that makes travel more difficult. Note, for instance, that friction *within* the system, provided by the bearings of the wheels, remains just as small on the rough surface as on the smooth, but the rolling costs increase (more energy is needed to maintain motion) because of energy losses between the wheels and the rough substrate.

The impression of increased internal friction in the skate wheel on a roughened surface, even though the cost actually arises from a loss of energy to the external

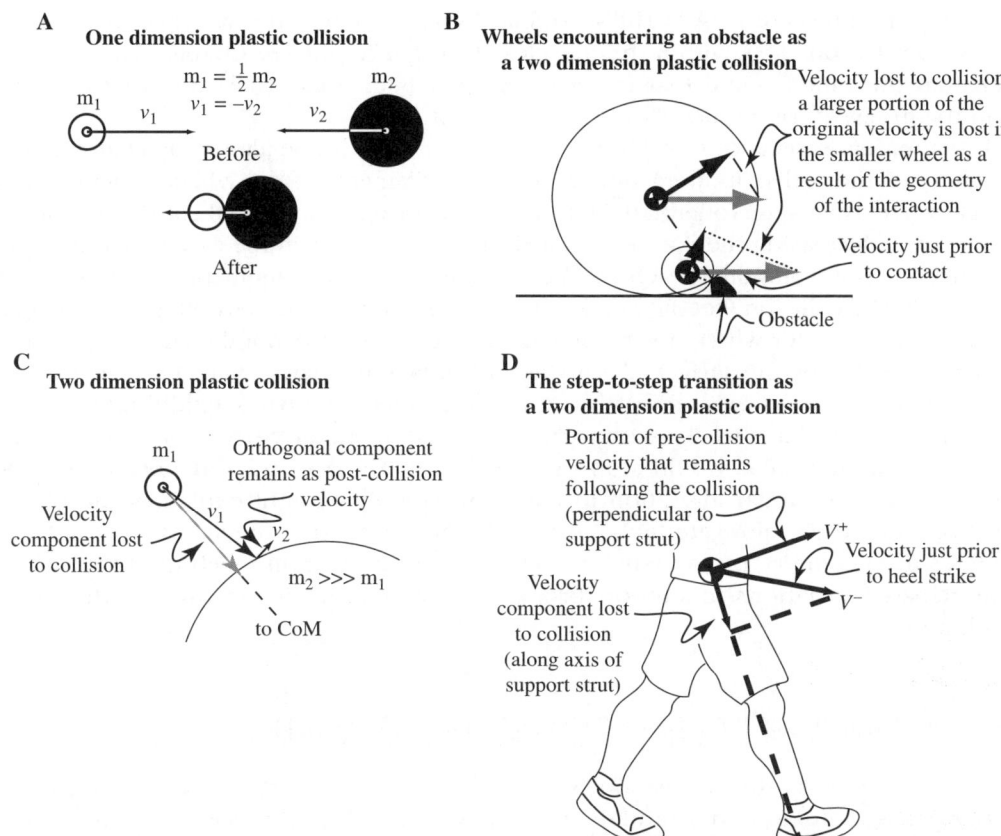

FIGURE 5.3

(**A**) Linear momentum balance in a one-dimensional plastic collision. Linear momentum of a body is the product of its mass and velocity. The total momentum remains constant through a collision. After a plastic collision, the two masses move with the same velocity, which is determined as the mathematical sum of the two momentums involved (one positive, the other negative). The kinetic energy of the system is reduced, however, because the collision will reduce the speed of the combined masses. (**B**) Momentum balance in a two-dimensional frictionless plastic collision. The small mass m_1 collides at a velocity v_1 with a much larger stationary mass m_2. The velocity component of m_1 in the direction of its common surface normal with m_2 will be lost to the collision, but the velocity component tangential to the surfaces is conserved. (**C**) The effect of collision geometry on momentum loss, considering wheels of different diameter. A given obstacle causes a much larger momentum loss for a small diameter wheel than for a larger one, due to the geometry of the collision. (**D**) Two-dimensional collision dynamics translated to human walking. Just prior to foot contact, the CoM trajectory will be perpendicular to the previous stance limb (on which it is falling forward and downward, as an inverted pendulum). As the new stance limb makes contact, the CoM will be diverted (a collision-like event). That component in line with the new stance limb will be lost to the collision, while that component perpendicular will remain, so that the CoM proceeds in an upward and forward direction, continuing the step cycle.

contact, is reminiscent of A.V. Hill's original hypothesis that the work of locomotion arises from viscous loss within the muscles (Hill, 1927; and see Chapter 1). It will be seen that the fundamental loss in terrestrial gait is also external, and arises largely from the inherent "bumpiness" of legged locomotion.

The apparent resistance to rolling on a rough surface can be altered by changing the geometry of the collision interaction. Making the diameter of the wheel larger, as in a bicycle wheel versus the rollerblade wheel, will increase the tangential component of the momentum, decreasing the component acting directly between the axle (CoM) and the obstruction. This component defines the portion of velocity (momentum and energy) involved in the collision (see Figure 5.3C). There is no particular advantage to having a larger wheel diameter when moving on a truly smooth surface, but the larger the irregularities presented by the surface, the greater advantage to a large diameter wheel.

This is taken to an extreme with the modern "monster" truck exhibitions, where the objective is to have trucks with extremely large diameter wheels pass over a variety of obstructions that would stop other vehicles. Usherwood and Bertram (2003) suggested that a similar alteration in collision geometry might explain some of the function behind the elongate limbs of the gibbon. Increasing the swing arc diameter changes the geometry of any tension collisions resulting from meeting obstructions, unanticipated variations in support properties, or inaccuracies in flight path during brachiation.

5.4 BASIC DYNAMICS OF THE STEP-TO-STEP TRANSITION IN BIPEDAL WALKING

Because the step-to-step transition is a key determinant of the mechanical cost in walking, it is useful to understand the various ways in which it is possible to transition from one support limb to the next, and to consider the optimal manner in which this can be accomplished in walking. To illustrate the dynamics of the transition, a 19th century vector visualization tool known as the hodograph is useful (Hamilton, 1847). A hodograph tracks the motion of the CoM velocity vector tip as it varies in response to an applied impulse (Adamcyzk and Kuo, 2009). The position of any point on the hodograph plot then has information on both the direction and magnitude of the vector. Three versions of the basic step-to-step transition options are depicted using a version of the hodograph in Figure 5.4.

In Figure 5.4A, the dynamics of a simple limb contact is depicted, showing the effect of the impact of a rigid stance limb on the motion of the CoM (comparable to a classic, stiff limb "compass" gait – Inman *et al.*, 1981). The center of mass velocity vector just prior to touchdown of the new stance limb is shown as a dark arrow, V^-, (falling forward as the new limb makes contact). At the instant of limb contact, the CoM velocity vector tip will traverse the path "*a*" (dashed) to reach its final orientation and magnitude after the collision (V^+). Note that path "*a*" is parallel to the contact limb and perpendicular to V^+; the contact limb determines the orientation of the interaction between the CoM and the substrate, and the collision cannot involve any component of the CoM velocity/momentum/energy perpendicular to that interaction. Due to the collision that results from the interaction (limb contact), the magnitude of V^+ is diminished, because of the momentum (and energy) absorbed along the axis of the contacting limb (strut). The remaining velocity (V^+) is perpendicular to the contact limb, because velocity

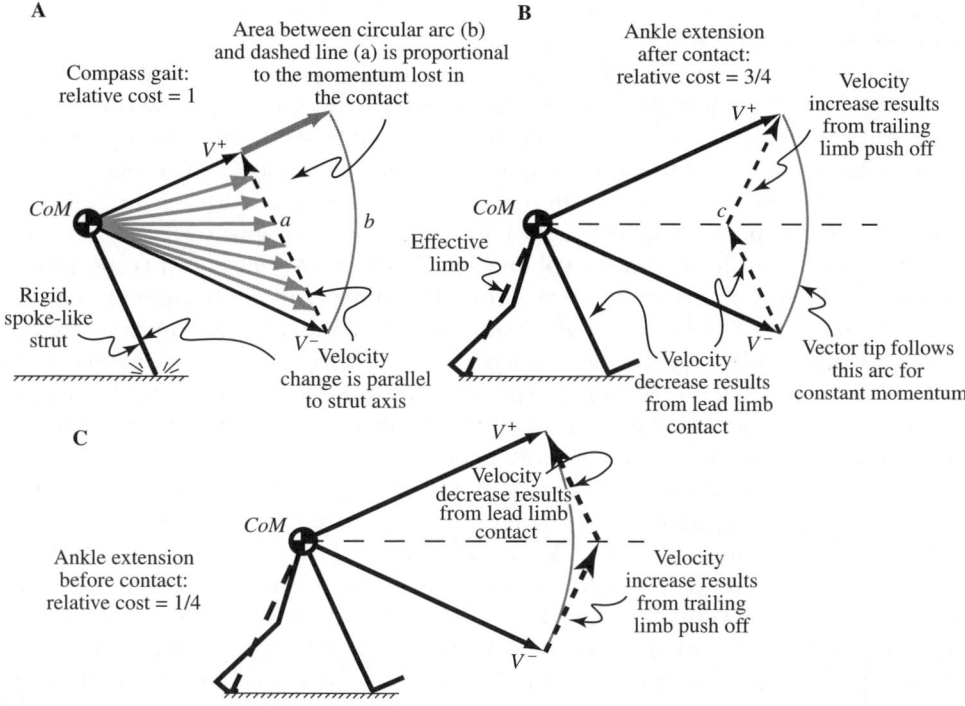

FIGURE 5.4

Hodograph-style depictions of center of mass velocity vector changes as a result of specific collision circumstances (after Ruina *et al.*, 2005). (**A**) The simple compass gait, in which the velocity of the CoM just prior to foot contact (V^-) changes (instantaneously) through path (*a*) to leave (V^+), following contact. The collision results in a loss of momentum (the difference in magnitude between V^- and V^+). This is compared to a constant momentum magnitude path (*b*). Note that path (*a*) is parallel to the contact limb and the velocity remaining after contact (V^+) is perpendicular, though much reduced (the shaded arrow shows how much velocity must be increased in order to return to the original velocity magnitude). (**B**) The compass gait model, extended to include an impulse from the previous stance limb that is applied following contact of the new stance limb. The momentum loss in the first half of contact is identical with that of A (parallel to new stance limb), but momentum is added by the impulse in the second half (parallel to effective former stance limb), resulting in the jointed path (*c*). The two applied impulses – one absorptive and the other generative – result in V^+ maintaining the same magnitude as V^-, but with a direction that continues progression. Much less total loss occurs, compared to the simple compass gait in A. (**C**) Changing the sequence of absorptive and generative impulses by having the former stance limb extend prior to new stance contact substantially reduces the collision loss. This is because the former limb impulse changes the direction of the CoM velocity vector, so that the total energy involved in the two "collisions" (one generative and one absorptive) is reduced.

perpendicular to the axis of the contact limb will not be affected by the collision, so will remain in spite of the contact transition.

If motion is to continue at a constant average forward speed (steady state locomotion), energy must be added to increase the velocity magnitude (gray arrow extending V^+) following the collision loss. For constant (average) speed, the velocity magnitude at the beginning and end of the transition must be equivalent, but the velocity orientation will be altered by the contact. This then allows for the next portion of the stride – the pendular exchange portion, as the CoM passively moves up and over the new stance limb. This velocity addition is not explained in Figure 5.4A, but could come passively from potential energy if this system is walking downhill (as in passive dynamic walking machines – McGeer, 1990; Collins *et al.*, 2001), or from some other means not considered in this simplistic model (turning on a jet pack at the proper time, for instance). Even with the retention of the momentum perpendicular to the limb, following the transition to the new stance limb, substantial loss is inevitable, due to the geometry of the CoM velocity-ground reaction force vector interaction.

Figure 5.4A also shows a path "*b*" – a hypothetical path where the velocity vector maintains its original magnitude as it changes direction through stance. Force application would be needed to support the CoM and alter its direction of travel, but velocity is neither lost nor added. This would be accomplished, in the ideal case, by having a smooth transition between the pendular fall (that ends with V^-) and the transition between support limbs (that ends with V^+ of magnitude equivalent to V^-). This is theoretically possible in brachiation, where contact is facilitated by the overhead support, and the ballistic path at the end of a flight trajectory could be precisely tangential to the swing arc of the next handhold (the vector of the support force and the CoM velocity are perpendicular). This is not possible in over-ground locomotion (where such a tangent cannot be maintained over the entire transition – Gomes and Ruina, 2005). Although it is not possible to generate in reality, the path "*b*" does serve to illustrate the absolute limit of an effective transition in walking.

The area between paths "*a*" and "*b*" is proportional to the velocity lost in the compass gait transition, and can be used to evaluate the effectiveness of alternative transition strategies. Even a simple, stiff, compass gait transition results in some remaining velocity (V^+), even though a substantial proportion of the original momentum must be lost.

The influence of the previous stance limb can be added to the simple spoke-like (compass gait) model of Figure 5.4A in order to provide a more realistic mechanism for replacing the momentum and energy lost in the transition between support limbs. The previous, or "trailing" limb in the context of the transition between supports, acts as an extensible strut (by extending the joints – particularly the ankle – and increasing the length of the limb) following contact by the new stance limb (Figure 5.4B). In this case, the velocity of the system will be subject to the effects of two separate collisions (applied impulses) – one absorptive, from impact of the new contact limb – and one generative, through extension of the former stance limb.

It is common, when considering the dynamics of inanimate objects, to acknowledge only absorptive collisions but, in biological systems, the muscles are able to both absorb and generate force. Recall that a collision is defined as a discontinuity in the path of the CoM, so a collision-like event can cause this discontinuity, either by passively absorbing energy, or by actively applying an impulse that increases the energy state of the system.

Just as in path "*a*" of the spoke-limbed compass gait, the velocity change resulting from the initial contact (collision) will parallel the orientation of the new stance limb (the collision acts along the axis of the effective stance limb). This will be followed by the second "generative" collision that parallels the effective orientation of the previous stance limb as it adds velocity back to the CoM (path "*c*"). Calculation of the energy balance, in this circumstance, indicates that this strategy would result in only 75% of the energetic loss, compared to that lost through the simple, stiff limb compass gait of Figure 5.4A (Ruina *et al.*, 2005).

The simplest and most obvious way to replace the velocity lost to the collision that results from contact of the new stance limb is to extend the previous stance limb. As described above, this replaces the energy that was lost and moves the CoM up and over the new stance limb. Counter-intuitively, however, it is actually far better to extend the previous limb and add thrust just *prior* to contact by the next stance limb. This is an illustration of how important it is to consider the whole-system dynamics when evaluating locomotion on limbs. Understanding why this is the case is an important insight into the role of collisions and mechanical energy management in legged loco-motion. It indicates the critical influence of relatively subtle changes in the geometry of the contact interaction between the organism and the support surface on which it moves (where, at its most fundamental level, the "geometry" refers to the relationship between the CoM velocity and ground reaction force vectors).

If the previous stance limb produces a thrust prior to contact by the next stance limb (Figure 5.4C), then the energy lost in the transition between limbs is only 25% of that lost in the simple spoke-like compass gait. This is one-third of the energy lost in 5.4B, where identical contacts are made, just reversed in sequence.

How could such a simple change in the sequence of two apparently equivalent events make such a profound difference in the mechanical cost of locomotion? The answer lies in analyzing the consequences of the geometry of the interaction between the CoM velocity vector and the strut-like limb contact (which determines the ground reaction force direction). Extension of the previous stance limb (application of a ground reaction force in line with the mechanical axis of the limb by increasing its length) prior to contact by the next stance limb diverts the CoM velocity vector from downward-forward to forward (from V^- to horizontal in Figure 5.4C). This is a genera-tive collision that increases the velocity of the CoM (note that the path of the velocity vector moves beyond the constant velocity arc along a path parallel to the mechanical axis of the previous stance limb, where the thrust originates).

When the next stance limb does make contact, however, it now diverts the CoM path far less than it would have if it made contact first, because it diverts the CoM path from horizontal-forward to upward and forward (from horizontal to V^+ – Figure 5.4C). Although CoM velocity again decreases along a path parallel to the mechanical axis of the new stance limb, the velocity change involved is substantially less than that of Figure 5.4B (area on diagram between constant velocity arc and velocity changes determined by limb contact). This simple change in geometry of the transition results in maintaining a substantial proportion of the available velocity, and energy, through to the next portion of the stride (McGeer, 1993; Kuo, 2002; Ruina *et al.*, 2005).

Through the use of a generative collision (push-off) by the previous stance limb, at a time when its orientation is perpendicular to the CoM velocity vector (V^-), com-bined with an absorptive collision (heel-strike) when the vector has been changed to horizontal, two CoM deflections are created during the transition, each of which is substantially smaller than the total deflection in the required transition. The energetic loss resulting from a collision is proportional to the square of the angle change it causes (Ruina *et al.*, 2005). Thus, even though this transition involves two collisions, their reduced angular deflection means that each only involves the square root of the energy of the original collision (i.e. the spoke-like compass gait). The sequencing of these collisions has a profound effect on the total momentum and energy lost in the transition between support limbs.

It has been well documented that, in normal human walking, the main plantar flexors crossing the ankle – the gastrocnemius and soleus muscles – are active prior to touchdown of the next stance limb (Anderson and Pandy, 2001; Rab, 1994). Such evidence fits well with the model described in Figure 5.4C and analyzed extensively by Kuo (2002). This indicates that the relative timing of triceps surae activity could be a definitive innovation in the development of the striding form of human bipedal walking. It is interesting to note that adding this energy through extension of the ankle does not appear to be equivalent to the same energy added through a torque applied to the hip (McGeer, 1993; Kuo, 2002), where energy is added through a lever-like motion of the limb. Such a mechanical difference is an indication of how strut-like action of limbs can be advantageous, and could be a meaningful insight in distinguishing functional capabilities in fossil post-crania, where hip extensor function useful in arboreal climbing would trade-off against the ankle function of a fully adapted terrestrial walker.

5.5 SUBTLE DYNAMICS OF THE STEP-TO-STEP TRANSITION IN BIPEDAL WALKING AND RUNNING

Bipedal walking and running have some important features in common. In both, progression is made through alternating limb contacts, and it is during contact that the path of the CoM is actively controlled. The limb functions to support the body against the action of gravity, and can do work to maintain progression in the face of energetic losses that result from moving across the supporting surface. Both gaits have a portion of the stride cycle that incurs relatively low energetic cost – the inverted pendulum motion of the stance limb in walking and the non-contact "flight" phase of running. The low-cost portion of the step alternates with a portion of the stride cycle that can involve substantial mechanical energy loss – the step-to-step transition in walking and the single limb stance in running. Understanding walking and running locomotion strategies depends on recognizing the interplay between energetic losses, costs arising from actively controlling the CoM and body motions, and how changes in either the low or high-cost portions of the stride influence the energetic investment required to produce locomotion.

Walking and running are often portrayed as being substantially different from each other. For instance, in human walking, at least one foot is always in contact with the substrate while, in human running, there is (usually) a portion of the stride where neither foot contacts the ground (see Chapter 4, Figure 4.4). One apparently fundamental difference between walking and running is the position of the CoM at midstance (when it is directly over the stance limb in a biped, or over the functional limb – the net effect of all limbs on the ground at one time in a quadruped). The CoM is at its *highest* point as it swings over the extended stance limb in walking, and at its *lowest* point as the stance limb is compressed in the running stance.

The relationship between the vertical position of the CoM and stance limb orientation is, of course, a true description of what occurs in each of these gaits. However, it might well be asked if this is important to understanding what is accomplished in each of these gaits, or explaining why each exists as it does? In both gaits, when the CoM is at its lowest point in the gait cycle, its path transitions, from travelling forward but downward, to forward and upward. The transition from downward to upward is mediated by double contact during transition between stance limbs in walking, and by a single limb in running. The alternate transition in the gait cycle, from the CoM moving forward and upward to forward and downward, is not much mediated by the limbs, but occurs as a result of the passive effects of gravity.

Many factors contribute to the mechanical cost of locomotion: support; relative motion of the limbs and trunk; stability of the whole system (preventing falling); and stability of the jointed skeleton (preventing collapse). Under most circumstances in legged locomotion, a large portion of the mechanical cost can be attributed to mechanical energy losses that occur in diverting the path of the CoM, and both walking and running have this in common. These two gaits can, then, be seen as using alternate limbs to progress forward through cyclic motions that have a phase of low mechanical cost, interspersed between phases of high mechanical cost. Of value in this perspective is analyzing how the low mechanical cost phase can be emphasized, and the high mechanical cost phase de-emphasized – essentially, the strategies employed to minimize the mechanical cost of moving the CoM across a solid substrate on limbs.

Bipedal running has largely been characterized as a "bouncing" gait, where much of the momentum loss that comes from decelerating the body in both the downward and forward direction (with concomitant kinetic energy loss) is converted to strain potential energy in the muscles and connective tissue of the limb and is returned through recoil of these elastic structures (Cavagna *et al.*, 1964; and see Chapter 4). In its simplest form, then, running is likened to a mass supported by an extensible strut having spring-like properties (Blickhan, 1989; Seyfarth *et al.*, 2002). Such characterizations are referred to variously as the mass-spring, spring-loaded inverted pendulum (SLIP) or "pogo stick" models.

For bipedal running and quadrupedal trotting, some form of mass-spring system appears to be a very good model of support limb function and whole-body motion. But, one might ask, is the elastic recoil, and the energetic return involved, a critical component of the function in running? If so, how much value does it provide? And what of pendular exchange in walking? Is it a critical feature that must be "designed" into the system, or simply one aspect among others that combine to allow the walking gait? To understand what makes these gaits what they are, it is important to identify the factors involved with determining the effectiveness of each gait, and the strategies employed to implement them.

Momentum and energy loss in the walk could be essentially eliminated by having the CoM follow a level path, where it does not move up or down as it progresses forward. There will be no discontinuity in its path, and no opportunity for a collision interaction between the CoM and the substrate (even though some collision loss might occur when the mass of the foot or lower limb segments interact with the substrate, although these can be eliminated if the velocity of the foot is zero at contact). Such a smooth and level motion is much like the path of a system rolling on wheels over a smooth surface.

To accomplish this on legs, however, it is necessary for the limb to flex and extend substantially over the course of the contact, so that the CoM does not vault over the stance limb. Although collision loss *per se* might be eliminated, the muscles of the limb need to produce more work to control this action than would have been required to replace the original collision loss, had some portion of the collisions been allowed. This has been determined analytically (Srinivasan and Ruina, 2006) and confirmed experimentally in the bipedal human (Ortega and Farley, 2005; Gordon *et al.*, 2009), where the bent limb walk with little or no vertical oscillation of the center of mass is found to be metabolically strenuous – much more so than our preferred walking gait.

The above description points to two important factors involved with legged locomotion that trade off against one another to provide functionally cost-effective locomotion. One is velocity loss resulting from CoM deflection caused by the strut-like stance limb. The other is the active work performed by the stance limb. This latter is required to flex and extend the limb as the CoM passes over the support contact point. In order for the gait to be energetically effective, the stride will represent a good compromise between these competing factors.

Alternatively, collision loss could be virtually eliminated (the CoM could move on a nearly level path across the substrate) if steps were short enough. With extremely short steps, the limb "spokes" would be so close together that it would emulate a rimmed wheel. There would be virtually no downward "fall" of the CoM between contacts, so no energy-dissipating redirection of the CoM would be necessary. Although this would eliminate CoM collision loss, it would require substantial investment in swing limb work to move the next limb into place so often (and so rapidly, if any reasonable forward speed is to be maintained – Doke and Kuo, 2007).

The most fundamental energetic problems and opportunities of legged locomotion can, therefore, now be identified. The major loss is derived from the portion of the gait, where the CoM transitions from downward to upward, regardless of gait. At the transition from downward to upward motion of the CoM (mediated by the transition between limbs for the walk, or for limb contact in the run), the trade-off is between energy loss due to collision-like diversion of the CoM path (where the limb acts as a rigid strut) versus muscle-mediated stance leg work (which applies an appropriately timed collision-reducing thrust, and/or allows the limb to flex and extend) and swing limb work (which influences step length optimization).

These high "cost" transitions are interspersed between low-cost (relatively passive) portions of the gait – the inverted pendulum-like motion of the CoM over the support limb in the walk, or the non-contact ballistic "flight" phase of the run. Cost-effective legged locomotion, then, involves optimizing between the low-cost portion of the step and the high-cost portion. This optimization will itself depend on the implementation of movement strategies that mitigate the energetic costs involved with the high-cost transition. This must all occur within the constraints of the physical limitations of the organism, and any complexities imposed by the physical environment. For instance, maximum speed possible in the walking gait is determined by gravitational and centripetal accelerations at the end of limb contact (Usherwood, 2005; and see Chapter 6), a limit determined by limb length, the mass of the organism, and the mass of the planet it moves on (i.e. gravitational attraction). Organic features of the organism, such as maximum muscle and joint forces, also set constraints on the strategies possible. Having identified the fundamental problems of legged locomotion in a general sense, it is possible to evaluate the more subtle strategies for optimizing gait.

In the bipedal walk of humans, the plantigrade feet allow the orientation of the effective limb to move during the course of contact (Ruina *et al.*, 2005). Since the momentum loss associated with the step-to-step transition in walking is largely dependent on the geometry of the strut function of the limb, shifting contact from the heel to the toe of the foot provides a means to adjust and manage the geometric interaction of the limb and the CoM trajectory (Figure 5.5). In the analysis of low-cost walking strategies available for a simple model, Ruina *et al.* (2005) calculated that the optimal foot length (heel to functional push-off point, the metatarsal-phalangeal joint) for a human-like bipedal walker should be one-quarter of the stride length.

Similarly, investing some active leg work (length change) for a reduction in collision-based momentum loss can result in a net decrease in locomotion cost (Ruina *et al.*, 2005; Srinivasan and Ruina, 2006). This would account for some of the compliance observed in the human limb during walking (see Chapters 6 and 7). The optimization of this depends on features determined both morphologically (body and limb mass,

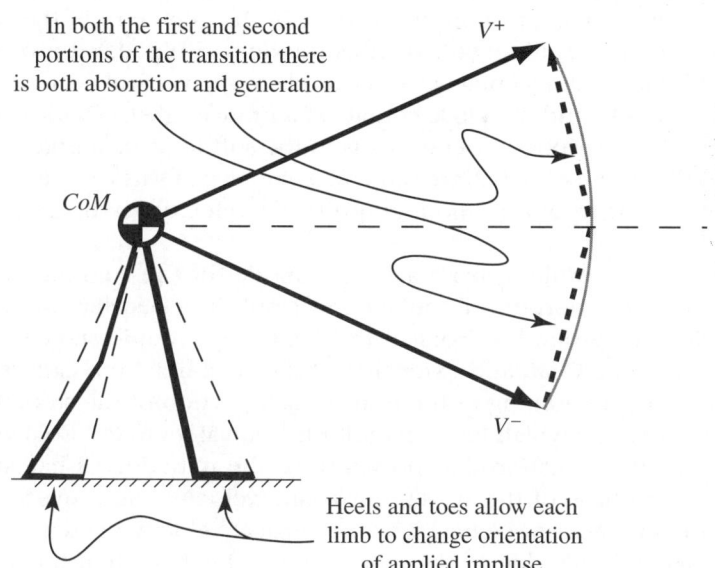

In both the first and second portions of the transition there is both absorption and generation

V^+

CoM

V^-

Heels and toes allow each limb to change orientation of applied impluse

Feet and spring-like leg muscles:
cost = 1/8
(relative to simple compass gait)

FIGURE 5.5

A hodograph illustration of the momentum changes possible in the walk when exploiting the subtleties of plantigrade feet and spring-like (force proportional to length change) limbs (after Ruina *et al.*, 2005).

limb and step length, etc.) and physiologically (cost of positive and negative muscle work, investment in rigid supporting tissues, etc.).

The movement strategies that mitigate collision loss in the run differ from those of the walk. In the run, energy loss associated with limb contact could, in theory, be eliminated by having the limb contact with the ground with a perfectly vertical orientation, and with the CoM velocity horizontally directed. In this circumstance, the limb is perpendicular to both the surface and the CoM velocity vector, meaning that there would be no collision or loss associated with contact of the limb and motion of the body across the surface. However, eliminating the high-cost portion of the step (the redirection of the CoM path) would also have the consequence of eliminating the low-cost portion – if there is no vertical motion of the CoM velocity vector, there can be no flight phase. With no flight phase, steps would be short, and the change from one limb to the next would have to occur very rapidly. The physiological limitations of moving the limbs rapidly into position for extremely short steps precludes eliminating the collision cost entirely. The optimization problem becomes reducing the energy loss and leg work associated with contact in a manner that emphasizes the low-cost (non-contact) portion of the step, while not allowing the high-cost portion of the step to overrule the advantages provided by the low-cost portion.

The ballistic path of the non-contact portion of the step, then, determines that some energy will be lost during the contact. The flight phase of the running step also means that the strategy used in walking, where sequencing a generative impulse prior to the absorptive impulse of contact, in order to make the geometry of the contact favorable, is not available in a running gait, such as the human bipedal run or the quadrupedal trot. The limb must make contact at some angle to the vertical (where the foot is placed in front of the CoM), and then lose contact at a finite angle beyond vertical. The contact must begin with absorption of energy (as in the stiff-legged, simple compass gait walk depicted in Figure 5.4A). What are the consequences of this on the velocity changes of the contact – what would the hodograph (CoM velocity vector tip path) of a bipedal run look like?

The hodograph resulting from a simple model of the running stride is shown in Figure 5.6. The first portion of contact will result in a deceleration as the initial foot contact strikes the ground in front of the CoM. Beyond mid-stance, the contact point will move under the CoM, and extension of the strut-like limb causes a reacceleration of the CoM. At the beginning of the non-contact portion of the step, the CoM achieves the velocity vector V^+, which has a magnitude equivalent to the CoM velocity at contact (V^-), but with an orientation that provides for the initiation of ballistic flight. During the non-contact phase of the stride, horizontal velocity will remain virtually constant (the largest component of velocity), while vertical velocity decreases and increases as the CoM rises and falls due to the action of gravity. Note that the velocity vector tip defining the hodograph path moves in a clockwise direction in the run. This is fundamentally different from that of a walking gait, where the best collision mitigation strategy drives it in a counter-clockwise direction (also see Chapter 6, Figure 6.3).

In comparison to the walk, the velocity changes involved in a running step will be quite large. There are no means available in the run to initiate contact with a horizontal acceleration (which would alter the path of the CoM velocity vector, and mitigate the loss from the initial contact by the strut-like stance limb). This loss appears acceptable, however, due to minimized leg work and substantial energetic "payback" provided by

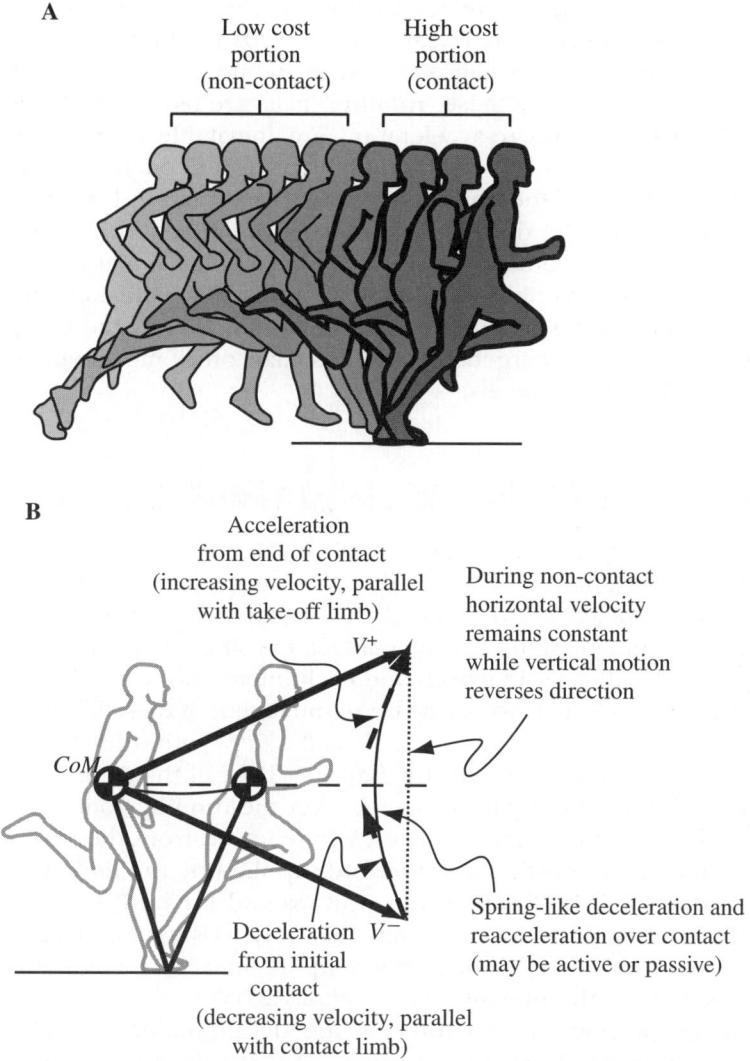

FIGURE 5.6

(**A**) The bipedal run (and other running gaits, such as the quadrupedal trot) has a high-cost portion, where the CoM is actively diverted from downward to upward (impulse supplied by the limb), and a low-cost portion, where the CoM is passively diverted from upward to downward (impulse supplied by gravity). (**B**) A hodograph of the running contact. At initial contact, the limb acts as a strut that decreases velocity. As the limb realigns with the CoM, it can generate thrust. If the deflection of the limb is proportional to force magnitude, the CoM will travel in the most beneficial path, even if no strain energy exchange occurs. This path is equivalent to that of a system deflecting an elastic support, so elastic strain (and energy return) through passively deforming tissues is a low-cost, complimentary way of providing this motion.

the distance covered during the relatively low-cost, non-contact flight phase of the stride (Srinivasan and Ruina, 2006; Hasaneini *et al.*, 2013). The energetic advantages of the gait are increased if more distance is covered while in the low-cost flight portion of the gait, contributing to why these "running" gaits are used at higher speeds.

Although deceleration and re-acceleration are inevitable during the contact, leg work is reduced by using a relatively stiff limb that makes contact at an acute angle to the ground. The stiffness of the limb limits work performed in length changes of the leg, and also means that the duration of contact is relatively brief. The near-upright orientation of the limb at contact ensures a geometry that limits the involvement of horizontal velocity. Thus, in spite of substantial impulse-like collision interaction in the run, the effect of these collisions on the horizontal momentum and energy changes is tolerable – that is, the net energetic cost is less than other movement strategies that could be employed at these speeds.

5.6   PSEUDO-ELASTIC MOTION AND TRUE ELASTIC RETURN IN RUNNING GAITS

Although the human run involves a single limb striking the ground and mediating the motion of the body through the contact, there are strategies available to minimize the energy loss during contact and to optimize the way that the stance limb alters the path of the CoM. By changing length over the contact, the single limb acts equivalently to a series of limbs, each differing in length and each interacting with the body mass over an instant of the contact. What would the optimum length of these limbs be to minimize the momentum loss over the contact and deflect the CoM path from downward-forward to upward-forward in preparation for the ballistic flight of the step?

If the manner that the CoM path is altered over the running contact is optimized, a minimum amount of momentum and energy is removed from the system. If support limb length change is proportional to the force applied by the limb, the transition of the CoM from downward and forward to upward and forward will be as smooth as possible, and the minimum velocity change will occur. Using a deflection of this type can be considered equivalent to using the support limb to create a curved "surface" that deflects the CoM path optimally (Ruina *et al.*, 2005).

Length change proportional to force is the definition of a mechanical spring. However, the advantages of using the support limb in this manner do not depend on the storage and return of strain energy directly, but simply upon the manner in which the path traveled by the CoM is affected by the supporting limb. So, to minimize the energy loss during the stance of a running step, it is necessary for the stance limb to *act* like a spring, even if it does not have any true spring-like energy storage and return. The change of length required of the limb can be controlled by active muscle length changes of the jointed limb, with negative work in the absorptive portion of the contact, and positive work in the generative portion. Even in this actively mediated case, the combined major loss and costs (collision loss, stance leg work and swing leg work) of this contact strategy results in a reduction of the total cost (Srinivasan and Ruina, 2006).

It should be stated unequivocally, that this description of CoM motion in a running contact does *not* preclude the advantages of using passive energy storage and return springs in the limbs of people and animals. Rather, active muscle-mediated spring-like motion and passive strain energy storage motion are co-incident. As a consequence, passive spring storage and return can compliment the energetic advantages of using limbs to guide the CoM through the most energetically advantageous path.

This analysis substantially alters our consideration of the role of spring-like elasticity in running. The well-documented spring-like behavior of the CoM-substrate interaction has previously been considered to exist as the only means to reduce the cost of bounce-like running gaits (Cavagna et al., 1964; Blickhan and Full, 1993; Taylor, 1994). Instead, mechanically effective limb use in the run requires spring-like motion of the limbs, whether this spring-like function is actively or passively generated. In order to distinguish the spring-like deformation of the limb, predicted through consideration of collision-like deflection of the body mass, from the conventionally inferred strain energy exchange, this type of deformation of the limb (during contact) has been referred to as "pseudo-elasticity" (Ruina et al., 2005).

Passive, spring-like storage and return of strain energy does occur in the limbs of mammals (Alexander, 1984; Roberts, 2002). This finding does not contradict the concept of managing the CoM path during the contact of a running stride (trot or bipedal bounce), or the value of pseudo-elastic behavior of the limb. Analysis of the momentum balance over the course of contact requires load-proportional deformation of the stance limb, and the energetic advantage is maintained, whether the load-deformation relationship is determined actively through muscle length changes, or passively through connective tissue changes. Passive strain energy storage and return is synergistic with the requirements of CoM energy management during the running step. Given this synergy, it is not surprising that running organisms (whether bipedal or quadrupedal) have exploited the energetic advantages of passive and near-passive elastic tissues.

It must be recognized, however, that the running gaits, and CoM motions, are functionally effective with or without passively elastic tissues, making their origin eminently more sensible from an adaptation point of view. If the primary function of the bounce-like run were to exploit elastic strain energy storage and return, development of the gait would be dependent on the serendipity of an organism developing appropriate elastic tendons, then coordinating movement that would exploit their properties. Alternatively, the value of pseudo-elastic limb behavior provides a mechanism for a cost-effective fast-moving running gait – a mechanism that could later take advantage of adaptive modification of the passive elastic properties of existing connective tissues to enhance its effectiveness.

5.7 MANAGING CoM MOTION IN QUADRUPEDAL GAITS

Effective locomotion on limbs is largely an issue of managing the interaction of the mass with its support surface, while minimizing the work that the supporting limbs have to contribute to effect this management. The strategies that provide effective energy management are depicted relatively easily in a biped model (and even more so for brachiation). However, the same principles apply to the quadrupedal case, albeit with more strategies available due to a greater number of limbs. For this reason, the quadrupedal gaits are poorly understood, particularly at a quantitative level. Some qualitative insights have arisen recently, which will, hopefully, form the basis of analyses that will eventually lead to more complete understanding of the dynamic opportunities and constraints that influence quadrupedal locomotion and the morphological features it depends upon.

5.7.1 Walk

As in the bipedal walk, the quadrupedal walk involves diverting the CoM as it falls forward to contact of the next (pectoral) stance limb. However, at this point in the gait, both the previous pectoral stance limb and the ipsilateral pelvic limb are available to provide forward thrust to shift the direction of the CoM to a more horizontal path, reducing collision loss (Figure 5.7). This strategy is equivalent to thrust from the previous stance limb in bipedal walking, preceding the next stance limb contact.

As the CoM is lifted over the forelimb strut in the quadrupedal walk, the contralateral rear limb contact is nearly vertical, moving through the relatively low-cost portion of its motion. As the next forelimb falls forward toward its contact, the contralateral forelimb and ipsilateral hindlimb are again available to add thrust, which shifts the CoM velocity vector to a more horizontal orientation, reducing collision loss as the new stance limb begins to divert the CoM. Like the bipedal walk, the quadrupedal walk is a gait strategy that utilizes the limbs largely as struts, with the main objective of limb sequencing to reduce energy loss at the transition between stance limbs. This is accomplished by maintaining the CoM velocity vector as perpendicular to the net ground reaction force as functionally possible (Lee *et al.*, 2011). With four limbs available to mediate the contact, it is possible for the quadruped to maintain a much more consistent vertical position of the CoM, and the "bumpiness" of the quadrupedal walk is much reduced from that of the human biped (hence, the apparent reduction in energy "recovery" in the quadrupeds, relative to human bipedal walking).

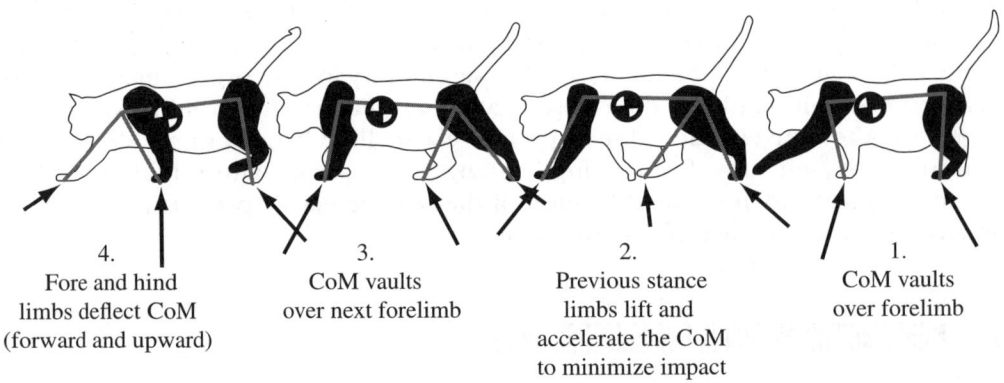

4.
Fore and hind limbs deflect CoM (forward and upward)

3.
CoM vaults over next forelimb

2.
Previous stance limbs lift and accelerate the CoM to minimize impact

1.
CoM vaults over forelimb

FIGURE 5.7

Tracings of a cat walking across a series of evenly spaced pedestals. The pedestal spacing is further than the animal's normal step length, which causes an exaggeration of the normally subtle changes in CoM height. Ground reaction force on each limb is indicated. The net ground reaction force is maintained as close to perpendicular with the CoM velocity vector as possible, so that collision-like loss is minimized (see Lee *et al.*, 2011).

5.7.2 Trot

Walking speed is increased largely through increased step length. At higher walking speeds, the geometry of the transition between limbs makes substantial momentum and energy loss inevitable, and walking becomes an ineffective gait. In parallel with the bipedal walk, higher speeds in the quadrupedal walk mean that the previous stance limb(s) will not be physically capable of maintaining contact at the end of contact, when thrust is needed to shift the CoM path (see Chapter 6). Since large energy losses occur at these speeds, the most "energy cost-effective" strategy changes as it did for the biped, and the trot becomes the more effective gait strategy.

The trot is directly parallel to the bipedal run – relatively stiff, strut-like limbs contact as the CoM falls from a low-cost, non-contact phase of the step. Energy loss during contact is managed through elastic-like deformation of the support limbs, now involving a diagonal pair of front and rear contacts (or ipsilateral, in the case of the pace). This provides opportunity for the adaptation of highly effective, energy-saving elastic structures in the distal limbs of those species that exploit this gait for cost-effective travel at moderately fast speeds.

The quadrupedal trot is characterized by a vanishingly small non-contact phase in some species. In the domestic dog, for instance, there can even be substantial overlap of diagonal pairs, removing entirely the non-contact phase of the gait (Lee *et al.*, 2002). The relatively small non-contact phase of some quadrupedal trots differs from the case of the bipedal human run, where a substantial non-contact phase occurs and contributes meaningfully to the energetics of the gait (see Chapter 4, Figure 4.4). The quadruped has the advantage of supporting the front and rear of the body simultaneously, allowing the CoM trajectory to be relatively flat in the trot.

Although some advantages of the low-cost non-contact phase of the gait are sacrificed by maintaining a relatively low profile to the rise and fall of the CoM, this also diminishes the energy loss caused by substantive vertical deflection of the CoM during limb contact. The gait still remains as an optimization between a relatively low-cost ballistic phase, interspersed between higher cost phases, where the CoM path transitions from downward to upward. During the "bounce" of the trot, the CoM still follows a largely ballistic path, and the mechanical cost of this portion of the stride is relatively small, even if the time that the animal's feet completely lose contact with the substrate is small or non-existent.

5.7.3 Gallop

As speed increases in the trot, two aspects of the gait become limiting (just as they do for the bipedal run). Stride frequency increases as speed increases. In order for the limbs to move forward to the next contact, and to move backward as swiftly as the CoM passes over the substrate, the angular velocity of the limbs becomes high. Eventually, an angular velocity will be reached, where the contractile speed or power output of the muscles activating the swing limb will be limiting (Doke *et al.*, 2005; Hildebrand and Hurley, 1985; Figure 5.8A).

Also, the absolute time of contact of the support limbs will decrease, which is a function of the geometry of the system. The angles of protraction (when the limb contacts the ground) and retraction (when the limb leaves the ground) will be

A

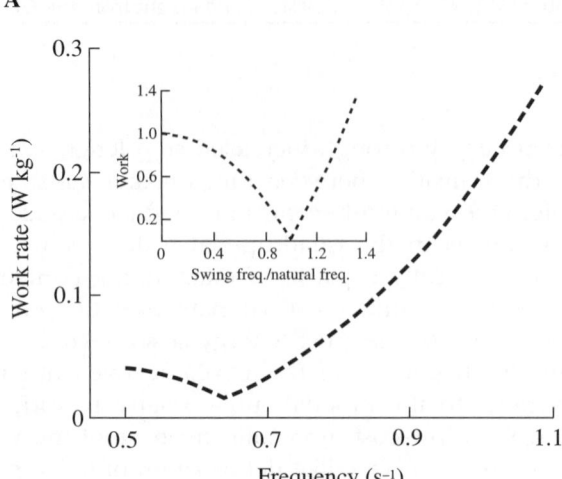

B

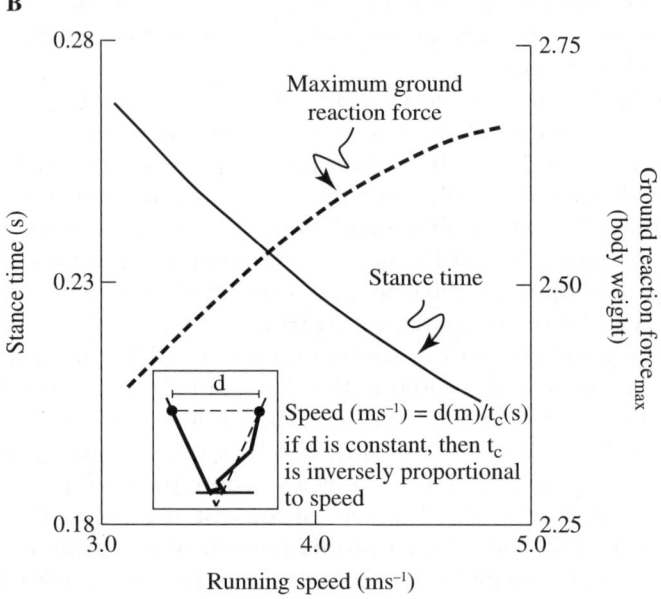

FIGURE 5.8

(A) The changes in power required to swing the human leg at different frequencies (after Doke *et al.*, 2005). Note that the power requirement increases substantially at higher frequencies. Inset: Comparative model of limb swing cost proposed by Hildebrand (1985). A work level is calculated as $2mgD\theta^2 = 1$, where m is the mass of a rectangular solid representing the proximal and distal components of the limb, g, is gravitational acceleration, D is the radius of gyration of the limb (from the proximal pivot), and θ is the amplitude of a half swing (from highest swing position to the bottom of the swing arc. Power (work rate) and swing frequency can be generated from the Hildebrand model by dividing both axes by the limb natural frequency for the particular species of interest. (B) The change in vertical ground reaction force that must accompany changes in contact time as speed increases. Inset: a diagram of the geometry involved with contact, illustrating how time of contact and speed are linked if protraction and retraction angle of the limb remain constant.

relatively constant. The faster the animal moves across the substrate, the shorter will be the time available for the limb to move through this angle. However, over the brief contact period, the limb must support the weight of the animal when the limbs are in contact with the ground, and compensate for when ground reaction force is less than body weight. As a result, the vertical ground reaction force over the middle part of the stance (when vertical forces are greater than body weight) must increase (Figure 5.8B). Eventually, the vertical force capacity of the limb will reach a limit.

Note that, contrary to what might be expected, it is *vertical* force involved with supporting the mass of the animal against gravity that is a major limitation as speed increases, not horizontal force that propels the mass along the surface. In theory, increases in vertical force requirement could be mitigated by decreasing step length, so that the period over which vertical support is provided by each contact is decreased. However, this requires a concomitant increase in frequency, which is itself limited by the angular velocity and power required to swing the limb. When the ground reaction force reaches the limit that the limb can support, and/or no more power is available to swing the limb, the bipedal runner has achieved its fastest running speed. This will also correspond to the fastest trotting speed of the quadruped. However, with double the number of limbs available, the quadruped has an alternative strategy for mediating the interaction between the CoM and the substrate as it moves across the substrate at high speeds – the gallop.

The gallop (which also includes the canter – simply a slower version of the gallop, see Chapter 2) is an asymmetric gait where the fore and hind limbs contact in sequence. What advantage is there to sequencing the limb contacts?

One common form of the gallop uses a relatively stiff torso, and foot contacts occur in a transverse pattern (Hildebrand, 1959). Traditionally, the convention has been to denote the sequence of limb contacts, starting with the first forelimb contact. Selection of this starting point as the convention probably derives from a similar convention applied to the quadrupedal walk. The walk is a cyclical, symmetric gait that has no non-contact phase, so selection of the beginning and end of the contact cycle is functionally arbitrary – any convention would suit the mechanics involved in the walk equally well. The gallop does have a non-contact phase, however, and selection of an inappropriate beginning of the cycle may well obscure the function of the limb contact sequencing strategy.

As with the walk and trot, the portion of the gait cycle of the gallop where substantial collision loss is possible is during the transition of the CoM motion from downward to upward as the animal moves forward across the substrate. For a horse, which uses the transverse gallop with a relatively rigid spine, a rear limb is the first to make contact following the non-contact, collected flight phase of the stride cycle (Figure 5.9; see Chapter 2). The initial limb contact then diverts the CoM path from steeply downward to more forward (implying that it must accelerate the animal in the forward direction). Altering the trajectory of the CoM, toward a path more parallel with the substrate, means that strut-like contact of the next limb – the contralateral rear limb – is more perpendicular to the CoM path than it would have been if the path had not been altered by the action of the first limb. This is much the same way that the initial thrust of the previous stance limb in bipedal walking changes the geometry for the ensuing contact.

FIGURE 5.9

The sequence of limb contacts in the transverse gallop of the horse. Following the non-contact, ballistic flight phase, the first limb to contact is a hind limb. Due to the contact point lying behind the CoM, this contact causes a net propulsive thrust that begins to deflect the CoM path from downward to more forward. The next hind and forelimb contacts continue this transition forward. The final forelimb contact, referred to as the "lead" limb, must impart a net braking force in order to divert the CoM path upward into the next non-contact phase.

Following contact by the second rear limb, the CoM path is near-horizontal, in preparation for the first forelimb contact, which can begin the change to upward motion. The final limb to contact the substrate is the remaining forelimb (referred to as the "lead" limb, indicating its position in space rather than time, as it is actually the last limb of the sequence to make contact) (Figure 5.9). This limb must divert the CoM path from forward to upward, in preparation for the next non-contact phase of the stride. This means that it must apply substantial breaking force to divert the CoM path from horizontal to forward and upward. Counter-intuitively, this deflection by the last limb contact means that the flight phase of the transverse gallop, where horizontal speed cannot be changed, is one of the slowest portions of the stride cycle (Minetti *et al.*, 1999) – surprising for a gait supposedly developed for high running speed. The portion of the gait where the animal is actually traveling most rapidly is near mid-stance, completely opposite the condition for any spring-mass or pogo-stick model of the gait. This limitation is imposed, however, because the potential losses of energy at the transition of the CoM from downward to upward at these high running speeds would be prohibitive for (almost) any other limb contact strategy.

Why sequence the limb contacts? The advantage of sequencing limb contacts can be seen by considering what is accomplished over the contact portion of the stride. The CoM of the animal is falling downward as it moves forward when the first limb makes contact (a rear limb), and is rising as it moves forward following the final limb contact (a forelimb). The transition of the CoM motion from downward to upward is actively mediated by the limb contacts (while the transition from upward to downward in the flight phase of the gait is passively mediated by gravity).

In the highest speed gallop of the horse, this deflection is divided into loosely equal portions, where each limb is responsible for a portion of the transition. This could be done with all four limbs contacting the ground simultaneously (a gait referred to as a "pronk", performed by some ungulates as a display to demonstrate vigor – see Chapter 2), where simultaneous action of the limbs would mean that the effort required would naturally be divided between all four limbs. However, in terms of energy loss, sequencing the limbs to gradually shift the path of the CoM has substantial advantages.

If the CoM transition from downward to upward were mediated by all limbs acting in a single contact (as in the pronk), the CoM velocity loss involved would result in a change in energy, ΔE_s (the change in energy resulting from a given angular deflection of the CoM mediated by a single impulse, acting on the CoM like one "effective" limb). Alternatively, the same transition could be mediated by n limbs, each deflecting the CoM path by an equal portion of the total transition, resulting in ΔE_M, the decrease in energy for the same angular deflection spread between multiple limbs (Figure 5.10). This energy decrease must be repaid by positive work at some other portion of the stride, in order to maintain speed. Multiple "sub-collisions" are superior, because the

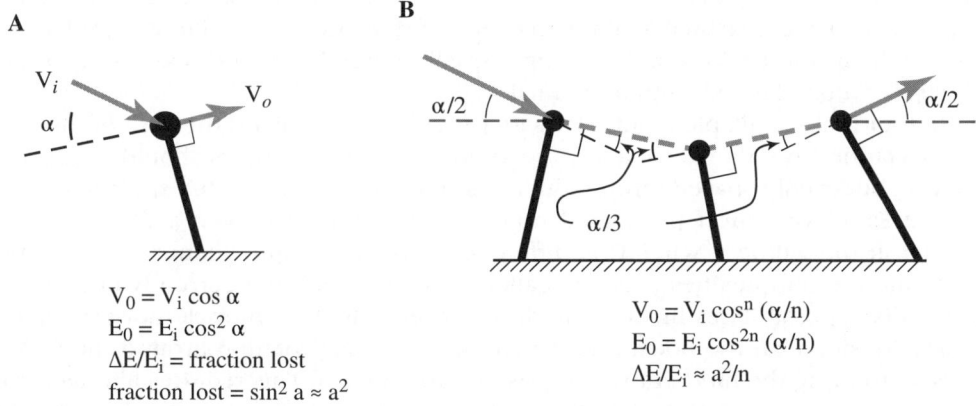

A

$V_0 = V_i \cos \alpha$
$E_0 = E_i \cos^2 \alpha$
$\Delta E/E_i = \text{fraction lost}$
$\text{fraction lost} = \sin^2 a \approx a^2$

B

$V_0 = V_i \cos^n (\alpha/n)$
$E_0 = E_i \cos^{2n} (\alpha/n)$
$\Delta E/E_i \approx a^2/n$

FIGURE 5.10

If α^2 dictates that one unit of energy is lost with a single limb contact (or multiple limbs acting simultaneously, shown in (**A**)), then 1/3 units of energy is lost when the contact is spread between three sequentially supporting limbs (indicated in (**B**)). That is, three times as many collisions occur, but each collision only contributes 1/9 the energy loss. Detailed explanation: When we have n sequential contacts, each deviating the velocity vector by α/n, then we have:

$$E_{o1} = E_{i1} \cos^2 (\alpha/n) = E_i \cos^2 (\alpha/n)$$
$$E_{o2} = E_{i2} \cos^2 (\alpha/n) = E_{o1} \cos^2 (\alpha/n) = E_i \cos^4 (\alpha/n)$$
$$\cdots$$

$$E_o = E_{on} = E_{in} \cos^2 (\alpha/n) = E_i \cos^{2n} (\alpha/n)$$
$$\Delta E_1 = E_{o1} - E_{i1} = E_{i1} \sin^2 (\alpha/n) = E_i \sin^2 (\alpha/n) \approx (\alpha/n)^2 E_i$$
$$\Delta E_2 = E_{o2} - E_{i2} = E_{i2} \sin^2 (\alpha/n) = E_{o1} \sin^2 (\alpha/n)$$
$$= E_i \cos^2 (\alpha/n) \sin^2 (\alpha/n) \approx (\alpha/n)^2 E_i$$
$$\cdots$$

$$\Delta E_n = E_{on} - E_{in} = E_{in} \sin^2 (\alpha/n) = E_{o(n-1)} \sin^2 (\alpha/n)$$
$$= E_i \cos^{2(n-1)} (\alpha/n) \sin^2 (\alpha/n) \approx (\alpha/n)^2 E_i$$
$$\Delta E = \Delta E_1 + \Delta E_2 + \cdots + \Delta E_n \approx n E_i (\alpha/n)^2 = E_i \alpha^2/n$$

energy involved with the transition, whether absorbing or generative, varies as the deflected angle squared (ϕ^2). As a consequence, halving the deflection angle through spreading the transition through two independent contacts ($n = 2$), then, decreases the individual limb contact energy loss by the root value.

Even by dividing the transition between two contacts (front and rear), the strategy results in a decrease in total energy loss – that is, each contact will involve one-quarter of the energy loss, but now there are two contacts, so the net result is ¼ + ¼ = ½ (Ruina *et al.*, 2005). Each sub-collision will involve $1/n^2$ energy decrease, but there will be n sub-collisions. Ultimately, $\Delta E_M/\Delta E_s \approx 1/n$, where the decrease in energy associated with multiple sequenced sub-collisions is related to the energy lost in performing the transition in a single collision, by the inverse of the number of sub-collisions. This extrapolates to the reasonable limit that an infinite number of limbs, emulating a rimmed wheel, should have a vanishingly small energetic cost (at least compared to each step of a one-legged running animal).

With four limbs available to interact with the substrate to divert the path of the CoM, the expectation from the above analysis would be that the contacts should progress in sequence, but evenly spaced throughout the stride, like evenly spaced spokes of a rim-less wheel and without a protracted non-contact phase. However, a flight phase is observed in the gallop – why? This difference between simple geometric interaction models and the complexities of actual galloping mammals is not currently understood. Some realistic possibilities include the distribution of limbs – they do not originate at the same position on the body, and it may be that some rotation around the CoM is necessary to bring the hind limbs into position for contact. This requires a longer non-contact phase intervening between the contact sequences that cause the downward to transition upwards.

The body itself is not symmetric; the horse, for instance, has a large head and neck projecting from the front of the torso. Although the head and neck can influence locomotion, its form is largely dictated by feeding requirements. Asymmetries of the contacts in the gait cycle may be required to contend with aspects of the distributed mass of the animal. The asymmetry may originate with fundamental asymmetries of the actuator system – the muscles themselves. Ground contact forces in human running are slightly asymmetric, and this has been attributed to differences in the action of the muscle during negative and positive work portions of the contact (Cavagna, 2006).

It may be that large asymmetries in the gallop originate with tradeoffs between collision-based costs, low-cost benefits of the non-contact phase, and metabolic costs associated with active motion and support needed to precipitate the gait. Resolution of this particular "controversy" will require developing more complex models which better identify and represent competing factors within the system. However, those models will most likely be successful if they begin by considering the function of the limbs in mediating CoM path transitions over the course of the stride cycle.

5.8 CONCLUSION

The complex morphology displayed by terrestrial mammals makes interpreting details of the locomotory machinery difficult. Attempting to evaluate the form of components without the context of their function in the animal as a whole can lead to much fruitless effort and misinterpretation, which can confuse the issue and hinder progress toward a full understanding of the system. Identifying the functional role of limbs, without becoming overwhelmed by consideration of the details of how they accomplish this function, provides a simplifying strategy that allows interpretation of the circumstances the system functions within. Putting this in the context of the dynamic trade-offs inherent in over-ground locomotion of the whole animal leads to insights into the problems faced in legged locomotion, and the strategies available to overcome them most effectively.

▌ REFERENCES

Adamcyzyk PG, Kuo AD (2009). Redirection of center-of-mass velocity during the step-to-step transition of human walking. *Journal of Experimental Biology* **212**, 2668–2678.

Alexander RMcN (1984). Elastic energy stores in running vertebrates. *American Zoologist* **24**, 85–94.

Anderson F, Pandy M (2001). Dynamic optimization of human walking. *Journal of Biomechanical Engineering* **123**, 381–390.

Bertram JEA (2004). New perspectives in brachiation mechanics. *Yearbook of Physical Anthropology* **47**, 100–117.

Bertram JEA, Chang YH (2001). Mechanical energy oscillations of two brachiation gaits: measurement and simulation. *American Journal of Physical Anthropology* **115**, 319–326.

Bertram JEA, Hasaneini SJ (2013). Neglected losses and key costs: Tracking the energetics of walking and running. *Journal of Experimental Biology* **216**, 933–938.

Bertram JEA, Ruina A, Cannon CE, Chang Y-H, Coleman M (1999). A point-mass model of gibbon locomotion. *Journal of Experimental Biology* **202**, 2609–2617.

Blickhan R (1989). The spring-mass model for running and hopping. *Journal of Biomechanics* **22**, 1217–1227.

Blickhan R, Full RJ (1993). Similarity in multilegged locomotion: Bouncing like a monopode. *Journal of Comparative Physiology A* **173**, 509–517.

Cannon CH, Leighton M (1994). Comparative locomotion ecology of gibbons and macaques: selection of canopy gaps. *American Journal of Physical Anthropology* **93**, 505–524.

Cavagna GA (2006). The landing-take-off asymmetry in human running. *Journal of Experimental Biology* **209**, 4051–4060.

Cavagna GA, Saibene FP, Margaria R (1964). Mechanical work in running. *Journal of Applied Physiology* **19**, 249–256.

Cavagna GA, Heglund NC, Taylor CR (1977). Mechanical work in terrestrial locomotion: two basic mechanisms for minimzing energy expenditure. *American Journal of Physiology* **233**, R243–R261.

Chang Y-H, Bertram JEA, Lee DV (2000). The external force and torques generated by a brachiating White-handed gibbon (*Hylobates lar*). *American Journal of Physical Anthropology* **113**, 201–216.

Collins SH, Wisse M, Ruina A (2001). A 3-D passive dynamic walking robot with two legs and knees. *International Journal of Robotics Research* **20**, 607–615.

Doke J, Kuo AD (2007). Energetic cost of producing cyclic muscle force, rather than work, to swing the human leg. *Journal of Experimental Biology* **210**, 2390–2398.

Doke J, Donelan JM, Kup AD (2005). Mechanics and energetics of swinging the human leg. *Journal of Experimental Biology* **208**, 439–445.

Fleagle JG (1974). The dynamics of the brachiating siamang (*Hylobates [Symphalangus] syndactylus*). *Nature* **248**, 259–260.

Gomes MW, Ruina A (2005). A five-link 2D brachiating ape model with life-like zero-energy-cost motions. *Journal of Theoretical Biology* **237**, 265–278.

Gomes M, Ruina A (2011). Walking model with no energy cost. *Physical Review E: Statistical, Nonlinear, and Soft Matter Physics* **83**, 032901.

Gordon KE, Ferris DP, Kuo AD (2009). Metabolic and mechanical energy costs of reducing vertical center of mass moving during gait. *Archives of Physical Medicine and Rehabilitation* **90**, 136–144.

Hamilton WR (1847). On theorems of hodographic and anthodographic isochronism. *Proceedings of the Royal Irish Academy* **3**, 465–466.

Hasaneini SJ, McNab CJB, Bertram JEA, Leung H. (2013). The dynamic optimization approach to locomotion dynamics: human-like gaits from a minimally-constrained biped model. *Advanced Robotics* (in press).

Hildebrand M (1959). Motions of the running cheetah and horse. *Journal of Mammalogy* **40**, 481–495.

Hildebrand M (1985) Walking and running. In: Hildebrand M, Bramble DM, Liem KF, Wake DB (eds). *Functional vertebrate morphology*. Harvard University Press, Cambridge, MA.

Hildebrand M, Hurley JP (1985). Energy of the oscillating legs of a fast-moving cheetah, pronghorn, jackrabbit, and elephant. *Journal of Morphology* **184**, 23–31.

Hill AV (1927). *Muscular movements in man: the factors governing speed and recovery from fatigue*. McGraw-Hill, New York.

Inman VT, Ralston HJ, Todd F (1981). *Human Walking*. Williams and Wilkins, Baltimore.

Jungers WL, Stern JT Jr (1980). Telemetered electromyography of forelimb muscle chains in gibbons (*Hylobates lar*). *Science* **208**, 617–619.

Kuo AD (2002). Energetics of actively powered locomotion using the simplest walking model. *Journal of Biomechanical Engineering* **124**, 113–120.

Lee DV, Bertram JEA, Todhunter RJ (2002). Force overlap in trotting dogs: A Fourier technique for reconstructing individual limb ground reaction force. *Veterinary and Comparative Orthopaedics and Traumatology* **15**, 223–227.

Lee DV, Bertram JEA, Anttonen JT, Ros IG, Harris SL, Biewener AA (2011). A collisional perspective on quadrupedal gait dynamics. *Journal of The Royal Society Interfacee* **8**, 1480–1486.

McGeer T (1990). Passive dynamic walking. *International Journal of Robotics Research* **9**, 62–82.

McGeer T (1993). Dynamics and control of bipedal locomotion. *Journal of Theoretical Biology* **163**, 277–314.

Michilsens F, D'Août KD, Aerts P (2011). How pendulum-like are siamangs? Energy exchange during brachiation. *American Journal of Physical Anthropology* **145**, 581–591.

Michilsens F, D'Août KD, Vereecke EE, Aerts P (2012). One step beyond: Different step-to-step transitions exist during continuous contact brachiation. *Biology Open* **000**, 1–11 (doi: 10.1242/bio2012588).

Minetti AE, Ardigo LP, Reinach E, Saibene F (1999). The relationship between mechanical work and energy expenditure of locomotion in horses. *Journal of Experimental Biology* **202**, 2329–2338.

Ortega JD, Farley CT (2005). Minimizing center of mass vertical movement increases metabolic cost in walking. *Journal of Applied Physiology* **99**, 2099–2107.

Parsons PE, Taylor CR (1977). Energetics of brachiation versus walking: a comparison of a suspended and an inverted pendulum mechanism. *Physiological Zoology* **50**, 182–188.

Preuschoft H, Demes B (1984). Biomechanics of brachiation. In: Preuschoft, H, Chivers, DJ, Brockelman, WY, Creel, N (eds). *The lesser apes: evolutionary and behavioral biology*. Edinburgh University Press, Edinburgh, pp. 96–118.

Rab GT (1994). Muscle. In: Rose J, Gamble JG (eds). *Human Walking*. Williams and Wilkins, Baltimore, pp. 103–121.

Roberts TJ (2002). The integrated function of muscles and tendons during locomotion. *Comparative Biochemistry and Physiology Part A: Molecular & Integrative Physiology* **133**, 1087–1099.

Ruina A, Bertram JEA, Srinivasan M (2005). A collisional model of the energetic cost of support work qualitatively explains leg sequencing in walking and galloping, pseudo-elastic leg behavior in running and the walk-to-run transition. *Journal of Theoretical Biology* **237**, 170–192.

Seyfarth A, Geyer H, Gunther M, Blickhan R (2002). A movement criterion for running. *Journal of Biomechanics* **35**, 649–655.

Srinivasan M (2007). *Bipedal running: "No muscle work and all tendon play" is energetically beneficial even with an energy cost for isometric force production.* SICB, Phoenix, Jan. 3–7, 2007.

Srinivasan M, Ruina A (2006). Computer optimization of a minimal biped model discovers walking and running. *Nature* **439**, 72–75.

Stern JT Jr, Larson SG (2002). Telemetered electromyography of the supinators and pronators of the forearm of gibbons and chimpanzees: implications for the fundamental positional adaptation of hominoids. *American Journal of Physical Anthropology* **119**, 92–94.

Stern JT Jr, Wells JP, Jungers WT, Vangor AK, Fleagle JG (1980). An electromyographic study of the pectoralis major in Atelines and Hylobates, with special reference to the evolution of the pars clavicularis. *American Journal of Physical Anthropology* **52**, 13–25.

Swartz SM (1989). Pendular mechanics and the kinematics and energetics of brachiating locomotion. *International Journal of Primatology* **10**, 387–418.

Taylor CT (1994). Relating mechanics and energetics during exercise. In: Jones JH (ed). *Comparative vertebrate exercise physiology: Unifying physiological principles*, Advances in Veterinary Science and Comparative Medicine Vol. 38A, Academic Press, New York, pp. 181–215.

Usherwood JR (2005). Why not walk faster? *Biology Letters* **1**, 338–341.

Usherwood JR, Bertram JEA (2003). Understanding brachiation: Insights from a collisional perspective. *Journal of Experimental Biology* **206**, 1631–1642.

Waters RL, Morris JM (1972). Electrical activity of muscles of the trunk during walking. *Journal of Anatomy* **111**, 191–199.

Wirkus S, Rand R, Ruina A (1998). How to pump a swing. *The College Mathematics Journal* **29**, 266–275.

CHAPTER SIX

Reductionist Models of Walking and Running

James R. Usherwood

Royal Veterinary College, University of London, UK

6.1 PART 1: BIPEDAL LOCOMOTION AND "THE ULTIMATE COST OF LEGGED LOCOMOTION?"

6.1.1 Introduction

The mechanics underlying bipedal locomotion are, from some viewpoints, exceedingly complex. They involve accurate timing of muscle actions, joint torques, powers and kinematics; they involve very different anatomical arrangements between species, especially between disparate species such as humans and dinosaurs (including birds); and, at some level, they vary from subject to subject, from patient to patient, and even from step to step.

Studying the detailed mechanisms within limbs down to the level of individual muscle dynamics (and even beyond) is certainly valuable, giving the potential to study muscle synergies, control and, potentially, the consequences of specific disease, surgery or prosthetic influences. However, the motivation of this chapter (and, indeed, this book) is to explore unifying principles in locomotion through admittedly

Understanding Mammalian Locomotion: Concepts and Applications, First Edition.
Edited by John E. A. Bertram.
© 2016 John Wiley & Sons, Inc. Published 2016 by John Wiley & Sons, Inc.

extreme reductionism – deliberately avoiding the level of detail traditional in clinical biomechanics. One aim of such reductionist approaches must be that they result ultimately in *predictive*, and therefore *testable*, models of the gross workings of a locomoting animal. Reductionism is far less powerful if it is merely *descriptive* although, perhaps, a "reduction in parameter space" or "compression in degrees of freedom" may be helpful in terms of reporting the key properties of a system or, perhaps, investigating the workings of the control/nervous system.

6.1.2 Reductionist models of walking

"If a man were to walk parallel to a wall in sunshine, the line described (by the shadow of his head) would be not straight but zigzag, becoming lower as he bends, and higher when he stands and lifts himself up." Aristotle (as translated by Farquharson, 2007).

Mechanically, what might this mean? An "inverted pendulum" appears to be a good analog representing this vaulting behavior, with interchange between gravitational potential energy and kinetic energy each step. This interchange is evident from force plate-derived measurements of center of mass motions (Cavagna *et al.*, 1977), and appears to be a fairly general phenomenon in walking – at least at some speeds – among animal bipeds (currently dominated by humans and birds). This mechanical description of walking has often been adopted as a "mechanical definition" of walking as a gait, leaving the traditional description of a symmetric gait without an aerial phase as a "kinematic definition".

While these two definitions broadly overlap for human gaits, this is not the case for many birds over a considerable range of speeds. For instance, domestic ducks can locomote terrestrially at relatively high speeds, using a mechanical run but a kinematic walk (sometimes termed a "grounded run"), with only athletic breeds achieving aerial phases at close to their top speeds (Usherwood *et al.*, 2008). Whether the human or the bird should represent the "typical" biped depends entirely on personal bias – though both cases should be remembered whenever general rules for bipedal locomotion are being developed.

Extreme reduction: the compass-gait

The extreme application of the inverted pendulum model of walking is the compass-gait, in which body motions are considered entirely passive and each leg completely stiff: the center of mass vaults up and over the leg, slowing as it approaches its highest point at midstance (kinetic energy, KE, is passively converted to potential energy, PE), and speeding up again as it re-accelerates due to gravity (the loss of PE powering the gain in KE). The mechanics during this vault of the body over a stiff leg are relatively easy to model (Figure 6.1A). They present a prediction for the maximum speed and step angle (or step length) that should be possible before the stance leg experiences tension forces. The stance foot either lifts off or drags along the ground, and standard "walking" becomes impossible at speeds greater than this (Figure 6.1B).

Interestingly, and perhaps counter-intuitively, the model predicts that higher walking speeds are achievable with shorter step angles (step lengths). This is due to two effects: high step angles result in high instantaneous speeds and, thus, high "centrifugal" forces; and the component of weight acting in line with the leg and providing the opposing "centripetal" forces is also reduced. Using this ultra-reductionist compass-gait model is

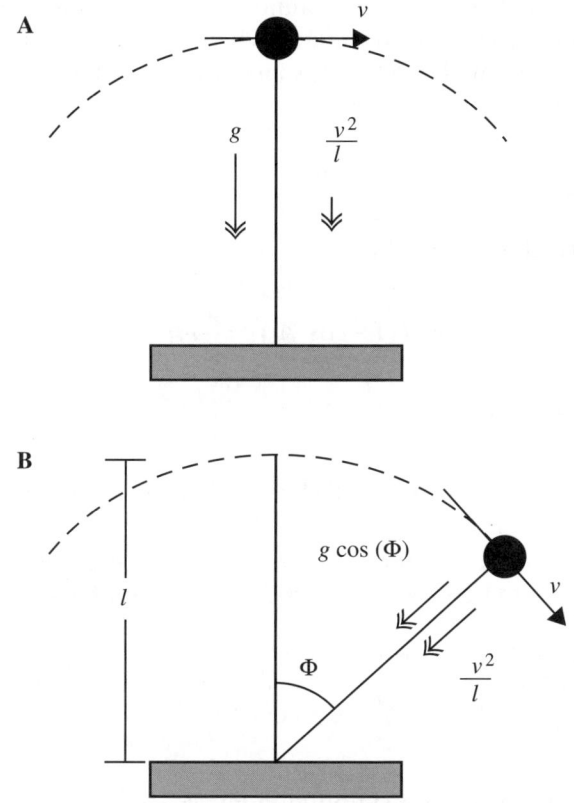

FIGURE 6.1

Walking with a passive, stiff-limbed stance leg imposes limits on speed and stance-length if leg tension resulting in "take-off" or "toe-drag" are to be avoided. The centripetal acceleration requirements for the center of mass to arc over the foot has to be provided by gravity. At midstance, or for very small step lengths, this provides the condition shown in **A**. However, the extremes of stance, especially for large step angles Φ, may provide the limiting conditions. Here, a smaller component of gravity acts in line with the leg and, as the mass is falling passively, its speed v is greater than at midstance. (Adapted from Usherwood, 2005).

surprisingly successful at predicting walk-run transition speeds (at least using the "mechanical definition" of walking) in humans and birds (Usherwood *et al.*, 2008). It leads to the conclusion that humans are able to walk relatively quickly (at least compared with ducks) because they are capable of relatively short steps, associated with the capacity to drive the swing-leg to achieve high step frequencies (see also Dean and Kuo, 2009).

The special case that appears particularly successful (try it!) is the prediction of the longest walking step length, from a bit of geometry and a couple of lines of math:

Consider a point-mass vaulting walker falling over a leg of length l as slowly as possible through a step angle Φ (this is a special case of a falling chimney or pencil – see Madsen, 1977; McGeer and Palmer, 1989). If falling as slowly as possible, the body starts at zero speed (at midstance, when the leg is vertical) and, as we are considering a passive system, the drop in potential energy equates to the rise in kinetic energy – and thus speed v – of the body:

$$\Delta PE = KE \tag{6.1}$$

With a little geometry, this means that:

$$mgl\left(1 - \cos(\Phi)\right) = \frac{1}{2}mv^2 \tag{6.2}$$

Thus,

$$v^2 = 2gl\left(1 - \cos(\Phi)\right) \tag{6.3}$$

The avoidance of take-off or toe-drag requires the centripetal acceleration to keep the body arcing over the leg to be provided by the component of gravity acting in line with the leg:

$$\frac{v^2}{l} \le g\cos(\Phi) \tag{6.4}$$

Substituting with equation 6.3 and simplifying gives:

$$\Phi \le \cos^{-1}\left(\frac{2}{3}\right) \tag{6.5}$$

And, again from simple geometry, as step length L_{step} relates simply to step angle and leg length:

$$L_{step} = 2l\sin(\Phi), \tag{6.6}$$

the constraints to step length can be expressed as:

$$L_{step} \le 2l\sin\left(\cos^{-1}\left(\frac{2}{3}\right)\right) \tag{6.7}$$

(Usherwood, 2005), or the longest possible step according to the compass-gait model is 1.49 times the leg length (Figure 6.2):

$$\frac{L_{step}}{l} \le 1.49. \tag{6.8}$$

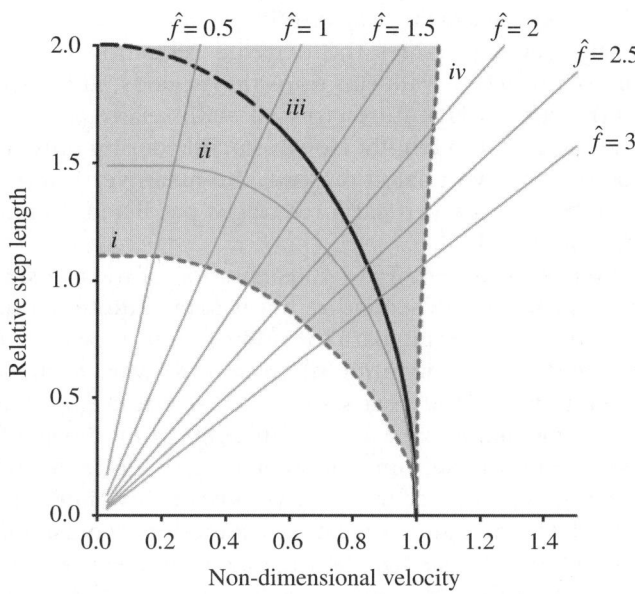

FIGURE 6.2

The shaded region (between contours *i* and *iv*) is that predicted by numerical optimization to favor (energetically) inverted pendular running (Srinivasan and Ruina, 2006). The gray contour (*ii*) indicates the maximum boundary for a passive, stiff-limbed vaulter, in which gravity provides sufficient centripetal acceleration of the CoM towards the foot to keep the leg under compression. The outer, black curve (*iii*) indicates the walk-run (run-walk) transition boundary, predicted from energy minimization derived from collision mechanics *while allowing limb tension*.

Despite the extreme reductionism of using compass-gait mechanics to model walking, this prediction appears remarkably effective, and very easy to test: try it on yourself!

Limitations of compass-gait constraints to prediction of top walking speeds
A couple of notes of caution should be expressed concerning the success of compass-gait mechanics in predicting the walk-run transition speed. First, the model says nothing about *why* a biped would choose walking over running at low speeds (just that it is possible to walk). Second, if the definition of walking is given in mechanical terms (as predominantly passive and stiff-limbed), then it should not be surprising – and is, in fact, circular – that "walking" cannot be achieved at higher speeds or longer step lengths than those possible with a passive, stiff-limbed model. This said, it does appear that the mechanical constraints of compass-gait walking can be successfully predictive. In uphill walking, the geometry of Figure 6.1 is slightly altered, and slower maximum walking speeds are predicted (Hubel and Usherwood, 2013), and this is quantitatively close to the 1% per 1% incline reduction in preferred walk-run transition that has been reported (Minetti *et al.*, 1994).

Introducing the foot-foot interchange to the compass-gait

So, the compass-gait model provides the extreme reduction of bipedal walking; it accounts for naturally observed walk-run transition speeds, and successfully predicts (admittedly only if the subject is required to walk with relatively stiff legs and without an aerial phase) the longest step length. However, considering only the vaulting phase is clearly unsatisfactory for two related reasons: the mean vertical force is insufficient to oppose body weight; and the vertical force traces predicted do not remotely match those of actual walking bipeds.

To account for the first issue requires a consideration of the transition from one foot to another, while to approach the second requires some additional biological realism. In order to transition from one arcing path to another takes a (or several) step change(s) in velocity, with the vertical velocity jumping from downward at the end of the stance to upward at the beginning of the next stance. It is worth noting that, in this respect, "inverted pendulum" mechanics are quite different from those of a conventional pendulum. If we stick with the stiff-limbed compass-gait model for the moment, these step changes in vertical velocity occur over a very brief (infinitely brief, if the legs are infinitely stiff) period of time, and can be considered as "collisions". As a new foot contacts the ground, the body jolts against the new stance leg, which may succeed, to some extent, in realigning the velocity upwards, following the new arcing path, but results in a loss of energy (see Chapter 5).

As Kuo (2002) demonstrates, an energetically effective way of overcoming this energy loss is to push with a telescoping leg extension towards the end of stance, just prior to the "lossy" new foot contact. This leg extension can be dominated by ankle extension in normal human walking, and similar force profiles are seen in high-heel, tiptoe or ostrich walking – cases where ankle extension is highly constrained (Usherwood *et al.*, 2012).

Such impulses are indicated in Figure 6.3 (top). With these, both horizontal and vertical force balances work (zero net horizontal force, an average vertical force matching body weight). Of course, if these step changes in velocity are supposed to act on infinitely stiff legs then, over infinitely small time periods (though providing finite and explicit impulses), infinite forces are required, and force traces remain quite dissimilar to those measured. This is clearly biologically unrealistic, but it can be simply modified with an addition of some degree of biological squishiness (modeled in Figure 6.3, bottom, with a sympathetically tuned Butterworth filter) to broadly replicate measured force traces.

Springy walking – a mechanical filter?

An alternative development of the inverted pendulum, which is also effective at accounting for the ground reaction forces of walking, is the spring-mass (or mass-spring) model of walking, with simple, perfectly elastic linear springs acting as legs. Such a system is sometimes termed a "spring-loaded inverted pendulum", or SLIP model and, unlike the pure compass-gait, it can include periods with both legs on the ground. Geyer *et al.* (2006) demonstrated that the mechanics of walking are broadly consistent with such a model. This demonstration (which may, today, be confirmed relatively easily with off-the-shelf mechanics simulators) was an important step. It results in a paradigm shift from the hard and fast distinction between walking (not spring-like) and running (spring-like).

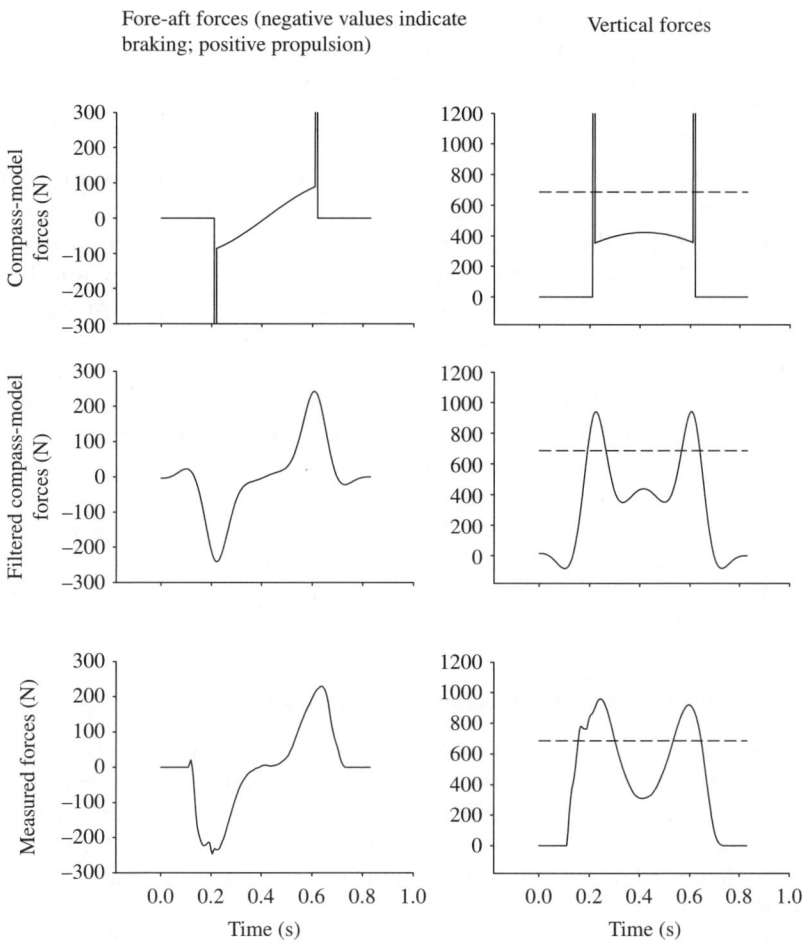

Fore-aft forces (negative values indicate braking; positive propulsion)

Vertical forces

FIGURE 6.3

Forces for a single leg during walking from the compass-gait model including powering impulses at foot-off and dissipative impulses at foot-on (top row). Second row: the modeled forces after an arbitrary Butterworth filter. Third row: actual forces for a walking undergraduate (at 1.97 m/s). Admittedly, getting a match this nice does require some tuning of both the numerical simulation of the compass-gait forces and filter parameters, but it remains a good match, given that the inputs are only the mechanics of an idealized compass-gait walker and a Butterworth filter.

One interpretation of the SLIP model applied to walking is to consider the spring effectively as a mechanical filter. Just as an arbitrary conventional filter (I used a Butterworth to generate Figure 6.3) can be applied to compass-gait forces to replicate observed forces, so can a mechanical filter consisting of a spring and gravity. The appeal of the mechanical filter is clear; thinking of a leg as spring-like is far more intuitive than thinking of forces filtered by whatever capacitors, inductors and resistors are modeled by computerized renditions of the Butterworth filter. In addition, viewing locomotion as some form of spring-mass system allows it to be both modeled and replicated with actual springs and masses. This gives a starting point for understanding issues of stability and control, and has inspired many walking, hopping and running robots.

However, a strong note of caution must be made, particularly relating to biological interpretation. By showing that walking *can* be modeled as a simple SLIP, it is sometimes inferred that the leg *is* somehow actually a mechanical spring, rather than merely behaving like one. This is certainly not the case, as Geyer would be the first to acknowledge; even though muscles and tendons may contain spring-like elements which can store and return "elastic strain energy", any real biological leg is actually very different from a spring. Another way of putting this is that *of course* locomotion at constant speed on level ground *must* average to be energetically conservative (and therefore broadly spring-like), but there are many combinations of inelastic mechanisms that might add to spring-like or "pseudo-elastic" behavior.

What is certain is that no animal biped leg acts as an obligate, perfectly elastic spring, so the benefits of considering them as such must be assessed with caution. The question then arises as to whether the SLIP is a useful reduction of the actual mechanics of walking (and, indeed, running), or is simply a way of compressing measurements of force and deflection into a single parameter, leg stiffness.

6.1.3 The benefit of considering locomotion as inelastic

One advantage of considering locomotion as largely inelastic is that you can begin to investigate the power requirements of legged locomotion. Once springs have been invoked, why should there ever be cost to locomotion? Granted, springy tendons (and, to some extent muscles, viscera, etc.) can reduce the cost of a given cyclical fluctuation in mechanical energy, but the reduction of any gait to a true, completely elastic bounce removes any need for energy input.

From walking to running
Geyer's demonstration that walking can be expressed as a SLIP somewhat blurs the distinction between walking and running – both can be modeled as actual springs. However, the opposite is also true – both can be viewed as only *appearing* like springs, actually being *pseudo-elastic*. Srinivasan and Ruina (2006, 2007) consider this possibility and show that walking and running, including both the vaulting-walking/bouncing-running transition with speed and the spring-like behavior of the leg, is predicted from simple mechanical energy cost minimization. This is achieved *without* elastic springs – the modeled mechanics merely "*look*" elastic as, indeed, is somewhat inevitable if a stable, steady-speed gait on level ground is to be considered.

So, is this the answer? Does Srinivasan's analysis subsume both compass-gait approaches to walking (to which it converges at low speeds) and SLIP models? After all, the compass gait never accounts for *why* running is not appealing (it is clearly mechanically a possibility) even at low speeds – it just highlights that vaulting walking mechanics are impossible at high speeds – and the SLIP spring-mass development, again, provides little insight into *why* particular spring-like properties should be selected.

Srinivasan's simulations do start to demonstrate the *why* behind selected leg properties. What is more, the simulations have had some success in predicting a novel gait between walking and running. At intermediate speeds, particularly at relatively large step lengths, the "inverted pendular run" is found to be energetically optimal. This gait involves periods of arcing, stiff-limbed walking-like vaulting, separated by ballistic, running-like flight phases. If this gait exists, it would be expected in birds with relatively low step frequencies and long step lengths. However, at these speeds, birds remain "grounded", showing no ballistic phase. But is some form of compliant, grounded "inverted pendular run" present? If so, then how might it be identified?

In order to distinguish between walking and running gaits, Kuo (e.g. Adamczyk and Kuo, 2009) identified center-of-mass "hodographs" to be highly informative. These follow the horizontal and vertical velocities of the center of mass through the gait cycle. Kuo describes an elegant, objective distinction between walking and running through the "sense" of such hodographs: walking hodographs are anticlockwise, while running one are clockwise (Figure 6.4). This framework for identifying gaits also reveals a grounded form of the inverted pendular run predicted by Srinivasan's work: at around the predicted speeds and step lengths, pheasants and guinea fowl locomote with figure-of-eight hodographs (Usherwood, 2010), where portions of the stride cycle follow different aspects of the hodograph trace expected for walking and running gaits (Figure 6.4).

Do the energy-minimizing models pioneered by Srinivasan and Ruina provide all the answers? Not yet. Unless Srinivasan's models are arbitrarily constrained from certain regions in parameter space, some very unrealistic optimal gait strategies are found, which require infinite limb forces. At low speeds, a stiff-limbed compass-gait, with infinitely short steps, is found to be optimal and, as discussed above, this requires infinite forces at each vault-vault interchange. At high speeds, "impulsive running" is predicted, with infinitely stiff spring-like bouncing. Clearly, any gait requiring infinite forces departs to a certain extent from that observed in biology.

In the case of walking, there is an obvious downside to this prediction of very short steps: small (upright) stance angles require short steps and hence, at anything above the lowest speeds, high step frequencies; and the swinging of the leg might, in itself, be costly. Indeed, driving the protraction of the swing leg, even during walking, can be seen to impose a considerable metabolic cost (e.g. Doke *et al.*, 2005; Marsh *et al.*, 2004). As a consequence, there appears to be a compromise between the benefits of short steps and high step frequencies (both in terms of locomotor efficiency and maximum walking speed) and the cost of driving the protracting swing leg. Perhaps unsurprisingly, this compromise does not appear to be balanced identically amongst all species: walking ducks maintain step frequencies around 1.2 times above that of a passive leg length pendulum ($\hat{f} = 1.2$), while humans drive their legs at up to $\hat{f} = 3$ when walking. Although elastic mechanisms improving the efficiency of swinging the leg forward may allow smaller steps, smaller collision angles and greater locomotor efficiency (Dean and Kuo, 2009), the dangers of including ideal springs in locomotion models must again be remembered: *if* springs between the leg could be swung very quickly and for free, walking should be costless as the center of mass approaches a smooth, level path, analogous to the axle of a wheel.

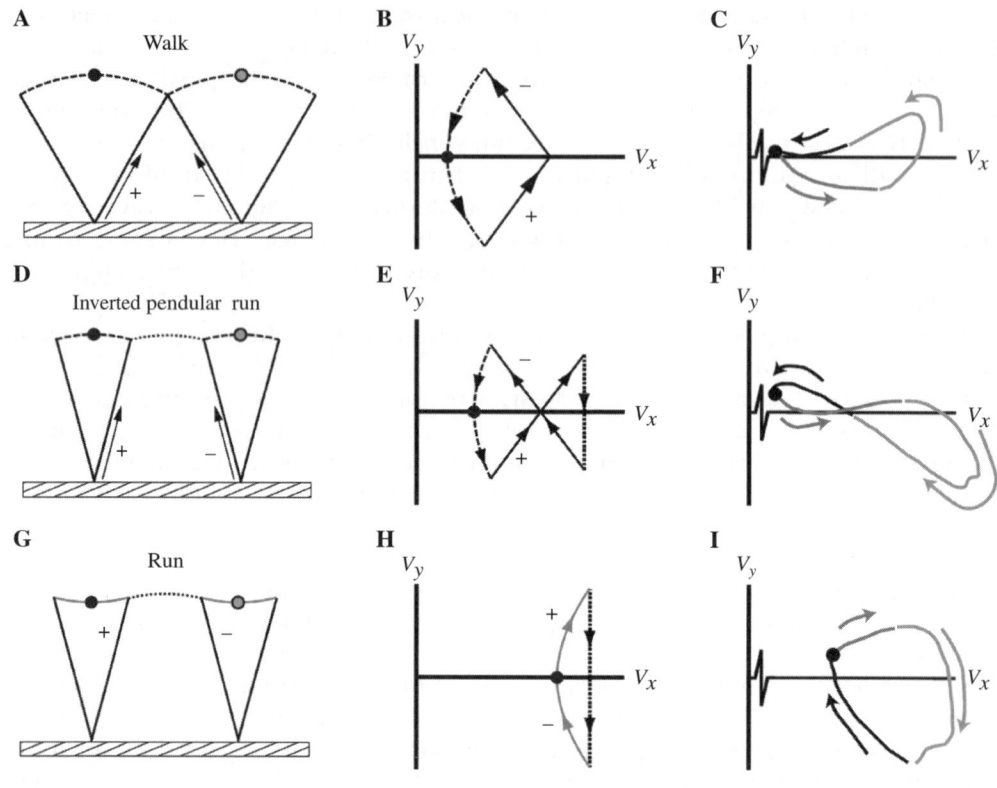

FIGURE 6.4

Idealized cartoons and hodographs, and example empirical (pheasant) hodographs of walking (**A-C**), Inverted Pendular Running (**D-F**) and running (**G-I**). Black arrows relate to momentary impulses; dashed lines to passive vaulting; dotted lines to ballistic falling. Note that while the cartoons and hodograph sketches for walking and inverted pendular running approach those for ideal, impulsive gaits, the running figure includes a qualitative realism of finite limb forces and stance paths (gray lines). Running is not impulsive, stance periods are finite, and the hodograph is not vertical during the leg impulse. In each case, the step cycle begins at the black circle, and the cartoons progress from left to right. +/– symbols indicate gain and loss of energy respectively. Shading in the empirical hodographs indicates different segments of the cycle.

This compromise between swing-leg driving and stance angle appears convincing in walking, but is unsatisfactory for running. According to Srinivasan's approach, running should also be most efficient with very small stance angles. This limits the fluctuations in energy to only those associated with vertical motions, whereas any finite step angles also include apparently wasteful or costly fluctuations in kinetic energy, due to changes in horizontal velocity. Also, at high speeds in real running bipeds, these costs vastly dominate those due to vertical motions. Unlike walking, there need not be a compromise between small stance angles and high swing-leg speeds; with an aerial phase, these two kinematic parameters become decoupled. So, why do animal bipeds not use near-vertical stance legs while running?

The cost of force

Kram and Taylor (1990) noted that metabolic cost was remarkably closely associated with the inverse of foot contact time, and not fluctuations in mechanical energy. In fact, this expression is so important, and apparently counterintuitive from a traditional engineering viewpoint, that it should be explicitly repeated:

$$\frac{\text{Metabolic power}}{\text{Body weight}} = \frac{0.18}{\text{Foot contact time}}$$

Kram and Taylor attributed this to a cost due to some "rate of force application". More recently, Pontzer has highlighted that this inverse of foot contact time should also be related to the forces experienced by the legs. Consider the following argument:

$$\frac{\text{Metabolic power}}{\text{Body weight}}$$

$$\propto \frac{\text{Volume of muscle activated}}{\text{Body weight}} \text{Step frequency}$$

$$\propto \frac{\text{Limb force}}{\text{Body weight}} \text{Step frequency}$$

$$\propto \frac{\text{Step frequency}}{\text{Duty factor}} = \frac{\text{Step period} \times \text{Step frequency}}{\text{Foot contact time}} = \frac{1}{\text{Foot contact time}}$$

Pontzer has also had considerable success demonstrating a close relationship between models for muscle force and measured metabolic costs of locomotion in both bipeds and quadrupeds (Pontzer, 2005, 2007a,b). Thus, it appears empirically that the metabolic cost of transport relates to the volume – and frequency – of muscle "active" during locomotion. It seems likely that this is a very good account for the *proximal* cause of costs during limbed terrestrial locomotion.

However, it does raise an interesting question: why do animals – through evolution or simply posture – not alter the mechanical advantage or gear ratio, in order to reduce the loading on the muscles? Why "*should*" the muscles expose themselves to such forces, and the animal to such metabolic cost? Alternatively, if low duty factors and high limb forces (as opposed to muscle forces) are so costly, why do animals not increase their duty factors by extending their stance periods? Therefore, this "cost of force" viewpoint fails to account for the *ultimate* energetic requirements of locomotion, and its predictive capability is limited.

The muscle-force view appears to be equivalent to measuring the rate of fuel consumption in the engine of a car: it provides an accurate measurement of the *proximal* power requirements (how much gas is required), but it does not really tell you *why* the fuel is needed – the *ultimate* costs. In the case of a car, we are familiar with these ultimate costs – aerodynamic drag, rolling drag, transmission losses, and so on.

However, these do not apply to legged locomotion; you cannot run *that* much faster on a treadmill (avoiding aerodynamic drag), and relatively little energy is lost heating up the shoe soles or running surface. So, why is it that we get tired when we run?

Inevitable constraints of biology?

The most common response to this question appears very reasonable: that real, biological legs are simply not made to withstand high (especially infinite) forces. Muscle simply gets in the way of the ground reaction forces (via some degree of gearing, due to the levers and pulleys of joints, tendons, etc.), and so must end up activating and deactivating, with the associated metabolic costs connected to both the mechanical work loop cycle and such issues as calcium pumping. However, this would suggest that all that is needed to reduce the muscle activation in response to an applied ground reaction force is an appropriate change in gearing (or effective mechanical advantage – see Biewener, 1989; Carrier *et al.*, 1994). This could be achieved either through a simple behavioral or evolutionary modification – run with more straightened legs, or evolve different bone geometry. There is something clearly wrong with this argument, though. Imagine trying to run with stiff, locked-out legs – you are certainly not going to run any faster, nor reduce the metabolic costs of locomotion. In fact, the experience is jolty and painful, and steady-state locomotion cannot be maintained.

Similarly, is it not surprising that the cost of transport relates so nicely with both some aspect of limb force (due to the inverse of either contact time or duty factor) and muscle recruitment, for animals with considerably different gear ratios (mechanical advantages) – unless, that is, we can find an underlying mechanism that might link these relationships through an *ultimate* cost.

One potential account for the ultimate energetic cost of legged locomotion

A simple, and perhaps very general, account for the ultimate cost of locomotion is the hysteresis losses due to loading of animal "viscera", consisting of the non-locomotor organs that have to be carried around by any successful animal – sometimes termed the "guts and gonads" (Daley and Usherwood, 2010). Such organs are situated above the legs, and therefore deflect (absorbing mechanical energy) in the direction of the dominant – vertical – force. So, unlike the mechanical work attributable to leg deflections, which are dominated by energies associated with fluctuations in horizontal velocity, the "guts and gonads" hysteresis losses would be more closely related to vertical forces, and thus the inverse of contact time or duty factor. In effect, this is one possible energetic argument underlying a "cost of force" which, at least in idealized engineering terms, should be energetically costless (consider that a table is very effective at supporting its own weight without requiring any energetic input).

A model is developed (Appendix A; refer also to Daley and Usherwood, 2010) as a first approximation for this mechanical system, shown in Figure 6.5. The loading on the legs is assumed to be largely compressive, and follows a sinusoidal vertical force profile. Energy fluctuations (expressed as a "cost of transport", or mechanical energy fluctuation per mass per distance) closely match those of a full numerical SLIP model (deviating by less than 5% at a stance angle of 45°). The loading on the viscera is assumed to be exactly that transmitted along the leg, and is maximum at midstance (Figure 6.5A, B*ii*).

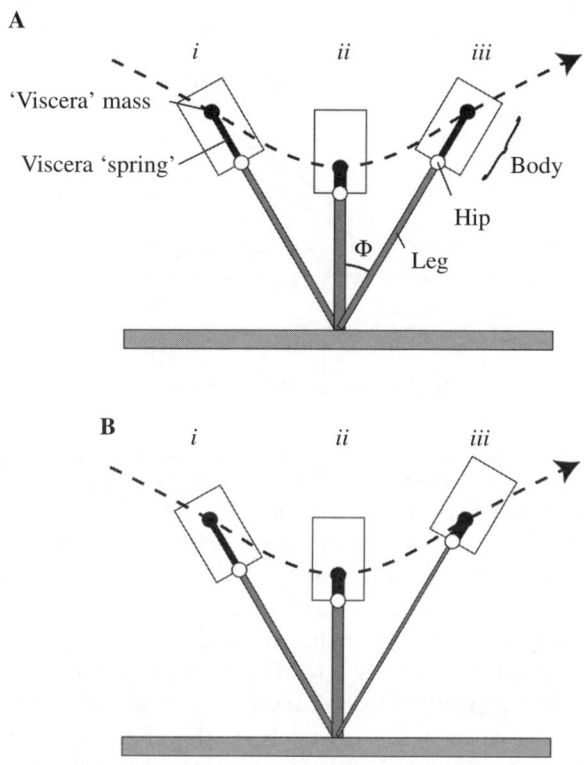

FIGURE 6.5

The mechanical system incorporating the "viscera" to the bouncing, compliant leg model of running. The forces along the legs are assumed to result in a sinusoidal vertical force profile (of sufficient magnitude to counteract body weight), and horizontal forces resulting in pure leg compression. The forces applied to the viscera are taken to be due to accelerations at the hip, calculated from the leg forces. In (**A**), both viscera and leg are considered to be spring-like, deflecting symmetrically from foot-on (*i*), through midstance (*ii*), to foot-off (*iii*). The more realistic case (**B**) is that plastic deflections of the viscera are "paid for" by net positive work, due to leg extension. In either case, smaller (more upright) stance angles (requiring stiffer legs) results in less energy being put into deflecting the leg, but more into deflecting the viscera – and vice versa with larger (longer stance length, lower leg stiffness) stance angles. The compromise provides one potential, testable account for the stance angles selected by biological runners and hoppers (and will influence design considerations for robotic runners).

Treating these two components of the mechanical system independently is clearly a little presumptuous. However, experience appears to show that "viscera" and "leg" bouncing occur over roughly the same period (stance) in actual running. The cleverness behind this should not, though, be overstated: if the viscera is actually largely dissipative, and the lost mechanical energy is largely returned by the leg (as shown in Figure 6.5 B), very little tuning is required, at least compared with the situation when both legs and viscera act as elastic, obligate springs.

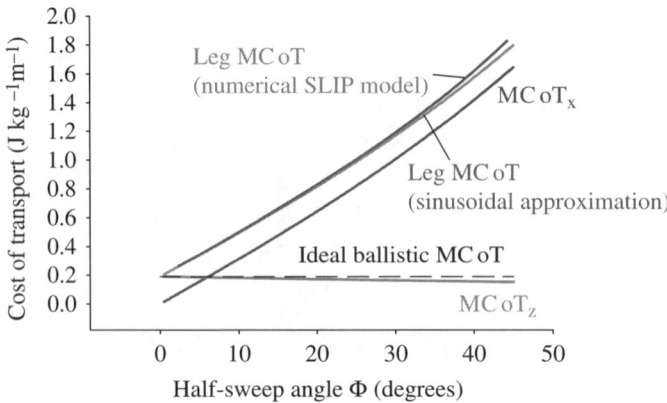

FIGURE 6.6

The effect of stance angle Φ on energy fluctuations presented as mass-specific costs of transport for a human sprinter (mass = 80 kg, leg length = 1 m, protraction period = 0.315 sec, speed=10 m/sec). With an infinitely stiff leg, the leg is vertical during an infinitely brief stance, and the only energy fluctuations are those associated with changing from falling to ascending ("CoT ideal ballistic"). Those costs of transport relating to vertical motions (CoTz) fall slightly with increasing stance angle. However, horizontal velocity (and, hence, energy, leading to CoTx) fluctuations increase rapidly with stance angle. This approximation, based on sinusoidal vertical forces and compression leg forces, provides a close match to the maximum energy stored in the "spring" of the full numerical "spring loaded inverted pendulum", even at relatively high angles.

The model results are shown for typical human sprinting (mass = 80 kg; leg length = 1 m; speed = 10 ms^{-1}; gravity = –9.81 ms^{-2}; protraction period = 0.315 sec), and are largely intuitive (Figure 6.6). With low-swept stance angles (relatively upright, stiff legs), there are very small fluctuations in horizontal velocity, "impulsive running" is achieved, and the energy fluctuations are largely limited to having to reverse the downward velocity (which can be calculated from simple ballistics – see Rashevsky, 1948). With larger swept stance angles, the time spent in the air declines a little, and so the vertical velocity reversal (and associated energies) decrease slightly. However, vastly dominating this effect is the increase in horizontal velocity (and associated energy) fluctuations, highlighting the apparent energetic "waste" of running with anything other than infinitely stiff legs (Figure 6.7). When combined, the energies associated with horizontal and vertical velocity fluctuations match those put into maximum spring "strain energy" in the full numerical SLIP model.

The energetic losses predicted due to deflections of the viscera (Figure 6.5B) depend on both the properties of the "guts and gonads" (how stiff, elastic and massive they are), and of the legs (how stiff, and so how upright, they are, and how great the peak vertical force applied to the viscera is). If the viscera are relatively massless, elastic

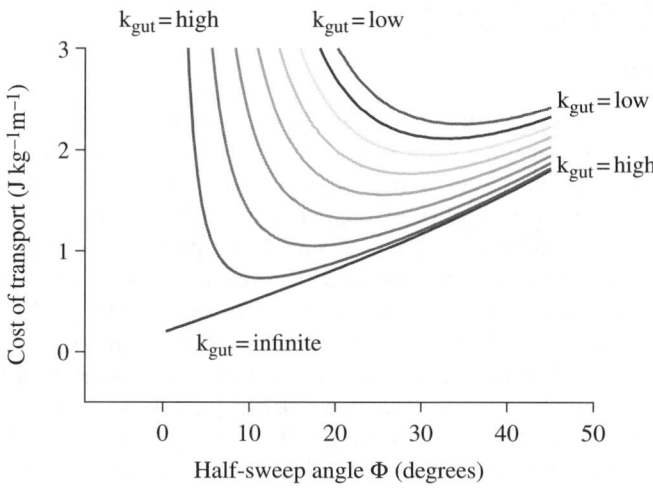

FIGURE 6.7

Energetic consequences of stance angle (directly related to leg stiffness) and stiffness of the viscera (or "guts and gonads"). With a completely stiff (or totally elastic) viscera, the costs are exactly those for the legs (see Figure 7.5), and a zero stance angle (infinitely stiff leg, infinitely high limb forces) is predicted to be energetically favorable, matching Srinivasan's modeling. Energetically optimal stance angles increase with decreasing viscera stiffness (or elasticity); the model suggests that stance angles (leg stiffnesses) may represent a compromise between leg and viscera losses.

and/or stiff, then their losses are relatively low, and small stance angles, high leg stiffnesses and high peak forces are predicted as favorable. This model suggests that some middle ground in viscera and leg properties results in a compromise in leg angles (i.e., neither $\Phi = 0$ nor $\Phi = 90$).

Thus, compliant, somewhat inelastic viscera may present the ultimate "cost of force" that the legs both have to ameliorate (with finite stance angles), withstand and, what is more, perhaps repay (with the legs performing net positive work – Figure 6.5B). This cost of force is appealing, as it effectively becomes again a cost of work (and those with physics/engineering training are more familiar with mechanical work requiring an energy input), is not avoided easily with simple changes in gear ratio, and results in a range of explicit, testable predictions (an exciting direction for future research).

However, it must be remembered that this presents only one possibility to account for the apparently inefficient exposure of leg muscles to forces during steady locomotion, and finite leg stiffness. Others include benefits relating to stability and control (see Chapter 11), and further mechanical (how strong can the bones be? What are the costs of very strong legs?), evolutionary and developmental constraints. Of course, as is so often the way in biology, the truth may well fall within a number of accounts – but simple, reductionist mechanical models provide great, intuitive, and testable starting points to begin the study.

6.2 PART 2: QUADRUPEDAL LOCOMOTION

6.2.1 Introduction

Just as with bipeds, the complexity of legged locomotion can be overwhelming. In animals, a huge variety of muscles, bones, nervous impulses, etc. combine to produce effective acceleration, maneuverability, high speeds and high efficiencies. In some respects, having double the number of legs simply doubles the complexity, and so also increases the difficulties involved in analysis and interpretation of the system. However, having four legs does provide the potential for considerably simplifying stability – at least at slow gaits – because the body may be permanently held over a "tripod of support" produced by three legs, while the fourth repositions itself. This allows repositioning of the legs to happen in its own good time, unconstrained by the dynamics of a falling body (as is clearly the case in a vaulting, walking biped).

Interestingly, however, this is very rarely the strategy actually adopted by animal quadrupeds. Even tortoises make use of the dynamics of toppling (Jayes and Alexander, 1980) – they are not "statically stable" throughout their gait cycle. While some slow, stealthy stalking by predators does, presumably, achieve locomotion with continuous static stability (see, for instance, Bishop *et al.*, 2008), this appears to be the exception, rather than the rule.

6.2.2 Quadrupedal dynamic walking and collisions

Walking quadrupeds generally allow some of their motions to appear passive, with fluctuations in height countered by changes in speed (Cavagna *et al.*, 1977). This principle appears to hold for quadrupeds ranging from running frogs (Ahn *et al.*, 2004), through the more traditional medium-sized cursors such as dogs (Griffin *et al.*, 2004), right up to elephants (Ren and Hutchinson, 2008). If these motions are viewed as the result of relatively stiff-limbed vaulting mechanics, a four-bar linkage model – the linkages consisting of one back leg, the back, the front leg, and the ground connecting back and front foot – represents the quadrupedal analogue of the inverted pendulum (Figure 6.8).

With appropriate empirical inputs (including speed, step frequency and fore-hind phasing), such a model is reasonably effective in predicting the forces, motions and energy fluctuations of walking dogs (Figure 6.9; Usherwood *et al.*, 2007). But what insight – apart from identifying that walking is moderately stiff-limbed, and that forces act predominantly in line with the legs – does such a viewpoint allow? Smith and Berkemeier (1997) extended the passive dynamic walking analyses pioneered in bipeds (McGeer, 1990) to a pair of rimless wheels connected by a stiff "body" – effectively the same as the four-bar linkage, providing a side-view of a walking quadruped (Figure 6.10), assuming the segments to be stiff (and so exactly one "front" and one "back" leg in contact with the ground at any time).

The key insight from this work is that, for a given angle between the leg "spokes", the "quadruped" was more efficient, and so was able to "walk" passively (powered only by gravity) down a shallower slope. In retrospect, this finding is obvious. In effect, having two sets of legs allowed each foot-foot transition to result in a smaller step change in velocity of the center of mass; quadrupeds are able to act in a more

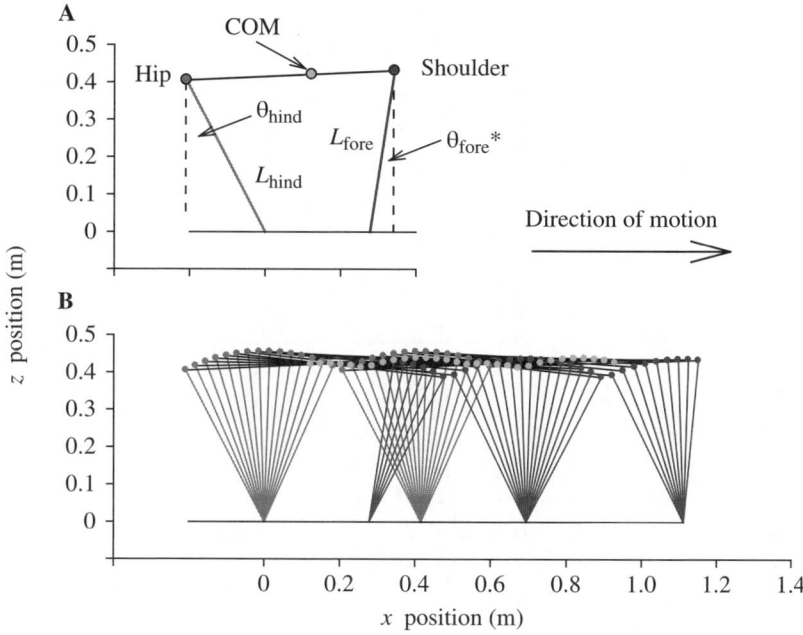

FIGURE 6.8

The motions of the center of mass of a quadruped (black dots) can be modeled with a four-bar linkage to give the fluctuations in potential and kinetic energy, and vertical and horizontal accelerations, that would occur with stiff-limbed, passive walking.

wheel-like manner than bipeds. However, it must be emphasized that this is *for* a given leg length and angle between legs; if bipeds are capable of swinging their legs faster and taking shorter steps, or *if* bipeds have proportionally longer legs, the "quadrupedal locomotion is efficient" story becomes muddied. At this stage, it is simply interesting to note that bipeds are often relatively long-legged (for their weight) or that, equivalently, quadrupeds are relatively heavy (for their leg length).

Another insight that can be gained from the passive/dynamic/collisional perspective is the manner and timing of powering in quadrupedal walking. As Kuo (2002) highlighted for bipeds, this may have significant consequences in terms of efficiency. Reducing Kuo's observations to "powering with a mechanism that smoothes the major collision" allows it to be applied to quadrupeds (Figure 6.11).

Many cursorial quadrupeds have, through evolution, extended their legs and even lost their "toes" (or "fingers"), in extreme cases (such as the horse) standing on only the toenail of a single toe. Compared with humans and most birds, then, these quadrupeds appear to have little opportunity to power walking with a functional ankle-extension just prior to the next footfall – the mechanism identified by Kuo as effective in smoothing out the collision, resulting in smaller power requirements. With four legs, and a bias of the weight support towards the front limbs (as appears to be generally the case in large mammals), powering with the hindlimbs just before the front legs hit the ground potentially results in the same center of mass path-smoothing and efficiency gains as Kuo's ankle extension (Usherwood *et al.*, 2007).

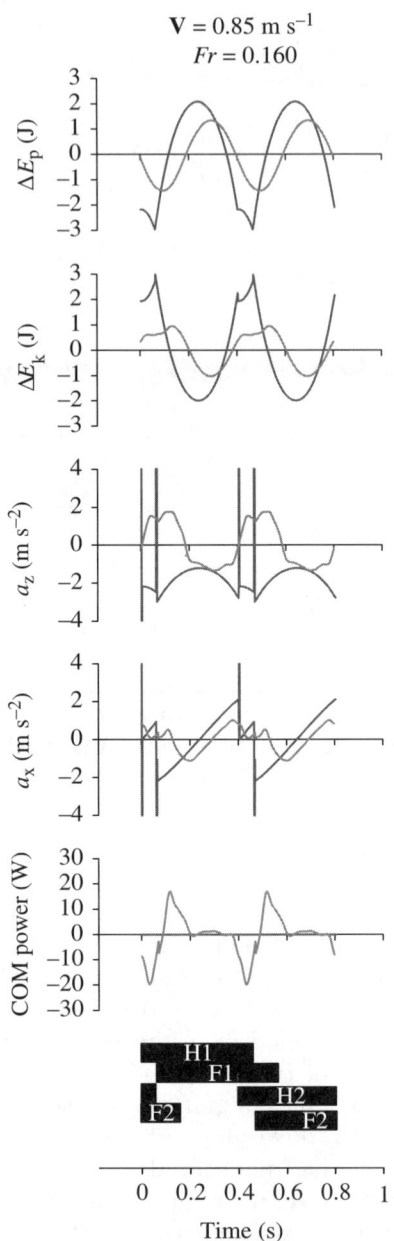

FIGURE 6.9

Fluctuations in potential and kinetic energies, along with net vertical (a_z) and horizontal (a_x) accelerations modeled with a stiff-limbed, passive four-bar linkage, using the observed leg timing (black), and the actual values derived from force plate measurements (grey) for a dog walking at intermediate speed (0.85 m/s; a Froude number of 0.16).

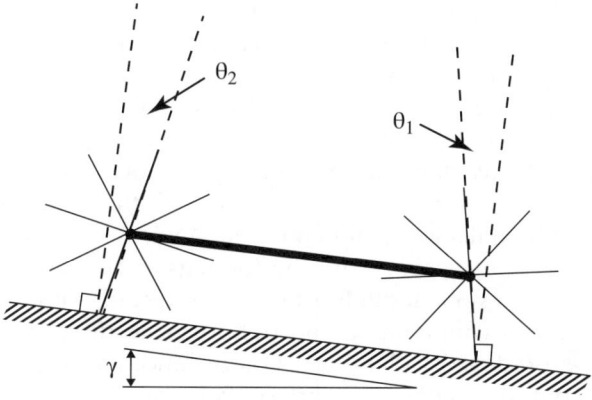

FIGURE 6.10

Smith and Berkemeier (1997) model of quadrupedal walking as a pair of linked rimless wheels.

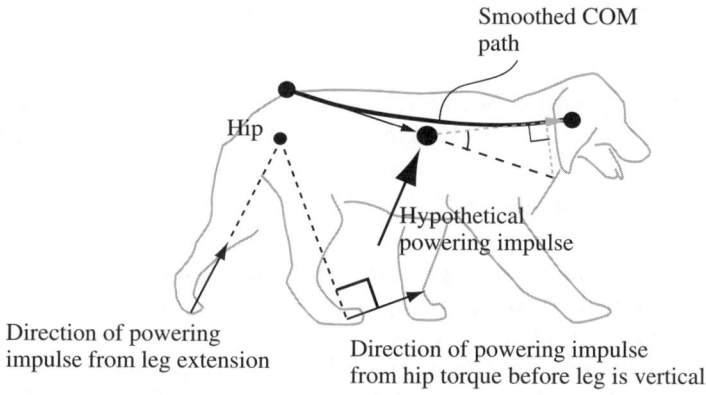

FIGURE 6.11

Potentially beneficial deviation from passive, stiff-limbed quadrupedal walking. With appropriate powering from the hindlimbs just as the forelimbs (supporting around 60% of body weight) hit the ground, the path of the centre of mass can be smoothed in a manner analogous to the ankle extension powering observed in human (bipedal) walking.

This, then, provides one nice account for why many animals use a "clip-clop" footfall pattern, with the hindfoot landing and powering just before the weight-supporting front foot hits the ground. Potentially consistent with this, in cases where the bias of weight support is reversed, often so is the timing of footfalls; both Komodo dragons and crawling babies tend to time the landing of their front feet just before their hind.

6.2.3 Higher speed quadrupedal gaits

The principle that more foot contacts allows a smoother path of the center of mass and, hence (if point-mass collision mechanics are applicable), smaller mechanical energy losses, has been applied to cantering and galloping (Ruina *et al.*, 2005). In this case, a smoother path need not mean wheel-like; the extreme, smooth, lossless path for transition from one ballistic flight phase to another would result in the mechanical analogue of an appropriately spinning oval. With the correct oval shape and spin rate, a perfectly lossless, yet inelastic "bounce" should be achievable (Haggerty, 2001).

Might the timing of footfalls in cantering and galloping quadrupeds be understood from this viewpoint? At this stage, predictions from a point-mass, plastic, momentary collision model match observations rather poorly – all things being equal, an exactly even spacing between footfalls would minimize energy losses. Adding the reality that the hind limbs might provide the power, and the forelegs are largely dissipative, also does not help. In this case, the reverse footfall pattern is predicted, with periods biased so that the higher energy (faster) condition should be longer, leading to a bunny-like footfall timing, with forelegs predicted to land just before hind. For a given average speed, it appears wasteful to put the energy in immediately before taking it back out again.

So, beyond the sensible point that footfalls should be distributed through time, collision mechanics are not yet able to account for the cantering and galloping footfall patterns typical of horses. Whether extensions of the collisional viewpoint, for instance towards distributed masses, or away from one-leg-one-momentary-collision models, will end up in accounting for these gaits remains unclear. We await the model that can truly account for:

1. different galloping styles between species, including "extended" (bunny-like), "gathered" (horse-like) and both extended and gathered (cheetah/greyhound-like) galloping; and
2. the changes in gait with speed.

In particular, the benefits of pacing (both legs on the same side hitting the ground at the same time) and trotting (diagonal opposites at the same time) are very difficult to understand. These gaits appear to ignore the apparently sensible collision-based guideline of spreading out the contacts to smooth the path. Also, there appears to be no particular mechanical constraint prohibiting either very slow cantering (common in dressage) or high-speed locomotion with a walking footfall pattern (termed variously "ambling" or "tölting"), observed in both special breeds of horse (see, for instance, Robillard *et al.*, 2007) and elephants (Hutchinson *et al.*, 2006). So what possible fundamental benefits are there to trotting and pacing at intermediate speeds? Stability? Elephants are not known for their tendency to trip. At this stage, it appears to be anybody's guess.

6.2.4 Further success of reductionist mechanics

Passive/dynamic and collisional reductions of locomotor mechanics are not the only ones capable of providing insight. Gray's work describes how the limbs of a quadruped might be viewed in terms of struts (supporting compression loads) or levers (applying, or resisting, torques about joints, experiencing bending moments; Gray, 1944; Bertram, 2007; and see Chapter 4)). When considered in terms of a powering mechanism, such limbs may equally be described as "telescoping" (extending while under compressive load) or "torquing" (in a wheel-like manner). It is important to distinguish between the actions of a whole limb and those of its constituent joints; a telescoping limb achieves this action through torques at a number of joints. So, where does such a reduction take us? Well, it allows us to account for the following observations:

1. Sprint speeds in humans appear to be constrained (amongst other things) by limb forces; sprint speeds around appropriately banked, tight bends (requiring high centripetal accelerations, and so an increase in effective bodyweight) are predictably worse that those on straight tracks (Figure 6.12; Usherwood and Wilson, 2006).
2. Sprinting greyhounds appear largely unaffected by increases in effective weight due to bend-running; limbs withstand an increase in force of around 65% (Usherwood and Wilson, 2005).
3. Peak accelerations in greyhounds are around 1 G.
4. Peak accelerations in polo ponies are around 0.6 G (Williams *et al.*, 2009).

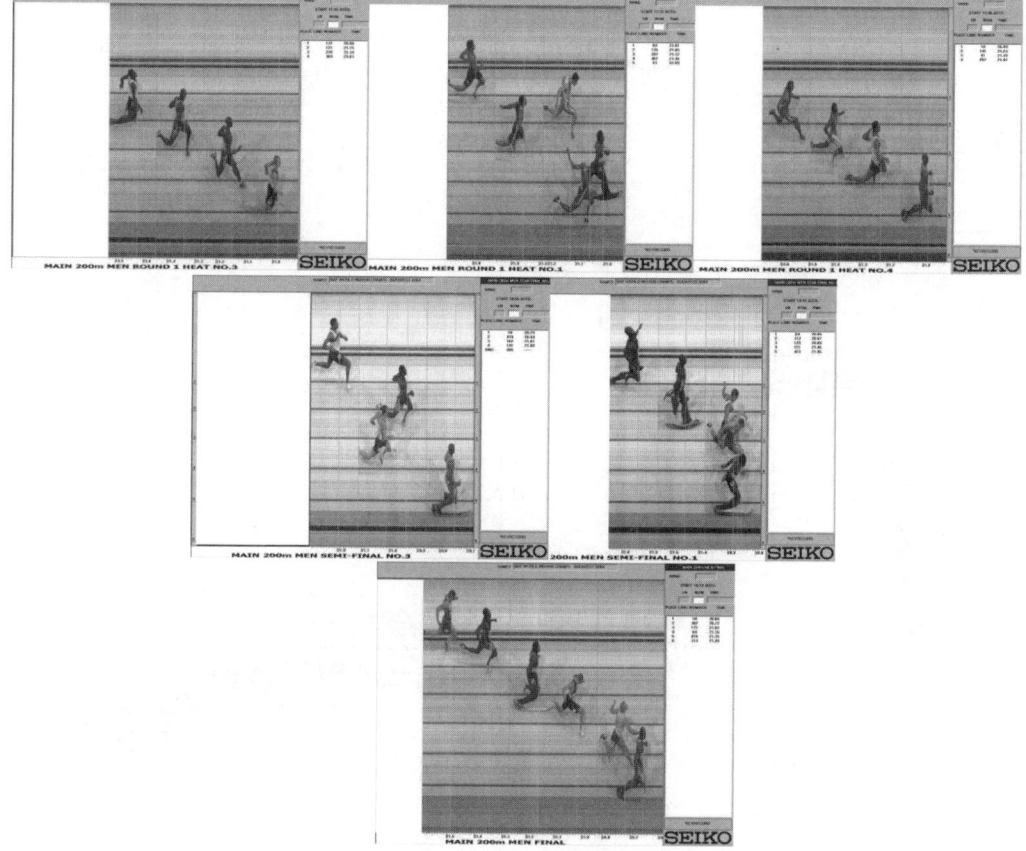

FIGURE 6.12

Finishing photographs of the last series of 200 m races to be held at the World Indoor Championships (2004). The athlete on the inner tracks (positioned at the top of each photo) face a notable disadvantage, quantitatively consistent with a simple limb force constraint model (Usherwood and Wilson, 2006).

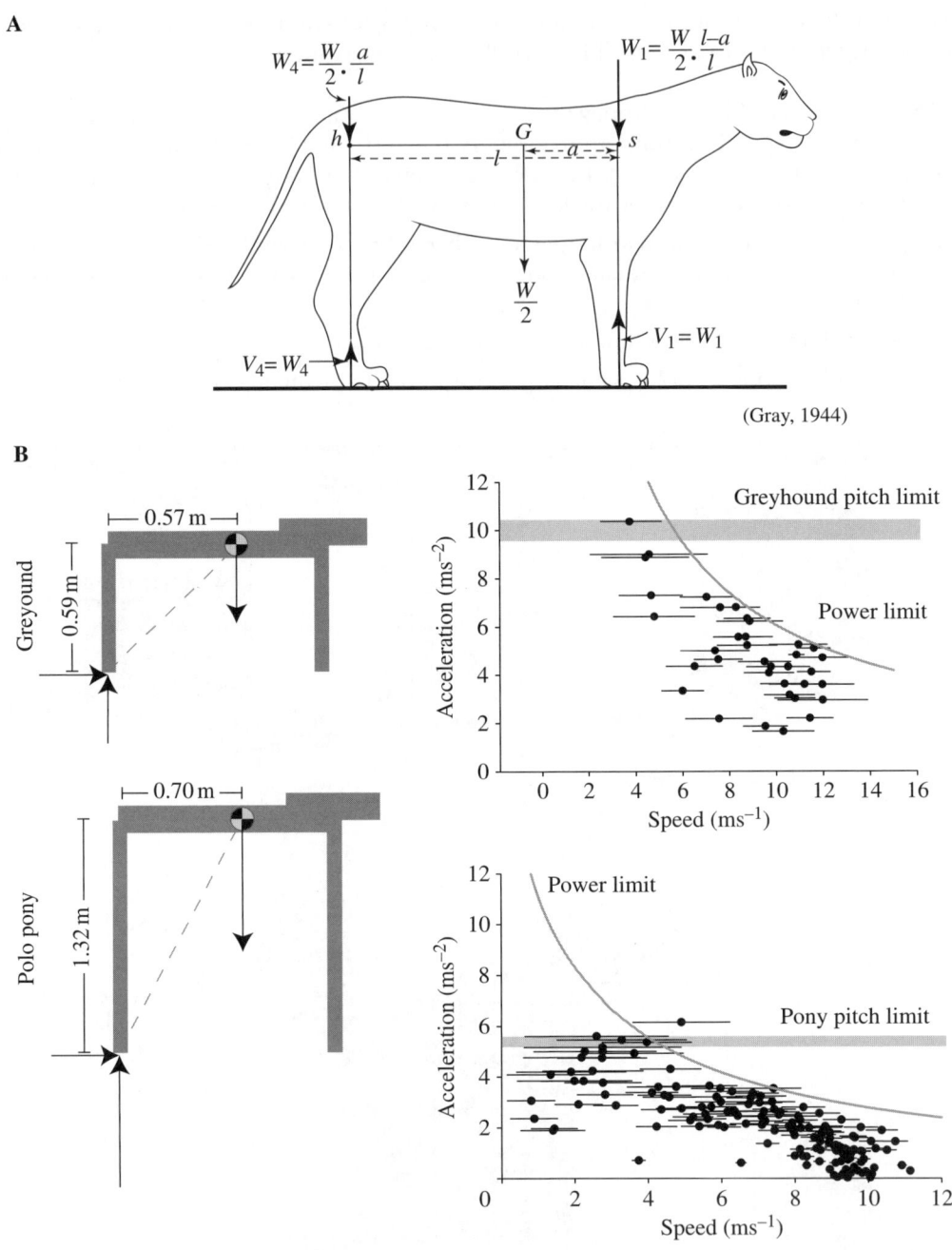

(Gray, 1944)

FIGURE 6.13

(**A**) The classic free-body diagram showing the forces acting on a standing quadruped was developed by Gray (1944), demonstrating the bias of weight support towards the front limbs. (**B**) When extended to maximally accelerating greyhounds and polo ponies, a pitch limit is revealed. At low speeds, acceleration is limited by the avoidance of flipping over backwards; at higher speeds, a simple power limit appears to apply.

The observations are consistent with the hypothesis (involving rather extreme reductionism) that humans power sprinting with telescoping legs, while greyhounds and horses employ torquing, wheel-like limbs. *If* humans power with leg extension, then those muscles powering sprinting are loaded by body weight (via some degree of gearing). Thus, it is not surprising that an increase in effective body weight due to running around a tight bend results in a reduction of sprint speed. In order to protect the muscles, the foot must spend a larger proportion of time on the ground (increase "duty factor") and, if the leg cannot be swung forward any more quickly (as is the case in a flat-out sprint), or the leg angle swept during stance increased, this cannot be achieved without slowing down.

If greyhounds power with the alternative strategy, of wheel-like torquing or levering of the leg, then speed need not be reduced with increased effective weight. Just as with a cyclist, the structures supporting body weight can be distinct from the muscles providing the power; increases in effective weight on a bend do not threaten to overload the powering muscles.

If we take the bicycle as the reductionist mechanical analogy of cursorial quadrupeds, constraints on maximum acceleration become apparent. *If* greyhounds and horses are powered predominantly with wheel-like torquing limbs, then the net forward acceleration over a stride cycle is constrained by the requirement to avoid pitching – or "wheelie-ing" – over backwards (Figure 6.13). Remarkably, with two simple measurements – rear leg length, and back length – and one further anatomical assumption (that the center of mass lies 60% along the way from the hips to the shoulders, as indicated by several force plate studies), peak accelerations can be successfully predicted. Greyhounds are relatively long, with the center of mass around a leg length forward of the hips, and achieve accelerations up to 1 G (10 ms^{-2}). Ponies are relatively tall, with centre of mass only 50–60% leg length in front of the hips, and they accelerate at 0.5–0.6 G.

While myriad other factors, from muscle dynamics to friction properties of the track surface, might come in to play, the "wheelie-avoidance" account (lifting the forelimbs off contact with the substrate) for peak acceleration performance is pleasingly parsimonious. With only one mechanical constraint – torque-dominated powering – and two tape measure measurements, a performance parameter of critical importance to predator, prey or athlete can be quantitatively predicted with some success.

The torquing nature of cursorial quadruped limbs (at least greyhounds and ponies) are also consistent with Biewener's (1989) observations concerning posture. Relatively large, cursorial quadrupeds have relatively upright limbs. One consequence of this is that the "mechanical advantage" or "gearing" between the powering muscles and the forces from the ground is altered in favor of the muscles – with upright limbs, the muscles experience relatively smaller proportions of the body weight. One consequence of this is that these upright limbs have little opportunity to extend, and so the capacity to power with a telescoping limb becomes limited. In effect, the observation that larger cursorial quadrupeds have relatively upright limbs, with consequences in terms of muscle gearing, is the same as noting more torquing or "wheel-like" powering mechanics, consistent with both the costs (limited acceleration) and benefits (ability to corner without slowing) of bicycle-like locomotion.

APPENDIX A: ANALYTICAL APPROXIMATION FOR COSTS OF TRANSPORT INCLUDING LEGS AND "GUTS AND GONADS" LOSSES

6A.1 List of symbols

CoT	Mass-specific cost of transport (J kg^{-1} m^{-1})
F	Force (N)
g	Gravity (ms^{-2})
h	Height (m)
l	Leg length (m)
m	Mass (kg)
S	Displacement (m)
t	Time (sec)
T	Period (sec)
U	Initial velocity (ms^{-1})
V	Velocity (ms^{-1})
W	Work (J)
ρ	Instantaneous leg angle from vertical
Φ	Stance angle swept from vertical (half stance angle)

Subscripts

gut	relating to body components – the "viscera" above the leg
leg	relating to the leg
step	for a step (half a stride)
stiff	infinitely stiff leg
stride	for a stride (two steps)
t	relating to an instant in time
x	horizontal axis in direction of travel
z	vertical axis

One key assumption made here is that "negative" work – specifically shortening while under compressive loads – "costs" in energetic terms. That is, while elastic components might be present, none is perfect (hysteresis is present), so a proportion of this negative work (which itself might be achieved relatively easily) must be "paid" with positive work.

6A.2 Period definitions for a symmetrically running biped

The horizontal displacement during stance $S_{x,\text{stance}}$ relates to the leg length l and stance angle (expressed here as the maximum angle swept from vertical) Φ:

$$S_{x,\text{stance}} = 2l \sin(\Phi) \tag{6.9}$$

Thus, the stance period T_{stance}, stride period T_{stride}, step period T_{step} and aerial period T_{aerial} are simply:

$$T_{\text{stance}} = \frac{S_{x,\text{stance}}}{\overline{V}_x} \tag{6.10}$$

$$T_{\text{stride}} = T_{\text{prot}} + T_{\text{stance}} \tag{6.11}$$

$$T_{\text{step}} = \frac{T_{\text{stride}}}{2} \tag{6.12}$$

$$T_{\text{aerial}} = T_{\text{step}} - T_{\text{stance}} \tag{6.13}$$

respectively, where $\overline{V}_x$ is the mean horizontal velocity and T_{prot} the protraction or swing-leg period. These are correct for a symmetrically running biped. If the development is to be applied to hopping or galloping, these should be adjusted appropriately.

6A.3 Ideal work for the leg

The minimum work that must be absorbed (and returned) by a leg is that associated with vertical motions for very brief stance periods (very stiff legs) – without any horizontal forces. The limiting case can be derived from simple ballistics. Given, for the special case where of Φ = zero (infinitely stiff limbs, negligible stance period):

$$T_{\text{aerial,stiff}} = \frac{T_{\text{prot}}}{2} \tag{6.14}$$

the initial upward velocity U_z at the beginning of the aerial phase must be:

$$U_z = -g\frac{T_{\text{aerial}}}{2}. \tag{6.15}$$

where g is -9.81 ms^{-2} (down). Thus, the change in height during aerial phase Δh_{aerial} is given by:

$$\Delta h_{\text{aerial}} = U_z \frac{T_{\text{aerial}}}{2} + \frac{1}{2}g\left(\frac{T_{\text{aerial}}}{2}\right)^2 \tag{6.16}$$

Therefore, the energy that must be absorbed and returned by the leg due to the transition from downward to upward motion equates to the change in potential energy during the aerial phase:

$$W_{\text{ideal}} = mg\Delta h_{\text{aerial}} \tag{6.17}$$

where W_{ideal} expresses the "ideal" or minimum work associated with running with a constrained protraction period for a body of mass m (see also Rashevsky, 1948). However, for real legs, stiffness is not infinite, forces are not infinite, both vertical and horizontal deflection are finite, and so additional work is required.

6A.4 Vertical work calculations for leg

Making the assumption that the vertical forces $F_{z,t}$ follow a half-sinusoidal profile during stance (broadly agreeing with empirical observations for a large variety of running animals):

$$F_{z,t} = F_{z,max} \sin\left(\frac{2\pi t}{2T_{stance}}\right) \qquad (6.18)$$

where: $F_{z,max}$ is maximum vertical force and t is time.

In order for the average vertical force over a step cycle to oppose body weight:

$$mgT_{step} = \int_0^{T_{step}} F_{z,t} \; dt = \int_0^{T_{stance}} F_{z,max} \sin\left(\frac{\pi t}{T_{stance}}\right) dt \qquad (6.19)$$

And so (broadly following Alexander, 1976):

$$F_{z,max} = \frac{-\pi m g T_{step}}{2T_{stance}} \qquad (6.20)$$

and we can calculate the vertical force profile from expressions in Equations 6.18 and 6.20.

The vertical velocity at the beginning of stance is (assuming symmetry during stance) the negative of that for initiation of the ballistic aerial phase (Equation 6.15), using Equation 6.13 to calculate the aerial period. With this, and the vertical force profile (Equation 6.20), the vertical force profile throughout stance $V_{z,t}$ can be determined:

$$V_{z,t} = -U_z + \int_0^t \left(\frac{F_{z,t}}{m} + g\right) dt \qquad (6.21)$$

The negative work, W_z, due to vertical forces and motions, occurs during the first half of stance:

$$W_z = \int_0^{\frac{T_{stance}}{2}} V_{z,t} F_{z,t} \; dt \qquad (6.22)$$

6A.5   Horizontal work calculations for leg

For small values of Φ, the angle ϕ phi from vertical through time t of stance approximates:

$$\phi_t \approx -\Phi + \frac{\overline{V_x}\, t}{l} \tag{6.23}$$

Assuming predominantly compression leg forces (broadly matching observations for the majority of vertebrate legs), again for small ϕ, the horizontal forces through stance $F_{x,t}$ can be determined:

$$F_{x,t} = F_{z,t}\tan\left(\phi_t\right) \approx F_{z,t}\phi_t \tag{6.24}$$

Continuing with the assumption that stance angles are relatively small, changes in horizontal velocity are also relatively small, thus the negative work (during the first half of stance) is approximately:

$$W_x \approx \int_0^{\frac{T_{stance}}{2}} F_{x,t}\overline{V_x}\; dt \tag{6.25}$$

6A.6   Hysteresis costs of "guts and gonads" deflections

Supposing that the deflections of components "above" the hip – for instance the guts – are due to the forces imposed through the legs, the negative work associated with these deflections can be approximated as:

$$W_{gut} = \frac{F_{gut,max}^{\;2}}{2k_{gut}} \tag{6.26}$$

where $F_{gut,max}$ is the maximum force experienced by the gut (and the – largely unimportant – factor of two is due to the assumption that the viscera act as a linear spring). If the leg is treated as being relatively low-mass, the forces experienced at the top of the leg (the hip, or leg-body junction) approximate those at the foot – the ground reaction forces modeled above with sinusoidal assumptions. Thus,

$$F_{gut,max} \propto F_{z,max} \tag{6.27}$$

Therefore, smaller (more upright) stance angles (requiring stiffer legs), which result in larger peak leg forces, also result in larger loading – and work done – on the viscera.

6A.7 Cost of transport

These expressions for work can also be expressed as mass-specific cost of transport. This is particularly convenient, as it makes comparison reasonable between steps of different length. The horizontal displacement over a step $S_{x,step}$ is:

$$S_{x,step} = \overline{V_x} \, T_{step} \tag{6.28}$$

The mass-specific mechanical cost of transport CoT relates to the negative work performed W and the degree to which this energy can be stored and recovered elastically η:

$$CoT = \frac{(1-\eta)W}{mS_{x,step}} \tag{6.29}$$

Realistic, reliable values for hysteresis $(1 - \eta)$ for whole limb or whole body (and especially "guts and gonads") are very hard to come by. While materials such as tendon or connective tissue may provide reasonable elasticity (93% appears to be the rule of thumb), whenever calculations are made of how much of the center of mass mechanical energy fluctuation can be directly attributed to such tissue loading, the view of even specialized cursorial legs as useful springs is usually questionable. As a first approximation, the case of zero elasticity is considered; the whole negative work for both legs and "guts" is assumed to be "paid" directly by positive, "costly" work.

▌ REFERENCES

Adamczyk PG, Kuo AD (2009). Redirection of center-of-mass velocity during the step-to-step transition of human walking. *Journal of Experimental Biology* **212**, 2668–2678.

Ahn AN, Furrow E, Biewener AA (2004). Walking and running in the red-legged running frog, *Kassina maculate*. *Journal of Experimental Biology* **207**, 399–410.

Alexander RMcN (1976). Mechanics of bipedal locomotion. In: Spencer Davies P (ed), *Perspectives in Experimental Biology*, **1**: 493–504. Oxford, Pergamon.

Bertram, JEA (2007). How Animals Move: Studies on the mechanics of the tetrapod skeleton. *Journal of Experimental Biology* **210**, 2401–2402.

Biewener AA (1989). Scaling body support in mammals: limb posture and muscle mechanics. *Science* **245**, 45–48.

Bishop KL, Pai AK, Schmitt D (2008). Whole body mechanics of stealthy walking in cats. *PLoS One* **3**, e3808.

Carrier DR, Heglund NC, Earls KD (1994). Variable gearing during locomotion in the human musculoskeletal system. *Science* **265**, 651–653.

Cavagna GA, Heglund NC, Taylor CR (1977). Mechanical work in terrestrial locomotion: two basic mechanisms for minimzing energy expenditure. *American Journal of Physiology* **233**, R243–R261.

Daley MA, Usherwood JR (2010). Two explanations for the compliant running paradox: reduced work of bouncing viscera and increased stability in uneven terrain. *Biology Letters* **6**, 418–421.

Dean JC, Kuo AD (2009). Elastic coupling of limb joints enables faster bipedal walking. *Journal of The Royal Society Interface* **35**, 561–573.

Doke J, Donelan JM, Kuo AD (2005). Mechanics and energetics of swinging the human leg. *Journal of Experimental Biology* **208**, 439–445.

Farquharson ASL (2007). *On the gait of animals by Aristotle.* (http://etext.library.adelaide.edu.au/a/aristotle/gait/complete.html) eBooks@Adelaide, Univ. of Adelaide Library.

Geyer H, Seyfarth A Blickhan R (2006). Compliant leg behaviour explains basic dynamics of walking and running. *Proceedings of the Royal Society B* **273**, 2861–2867.

Gray J (1944). Studies in the mechanics of the tetrapod skeleton. *Journal of Experimental Biology* **20**, 88–116.

Griffin TM, Main RP, Farley CT (2004). Biomechanics of quadrupedal walking: how do four-legged animals achieve inverted pendulum-like movements? *Journal of Experimental Biology* **207**, 3545–3558.

Haggerty P (2001). *Radiation-induced instability.* Ph D. thesis, University of Michigan, Ann Arbor.

Hubel TY, Usherwood JR (2013). Vaulting mechanics successfully predict decrease in walk – run transition speed with incline. *Biology Letters* **9**, 20121121.

Hutchinson JR, Schwerda D, Famini DJ, Dale RHI, Fischer MS, Kram R (2006). The locomotor kinematics of Asian and African elephants: changes with size and speed. *Journal of Experimental Biology* **209**, 3812–3827.

Jayes AS Alexander RM (1980). The gaits of chelonians: walking techniques for very low speeds. *Journal of Zoology, London* **191**, 353–378.

Kram R, Taylor CR (1990). Energetics of running: a new perspective. *Nature* **346**, 265–267.

Kuo A D (2002). Energetics of actively powered locomotion using the simplest walking model. *Journal of Biomechanical Engineering* **124**, 113–120.

Madsen EL (1977). Theory of the chimney breaking while falling. *American Journal of Physics* **45**, 182–184.

Marsh RL, Ellerby DJ, Carr JA, Henry HT, Buchanan CI (2004). Partitioning the energetics of walking and running: swinging the limbs is expensive. *Science* **303**, 80–83.

McGeer T (1990). Passive dynamic walking. *International Journal of Robotics Research* **9**, 62–82.

McGeer T, Palmer LH (1989). Wobbling, toppling and forces of contact. *American Journal of Physics* **57**, 1089–1098.

Minetti AE, Ardigo LP, Saibene F (1994). The transition between walking and running in humans: metabolic and mechanical aspects at different gradients. *Acta Physiologica Scandinavica* **150**, 315–323.

Pontzer H (2005). A new model predicting locomotor cost from limb length via force production. *Journal of Experimental Biology* **208**, 1513–1524.

Pontzer H (2007a). Effective limb length and the scaling of locomotor cost in terrestrial animals. *Journal of Experimental Biology* **210**, 1752–1761.

Pontzer H (2007b). Predicting the cost of locomotion in terrestrial animals: a test of the LiMb model in humans and quadrupeds. *Journal of Experimental Biology* **210**, 484–494.

Rashevsky N (1948). On the locomotion of mammals. *Bulletin of Mathematical Biology* **10**, 11–23.

Ren L, Hutchinson JR (2008). The three-dimensional locomotor dynamics of African (Loxodonta africana) and Asian (Elaphas maximus) elephants reveal a smooth gait transition at moderate speed. *Journal of The Royal Society Interface* **5**, 195–211.

Robillard JJ, Pfau T Wilson AM (2007). Gait characterisation and classification in horses. *Journal of Experimental Biology* **210**, 187–197.

Ruina A, Bertram JEA, Srinivason M (2005). A collisional model of the energetic cost of support work qualitatively explains leg sequencing in walking and galloping, pseudo-elastic leg behavior in running and the walk-to-run transition. *Journal of Theoretical Biology* **237**, 170–192.

Smith AC, Berkemeier MD (1997). Passive dynamic quadrupedal walking. *Proceedings of the 1997 IEEE International conference on robotics and automation* 34–39.

Srinivasan M, Ruina A (2006). Computer optimization of a minimal biped model discovers walking and running. *Nature* **439**, 72–75.

Srinivasan M, Ruina A (2007). Idealized walking and running gaits minimize work. *Proceedings of the Royal Society of London. Series A* **463**, 2429–2446.

Usherwood JR (2005). Why not walk faster? *Biology Letters* **1**, 338–341.

Usherwood JR (2010). Inverted pendular running: a novel gait predicted by computer optimization is found between walk and run in birds. *Biology Letters* **6**, 765–768.

Usherwood JR, Wilson AM (2005). No force limit on greyhound sprint speed. *Nature* **438**, 753–754.

Usherwood JR, Wilson AM (2006). Accounting for elite indoor 200 m sprint results. *Biology Letters* **2**, 47–50. doi: 10.1098/rsbl.2005.0399

Usherwood JR, Williams SB, Wilson AM (2007). Mechanics of dog walking compared with a passive, stiff-limbed, 4-bar linkage model, and their collisional implications. *Journal of Experimental Biology* **210**, 533–540.

Usherwood JR, Szymanek KL, Daley MA (2008). Compass gait mechanics account for top walking speeds in ducks and humans. *Journal of Experimental Biology* **211**, 3744–3749.

Usherwood JR, Channon AJ, Myatt JP, Rankin JW, Hubel TY (2012). The human foot and heel–sole–toe walking strategy: a mechanism enabling an inverted pendular gait with low isometric muscle force? *Journal of The Royal Society Interface* **9**, 2396–2402.

Williams SB, Tan H, Usherwood JR, Wilson AM (2009). Pitch then power: limitations to acceleration in quadrupeds. *Biology Letters* **5**, 610–613.

CHAPTER SEVEN

Whole-Body Mechanics: How Leg Compliance Shapes the Way We Move

Andre Seyfarth[1], Hartmut Geyer[2], Susanne Lipfert[3], J. Rummel[4], Yvonne Blum[5], M. Maus[6] and D. Maykranz[4]

[1]*Institute for Sport Science, Technical University of Darmstadt, Germany*
[2]*Carnegie Mellon University, Pittsburgh, PA, USA*
[3]*Human Motion Engineering, Lausanne, Switzerland*
[4]*Friedrich Schiller University, Jena, Germany*
[5]*Royal Veterinary College, London, UK*
[6]*Imperial College London, London, UK*

7.1 INTRODUCTION

The function of the human leg during movements is complex. One challenge lies in the segmented structure of the leg, which consists of thigh, shank and foot. The situation is further challenged by the parallel arrangement of muscles spanning a single or multiple leg joint(s). Depending on the coordinated activation of these muscles and the configuration of the leg segments, the leg is capable of generating the forces that are required to fulfill a desired movement task. This poses the question of how the leg function needs to be organized to make typical movements, such as walking, running or jumping, possible and easily accessible, despite the high complexity of the underlying musculo-skeletal system?

In this chapter, we argue that spring-like leg function is a commonly observed feature during movements, which can be described by characteristic leg parameters. By properly adjusting these global leg properties, selected movement tasks can be realized (Figure 7.1, upper panels). These tasks include maximizing jumping performance, and stabilizing cyclic tasks like running and walking. Surprisingly, to achieve a certain movement goal, the leg adjustment is usually not unique. This means that the same performance or level of stability can be achieved by different leg configurations (e.g., joint angles, muscle activations). For instance, this functional redundancy of the leg adjustment can be found in the long jump (Seyfarth *et al.*, 1999), where the same jumping distance can be achieved by different leg strategies (e.g., different combinations of leg stiffness and angle of attack[1]). This basic observation of redundant leg adjustments fulfilling the same movement goal is a key for understanding the organization of the body mechanics during movements.

In legged locomotion, such as walking or running, the optimal strategy is less obvious, as there is no single parameter (such as maximum jumping distance or height) which needs to be optimized. We propose that cyclic gait stability could be an appropriate movement goal, which can be used to identify successful strategies for adjusting compliant legs in human or animal locomotion. In the framework of deterministic biomechanical models, cyclic stability requires that a steady-state gait pattern (e.g.,

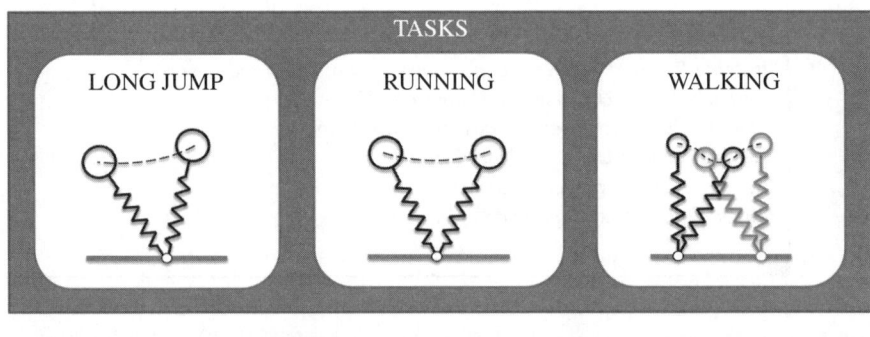

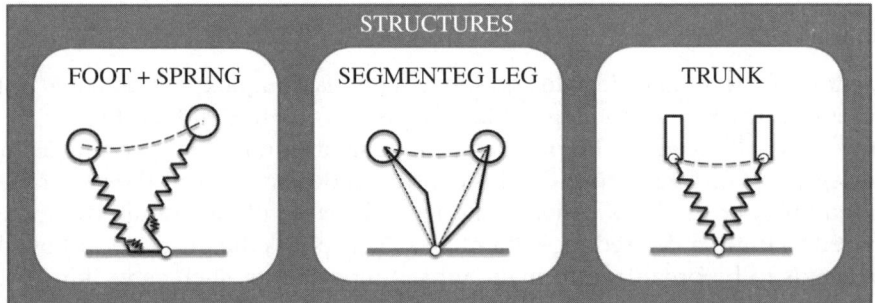

FIGURE 7.1

Movement tasks and structural elements used in this chapter to set up conceptual models for investigating the organization and function of the human body during locomotion. Upper panels: During different movement tasks (jumping, running, walking), spring-like operating legs are found in different configurations, redirecting the center of mass during ground contact. Lower panels: Different structures of the segmented body (foot, lower and upper leg, upright trunk) specifically modulate the function of the leg during locomotion.

same apex height[2] of the center of mass, CoM, from step to step) is approached by different initial conditions (e.g., perturbed initial CoM speeds and heights at apex). It is important to note that a perfectly periodic limit cycle, as predicted by the model, is generally not observed in human or animal locomotion. Hence, in a more relaxed way, steady-state gait patterns require achieving approximately the same system state over a high number subsequent steps, taking ongoing internal or external variations into account.

With this approach, it is possible to identify adjustments of a spring-like operating leg, which result in highly diverse but stable movement patterns covering different gaits, step lengths and speeds. Stable running is predicted for moderate and fast speeds (Seyfarth *et al.*, 2002). In contrast, walking patterns are predicted for lower speeds with ground reaction forces remarkably similar to those in human and animal walking (Geyer *et al.*, 2006).

The concept of deriving gait patterns based on stability analysis of conceptual models will be extended in more detailed models addressing:

1. the roll-over function of the foot;
2. the segmentation of the leg; and
3. the upright trunk posture (Figure 7.1, lower panels).

With the help of these models with increased complexity, we aim at better understanding the origin of movement patterns observed in legged locomotion under varying structural conditions. Comparing model predictions with experimental data allows these concepts to be validated, so they may become accessible for future technologies such as orthoses or prostheses.

This chapter is organized as follows: first, we will describe the leg function during the take-off phase in the long jump, to explain the interplay between leg stiffness and leg angle at touch-down in optimizing jumping distance. Then, we will explain how, for gait patterns like running or walking, the movement goal can be formulated. We will see that the requirement for stable cyclic locomotion can lead to similar leg adjustment strategies as those previously identified for the long jump. This proposed gait model can not only predict common gait patterns like walking and running, it also predicts gait patterns at low speeds, which are less stable and which were not described previously. Finally, we will discuss how legged locomotion can be represented in greater detail when the mechanics of the trunk, the upper and lower leg, or the foot are taken into account. We will conclude this chapter by pointing to possible future directions in the development of conceptual models for legged locomotion.

7.2 JUMPING FOR DISTANCE — A GOAL-DIRECTED MOVEMENT

The take-off phase in long jump – the last ground contact before the flight phase – is used to redirect the movement of the body from a high horizontal velocity to an optimal take-off condition for the subsequent flight phase. If all of the initial energy could be redirected, an optimal take-off angle of about $\beta = 45°$ (Figure 7.2A) would be expected. Given a run-up speed of 10 m/sec and a contact time of 100 ms, this would require extreme leg forces of more than eight times body weight (e.g., over 6400 N for

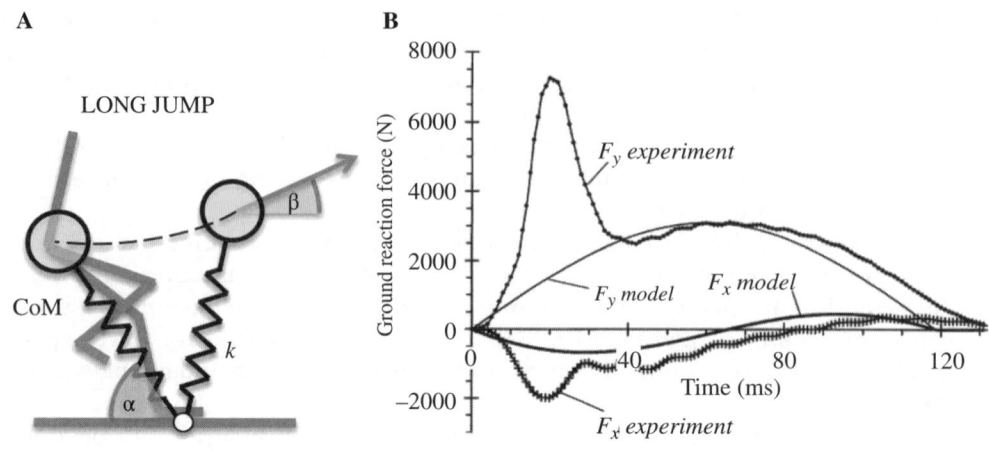

FIGURE 7.2

(**A**) The dynamics of the center of mass (CoM) during the take-off phase of a running long jump can be described with a simple spring-mass model (Blickhan, 1989). The leg is described by the angle of attack α and the leg stiffness k. The take-off angle of CoM is described by the velocity angle β. (**B**) With this model, the pattern of the experimental ground reaction forces after impact (after 40 ms) can be approximated in vertical (F_y) and horizontal (F_x) direction (Seyfarth *et al.*, 1999).

an athlete of 80 kg body mass). High forces require eccentric muscle operation, and both leg force and the capacity to redirect the body without energy losses are limited. Therefore, take-off angles in the long jump are clearly less than 45 degrees, and energy losses cannot be prevented during the take-off phase (Seyfarth *et al.*, 2000).

An optimal take-off strategy could be to redirect the body with take-off angles allowing for a substantial recovery of elastic energy stored in the leg. This is indicated by the observed patterns of the ground reaction force (Figure 7.2B). Shortly after touch-down, there is a high impact peak, followed by an almost sinusoidal second force peak – the active force peak. The shape of the second peak can be represented with a simple spring-mass model (Figure 7.2A), where a linear prismatic spring describes the function of the leg during contact.

Assuming a spring-like leg behavior in the long jump, possible strategies for optimizing jumping distance can be predicted by the model. Given a certain approach velocity $3v_0$, sufficient leg stiffness is required to redirect the body to achieve close-to-optimum jumping distance. However, there is no unique combination of leg stiffness and angle of attack for maximizing jumping performance (Figure 7.3A). A similar distance can be achieved by alternating strategies (i.e., by using flat angles of attack and soft legs, or by using steep touch-down angles and stiff legs). Therefore, there is no single optimal leg stiffness or angle of attack for the take-off phase in long jump. It is, rather, the combination of the two parameters, which determines the performance of the athlete. This means that leg stiffness does not differ between the better and the worse athletes (Figure 7.3A), but the location of the pair of leg stiffness and angle of attack are closer to the optimal region for the better jumpers.

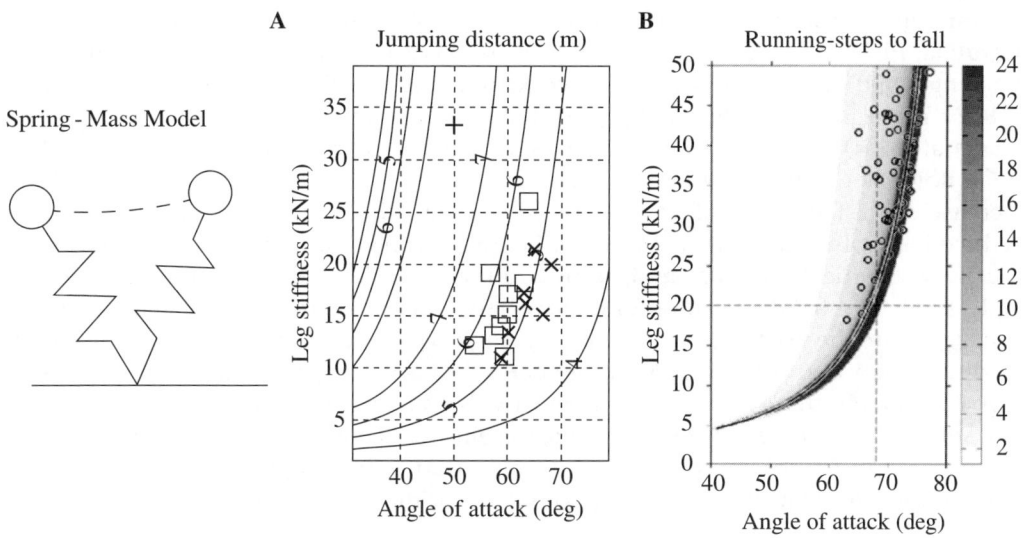

FIGURE 7.3

Predictions of the spring-mass model (left) dependent on the adjustment of the angle of attack and leg stiffness in comparison to experimental data: (**A**) predicted jumping distance for the long jump (Seyfarth *et al.*, 1999) and (**B**) predicted steps to fall in running (Seyfarth *et al.*, 2002). In both cases (**A, B**), appropriate initial conditions of the center of mass movement were given. The predictions of the model indicate that for good jumps (**A**) and stable running (indicated by at least 24 successful running steps, **B**), a proper combination of leg stiffness and angle of attack must be chosen. In both tasks, no preferred (optimal) leg stiffness is predicted. This is also reflected by the experimental data.*Experimental data:* (**A**) Leg adjustments for better jumps (squares: over 6 m) and worse jumps (crosses: under 5 m), (**B**) Leg adjustments in human running (circles) at about 5 m/sec.

7.3 RUNNING FOR DISTANCE — WHAT IS THE GOAL?

Compared with the long jump, the situation in running is somewhat different. The movement goal is not as simple as maximizing the distance travelled during the flight phase. In contrast to jumping for distance, running is a periodic motion, and periodicity is inherently coupled to stability in dynamic systems. Therefore, a consequent movement goal in running could be to ensure stability. But how need the leg parameters of a running system be adjusted to ensure stability?

At first glance, there seem to be unlimited possibilities to achieve a gait pattern for either walking or running (Srinivasan & Ruina, 2006). Assuming that the legs need to perform work during the contact phase, some of these trajectories are more costly than others. In fact, the cheapest way to travel on legs would be to keep them straight all the time (Alexander, 1976). Although this is kinematically quite similar to what we do during mid-stance in walking, it clearly provides no explanation for the human leg function in running.

Fortunately, there is a quite similar leg behavior in running as previously described in the long jump model (Figure 7.2A). The leg forces generated during the stance phase of running are remarkably similar to those a linear prismatic spring would generate (Cavagna, 1977). This, again, supports the spring-mass model (Blickhan, 1989; McMahon & Cheng, 1990) as a conceptual framework to investigate the dynamics of running.

For a given initial condition (e.g., defined by the running speed and center of mass height at apex) and depending on the selected leg parameters (e.g., rest length at touch-down, leg stiffness), forward speed may increase (e.g., steep angle of attack) or decrease (e.g., flat angle of attack). In both cases, the next apex height will adapt (reduced apex height or increased apex height, correspondingly). The energy is shuffled between the horizontal and vertical directions, depending on the selected leg angle. How can this distribution of energy between both directions be managed?

It seems to be tricky to calculate the angle of attack to match the steady-state condition with both constant running speed and constant apex height from step to step. Fortunately, there are two simple mechanisms that help to solve the problem. The first solution is offered by the mechanics itself; the second one results from the fact that we do not fix the leg parameters in preparation of the ground contact. We will explain both mechanisms with the help of the spring-mass model.

7.4 CYCLIC STABILITY IN RUNNING

For running with spring-like legs, two leg parameters need to be adjusted before landing – leg stiffness and angle of attack (Figure 7.2A). Given a certain apex condition (CoM height, forward CoM velocity), an appropriate set of these two parameters might lead to the same apex condition in the following flight phase. Interestingly, the combination of leg stiffness and angle of attack leading to a periodic gait pattern is not unique. Both leg stiffness and leg angle can be largely changed, still leading to a steady state running pattern (Blum *et al.*, 2007). It is, however, important how both parameters are adjusted to each other.

It turns out that, for running, similar to the long jump (Figure 7.3A), both parameters need to be aligned close to a 'J'-shaped curve when leg stiffness is plotted versus leg angle. This is illustrated in Figure 7.3B. Given initial conditions (running speed 5 m/sec, initial apex height 1 m), the number of running steps is counted until the model predicts a gait failure – that is, the CoM is hitting the ground (e.g., when the angle of attack is too steep) or the forward CoM velocity vanishes (e.g., when the angle of attack is too flat). This approach can be described as a step-to-fall method (Rummel *et al.*, 2008). Within a certain range, different leg angles may lead to continuous running for one single value of the leg stiffness (Figure 7.3B). For practical reasons[3], the maximum number of steps is limited to 24.

A more detailed analysis (based on a Poincaré return map) reveals that – within limits – even varying leg parameters may still result in stable running patterns (Seyfarth *et al.*, 2002). This indicates that the mechanical system represented by the spring-mass model is robust with respect to variations in the leg adjustment.

The simulation studies based on the spring-mass model indicate that the challenge of stabilizing periodic gait patterns may be facilitated by an attractive mechanical behavior of the body based on compliant leg function. However, the comparison with experimental data reveals that, in human running, the variation in the angle of attack (Lipfert, 2010) is much higher than is predicted by the model. Consequently, we need to question the assumptions made with the spring-mass model.

Both leg stiffness and angle of attack are not fixed parameters during locomotion (De Wit *et al.*, 2000). In fact, the leg angle is adapting during swing phase in order to reach the next landing configuration. The swing leg first protracts (passing the other leg), and finally retracts, in preparation for the ground contact (Lipfert, 2010). An adaptation of leg orientation before landing was used in previous control approaches to stabilize hopping and running robots (Raibert, 1986). In the following, we focus on swing leg retraction as a simple strategy that may further enhance running stability.

7.5 THE WHEEL IN THE LEG – HOW LEG RETRACTION ENHANCES RUNNING STABILITY

In the previous section, we demonstrated that both jumping for distance and stable running require a proper adjustment of leg stiffness and angle of attack. The model analysis, however, shows a considerable sensitivity of the system behavior to selected leg properties. For instance, for 5 m/sec running with a given leg stiffness, the angle of attack must be adjusted within a few degrees. This is hard to achieve in a biological or technical system under highly dynamic conditions. Therefore, we ask for strategies facilitating the adjustment of the leg parameters.

Shortly before ground contact, the leg angle is not constant in human running. Assuming a constant rotational velocity of the leg, similar to the spoke of a turning wheel, the leg angle would depend on an initial angle (e.g., preset at the apex of the flight phase: α_{APEX}) and the leg's angular velocity ω:

$$\alpha\left(t\right) = \alpha_{APEX} + \omega \cdot \left(t - t_{APEX}\right)$$

Depending on the duration of the flight phase, different touch-down angles are possible (Figure 7.4). For longer flight phases, the angle of attack would be steeper, resulting in a more forward orientation of the leg during contact. As a result, part of the vertical landing velocity would be transferred into horizontal CoM velocity after contact, resulting in a flatter flight phase. For shorter flight phases, the opposite is true; the landing angle is flatter, increasing the vertical CoM velocity at the beginning of the next flight phase. Therefore, swing leg retraction can influence running stability by finding an appropriate landing angle which can adapt to different landing conditions (e.g., ground levels).

The running stability can be further explored by selecting combinations of leg retraction, leg stiffening and leg lengthening during swing phase (Blum *et al.*, 2007, 2010). With these generalized swing leg strategies, running can be stabilized at speeds where spring-mass running with fixed leg parameters would become unstable (e.g., at slow running below 3 m/sec; Seyfarth *et al.*, 2003). In a study on bird running (Daley *et al.*, 2007), it was shown that swing leg retraction remains almost unchanged, even when the animal is running over a large, unexpected drop in the ground. This supports the idea of stabilizing running by simple leg strategies.

In the spring-mass model for running, the function of the leg is reduced to a linear prismatic spring, supporting a point mass. In the following, we add more structural details to the leg and body design, and analyze the effects on the dynamics of locomotion. These structures include a second contralateral spring-like leg, an upright trunk, a segmentation with upper and lower leg, and a foot with spring-like

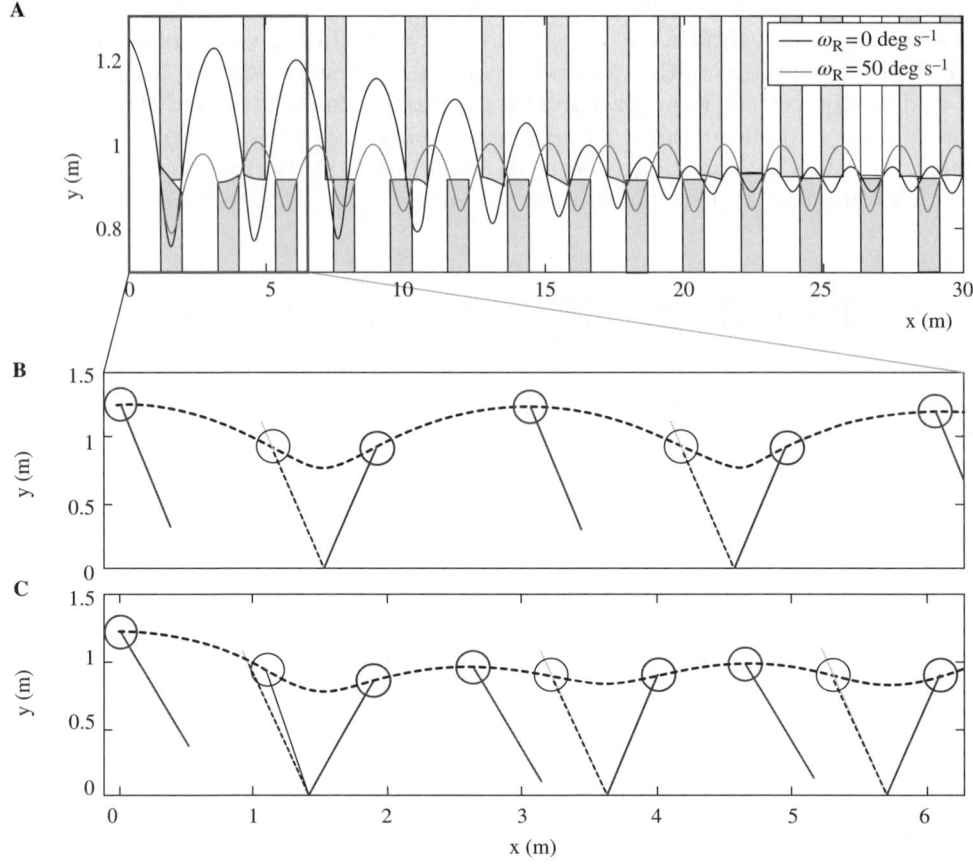

FIGURE 7.4

Influence of swing leg retraction (**A, B**: ω_R=0°/sec; **A, C**: ω_R=50°/sec) on the running pattern represented by the CoM (circles) trajectories predicted by the spring-mass model. As demonstrated for running down a 25 cm step (first step in **A, B, C**), with swing leg retraction (**C**) steady-state running is achieved within two steps (**A**, gray curve) whereas, without leg retraction (**B**), clearly more steps are required to settle down (**A**, black curve; Seyfarth *et al.*, 2003).

joint characteristics. With these extensions of the model, we demonstrate that the stabilizing properties of the underlying spring-mass model can be preserved, and additional system properties due to the systematic changes of the body mechanics can be obtained.

7.6 WALKING WITH COMPLIANT LEGS

The leg function in human walking seems to be fundamentally different from the leg function in running. At about mid-stance of walking, the knee joint is almost straight (Perry, 1992), in contrast to maximum knee flexion in running (Lipfert, 2010).

The opposite is true for the instant of take-off. Here, the knee joint is clearly bent in walking, and almost straight in running. Still, the similarity in leg function in human walking and running might be larger than expected if the leg forces are related to the leg length (Figure 7.5), defined as the distance between body's centre of mass, CoM, and the center of pressure (point of force application) on the ground.

Given the leg operating like a simple linear spring, leg force (i.e., the amount of the ground reaction force) would increase proportionally to the amount of leg shortening. In human running, leg force increases towards mid-stance and decreases until take-off. This force recording is mirrored by the trace of the leg length, with maximum leg shortening at mid-stance and leg extension until take-off. Hence, leg force appears linearly dependent on leg shortening (gray graph in Figure 7.5C). Only at the beginning of contact (after touch-down) is the slope of the force-length graph steeper than in the remaining ground contact. This almost spring-like leg function corresponds to a simultaneous maximum leg force (Figure 7.5A) and maximum leg shortening (Figure 7.5B) at about mid-stance.

In walking, the situation is slightly different. Typically, there are two maxima in leg force (Figure 7.5D) and two minima in leg length (Figure 7.5E). However, when plotted in one diagram (Figure 7.5C), the force-length dependency of the contacting leg in walking at normal speed is remarkably similar to running. As in running, the slope is

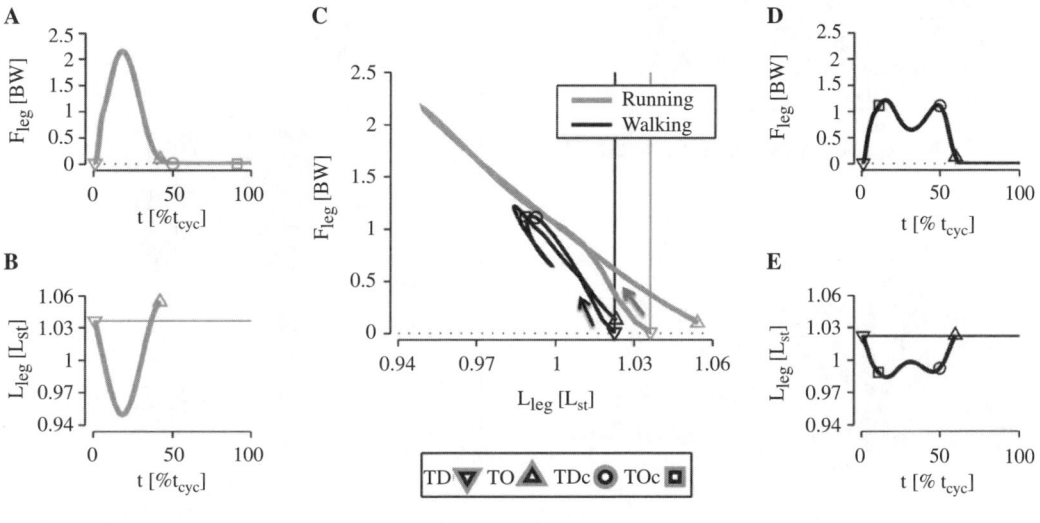

FIGURE 7.5

Average leg forces F_{leg} (**A, D**), leg lengths l_{leg} (**B, E**), and leg force vs. leg length curves (**C**) of 21 subjects running (gray lines) and walking (black lines) on a treadmill at 75% of the preferred walk-run transition speed (75% PTS, 1.55m/sec). Leg force is expressed in body weight (BW), and leg length is measured from center of mass CoM to center of pressure on the ground, and is expressed relative to the height of the center of mass above ground in upright standing (L_{st}). Touch-down and take-off are indicated by TD and TO, respectively. The index 'c' represents the contra-lateral leg. The time (**A, B, D, E**) is expressed relative to the cycle time (from TD to TD). The arrows in (**C**) indicate the progression in time. Data are taken from Lipfert (2010).

clearly steeper at the beginning of ground contact. Afterwards, the slope of the force-length curve remains rather constant. Between the two force maxima, the force-length tracings are slightly shifted towards smaller leg lengths or reduced leg forces. This phase of relative force depression in walking occurs within the single-support phase (i.e., during the swing-phase of the contralateral leg). It is therefore meaningful to extend the spring-mass model for running to a bipedal spring-mass model, in order to represent the dynamics of walking as predicted by Geyer *et al.* (2006). We will analyze the gait patterns predicted by this model and compare them with experimental data on human walking (Figure 7.6).

The bipedal spring-mass model is capable of generating both walking and running patterns (Geyer *et al.*, 2006). For fast speeds, running gaits are considered, as already identified with the original spring-mass model (Seyfarth *et al.*, 2002). At lower speeds, additional stable walking patterns are predicted (Figure 7.6A), which are characterized by double-support and single-support phases, as in human gait (Figure 7.6C). Not only does the model predict the kinematics of the center of mass

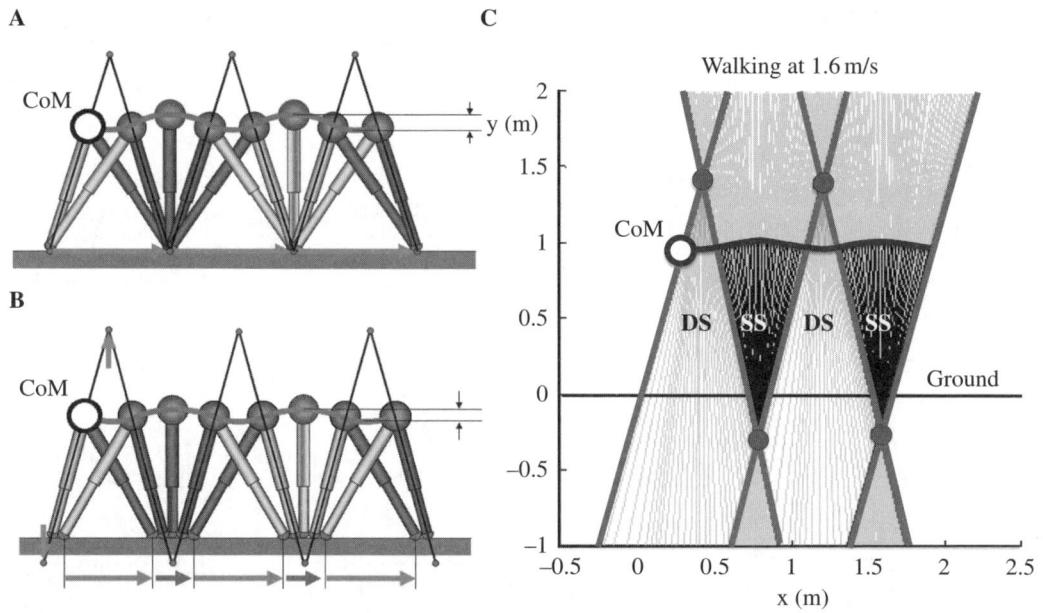

FIGURE 7.6

(**A**) In the bipedal spring-mass model for walking (Geyer *et al.*, 2006), the center of mass (CoM) is redirected upward during double-support (DS). During single-support (SS), the leg force only partially compensates for the body weight, and the center of mass is redirected downward. (**B**) The foot contact (center of pressure on the ground, CoP) is shifted from the heel to the toe (Bullimore, 2005; Maykranz *et al.*, 2009). (**C**) Leg axis as defined by CoP and CoM (gray lines) in human walking at 1.55 m/sec. During single-support and double-support, leg axes intersect in two virtual centers of rotation (filled circles), similar to the schematic drawing presented in (**B**).

quite realistically, it also provides a simple explanation for the double-humped shape of the ground reaction forces in human walking (Figure 7.5D).

At touch-down, the contacting leg is being loaded. This results in an increased leg force (Figure 7.5D) of the leading leg. At the same time, the trailing leg is unloaded until the end of ground contact. The leading leg is now almost fully compressed (Figure 7.5C, D), and redirects the center of mass upward. During single-support, the leg is lengthening until about mid-stance, with an associated drop in ground reaction force. Consequently, the center of mass is moving downward, leading to a second rise in leg force. In the following double-support phase, the forces generated by both legs generate the vertical momentum required to redirect the center of mass upward again.

This parallel action of leg force and leg shortening is not only a typical feature in hopping or running, it is also part of the leg function at preferred walking speed (Figure 7.5C). However, for slow walking speeds (1 m/sec and below – Figure 7.7), the leg compression during contact is quite small. This supports the concept of an inverted pendulum-like gait pattern (Srinivasan and Ruina, 2006). Also, at higher speeds, spring-like leg function is less obvious in walking than it is in running.

One common feature observed at higher speeds (Figure 7.7) in both human walking and running is the increase of leg length from touch-down to take-off. In the following section, we will discuss one mechanism that can explain this systematic change in leg length as a natural action of the foot during ground contact. This modification of leg function would also explain why the measured force-length curve during locomotion does not necessarily have to be a linear relationship, although the leg might operate in a purely elastic manner.

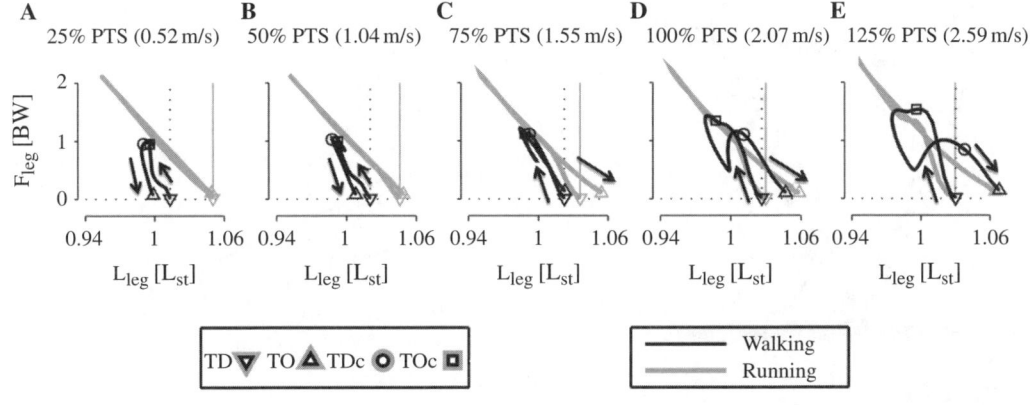

FIGURE 7.7

Influence of speed (panels **A-E**) on the leg force-length curves. In human running, spring-like leg function is indicated at all speeds. In walking, compliant leg function is present only at moderate speeds (1.55 m/sec). PTS refers to the preferred transition speed from walking to running, which is at about 2.1 m/sec. TD = touch-down, TO = take-off. The index 'c' indicates the contra-lateral leg. BW = body weight. L_{st} = height of the center of mass above ground in upright standing. Adapted from Lipfert (2010). Used with permission.

7.7 ADDING AN ELASTICALLY COUPLED FOOT TO THE SPRING-MASS MODEL

During the contact phase of human walking, the point of force application at the ground (center of pressure) is shifted from heel to toe (Rose and Gamble, 2005), as depicted in Figure 7.6B. As a result, experimental leg forces do not intersect in the foot – as assumed in the spring mass model – but below the ground (Figure 7.6C). In order to represent this shift of the center of pressure on the ground, Maykranz *et al.* (2009) introduced a foot segment attached with a rotational spring to the prismatic spring of the spring-mass model (Figure 7.8A).

The change in the force-length curve predicted by this extended spring-mass model is shown in Figure 7.8B. At touch-down the foot is flat on the ground. In the first phase (1), both the translational leg spring and the rotational foot spring are loading. After the maximum compression of the prismatic leg spring is reached, the second phase (2) starts. Here, the leg spring is lengthening, while the foot spring is further compressing. As a result, leg length is almost constant and leg force is decreasing. Finally, the joint torque in the heel suffices to lift the heel. This initiates the last phase, where both leg spring and foot spring are extending.

The relationship between rotational stiffness and translational stiffness influences the relative timing of maximum leg spring compression to maximum foot spring compression. In human running, these two instances occur simultaneously at about mid-stance, with maximum flexion of ankle and knee joint. For human walking, the

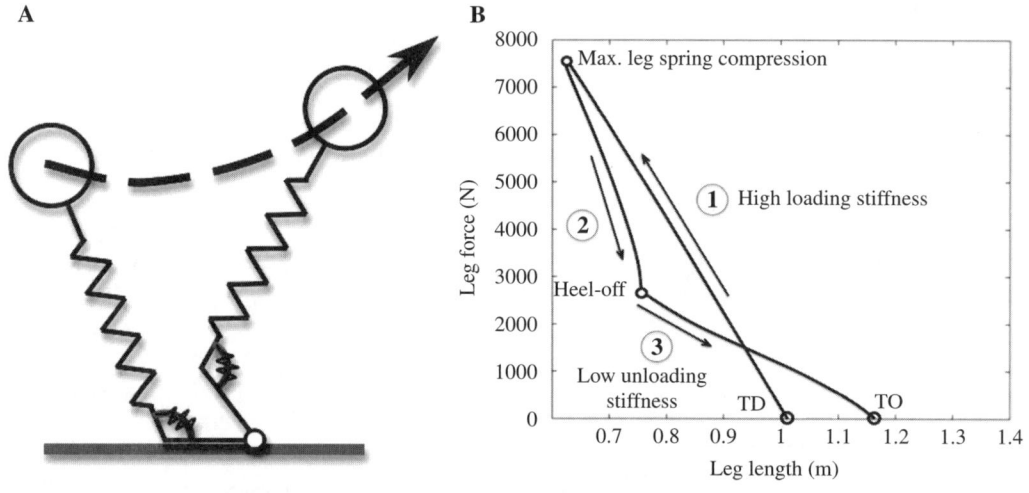

FIGURE 7.8

(A) Spring-mass model extended with a foot segment attached to the distal part of the prismatic leg spring by a rotational foot spring. (B) This combination of two elastic structures (linear and rotational spring) results in a change of leg stiffness and rest length of the leg during contact (Maykranz, 2008; Maykranz *et al.*, 2009). The model predicts high leg stiffness during initial loading of the leg (phase 1) and more compliant leg behavior after heel-off (phase 3). Dependent on the selection of prismatic and rotational spring stiffness, the leg spring may extend while the foot spring is still loading (phase 2). Here, large changes in leg force with little changes in leg length may occur.

ankle extension is delayed during contact. A similar situation can be observed in the model, as shown in Figure 7.8B (phase 2), where the leg spring extends before the heel is leaving the ground.

It is important to note that this model does not represent the anatomical knee or ankle joint. Therefore, leg compression and leg rotation cannot be directly attributed to human knee and ankle flexion, respectively. Only when the knee joint angle is constant and the foot is flat on the ground does leg rotation translate into ankle flexion, similar to foot spring flexion in the extended spring-mass model, shown in Figure 7.8A. This is approximately the case at mid-stance in human walking, where the knee joint is approaching an extended configuration, while the ankle joint is continuously flexing.

In conclusion, the combination of a linear leg spring and a rotational foot spring may lead to a modulation of the resulting leg stiffness. Such modulations of leg stiffness are also found in the force-length curves of human walking and running (Figure 7.7). The beginning of contact is characterized by high leg stiffness (with the exception of 25% PTS running, where no heel contact occurs). In walking at 100% PTS and 125% PTS, additionally, the leg force drops before mid-stance, with little changes in leg length, as predicted by the foot-extended spring-mass model (Figure 7.8A).

As found in both gaits for higher speeds, leg length is increasing until take-off. This increase in leg length, associated with a drop in leg stiffness, results in a higher center of mass position at the beginning of swing phase. This can be explained with a purely conservative model, which consists only of energy preserving, elastic elements. Hence, the interaction of the elastically coupled foot with the spring-like leg is a key feature of the human leg for understanding dynamic leg behavior during locomotion.

There are, of course, additional important functions of the foot, as described here in the context of the model, such as managing impacts and the shift of the center of rotation on the ground (Figure 7.6B). Another advantage is the reduction of leg forces and the prolongation of contact times. This is also predicted with the model in Figure 7.8A, compared with the spring-mass model. Finally, it was suggested (Bullimore & Burns, 2005) that a constant forward velocity of the center of pressure on the ground could reduce the effective forward speed in locomotion. This idea could not be supported by the model, which additionally considers the lifting of the heel with no further shift in the position of the center of pressure during this phase.

We have seen how an elastically coupled foot can tune the function of the leg and, therefore, manage the ground contact during locomotion. In the following section, we will demonstrate how the segmented structure of the upper and lower leg may further influence the dynamics of locomotion.

7.8   THE SEGMENTED LEG – HOW DOES JOINT FUNCTION TRANSLATE INTO LEG FUNCTION?

The structure of the human leg is characterized by a short foot, with a bent ankle joint, and two long segments – thigh and shank – linked by a straight knee joint. The foot, although small compared to the other segments, clearly shapes the leg behavior, as represented by the force-length curve (Figure 7.7 and Figure 7.8B). The coupling between translational and rotational elastic effects can explain part of the observed leg functions.

In this section, we aim to investigate how the segmentation of the upper leg segments may influence leg function during locomotion. For this, we again modify the spring-mass model, by replacing the prismatic leg spring with two segments, coupled by a rotational spring at the joint in between (Figure 7.9A). In this model, leg shortening translates into joint flexion. The resulting torque of the spring-like operating joint further determines the leg force. These two transformations within the leg (leg length to joint angle and joint torque to leg force) lead to a modified leg function. This change is reflected in the predicted regions of leg parameters resulting in stable running (Figure 7.9B).

Assuming a rotational joint spring with linear torque-angle characteristics, the leg force-length characteristics are becoming nonlinear with increased leg stiffness at more extended joint angles. Therefore, the resulting leg stiffness drops with increasing joint flexion during contact. During running, the adaptation in leg function predicted by the segmented model is comparable to the previously described mechanism of swing leg retraction (Seyfarth et al., 2003); the longer the flight phase, the more the leg redirects the center of mass forward during the next contact. Whereas, in the leg retraction model, the angle of attack is a function of flight time, here the resulting leg stiffness is decreasing with larger leg compression.

As a result, both leg retraction and leg segmentation can reduce the predicted minimum speed required for stable running (Figure 7.9B). Also, the robustness with respect to changes in the angle of attack can be dramatically increased with both strategies. In the segmented leg model, this comes at the cost of a reduced tolerance of speed changes. This issue could be resolved by adapting joint stiffness to running speed. In agreement with the model predictions, knee stiffness increases with increasing speed in human running (Kuitunen et al., 2002; Rummel & Seyfarth, 2008).

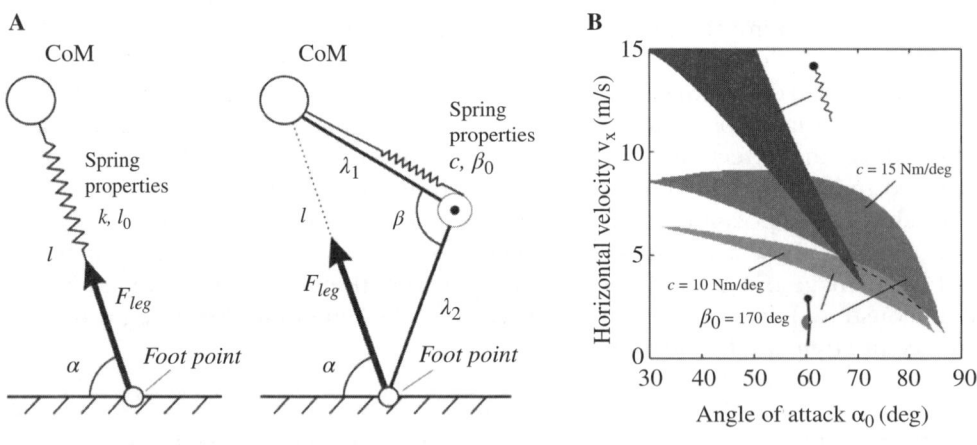

FIGURE 7.9

(**A**) In the segmented leg model (right) the prismatic leg spring (left) is replaced by two mass-less segments (lengths λ_1, λ_2), coupled by a rotational spring (rotational stiffness c, rest angle β_0). (**B**) The predicted regions of stable running for the model with segmented legs (**A**, right) include a larger range of leg angles, compared with the spring-mass model with a prismatic spring (**A**, left). With segmented legs, the speed range for stable running is smaller, and depends on the joint stiffness c (Rummel & Seyfarth, 2008).

For crouched leg configurations with more flexed nominal joint angles, the leg stiffness becomes less sensitive to the amount of leg shortening. In humans, the almost straight knee angle at touch-down may support running stability at low speeds. With increasing speed, the knee joint starts in a more bent landing configuration. Here, the predicted running stability provided by linear spring-like leg behavior is sufficient, and tolerates a large speed range, including maximum human running speeds.

In summary, leg segmentation offers a changed gearing between joint function and global leg function by adjusting the nominal joint angles. This is used in human locomotion and may help to guarantee gait stability at different speeds. Joint stiffness behavior can be adapted to speed and the amount of joint flexion. This further enhances the capability of the leg to adapt to speed.

7.9 KEEPING THE TRUNK UPRIGHT DURING LOCOMOTION

In the last two sections, we focused on the role of the foot and leg segmentation on gait and gait stability. Similar to the spring-mass model, the supported body is reduced to a point mass. In this section, we aim to extend the model by an upper body, in order to study mechanisms for stabilizing the upright trunk posture in human locomotion. Here, we use a rigid segment as a replacement of the point mass in order to describe walking and running (Figure 7.10A).

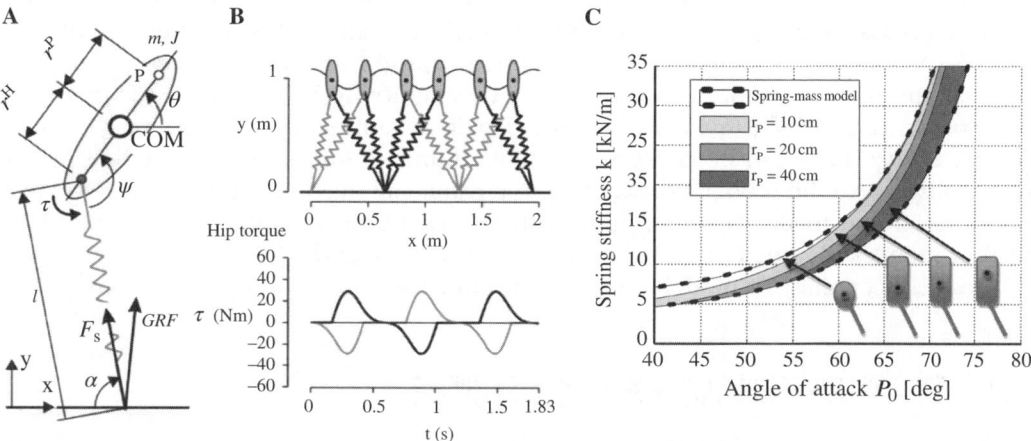

FIGURE 7.10

(**A**) Extended spring-mass model with a rigid trunk. (**B**) During walking, an upright trunk posture can be achieved by applying hip torques τ, redirecting the ground reaction force *GRF* to a virtual pivot point P located at the trunk above the center of mass (Maus *et al.*, 2008). The predicted hip torque patterns (lower panel) are similar to those observed in human walking (Maus *et al.*, 2010). (**C**) The region of stable running is a subset of the region in spring-mass running (Seyfarth *et al.*, 2002). The area is reduced as the position of P increases, as indicated by the black dot in the trunk (for definition of r_p see A).

Without any additional means of support, the upper body will not spontaneously stay in an upright posture, as the center of mass is located above the hip joint. In order to align the trunk during locomotion, appropriately controlled hip torques, acting between the prismatic leg spring and the trunk, are required. Such a control could rely on sensing the vertical axis, i.e., the direction of gravity, as previously applied in legged robots that can achieve active balance (Raibert, 1986). However, in highly dynamic conditions, this may be a challenging task, and sensory signals may become incorrect. An alternate approach would be to rely only on internal sensory signals of the body. Therefore, we ask whether there is a simple strategy to stabilize the trunk without the need of external information (e.g., direction of gravitation).

One solution would be that the hip joint would be located above the center of mass, as denoted by the location of P in Figure 7.10A (Maus *et al.*, 2008, 2010). In order to achieve such a functional point of rotation, the leg forces are redirected from the leg spring to this point P by applying appropriate hip torques. To some extent, this mimics the situation in which the leg spring would be attached to P instead of the hip joint. The required hip torque is calculated based on the spring force F_s, and the inner hip angle ψ (Figure 7.10A). Introducing this hip torque results in a non-conservative model, as the total system energy is no longer necessarily constant.

By implementing this simple concept, it is possible to predict both walking (Figure 7.10B) and running with a stabilized upright trunk posture. For this, P needs to be located in a certain range above the center of mass (Figure 7.10C). The predicted torque patterns (Figure 7.10B, lower panel) are in good agreement with experimental data on human walking (Lewis and Ferris, 2008; Maus *et al.*, 2010). The region of stable running with respect to leg stiffness and angle of attack is a subset of the region of spring-mass running, and varies with the vertical location of P (Figure 7.10C).

The gait model presented, complete with an upright trunk, demonstrates that the bipedal spring-mass model is not restricted to simplifying body representations such as a point mass. The previously identified gait dynamics for walking and running (Geyer *et al.*, 2006) can be preserved, while adding additional functionality, such as trunk posture, to the model. For walking, the trunk can even be used as an alternative reference frame to define the angle of attack resulting in stable gait patterns. This further reduces the need of global sensory information, and may facilitate the transfer of these conceptual gait models to technical implementations (e.g., legged robots).

7.10   THE CHALLENGE OF SETTING UP MORE COMPLEX MODELS

In this chapter, we have introduced a number of minimalistic models to systematically study the organization of the human body in selected movement tasks. With each model level, a new feature has been introduced, and its effect on the system dynamics analyzed. These features have comprised movement patterns such as jumping, running and walking and structural components such as foot mechanics, leg segmentation or trunk posture (Figure 7.1).

With this systematic approach, we have aimed at establishing a framework consisting of models with increasing complexity, which inherit fundamental properties of the underlying simpler models. This follows the very general idea of deriving more complex and more detailed models based on more fundamental conceptual *template* models (Full and Koditschek, 1999). With all structural extensions demonstrated here (bipedalism, foot, leg segmentation, trunk), some general properties of the spring-mass model, including cyclic gait stability and adaptability of leg parameters with corresponding gait patterns, can be preserved or even further exploited.

In the future, these elementary extensions of the spring-mass model (foot, leg segmentation, trunk) need to be combined and tested with experimental data. Both the foot model and the upright trunk model can already be successfully applied for both walking and running (Figure 7.10B). One challenge which needs to be addressed in future research is the gait-specific interplay of the individual leg joints (ankle, knee, hip) in walking and running (Lipfert, 2010).

In both gaits, leg shortening during initial contact is associated with knee flexion (Figure 7.9A). With the heel and forefoot contacting the ground, knee flexion is directly coupled to ankle flexion. Leg rotation additionally contributes to ankle flexion (Figure 7.8A). Hence, foot function is influenced by leg segmentation and the sequence of knee flexion and extension during contact. As knee kinematics are different between walking and running, so may the influence of leg segmentation on foot function depend on gait.

A similar coupling can be expected between leg segmentation and hip function. The rigid trunk model presented here (Figure 7.10) is able to predict hip torque patterns in good agreement to data on human walking (Moritz *et al.*, 2010), but it fails to predict the hip function in human running. One reason could be the larger knee flexion in running, compared to walking. This suggests a more complex interplay between leg segmentation and trunk stability in running than in walking.

All models and model extensions described in this chapter – except for the trunk extension – are conservative, and do not consider energy fluctuations. In the real world, energy losses exist and must be compensated for. For this, the spring elements need to be replaced by more physiological structures, such as visco-elastic structures and muscles. Energy losses could then be compensated for, based on active muscle control (van der Krogt *et al.*, 2009), including sensory feedback (Geyer *et al.*, 2003). Interestingly, local muscle-reflex systems are capable of mimicking spring-like joint function by taking advantage of the intrinsic properties of the muscle (e.g., force-length and force-velocity function). This not only allows for exploiting the mechanical advantages provided by spring-like leg function, but also contributes to stable locomotion in unknown environments, such as against an external force (e.g., wind), or on slopes.

Another important simplification made in the models presented here is the neglect of leg masses. With this simplification, the analysis of the models has been greatly facilitated and the parameter space reduced to a minimum. However, legs do not just produce forces to support the upper body and to perform work on the body's center of mass. With consideration of the inertial properties of the legs, hip torques are required for both keeping the trunk upright and for protracting and retracting the leg during the gait cycle.

The focus on the prismatic leg function, as suggested by the spring-mass model (relating leg force to the length of the leg) must therefore be extended to rotational leg function. This includes considering forces deviating from the leg axis. The foot model and the trunk model provide examples of how the orientation of the leg axis can be tuned during contact. In future, the resulting moment of inertia due to the inertial properties of the segmented leg also needs to be considered. A simple description of the leg function in leg axis and perpendicular to the leg axis (e.g., bending stiffness) could be a next step toward understanding the leg dynamics during locomotion.

Extending the conceptual models to better represent the dynamics of the human body during selected movements is a challenging task. However, the better description achieved with more complex models is, itself, not yet sufficient to understand how the body generates its movements. One equally important question is how the neuromuscular system can access the advantageous movement patterns offered by the mechanics of the body. For example, when simulating the spring-mass model on a computer, it is easy to keep leg length or angle of attack constant during swing phase. However, this is hardly possible within the biological system. Therefore, it is necessary to replace leg parameters assumed in the model by control strategies that tune these parameters in a real-world physical system. For this, the construction of technical models like simple legged robots can be a good framework to investigate the interaction of a control system with a real physical model of the body.

The extension of the conceptual models with corresponding control approaches was successfully demonstrated in the swing leg retraction model (Seyfarth *et al.*, 2003; Blum *et al.*, 2010) and in the trunk model (Maus *et al.*, 2008). By adding more structures to the model (e.g., trunk), new and unexpected options for movement control can appear. From this, the dependency on fixed leg adjustments (e.g., angle of attack defined with respect to the ground) can be resolved, and the body itself could provide a reference system to align the leg with respect to it (Maus *et al.*, 2010). This will reduce the demands on the required sensory input on the environment (e.g., vision), and can further enhance *self*-stability during movements.

▌ NOTES

1 The angle of attack describes the leg orientation at touch-down (Seyfarth *et al.*, 2001).
2 The apex is the highest point during the flight phase.
3 Fewer steps would reduce the size of the region with the maximum number of steps. However, the region for 1000 and more steps would be almost identical to the region shown for 24 steps.

▌ REFERENCES

Alexander R (1976) Mechanics of bipedal locomotion. In: Davies PS (ed). *Perspectives in Experimental Biology*, pp. 493–504. Oxford, UK: Pergamon Press.

Blickhan R (1989) The spring-mass model for running and hopping. *Journal of Biomechanics* **22**, 1217–1227.

Blum Y, Rummel J, Seyfarth A (2007) Advanced swing leg control for stable locomotion. In: Berns K, Luksch T (Eds.) *Autonome Mobile Systeme 2007*. Springer, 301–307.

Blum Y, Lipfert SW, Rummel J, Seyfarth A (2010) Swing leg control in human running. *Bioinspiration & Biomimetics* **5**, doi:10.1088/1748-3182/5/2/026006.

Bullimore S, Burns Y (2005) Consequences of forward translation of the point of force application for the mechanics of running. *Journal of Theoretical Biology* **238**, 211–219.

Cavagna GA, Heglund NC, Taylor CR (1977) Mechanical work in terrestrial locomotion: two basic mechanisms for minimizing energy expenditure. *American Journal of Physiology* **233**, 243–261.

Daley MA, Felix G, Biewener AA (2007) Running stability is enhanced by a proximo-distal gradient in joint neuromechanical control. *Journal of Experimental Biology* **210**, 383–394.

De Wit B, De Clercq D, Aerts P (2000) Biomechanical analysis of the stance phase during barefoot and shod running. *Journal of Biomechanics* **33**, 269–278.

Full RJ, Koditschek DE (1999) Templates and anchors: neuromechanical hypotheses of legged locomotion on land. *Journal of Experimental Biology* **202**(23), 3325–3332.

Geyer H, Seyfarth A, Blickhan R (2003) Positive force feedback in bouncing gaits? *Proceedings of the Royal Society B: Biological Sciences* **270**, 2173–83.

Geyer H, Seyfarth A, Blickhan R (2006) Compliant leg behaviour explains basic dynamics of walking and running. *Proceedings of the Royal Society B.* doi: 10.1098/rspb.2006.3637.

Kuitunen S, Komi PV, Kyröläinen H (2002) Knee and ankle joint stiffness in sprint running. *Medicine & Science in Sports & Exercise* **34**(1), 166–173.

Lewis CL, Ferris DP (2008) Walking with increased ankle pushoff decreases hip muscle moments. *Journal of Biomechanics* **41**(10), 2082–2089.

Lipfert SW (2010). *Kinematic and dynamic similarities between walking and running.* Hamburg: Verlag Dr. Kovac, ISBN: 978-3-8300-5030-8. Dissertation, Friedrich-Schiller-Universität Jena.

Maus HM, Rummel J, Seyfarth A (2008) *Stable Upright Walking and Running using a simple Pendulum based Control Scheme.* Advances in Mobile Robotics: Proc. of 11th CLAWAR, Marques L, Almeida A, Tokhi MO, Virk GS (Eds.), World Scientific, 623–629.

Maus HM, Lipfert SW, Gross M, Rummel J, Seyfarth A (2010) Upright human gait did not provide a major mechanical challenge for our ancestors. *Nature Communications* DOI: 10.1038/ncomms1073.

Maykranz D (2008) *Einfluss eines elastisch modellierten Fusssegmentes auf die Kinematik des Feder-Masse-Modells.* Diploma thesis, Jena University, Germany.

Maykranz D, Grimmer S, Lipfert SW, Seyfarth A (2009) *Foot function in spring mass running.* Autonome Mobile Systeme, Karlsruhe, Germany.

McMahon TA, Cheng GC (1990) The mechanics of running: how does stiffness couple with speed? *Journal of Biomechanics* **23**(1), 65–78.

Perry J (1992) *Gait analysis: normal and pathological function.* Thorofare, New Jersey: SLACK Inc.

Raibert MH (1986) *Legged Robots that Balance.* Cambridge, MA, MIT Press.

Rose JG, Gamble J (2006) *Human Walking. Lippincott Williams & Wilkins*, Third Edition.

Rummel J, Seyfarth A (2008) Stable running with segmented legs. *International Journal of Robotics Research* **27**(8), 919–934.

Seyfarth A, Friedrichs A, Wank V, Blickhan R (1999) Dynamics of the long jump. *Journal of Biomechanics* **32**, 1259–1267.

Seyfarth A, Blickhan R, van Leeuwen JL (2000) Optimum take-off techniques and muscle design for long jump. *Journal of Experimental Biology* **203**, 741–750.

Seyfarth A, Geyer H, Günther M, Blickhan R (2002) A movement criterion for running. *Journal of Biomechanics* **35**, 649–655.

Seyfarth A, Geyer H, Herr H (2003) Swing-leg retraction: a simple control model for stable running. *Journal of Experimental Biology* **206**, 2547–2555.

Srinivasan M, Ruina A (2006) Computer optimization of a minimal biped model discovers walking and running. *Nature* **439**(7072), 72–75.

Van der Krogt MM, de Graaf WW, Farley CT, Moritz CT, Casius LJR, Bobbert MF (2009) Robust passive dynamics of the musculoskeletal system compensate for unexpected surface changes during human hopping. *Journal of Applied Physiology* **107**, 801–808, 2009.

The Most Important Feature of an Organism's Biology: Dimension, Similarity and Scale

John E. A. Bertram

Department of Cell Biology and Anatomy, Cumming School of Medicine, University of Calgary, CA

8.1 INTRODUCTION

What is *the* most important feature of an organism's biology? Certainly, the acquisition and utilization of energy has to rank high on any list, because this is the fundamental process of life. Reproduction would also be prominent, because reproduction is required to sustain life and is, in fact, a character that distinguishes life from non-life. Although locomotion is not absolutely necessary for many organisms, some type of internal or external motion is necessary in order to facilitate the more fundamental activities of life. Within the range of organisms in existence, a variety of approaches to energy acquisition, metabolism and reproduction are utilized, as are a wide range of methods of internal and external transport. With such a myriad of organismal body "plans" and life strategies, is there a single *most important* feature common to all organisms?

Understanding Mammalian Locomotion: Concepts and Applications, First Edition.
Edited by John E. A. Bertram.
© 2016 John Wiley & Sons, Inc. Published 2016 by John Wiley & Sons, Inc.

If one underlying feature does exist, I would argue that it is size – because absolute size influences *all* other fundamental features of a living organism, and does so in profound ways. Whatever the way an organism operates to sustain itself within its environment, its absolute size will determine the constraints and opportunities available, so will be the most fundamental and important feature of its biology. With this view, I join such luminaries as Bartholomew (1981): "*It is only a slight overstatement to say that the most important attribute of an animal, both physiologically and ecologically, is its size.*" (p.46) and Schmidt-Nielsen (1984): "*We shall see that body size has profound consequences for structure and function and that size is of crucial importance to the question of how it manages to survive.*" (p.1).

If this seems surprising, consider the role that size plays in defining how the organism interacts with its physical environment. Organisms of different sizes are really operating in different physical "environments", because the relationship between the organism and its environment is highly dependent on absolute size. This is because almost all functional features of the organism are ultimately determined by physical interactions and these, in turn, will dictate specific consequences that are dependent on absolute size. Understanding these consequences requires an understanding of how absolute size can affect an organism.

8.2   THE MOST BASIC PRINCIPLE: SURFACE AREA TO VOLUME RELATIONS

For decades, science fiction movies have depicted numerous giant organisms, most of which are simply scaled-up versions of often quite diminutive creatures. They are depicted as functioning equivalently to their smaller model systems – which, on closer inspection, is physically impossible. An early analysis of a similar "creative musing" was performed by D'Arcy Thompson (1917, 1992), evaluating the physical consequences of size differences between Lemuel Gulliver and the Lilliputians from Jonathan Swift's famous 1726 novel, *Gulliver's Travels*.

A more modern treatment of this was discussed by Harris (1973) and summarized by McMahon (1984). In these discussions, the difference between metabolic rate at different sizes, or size-dependent influences of locomotion, are discussed, with many of the conclusions being surprisingly counter-intuitive in the absence of an understanding of the influence of size on function. Similar "surprises", sometimes with catastrophic results, arise from simple proportional increases in medical dosages between animals of different sizes, where a dose appropriate for a small body size animal may be totally inappropriate when increased in proportion for the body weight of an animal of large body size (West *et al.*, 1962). The reason for these differences lies with a fundamental consequence of size – the relationship between surface area and volume.

Volume, composing the three linear dimensions of our physical world, is a function of length cubed (l^3), while area is a function of length squared (l^2). Thus, for any size change without shape change, all else being equivalent (a loaded condition, itself), those features related to volume will change out of registry with those features related to area. This is the case whether the size change originates ontogenetically, as size changes between growth stages of an individual, from variation between adult members of a single species, or between species spanning a size range.

The volume-area relationship can most simply be appreciated by considering inter-specific changes in body mass, and making the assumption that small differences in shape that distinguish individual species are of negligible consequence for similar types of mammals. Depending on the phylogenetic group considered, mass differences between adult mammals can be several orders of magnitude (Table 8.1).

Given that the density of mammalian tissue is reasonably constant across species, mass differences will indicate volume differences. However, if we consider two species that are identical in shape, but one is three orders of magnitude more massive than the lesser, then the cross-sectional area of any comparable surface on the larger species will only be $(1000)^{2/3}$ – 100 times greater than the smaller species (Figure 8.1). In other words, the larger form has to hold up 1000 times the load with only 100 times the cross-sectional area. Cross-sectional area largely determines the structural integrity of the supporting skeleton and the force-producing capacity of muscles. If proportions of the smaller species are appropriate for effective function at that size, those proportions cannot be appropriate for effective function at the larger size.

As has been fully described and discussed in numerous previous publications (where the proliferation of such texts by renowned individuals indicates the value placed on this aspect of physical biology: Alexander *et al.*, 1979; Brown and West, 2000; Hill, 1950; McMahon, 1973, 1977; McMahon and Bonner, 1983; Pedley, 1977; Peters, 1983; Schmidt-Nielsen, 1984), maintaining a constant shape over a large size range is functionally untenable, due to structural consequences resulting from the volume-area relationship with absolute size. Regardless of the original design, at some stage, some feature of the system's functional capacity will be compromised as size changes.

This contrasts with the scaling of cetaceans, a group that does not contend with similar gravity effects in their aquatic environment. Over a mass range from 30–200,000 kg (almost four orders of magnitude) Economos (1983) found geometrically similar body shape in cetaceans, indicating that their environment relieves them of some of the support-related scaling issues that terrestrial forms must contend with. This does not mean, however, that other scale-dependent features do not influence their physiology as well.

The analysis of scale effects involves evaluation of the compensations that are used to allow for a large range of size in functionally "similar" forms. There are various possibilities for such compensations, depending on the function being investigated (e.g., issues of heat balance differ from those of structural support) but, in a terrestrial environment, some of the main ones might be classified as:

(i) internal structural, where properties of tissues change (see Bullimore and Burn, 2006);

(ii) external structural, where proportions and/or shape are altered to offer reinforcement (Carrano, 2001; Doube *et al.*, 2009);

(iii) functional, where the strategy of use is changed (Biewener, 1989); this could involve range of motion, timing of motion, or some combination, such as work or power output, or changes in gait parameters such as stride frequency or duty factor.

In most cases, it does not seem possible to alter material properties very much, probably because the basic properties of the structural components are approaching their optimum, at least within the functional environment of a group such as the terrestrial mammals. If room for improvement in tissue functional properties exists for larger mammals, why would smaller forms not take advantage of this as well?

TABLE 8.1 ■ **Orders of mammalian size range ontogenetically, intraspecifically and interspecifically**

Ontogenetic

Species (kg)	Birth mass (kg)	Adult mass	Increase (multiples of birth mass)
Elephant (Asian)	107	4200	39.3
(African)	105	5500	52.4
Giraffe	58.5	1000	17.1
Rhinoceros (White)	52.5	3000	57.1
Moose	12.8	750	58.6
Bighorn sheep	4.4	120	27.3
Gorilla	2.1	225	107.1
Lion	1.3	220	169.2
Binturong	0.32	12	37.5
Coatimundi	0.14	5	35.7
Black-tailed jackrabbit	0.084	4.5	53.6
Golden lion tamarin	0.055	0.65	11.8
American marten	0.030	1.0	33.3
Bushy-tailed wood rat	0.014	0.40	28.6
Red tailed squirrel	0.01	0.25	25
Rat	0.005	0.35	70
Meadow vole	0.002	0.02	10
Mouse	0.001	0.02	20

Intra-specific

Species	Smallest adult mass	Largest adult mass	Difference
Elephant (*Loxodonta africana*)	2200 kg	7500 kg	3.4
Horse (*Equus caballus*)	50 kg	1500 kg	30
Dog (*Canis domesticus*)	1.5 kg	90 kg	60
Rat (*Rattus norvegicus*)	0.20 kg	0.50 kg	2.5
Mouse (*Mus musculus*)	0.012 kg	0.035 kg	3

Inter-specific	no.	Smallest	Largest	Difference
All terrestrial mammals	4100	0.0015 kg	5500 kg	3.7×10^6
Carnivora	238	0.060 kg	800 kg	1.3×10^4
Rodentia	1680	0.01 kg	80 kg	8.0×10^3
Artiodactyls	192	0.80 kg	4500 kg	5.6×10^3
Ceratomorphs	9	225 kg	3600 kg	16
Proboscidians	3	500 kg	5500 kg	11

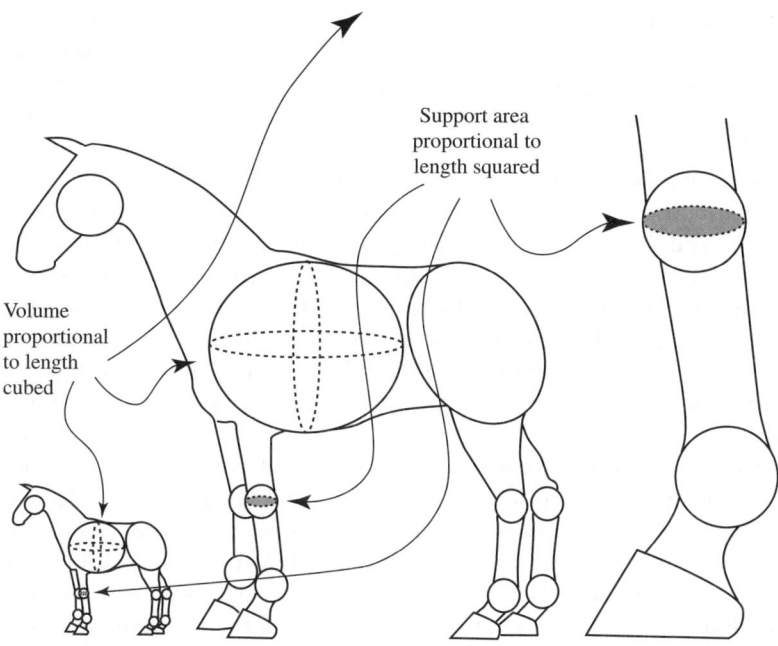

Support area
proportional to
length squared

Volume
proportional
to length
cubed

FIGURE 8.1

Illustration of relationship between surface area and volume for geometrically similar forms that differ in size. Surface area increases in relation to linear dimension squared, while volume increases as linear dimension cubed. Those features that depend on area – such as structural support or heat transfer – will change in relation to area, while those dependent on area – such as mass or heat production – will increase in relation to volume.

More likely are variations in external structure, to provide compensation for the consequences involved with the dynamic aspects of function. However, it must be noted that the division between "material" and "structural" features can become blurred in the integrated organization of the biological system. For instance, the density and orientation of the trabecular network within a bone can have an effect on the functional properties of the supporting element, in spite of the fact that the true material properties of the bone from which the trabeculae are constructed do not differ from those of the rest of the bone. In this case, internal structural features that reside at the boundary between material and structure can alter the functional capabilities of the structural element.

8.3 ASSESSING SCALE EFFECTS

Variations in form are often tested for by comparing species of different sizes against the assumption of constant shape, also referred to as isometry or geometric similarity. The strategy in such evaluations is to note that size has a powerful effect on function, and to utilize observed changes in shape or function to focus attention on the features that appear most important to the organism (i.e., those that have adapted to deal with the consequences of the organism's size, often at the expense of other features). There are examples of systematic changes in both the actuators and the supporting skeleton (Biewener, 2005; Prange *et al.*, 1979; Prothero, 1995), or specific features of them, such as curvatures in long bones (Bertram and Biewener, 1988).

We may find it odd that mammalian species as different as mice and elephants might be considered to have a "similar" shape, as they appear substantially different on first comparison. But are the observed differences functionally important? If so, how might we determine which are key to the success of the species and which are of lesser importance? In evaluating the skeletons of a wide range of mammals, literally from mice to elephants, Alexander *et al.* (1979) found a remarkable consistency in basic structural proportions across approximately six orders of magnitude in size (0.005–5000 kg).

The consistency of limb bone form is surprising, of course; such a wide size range should have substantial consequences for the structural integrity of the support skeleton. In an important illustration of how scaling studies can direct attention to features of functional importance in locomotion analyses, Biewener (1989) identified a systematic change in limb posture that accounts for much of the lack of individual bone proportional change, where the postural change alters the moment arm of the muscle, relative to the ground reaction force moment arm.

Much of the load applied to the skeleton arises from the reaction force generated by the muscles supporting the joints (Figure 8.2). These forces are often many multiples of the load applied by the animal's body weight, due to the generally small muscle moment arm acting around most joints. Smaller mammals tend to have a more crouched posture, likely facilitated because they posses a relatively high ratio of muscle area to body volume (body weight).

Although Biewener's observation is a compelling and satisfying explanation to how limb skeletal elements can maintain similar proportions over a relatively large body size range, it leaves unanswered questions about the functional consequences of this postural shift. For instance, if large mammals can function with an upright posture, why do small mammals not take advantage of this strategy to decrease their required investment in skeletal and muscular tissues (and the metabolic demands imposed by supplying and maintaining those tissues)? Is there a functional advantage to flexed posture that is unavailable to larger forms, or is the large range of motion necessary for smaller forms to deal with the "roughness" of the surface they have to interact with? Such questions do not currently have an answer, but are examples of issues that the scaling approach is suited to analyze.

8.4 PHYSIOLOGY AND SCALING

Even though the surface area-volume (SA-V) relationship is one of the fundamental and defining principles of scaling, the consequences of size are not restricted to structural issues. This is because many functional aspects of the organism and its physiology are determined by the relationship between volume, area, length, and even time. The easiest example to illustrate this fact is with regard to body temperature.

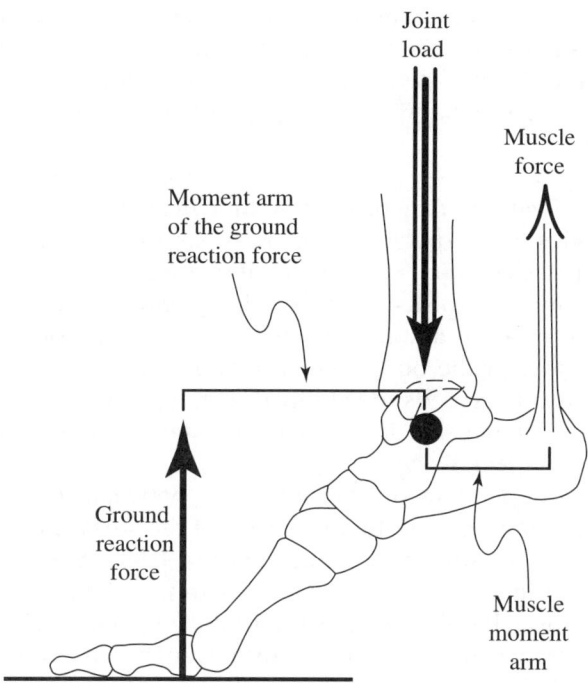

FIGURE 8.2

Illustration of the effect of posture on load bearing, the human ankle as an example. In order to balance the moment arm of the ground reaction force, the muscle force has to be larger than the ground reaction force due to the smaller muscle moment arm. The muscle force is generated through an attachment higher on the skeleton, which is resisted by the skeleton. The joint load will result from a combination of the muscle force (stabilizing the joint) and the body weight being supported. A more upright position of the foot would reduce the ground reaction force moment arm, so would require lower muscle forces to resist (though such a change may place the muscle contractile machinery in a less optimal position).

The metabolic processes of the body's cells produce heat. In order to maintain the homeostatic characteristic temperature equilibrium found in mammals and birds, the heat generated by the cells of the body (related to the volume of tissue) must be released through the body's surface (related to the available area). Thus, for any given size, there will be a limit imposed on the cellular metabolism that is related to the capacity of the organism's surface area to dissipate heat.

Of course, mechanisms exist to adapt either heat production or heat dissipation to varying environmental or functional circumstances. Shivering in the cold is a strategy for using mechanically futile muscle activity (activity that does not perform productive work) to convert metabolic energy into body heat, if the heat dissipation across the body surface outpaces normal heat production. Panting (respiratory cooling) and sweating are evaporative strategies for increasing heat loss when heat production surpasses the normal rate of heat dissipation. These strategies allow modification of heat transfer to suit the changing environmental circumstances in which the organism finds itself. Such mechanisms are crucial to maintaining function over a reasonable range of environments, but the bounds determined by such strategies are relatively small in comparison to the changes in SA-V ratio across the range of body mass represented by terrestrial mammals.

Some compensation for SA-V changes with size can be accommodated with morphological specializations that promote either heat dissipation or retention. Like physiological strategies, such specific compensations can only extend so far, and are commonly observed as adaptations of individual species to specific environments – for instance, small ears in arctic forms, or large, highly vascularized ears in large-bodied tropical mammals, or retae in aquatic forms, to cool specific organs while thermally protecting the remainder of the body. The very large range of body sizes represented by terrestrial mammals far exceeds the ability of either morphological or limited physiological strategies to compensate in order to maintain metabolic rate constant over even a partial range of species sizes.

Heat dissipation is dependent on the difference between the temperatures of the organism, and its environment and the properties of the surface the heat is transferred through. This is referred to as thermal conductance (C), which is measured normalized to area and depends primarily on transfer of heat to the vicinity of the exchange surface, and structural obstacles to heat transfer, such as insulation. For a homeothermic mammal, body temperature remains fairly constant, such that $H = C(T_b - T_e)$, where H is rate of heat flow, T_b is body (surface) temperature, T_e is environmental temperature and C is conductance.

Environmental temperature (T_e) can vary substantially, so the organism requires a mechanism to adjust the rate of heat flow (H). This can involve either or both of conductance (C) and body temperature (T_b). In the latter, the body temperature does not change substantially, but increased metabolic heat production is released through an increased heat flow. Metabolic heat sources can involve changes in general metabolic rate, activity levels such as movement and shivering, or even the fermentation process of the digestive apparatus in herbivores. Behavioral, metabolic and adaptive factors can help modify the heat exchange process in order to maintain temperature homeostasis or adapt a mammal to a more extreme environment, but such "adjustments" have a limit, and are not able to compensate for changes in surface area-volume relationship over a large range of sizes.

We are led to the apparently inevitable conclusion that, for a large mammal with approximately similar shape to a smaller one that is, itself, in thermal equilibrium between its heat-generating and heat-dissipating components, the metabolic rate (rate of heat production, which is ultimately dependent on sub-cellular processes) will have to be changed in a manner that allows the larger species also to be in thermal equilibrium.

Consider, for example, the metabolic rate of a gestating fetus. While it is within the womb, the fetus is essentially part of a organism of larger body size – the mother. Although the processes involved with growth require a somewhat elevated metabolic rate, the metabolism of the fetus pre-partum roughly matches that of the mother (Figure 8.3). Certainly the excess metabolic rate of the growing fetus can alter the heat balance of the mother; it is common during pregnancy for human mothers to be less sensitive to cold, or to be more sensitive to heat, than they would normally be.

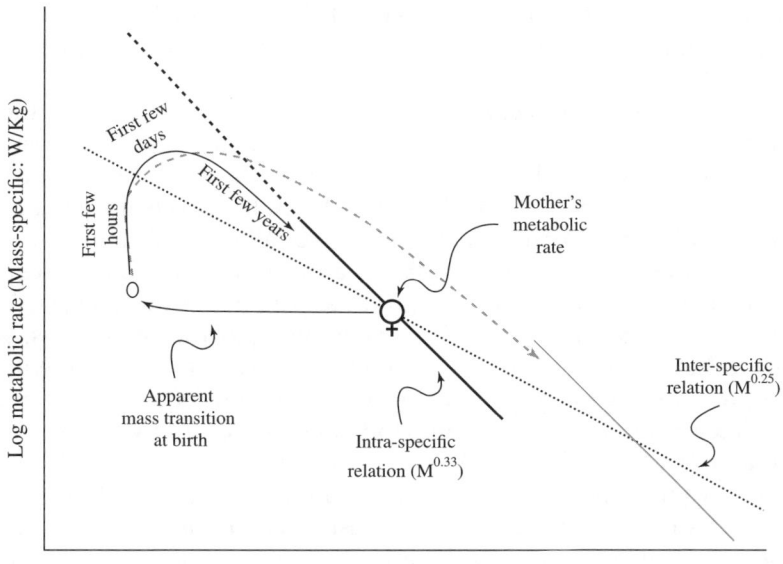

FIGURE 8.3

An overview of mass-specific metabolic scaling, from prenatal through the intraspecific relation for a species, indicating a range of adult body mass. Prior to birth, the metabolic rate of the fetus nearly matches that of the mother, indicating that the fetus must act, physiologically, almost like an organ of the mother, in spite of being a distinct organism. At birth, the apparent mass of the neonate changes drastically, creating a miss-match between the mass and the mass-specific metabolic rate. In the first few hours, days and months following parturition, the metabolic rate of the neonate increases so that it is appropriate for the independent body size. As the offspring approaches the adult body size for the species, mass-specific metabolic rate comes in line with that of the species (black line). Such changes at birth provide a mechanism through which inter-specific metabolic rates can shift adaptively to an alternative species body size (over some number of generations). A species of larger adult body size than the original (gray line) will have a mass-dependent metabolic rate that parallels the original (where isometric considerations dominate), but will be systematically displaced along the inter-specific relation (dashed gray line; Feldman and McMahon, 1983). Metabolic changes following birth stylized after Hill and Rahimtulla (1965).

However, heat equilibrium is a critical challenge for the neonate. At birth, the apparent surface area to volume ratio of the newborn changes dramatically, from acting much like an organ of the relatively large body size mother, to being an independent system with a mass a fraction of that of the mother. The tissue of the neonate itself must produce adequate heat to offset the loss through the proportionally large surface area it finds itself with. In the minutes and hours following birth, the metabolic rate of the newborn raises substantially (as body mass declines slightly, through changes in hydration and metabolic energy expenditure). Substantive elevation in metabolic rate, compared to adults, are evident through much of childhood, until it eventually falls in line with that expected of a human of full body size (Figure 8.3; Hill and Rahimtulla, 1965).

The dramatic changes that occur at birth are an extreme example of the influence of the surface area-volume ratio on physiological processes and metabolic rate. They do clearly indicate, however, that size – and basic form – can be a major (the most important?) determining factor of the activity of the cellular constituents and metabolic reactions, basically governing the general activity of all cells of the body.

Once the small body size infant establishes a metabolic rate that relates directly to the relationship between the surface area and the volume for the species, the relationship between surface area and volume, or metabolic rate and body mass, will remain consistent with the rest of the species. Within a species, metabolic rate will be related by the 2/3 power of body mass ($E_{metab} \propto M^{2/3}$), the relation required to compensate for the changes determined by the relationship between area and volume to mass (Area $\propto M^{2/3}$; Volume $\propto M^{3/3}$).

Due to the obvious expectation of a relationship between body size and metabolic rate, this issue has long been a topic of interest in comparing mammalian species of different body sizes, where body size is a basic and striking feature of each species' morphology. Interestingly – and such discoveries indicate the value of scaling analyses – interspecifically, the metabolic rate does not appear to vary with the surface area of the organism, as expected. Instead, it maintains a specific relationship throughout the range of mammalian body sizes that differs from the predictions of the area-volume relation as a constraint. Max Kleiber determined in 1932, and described in his book *The Fire of Life* (1961), that the relationship between animal body weight and metabolic rate is well described by a 3/4 power function, rather than the expected 2/3 of the area-volume relation. This is an example of a system apparently not explained by constant shape considerations (isometric expectations), and is therefore referred to as an "allometric" relationship (Huxley, 1932).

Numerous proposals have been made to explain the 3/4 mass exponent (also sometimes referred to generally as the 1/4 power model, where the power is determined as a mass-specific value, $M^{3/4}/M^{1.0} = M^{-1/4}$), but none are entirely convincing (see discussion in following paragraphs). The scaling of metabolic rate to body mass in mammals remains one of the more fundamental enigmas of modern biology.

A variety of controversies have developed regarding the overall relationship between body size and metabolic rate. Although the debate continues, scrutiny of the 3/4 power law has led to numerous insights into the complexities of metabolic scaling in mammals. Huesner (1982) analyzed seven mammalian species, ranging in size from 0.016 – 922 kg (house mouse to domestic cattle), and found that each species had a metabolic rate that scaled as a 2/3 power of mass, but the scaling coefficient (a in the power function $Y = aX^b$) changed systematically between species.

Although the argument was made that such results indicated that the 3/4 mass exponent was simply a statistical artifact, further analysis (Feldman and McMahon, 1983) indicated that the "systematic" change in the scaling coefficient (a) led to an interspecific relationship that matched the general 3/4 mass exponent. Thus, it is evident that intraspecific and interspecific scaling relationships differ – not surprising, considering the genetic consistency that defines each species, and the requisite differences between species. Within each species, metabolic rate scales in a manner consistent with surface area-volume considerations while, over a range beyond the species, a systematic shift in metabolic rate leads to a general scaling relationship that differs from this.

Two major questions arise from the above observations:

(i) *How* is the interspecific shift accomplished?

(ii) *Why* does it occur?

The mechanism of the transition likely depends on the early changes involving transient scaling exponents, as metabolic rate is adjusted to the surface area-volume relationship of the growing individual – a potential adaptive mechanism that allows for interspecific metabolic rate scaling (Figure 8.3; gray trajectory – and see also Wieser, 1984). Some evidence indicates that, even for a single species that has an artificially wide distribution of adult body size (such as domestic dogs), there may be some disparity between smaller and larger breeds (see Heusner, 1982, Figure 3, p. 9; although it was later argued that even this relationship was more complex than first indicated – Heusner, 1991).

Are such differences in the relationship between body mass and metabolic rate "real"? This is an important issue when addressing question (ii) above – why do such differences exist? With detailed analysis, and an understanding of differences between organisms down to the individual level, it comes as no major surprise that variation between groups exist; however, the variation with body size appears compellingly systematic. Intraspecific scaling likely represents the consistency in form (at numerous levels) determined by the closely shared gene compliment of the group.

As comparisons are made over a broader range of body sizes, it will require the inclusion of species where phylogenetic, functional and ecological features also contribute to the relationship (Cheverud, 1982). Yet again, it is possible that the simple extrapolation from basic geometry to metabolic scaling is not adequate to account for important determinants of metabolic rate across terrestrial mammalian species. Note, for instance, recent suggestions of fractal-based scaling rules imposed by material supply processes dependent on the organization of fundamental transport networks (West *et al.*, 1997, 1999; West and Brown, 2005).

If, however, the metabolic relationships observed are related to basic form of the organism and the underlying surface area-volume relationship, then other structural scaling relationships probably show similar intra- and inter-specific complexities. In a manner analogous to the heightened metabolic rates of the growing individual, many structural features, such as long bones and joints, appear over-sized during growth, in anticipation of structural requirements of the system when full body size is achieved (Main and Biewener, 2004, 2006), an aspect of growth and size change identified quite early by Huxley (1932).

8.5   THE ALLOMETRIC EQUATION: THE POWER FUNCTION OF SCALING

In the above discussions, variables X and Y are related by the relation, $Y = aX^b$. This power function is often referred to as the allometric equation, where the term "allometry" derives from the study of relative growth of body parts (Huxley, 1932). It has since been expanded in sense to also include comparisons between individuals within a species, and between species across groups.

The power function has substantial impact in assessing and evaluating scaling relationships. It is also built on several important assumptions, which will be included in the following discussion. The allometric equation can be used to describe size scale relationships for things as diverse as pharmacokinetics (Boxenbaum, 1982) through the patterns of modern human travel (Brockmann *et al.*, 2006 – an interesting study that analyzed the distribution patterns involved with the circulation of paper currency).

In the allometric equation $Y = aX^b$, the variables Y and X are related through the parameters a, a coefficient of equivalence, and b, the function determining power. Although a phenomenal number of morphological and physiological variables are verifiably related through the allometric relation, some caution needs to be taken in applying this extremely powerful analytic tool. This is because, as argued above, the effects of size are so ubiquitous that a statistically significant power-law relation may be found between two unrelated variables as each may be related in its own right, in some undetermined manner, to size. Under certain circumstances, the size relationship may become an implicit variable in the comparison of two features that are related only by association. Not to be too facetious, but it is highly likely, for instance, that across a broad enough range of terrestrial mammals, nose hair number can be related to scat length through an allometric relationship that shows statistical significance. Such a relationship, however, will not be functionally or biologically meaningful.

Body size not only influences metabolic processes, but can also have profound, and sometimes surprising, effects on functional capabilities. In human athletic performance, for instance, differences in size account for a number of functional advantages inferred by size. Diminutive stature tends to favor heat release (due to a relatively large surface area to volume ratio), so smaller individuals have an advantage in some long-distance races, especially in hot, humid circumstances (Dennis and Noakes, 1999; Godek *et al.*, 2008). A similar situation exists in weightlifting competitions. Although absolute strength generally goes to the largest competitors, the smaller weight classes always appear stronger than their larger counterparts on a weight-specific basis (Figure 8.4). This is because the smaller individuals have a relatively greater cross-sectional area of muscle (plus other size-related features of their joint leverage), which allows them to lift relatively larger loads. The same phenomenon accounts for the apparent strength of very small animals such as ants, which results from an issue of scale, rather than any particular facility at lifting and carrying (see explicit discussion of this topic in Chapter 9).

The allometric (power-law) relation of mammalian metabolic rate found by Kleiber (1932, 1961) and verified by others (Brody, 1945; Bartels, 1982; although there has certainly been debate over the decades) can be plotted on arithmetic axes to yield a curved plot indicating $b = 3/4$ (the power-law relation is rectilinear (straight) for $b = 1$; it is flat for $b = 0$). This can be compared to the alternate values of $b = 2/3$, the expectation for isometrically similar forms, in which the area-volume ratio of 2 : 3 applies (as in within-species comparisons), or of $b = 1$, where metabolic output remains a constant proportion to body mass (volume; Figure 8.5).

Systematic changes in proportion, as in having proportions get longer and thinner, or shorter and fatter, will have consequences on the SA/V ratio. Although modest

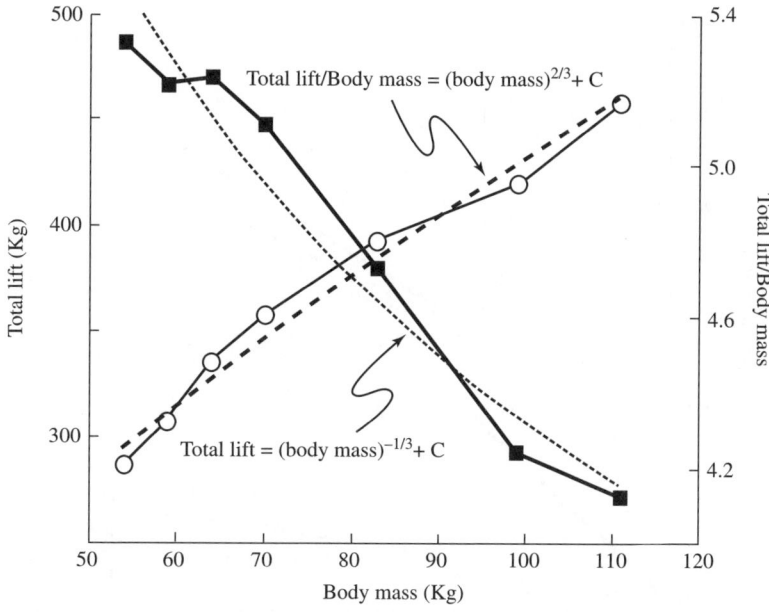

FIGURE 8.4

Illustration of size effects in human athletes – values from list of records of Olympic weightlifting (combined total, data from 2002). The total weight lifted increases with the body mass of the lifter (open circles), in a manner that is well represented by the power function with a mass exponent of 2/3. Relative lift weight will therefore decrease with body mass, in agreement with a power function with a mass exponent of $-1/3$ (relative lift weight $\propto M^{2/3}/M^1 \propto M^{-1/3}$).

changes in general proportion can have substantial effects on heat dissipation, there are no realistic proportions that allow volume and surface area to increase in a parallel fashion. Regardless of the scaling of proportions, it can be expected that processes such as heat dissipation would be severely limited by available surface area in large forms, so heat should accumulate over time (the system will not be expected to maintain homeostasis).

The curvilinear nature of the allometric function for almost all power values has been traditionally considered inconvenient, largely because statistical analyses were complex before the widespread availability of personal computers (and probably because evaluating curvilinear relationships by eye is difficult, especially over limited size intervals, where a small amount of data scatter can obscure differences between relationships).

It has, therefore, been common to apply the law of logarithms to the power function $Y = aX^b$, so that it can be rewritten $\log(Y) = \log(a) + (b)\log(X)$. This form of the relation indicates that it can be plotted as a linear relation, either on arithmetic axes after transformation of the data to log values (example: a bone length of 0.6 m translates to log -0.2219 (ln -0.5108), and a body mass of 720 kg translates to log 2.857 (ln 6.579)), or by plotting the values directly on logarithmic axes. After such manipulation, the function determining exponent b becomes the easily viewed "slope" of the

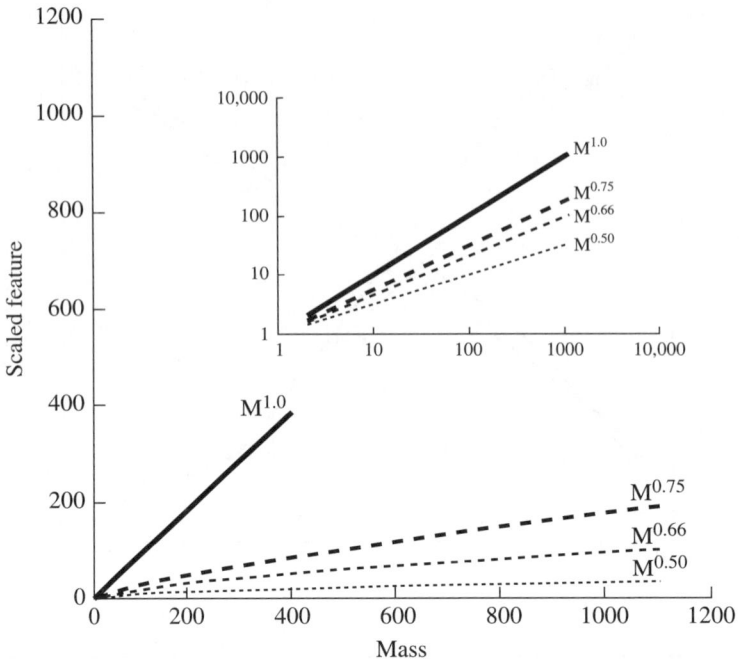

FIGURE 8.5

Comparison of scaling models plotted on arithmetic and logarithmic axes (inset). Plotting on logarithmic axes (or conversion to logarithmic representation) can substantially alter the apparent relationship. Note that the assumption of the allometric relationship, $Y = aX^b$, requires that the relationship passes through the origin. In many cases it is not reasonable to assume *a priori* that this is the case. This assumption is obscured in the plot using logarithmic axes, because the intercept is not indicated.

relation (but note that such transformations also have consequences for the data distribution that can affect the statistical inference – see Swartz and Biewener, 1992).

In analogy to a standard linear relation, the coefficient "*a*" is often described as indicating the "Y intercept". This, however, is *incorrect*. As a power function, the intercept must be at the origin (0, 0), as there is no value of X except 0 that will satisfy $0 = aX^b$ (if "a" is 0, then Y can only be 0 for all cases of X). Note that this is one of the most important assumptions implied by the allometric equation, essentially forcing the relationship to run through the origin (almost like adding this point to the plot). In many cases (for instance when considering structural variables like length and width of long bones), it may be justified to assume that, in theory at least, each dimension should decline systematically as size decreases to zero. However, the allometric equation is often applied to the size relationship between variables whose value, as they approach zero, cannot so easily be assumed. For example, it is not easy to predict what many physiological variables should be as animal size approaches zero.

It generally goes unnoticed that the range of the data used in allometric analyses is often well removed from the implied zero point (Figure 8.6). In such cases, it is possible that a simple power relation may provide a statistically verifiable, but entirely

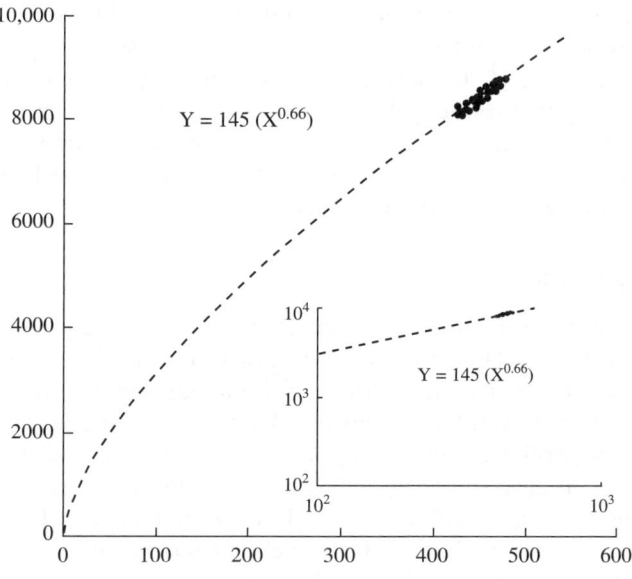

FIGURE 8.6

Plot of a data set substantially removed from the origin and fit with the allometric relationship. Although the data are well represented by the allometric relationship over the range available (inset), other relationships may also represent the data (Albrecht and Gevin, 1987; Albrecht, 1988).

spurious, fit. It is important to evaluate the underlying assumption (that the power function describes the data distribution) by subtracting the value of the smallest data point from the entire data set (essentially placing the smallest data point at the origin), and re-evaluating the relationship. If this manipulation does not alter the apparent power (b), then the allometric equation likely applies. If not, the relationship may be better served by another relationship ($Y = aX^b + C$ is one of several alternative possibilities – Albrecht and Gevin, 1987; Albrecht, 1988).

8.6 THE STANDARD SCALING MODELS

Size has important consequences for function, but it is not possible to interpret differences in form, mechanical function, or physiology, without evaluating organisms within the context of their relation to animals of different size; size effects are meaningless without considering the manner in which features change across a broad enough size range. Such a comparison is facilitated by comparing the relationship between forms, using models based on the expected consequences of the model parameters. That is, the models represent hypotheses that predict the effects of size differences based on standard presumed relationships between dimensional parameters.

The geometrical models most commonly utilized can be thought of in terms of distribution of size increases – if volume becomes larger, that volume can be apportioned equally in all directions (shape remains constant), or one dimension can increase out of proportion to another, as in trees becoming more robust as they increase in height (McMahon and Kronauer, 1976; Bertram, 1989). Although a near-infinite number of possibilities exist for partitioning volume into the constituent dimensions that define it, there are three models, based primarily on mechanical considerations, that warrant introduction, as they are generally involved – to some degree, at least – in all discussions of size effects in the terrestrial environment.

8.6.1 Geometric similarity

Structures of different sizes, including mammals, are geometrically similar if each of their linear dimensions can be related by a common proportion. That is, each linear dimension of the larger one is a constant multiple of the linear dimensions of the smaller. Put most simply, geometrically similar structures have the same shape. This is also referred to as "isometry", and implies a consistency in proportion. As discussed above, however, the assumption of geometric similarity has substantial implications for structural performance, for physiology and, as will be seen below, for activities such as locomotion.

Maintaining shape requires that equivalent changes occur in each dimension as size changes. That is, for increases in size, each of the linear dimensions will increase in proportion to (volume)$^{1/3}$ or M$^{1/3}$ (where M is mass, provided density remains constant) – because L $\propto$ D (length will be proportional to diameter) and M $\propto$ L^3 $\propto$ D^3 (Economos, 1982). Galileo (1638) noted that, with increasing size, objects that maintain similar overall shape will be more likely to fail, because the load to be supported (volume) increases to a greater extent than the cross-sectional area of supporting tissue. Since most supporting limb bones are loaded predominantly in bending (even if apparently upright: Bertram and Biewener, 1988; Biewener *et al.*, 1983; Lanyon and Baggot, 1976; Rubin and Lanyon, 1982), a proportional change in length and diameter can be proposed that will maintain a constant bending stress in the supporting elements.

Deformation of the supporting structure (strain of the constituent materials) results from applied loads. Loading in bending produces an uneven distribution of stress across the beam, such that the stress is zero at the neutral bending axis, and greatest at the external surface of the beam. The uneven distribution of load means that, in bending, beams will be much more vulnerable to failure than in pure compression or tension. Along with increases in stress levels, bending also involves greater deflection of the beam in comparison to pure tension or compression, again because the uneven distribution of load means that not all of the material of the beam effectively resists deformation under the applied load.

A cantilever is a simple example of a structural system loaded in bending. Simplifying the structure of skeletal elements to prismatic cantilevers can obscure many details of form and function, but it can also reveal fundamental features of the mechanics that are lost within considerations of specific detail. If we assume that the force (F) applied to the end of a cantilever is proportional to the volume of the beam (modeling the loads applied to the skeleton as proportional to mass and, consequently, volume), the consequences of size changes can be predicted from the equations governing beam loading:

$$\sigma = FLC / I \qquad (8.1)$$

and

$$\delta = FL3 / 3EI, \tag{8.2}$$

Where: σ = stress, δ = deflection, E = modulus (stiffness), F = force, L = length, I = second moment of area, C = position of stress relative to neutral axis.

Given the assumption that force (F) remains proportional to volume (or mass) in this simple model, F can also be considered proportional to mass (M). Second moment of area (I) is related to cross-sectional shape, but will vary as the fourth power of cross-sectional radius (or diameter, D), if shape remains constant (Wainwright *et al.*, 1982), so $I \propto D^4$. Stress will be greatest at the surface of a beam in bending, so C is maximum when $C \propto D$. The variable L is the length of the beam. Combining proportional changes in L and D define the proportions of the beam as it scales over any given size range. For geometric similarity, proportions remain constant, so $L \propto D$. Since $M \propto LD^2$, then $L \propto D \propto M^{1/3}$. By substitution in Equation 8.1, for geometrically similar beams, $\sigma \propto M^{1/3}$, indicating that stress will inevitably increase as size increases (or will inevitably decrease for decreases in size – stress would only be constant for M^0).

8.6.2 Static stress similarity

The above arguments indicate that, for geometrically similar beams, stress increases as the 1/3 power of volume (or mass). In order to maintain consistent stress in beams of different size loaded in bending, then, it is necessary for shape (proportions) to change systematically with size (increasing or decreasing). To reinforce the support area for increases in size, diameter must increase much more rapidly than length. This condition is satisfied for Equation 8.1 – that is, $\sigma \propto M^0$ (constant) when $L \propto D^{1/2}$, so that $L \propto M^{1/5}$ and $D \propto M^{2/5}$ (Note: the volume relation $M \propto D^2L$ must be maintained; adding a proportion of the mass to one dimension means that it must be removed from the other dimensions. In this case, $D^2L \propto (M^{2/5})^2 M^{1/5} \propto M^{5/5} \propto M$).

8.6.3 Elastic similarity

McMahon and Bonner (1983) recognized that the 3/4 power relation between metabolic rate and body mass (Kleiber's rule) could result from the surface area-volume ratio if shape did not remain consistent over the entire size range. If shape changed systematically with size, the relationship between volume and surface would also shift. If shape changed systematically over the range of mammalian body sizes, then the surface area-volume relation would be affected and could result in the observed metabolic scaling, simply because the surface area to volume ratio did not remain constant. However, the consistency of the observed relation over a broad size range suggested that the shape change was "governed" by some general feature of the system.

Expecting that such a shift in general proportions would be likely to originate with features of the structural skeleton, McMahon proposed the elastic similarity model (McMahon, 1973), where limb bone proportions change in such a way that deflection in bending remains proportional to the bone's length. This requires proportional changes in dimensions intermediate between geometric and static stress similarity, where $L \propto D^{2/3}$, $L \propto M^{1/4}$ and $D \propto M^{\sqrt{3}/4}$ ($D \propto M^{3/8}$).

McMahon found some interesting evidence supporting elastic similarity in the limb bones of a variety of ungulates (McMahon, 1975), and even in the branches of trees (McMahon and Kronauer, 1976). This was followed by a range of data that suggested elastic similarity may, indeed, be an important feature of mammalian scaling. However, there is no structural reason why supporting bone deflections should remain proportional to limb bone length. For example, even if resistance to deformation (keeping the joints apart) is the main function of the long bones (Currey, 1984), one might suppose that a deflection limit relative to bone width might be even more functional (Bertram, 1988). Elastic similarity proportions will result in increased stress in bending (Alexander, 1977), albeit to a lesser degree than geometric similarity (from Equation 8.1, $\sigma \propto M^{1/4}$ – Figure 8.5). Ultimately, this will lead to diminished support capability, just as geometrically similar proportions should.

8.7 DIFFERENTIAL SCALING — WHERE THE LIMIT MAY CHANGE

None of the above "standard" models (or their correlates) are able unambiguously to explain the structural (shape) scaling of terrestrial mammals. Given the diversity of forms and lifestyles represented, this may not be surprising. However, given the importance of size in functional considerations, it is problematic that these logical models cannot account for function across the size range of species represented by terrestrial mammals.

One issue may be that the group being analyzed covers too broad a size range, even though, by only considering terrestrial mammals, the group has already been restricted to a single physical environment. Even recognizing that substantial variability is anticipated, given the range of influences acting to determine form, the apparent reliability of the scaling relationship expressed over such a large size range as the shrew to elephant (Alexander *et al.*, 1979; or even larger – see Full, 1989) challenges us to expose the fundamental influences determining this aspect of mammalian form and function. However, it is possible that the accepted level of variability may obscure important aspects of the scaling relationship that could point to the process through which the scaling relationships are produced.

We might ask the question, *are* all mammals, indeed, geometrically similar? Is it reasonable to combine the limb bones of insectivores, carnivorans, ungulates and elephantidae in the same scaling analysis? This would be valid only if scaling considerations are so powerful that they overshadow the variability imposed by combining such a broad range of species. Note that, even over such a wide size range, the scaling predictions of each of the above models is rather subtle, even though the functional consequences may be great. Added variability will tend to obscure these relationships, and the system may well default to the pending null hypothesis – in this case, it is usually taken to be geometric similarity.

It is important to be aware that scaling considerations can be applied at three easily confused, partially related levels that require distinguishing. These are:

(i) ontogentetically (through the growth of an individual);

(ii) intra-specifically (between members of the same species); and

(iii) inter-specifically (between different species comparing an individual or characteristic dimensions representing each species).

Each of these comparisons is a valid approach to evaluating size influences on an organism's form, but there is a risk in mixing comparisons at these different levels. Statistical complications can easily obscure the true characteristics of the scaling relationship if data from the different levels of comparison are combined. This results from inappropriate weighting of data set components. For example, if some species are represented by a single point, while a number of individuals are depicted for other species, then intra and inter-specific relationships are confounded in the apparent result. A similar logic underlies the consideration of phylogenetic influences – closely related species will more likely resemble each other and will produce a bias similar to that of mixing intra/inter-specific groupings. For this reason, the phylogenetic distribution of the species included in the analysis must be considered and accounted for (Garland *et al.*, 2005).

Comparing forms with comparable functionality is also important. To some extent, the consideration of only terrestrial mammals accounts for this, but it may not be adequate to produce a reliable analysis of size effects without the complication of other functionally related factors. For instance, some terrestrial mammals are fleet-footed cursors (e.g., Pronghorn antelope, *Antilocapra americana*) while others are slow and lumbering, and utilize an entirely different method of procuring food and defending themselves from predation (e.g., porcupine, *Erethizon dorsatum*). Formulating scale effect comparisons between species assumes that size effects will outweigh either the phylogenetic or ecological differences.

Parsing out the true size effects becomes less confused if groups of functionally similar organisms are compared. In some regards, it may be more reasonable to evaluate structural consequences of size in locomotion through restricting the analysis to a group that utilizes a similar form of locomotion, such as cursors, whatever their phylogenetic origin, than to mix species of widely differing functional habits. In many cases, however, restricting the species also severely restricts the size range over which comparisons can be made.

Evaluation of size consequences, then, involves complications that can arise from both functional and phylogenetic relationship of the species included. One feature that has arisen from such evaluations is the possibility that a single scaling relationship does not hold across the full range of body sizes of terrestrial mammals (Bertram, 1988; Bertram and Biewener, 1990, 1992; Christiansen, 1999; Iriarte-Diaz, 2002). This was originally suggested by Economos (1983), who noted that body lengths of a wide range of mammals diverged from the expectation of geometric similarity when plotted against body mass. He noted that both smaller and larger body size species tended to lie below the prediction for the entire group. From this, he suggested that the data were better represented by two scaling relationships, and proposed this as a solution to the competing expectations of geometric and elastic similarity – these were the two prevalent models of terrestrial mammalian scaling at the time, each with its own supporting data (Alexander, 1977; Alexander *et al.*, 1979; McMahon, 1975).

With no statistical verification of the intersecting relationships, Economos (1983) suggested a relatively arbitrary break point at 20 kg body mass (this may represent the mid-point of species distribution in the sample, or an estimate of the point at which the data tend to diverge from the common regression, but it may also be the intermediate value that gives individual regression results that match the prevailing scaling hypotheses at the time). This suggestion was reiterated by Bou and Casinos (1985), who demonstrated that the relationship between bone mass and body mass in rodents and insectivores did not match that observed for large body mass species.

An allometric model with different exponents covering different size ranges, or as it has come to be known – "differential" scaling, has substantial consequences for the interpretation of scaling, far beyond simply as a means to reconcile prevailing models. Rather, such a situation would imply that the basic physical "rules" limiting the function of terrestrial mammals differs depending on general body size. If demonstrated to be the case, this would be an important consideration in evaluating functional limits and the influence of mechanics on form.

The Economos suggestion of different scaling in small and large body size ranges was directly tested by Bertram (1988) and Bertram and Biewener (1990, 1992). Much of this analysis was based on limb bone structural form from the terrestrial carnivorans, considered a well-represented monophyletic group spanning a wide range of body sizes (0.1 kg (*Mustela nivalis*) to 500 kg (*Ursus maritimus*)), all having relatively consistent locomotor behavior. Major long bones of 118 species, drawn from seven terrestrial families of the order, were analyzed. It was found that limb bones were demonstrably more robust in families with larger body size than those with smaller body size, particularly for distal long bones. Long bone form shifted systematically, as body size of the constituent groups became larger. For small body size families, limb bone shape could not be distinguished from geometric similarity but, in larger forms, the limb bones became more and more robust as size increased.

Following from this finding, Bertram and Biewener (1990) extended the body mass range by adding comparable limb bone dimensional measurements from other groups, including the Bovidae and the Ceratomorphs (the Ceratomorphs are composed of the tapirs and rhinoceroces; Prothero and Sereno, 1982). A reanalysis of the bovid data that McMahon (1975) originally used to confirm elastic similarity in limb bone scaling indicated that scaling, within even this group, was distinctly non-linear when on logarithmic axes. The apparent shift in bovid limb bone scaling mirrored that seen over the same size range in the carnivoran sample, but the relationship was displaced along the length axis, indicating that, for the same body size, bovids, as a group, have longer limb bones than carnivorans. The scaling exponent, however, appeared to change in a parallel manner to that of the carnivorans. Could this indicate the formative features of an emerging general pattern of skeletal scaling in terrestrial mammals?

Although some bovids become quite large, there are very few species in the extreme body size range of mammals. The only group of very large body size terrestrial mammals that are represented by an adequate number of species to allow evaluation of their scaling relationship are the Ceratomorphs – the tapirs and rhinoceroses.

An analysis of the limb bones of this group indicate that their scaling exponent diverges substantially from that of smaller body size groups, and indicates limb bone scaling with an exponent consistent with static stress similarity ($L \propto D^{1/2}$, $L \propto M^{1/5}$, $D \propto M^{2/5}$), as originally proposed by Galileo.

Familial and ordinal scaling differences in structural dimensions have been confirmed in mammals (Christiansen, 1999; Garcia and da Silva, 2004; Silva, 1998), and similar findings are reported for metabolic scaling (da Silva *et al.*, 2006; Kolokotrones *et al.*, 2010). Together, the data appears to indicate a two-phase scaling relationship, with structural scaling in small body size forms generally matching geometric scaling expectations, while the largest body size forms have structural scaling that matches the static stress model. In between, scaling can appear to match any relationship between these two, depending on the body size distribution of the members included within the data set analyzed. Performance follows from form, and analysis of locomotor performance also supports a transition in scaling relationships between small and large forms (Iriarte-Diaz, 2002).

If terrestrial mammals have two-phase, or differential, scaling, where does the transition occur, and why? The latter question is particularly important, because it would indicate that the consequences of size changes differ, depending on absolute size. The former question will need to be answered, however, as a key piece of evidence helping to identify what factors are at play influencing the integration between size, form, and function in mammals.

Economos originally estimated the change point at 20 kg (examples of species at about this size include the Canadian beaver (*Castor canadensis*), collared peccary (*Tayassu tajacu*), muntjac deer (*Muntiacus muntjac*), common duiker (*Sylvicapra grimmia*), wolverine (*Gulo gulo*), and coyote (*Canis latrans*)). However, this value was derived from estimating the apparent transition between smaller body size forms displaying geometric similarity, and those that began to diverge from geometric similarity. Using phylogenetically and functionally consistent groups suggests a change point in the range of 100–200 kg (e.g., North American elk (*Cervus elephus*), Okapi (*Okapia johnstonia*); Bertram and Biewener, 1990, 1992), while the effects of systematic changes in limb orientation to modify muscularly applied supporting loads (see below) suggests a value in the range of 300 kg (Biewener, 2005).

Geometric scaling over any range of body size greater than that of a given species (and potentially even within those species, such as humans and domestic dogs, that have excessive size variation) should have functional consequences, particularly with regard to structural issues such as limb bone and muscle stress (Figure 8.2). How, then, could mammals in the small body size range (whether less than 20 or 200 kg) have geometrically similar skeletons?

Biewener (1989) noted that, as terrestrial mammals become larger, they assume a relatively more upright limb orientation. This reduces the relative contribution of muscular load supported by the limb skeleton, through decrease in the lever arm of load that originates with the body weight. Note that a substantial proportion of the load applied to the skeleton results from stabilizing joints, rather than simply supporting body weight (though the motivation for this results from the body weight acting about the joint, so should be proportional to body weight for structural forms with equivalent shape and limb orientation; Figure 8.2). As effective as this strategy appears to be, there is a limit to how straight a limb can become. Presumably, the change in structural scaling of limb bones in larger body size forms could result because the option for adjusting overall limb orientation has reached its functional limit, leaving only modifications to structural proportions as an option to maintain functional integrity.

In this scenario, relatively small body size mammals would have a crouched posture and parasagittal limb movement. As body size increases within a lineage, increases in skeletal loading would be compensated more by changes in limb orientation than by limb bone proportions. Eventually, a body size limit is reached, in which the limbs are effectively as straight as functionally allows. At this point, a shift in limb proportion is required in larger body size forms, providing proportions that approach those which maintain stress levels, safety factor, and prevent undue risk of failure (static stress similarity).

We know that some forms, particularly the elephantidae, but also some very large body size extinct mammals like Baluchitherium (Paraceratherium – Fortelius and Kappelman, 1993), have remarkably vertical limbs, and their posture is referred to as "graviportal". It is likely that such a physical limit to the vertical orientation of limbs differs from the apparent limit seen at the change-point in general mammalian skeletal scaling. These extreme body size mammals compromise some functionality, and employ their large bulk as part of their survival strategy, relieving them from some of the performance requirements of moderately large, but not extreme, body size species.

Taken as a whole, the accumulated data on structural shape changes over the size range of terrestrial mammals may indicate a complex of scaling relationships, each serving a different size or phylogenetic group. How can discontinuities between these relationships be explained?

As has been demonstrated, mammals undergo substantial metabolic changes during ontogeny (Figure 8.3). In like manner, structural form undergoes substantial transition during the growth process, usually involving the development of an overly robust skeleton that becomes more slender as growth progresses (Bertram *et al.*, 1997; Main and Biewener, 2004). Such variation in form and physiology during growth provides a mechanism for "adjustment" in structural form and physiological activity with size changes in a clade. Whether the consistency of "general" form, structure and physiology in smaller body size mammals persists for functional reasons (i.e., from an advantage to these characteristics over the physical challenges faced by smaller terrestrial forms), or whether it is an anachronism rooted in the fact that mammals expanded from groups that were of relatively diminutive size, is yet to be determined.

The overwhelmingly powerful influence of size on function and physiology means that recognizing the adjustments in biological features over a range in size can direct attention and investigation into those aspects of function that are of critical value to the interaction of the species with the physical environment it inhabits. Size-dependent differences in metabolic rate can sometimes be identified even within a species (Feldman and McMahon, 1983), but are often much more obscure in structural features of muscle, connective tissue and the skeleton. This difference probably results from our ability to conveniently characterize metabolic rates with precision, where metabolic rate represents the accumulated consequences of all the metabolic processes supporting the organism's function, while the complex features of structural competence are not so cleanly characterized by a single value (though many try).

8.7.1 Assessing the assumptions

The models discussed above provide reasonable hypotheses from which to explore the possibilities of size effects on function and performance in terrestrial mammals. However, they all involve substantial assumptions that are violated, to some extent, in reality. In evaluating each model, it is important to also consider the assumptions involved. At first, it may seem intuitive that all mammals are constructed from essentially the same tissues, so the assumption of constant density is reasonable. However, it has been demonstrated in both birds and terrestrial mammals that the proportion of tissue dedicated to the skeleton increases with body size (Prange *et al.*, 1979; Bou and Casinos, 1985).

Skeletal scaling models also assume that the basic internal shape of bones remains equivalent over the body size range. Recent evaluation quantifying cross-sectional shape using CT scans (computed tomography) indicates that, even within a restricted lineage, relatively subtle features of cross-sectional shape, not evident on the surface of the bone, may contribute in an important manner to the structural capacity of the support skeleton as size changes (Doube *et al.*, 2009). The availability of such techniques may well lead to novel understanding of the strategies employed to circumvent the mechanical consequences of body size in the skeleton. Size effects still remain an important strategy for investigating and identifying the critical mechanical features imposed on the functional capacity of terrestrial mammals, even if our current understanding of the consequences of size may not be complete answers in themselves.

8.8 A FRACTAL VIEW OF SCALING

The fractal concept of dimension has arisen in the last 50 years – a relatively brief time span, considering some relatively current analytic techniques, such as wavelet analysis and Fourier transformation, originated one or more centuries ago. The essence of the fractal concept is that some objects may "inhabit" geometries that are not readily reducible to integer dimensions (where $LD^2 \propto$ volume), but are better characterized by "fractional dimensions" (fractals).

It is often said that no perfectly straight lines exist in nature – a statement of the observation that the surfaces and boundaries of natural (and biological) systems are dominated by irregularity, at least at some level. Any boundary can be measured by pacing out the length using x number of steps, each with y length so that $L = xy$. However, if the boundary is irregular, the magnitude of L will be dependent on step length y. This is because a larger step length will bypass the length of some irregularities, and underestimate the length of the boundary. A fractal analysis recognizes this, and uses it to compute a unique number that is not dependent on step length; this has been referred to as the "extent" of the boundary, and characterizes the apparent irregularity of the boundary (Pennycuick and Kline, 1986). Extent (E) can be calculated as:

$$E = xy^f, \qquad (8.3)$$

Where: y denotes the length of the interval between measurement points
x is the number of intervals
the exponent f represents the fractal dimension which, for a boundary, can vary between 1.0 (representing a smoothly curving boundary) to nearly 2.0 (a boundary so convoluted that its winding basically covers an area).

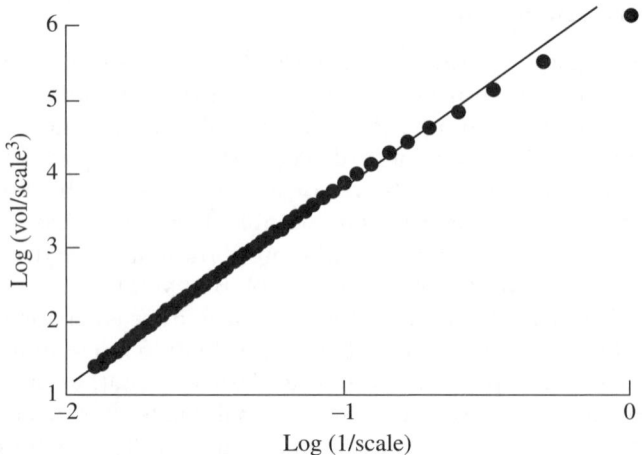

FIGURE 8.7

An example of a fractal analysis of biological form, illustrating the basic similarity of this approach with that of allometric analysis. The plot is on logarithmic axes, and is used to calculate the Minkowski dimension for a radiographic image of trabecular bone (re-drawn from Jaing *et al.*, 1999). The Minkowski dimension is a class of fractal dimension that can be used to characterize the textural orientation of image data. The dimension is determined as the regression of the plot over the linear (self-similar) portion.

In like manner, for forms that exist between a plane and a solid volume, as in many natural surfaces with substantial complexity, the value of f will reside somewhere between 2 (an area) and 3 (a volume).

If Equation 8.3 holds – that is, if the line has the special property of self-similarity (Mandelbrot, 1983), a plot of extent (E) against interval length (y) will yield a log-linear relationship (Figure 8.7), where the slope indicates the unique number that characterizes the irregularities of the boundary being measured. If the relationship asymptotes to a unique number, the system is "self-similar". The property of self-similarity distinguishes fractal from non-fractal systems.

As noted by Pennycuick (1992), no natural objects (including biological) can have self-similarity that continues to infinite scale (because they are based on building blocks with defined composition). However, many features of biological systems appear to scale in a manner that indicates a range over which self-similarity is maintained, at a fractal dimension that does not seem to conform to whole-integer dimension (i.e., the system appears fractal). Examples in nature include features as far removed as blood vascular patterns and coastlines (Iannacone and Khoka, 1996; Pennycuick, 1992).

The fact that the fractal dimension, f, is a power of the measurement step length, itself a key determinate of the power function $E = xy^f$, means that fractal analysis will have many features in common with allometric analysis.

Fractal theory was applied most thoroughly to the question of metabolic rate scaling by West *et al.* (1997). This group proposed that virtually all allometric relationships in complex organisms could be derived from a strategy for minimization of energy dissipation in internal transport, because all transport systems are a space-filling fractal network of branching tubes – with inherent limitations that will induce specific fractal influences on the solutions available. This has become commonly known as the WBE (West, Brown, Enquist) model.

Challenges to this model came immediately, where technical issues regarding the assumption of capillary invariance appeared incorrect on close inspection (Dawson, 1998), or other aspects were not a good representation of biological reality (Kurz and Sandau, 1998). These initial problems were dealt with fairly effectively (West and Brown, 2005). Recent challenges have been more fundamental, where it has been argued that the WBE model lacks self-consistency and depends on assumptions of homogeneity of form, rather than demonstrating them (Chaui-Berlinck, 2006), or that a more detailed analysis leads to predictions of the model that do not correspond to measured data (Savage *et al.*, 2008).

Certainly, the debate continues, while alternative models such as the "allometric cascade" (Hochachka *et al.*, 2003), a model suggesting that multiple mass-related scaling contributions are involved, each with its own coefficient of influence, are brought into the discussion. Final resolution of the scale effects in biological systems will provide substantial insight into an often neglected, but very important, aspect of the functional reality of organisms.

8.9 MAKING VALID COMPARISONS: MEASUREMENT, DIMENSION AND FUNCTIONAL CRITERIA

This volume focuses on issues related to the evaluation and interpretation of locomotion in terrestrial mammals. As such, it deals with a field that is fundamentally comparative in nature. The above discussions, however, have demonstrated that the simple process of comparison is fraught with complexities that can, at best, obscure the true relationship between variables or, at worst, lead to inaccurate portrayals of these relationships. How, then, should valid comparisons of form and function between animals be performed? A wide variety of approaches currently exist, and novel ones are being generated as new perspectives on function are established. In the process of making such comparisons, or evaluating those performed by others, it is important to keep some fundamental aspects of the process in mind.

8.9.1 Considering units

Just what temperature is it when it is twice as cold as zero? This question may appear a little facetious, but there is a good lesson to be learned in considering it. Measurement and analysis of physical systems, like moving animals, requires the use and manipulation of measurement units. The quantities with which both temperature and mechanical variables are measured are described with man-made units. In some circles, "man-made" is synonymous with "unnatural", but they are actually meant to be representations of the natural world. Misinterpretation of the manner in which the representation is made, however, can lead to important misunderstandings of the system. Since this holds for both temperature and mechanical quantities, our common familiarity with temperature units can be exploited to wrestle with this question, but the concepts discussed apply to other physical measures as well.

So what temperature *is* it when it is twice as cold as zero? The first response, and a reasonable one at that, is to ask if we are referring to Celsius (C) or Fahrenheit (F) degrees (at least currently, in most places – but the question could as easily involve degrees Delisle, Newton or Réamur, all of which exist)? This aspect does not much matter to the basic point, which is that applying standard mathematics to this particular question leads to a problem – it makes sense that twice as cold as zero Fahrenheit might be colder than twice as cold as zero Celsius (where 0° F is substantially below the temperature at which water freezes, but 0° C defines the freezing temperature of water) – but how is the value of "twice zero" determined? There are at least three "reasonable" answers to this question in either scale. The current convention in science is (usually) to use Celsius degrees for measuring temperature, so these will be emphasized here, but it should be an easy matter to convert the arguments to any scale.

So, what are the three reasonable answers to this question? They are:

(i) –26.7°C

(ii) –37°C

(iii) –136.5°C

… and the explanation for each of these options …

Mathematically, two times zero is still zero, but we know from experience that it can be colder than 0°C. So, one way to approach the problem is biologically, and this is where the first two answers originate. The surface temperature of human skin is approximately 26.7°C. Since this is the temperature difference above zero, then that many degrees below zero should "feel" twice as cold as zero. Alternatively, core body temperature is approximately 37°C, so this many degrees below zero would be twice as cold to the body as zero.

If these answers seem somewhat arbitrary, it is because they are. The third answer addresses this issue. The temperature normally referred to as zero degrees is arbitrary in the scheme of temperature measurement. For the Celsius scale, it is the temperature that pure water freezes (Fahrenheit was a physiologist so, in his scale, the freezing point of blood serum was used; thus, due to the salinity of serum, Fahrenheit's zero is a lower temperature than that of Celsius, a chemist).

Temperature is a measure of heat content of an object, so clear logic should lead to the conclusion that zero should be when the object contains absolutely no heat energy. This is referred to as absolute zero, 0°K (Kelvin, a scale using Celsius units that starts at absolute zero) or 0°R (Rankine, a scale using Fahrenheit units that starts at absolute zero) and it occurs at –273°C (–460°F). So, if 0°C is 273 degrees above absolute zero, then half way to absolute zero, –136.5 °C, is a good candidate for twice as cold as 0°C. Of course, twice absolute zero remains absolute zero, because it is impossible to have negative heat energy (an appealing result on many levels including mathematically, but one that requires understanding both the units and the physical phenomenon they represent).

The "take home" message in all this is simply that units are devised in sometimes arbitrary ways to assist us in measuring physical phenomena. Like the winding cattle path that eventually turns into a crooked thoroughfare through the city as the city grows and develops over the adjoining countryside, units can (and often do) bear the burden of their history. Delisle, Newton and Réamur degrees are all based on the temperature difference between freezing and boiling water, but each chose a different number of units between the two identifiable points. Delisle originally chose an unworkable 2400, but this was eventually reduced to 150 – an odd measure to us, but one that was in common use in Russia for a century. Newton chose 33 (suggesting an unusual arbitrariness to a mind we usually revere for its astute logic), and Réamur 80 – a scale currently in use in some parts of the cheese-making industry.

Units may occur in a range that is convenient for our day-to-day business (°Celsius) or be related to a feature of the physical world (°Kelvin). The same applies for other physical measurements, such as force, momentum and energy. As we will see, in some cases, the units applied to a problem can influence our interpretation of its apparent outcome (so, what is twice as cold as zero?).

The above illustrates that, through the historical development of units – often prior to understanding the physical principles on which the units are designed to measure – we are faced with the possibility that units could be numerically consistent and inconsistent. Numerically consistent units are linked in a meaningful way to the phenomenon they measure, as degrees Kelvin or Rankine are. As such, their manipulation through standard mathematical operations leads to a realistic conclusion. This is not the case for numerically inconsistent units, especially over contentious regions such as the transition from positive to negative.

The result of a measurement is always a number that expresses the ratio of a measured physical quantity to those of a unit selected to represent that measure (Duncan, 1953). Such a unit can be remarkably arbitrary, and still be useful. The temperature measures are examples, where the magnitude of the unit is defined only by selecting a number of divisions between physically reproducible events – making the unit numerically consistent requires only making the arbitrary scale coincident with the properties of the physical phenomenon they represent.

Another example of a fully serviceable, but notably arbitrary, unit is the measure of an arc of a circle. Degrees are a long-established measure related to the 360(ish) days it takes the earth to revolve around the sun. One would assume that, if Martian civilization ever existed, then their compasses would be divided instead into 670 degrees, corresponding to the 670-day Martian year (a Martian day, resulting from the rotation of the planet around its axis, is very similar to that of earth, but the Martian year, the period for the planet to revolve around the sun, is 1.8 times longer than that of Earth).

An attempt to be non-arbitrary in the measure of a circle is indicated by the use of the radian as an angular measure. The circumference of any circle is given by the product of the radius and 2π. Consequently, the arc of a circle can be divided into 2π angular units, each covering a partial circumference of the circle with a distance equal to the radius. These angle units are given the name "radian". Although derived from the unique geometric properties of circles, the use of the radian as a measure exists simply as a convention, and does not have any particular functional advantage, especially in locomotion studies, over the use of degrees (i.e., it is no more nor less meaningful than that of the degree).

8.9.2 Fundamental and derived units

There is an important feature of units that should be recognized, and which can inform analyses in important ways. This is the recognition of units as either fundamental or derived. Interestingly, there is nothing particularly "fundamental" about the difference between fundamental and derived units, but the function of each is a convention generally selected for the sake of convenience. For instance, if we select as a fundamental unit time (T) measured in whatever unit is at hand, then the measure of frequency (f), a derived unit, becomes the inverse of period (the time interval needed to complete a cycle), $f = 1/T$. Although it is common practice to select time as a fundamental unit, there may be cases where it is more convenient to select frequency instead, and define time as a derived unit, being $T = 1/f$ (Pennycuick, 1992).

In describing geometry, it is adequate to have a system of units consisting only of a single fundamental length unit, from which all others, such as area and volume, can be derived. If motion is of interest (kinematics), then a two-unit system is necessary; one for the measurement of length, and another for the measurement of time. For the mechanics of motion, kinetics are also needed, so a third unit involving mass, a measure of inertia, (or force, if that is more convenient) is required.

Even though a variety of units could qualify as *fundamental*, by convention and as a consequence of their utility, we commonly use the fundamental units of mass (M), length (L) and time (T) and, again by convention, utilize the SI system as a measure of these; kg, m, sec (kilogram, meter, second). It is obvious that not all features that could be measured in a physical system are included in the MLT system. Heat, for instance, is not represented by these fundamental units; if the analysis involves heat, or heat transfer, then the fundamental units would have to include a measure of heat, such as temperature.

Even with a restriction to mechanics, it would appear that important measures are neglected by the MLT system. Velocity, acceleration, force, work, power, etc. appear to be neglected. However, these all constitute *derived* units, where each can be derived by some combination of the fundamental units. For instance, velocity has the units LT^{-1}, acceleration LT^{-2}, force ($F = MA$) MLT^{-2}, work $w = F \times$ distance, $MLT^{-2} \times L$) ML^2T^{-2} and power (power = work/time; $ML^2T^{-2} \times T^{-1}$) ML^2T^{-3} (Table 8.2). In some

Table 8.2 ■ Dimensions of some important mechanical quantities used in locomotion analysis. Some dimensionless quantities from Hof (1996)

Quantity	Dimension	Dimensionless formula
Fundamental		
Mass	M	$M_{rel} = m/m_0$
Length	L	$L_{rel} = l/l_0$
Time	T	$T_{rel} = t/(l_0g)^{1/2}$
Derived		
Velocity	LT^{-1}	$V_{rel} = v/(l_0g)^{1/2}$ (relative velocity, a form of the Froude number)
Acceleration	LT^{-2}	$A_{rel} = a/g$
Jerk (rate of acceleration change)	LT^{-3}	$J_{rel} = j/gt$
Momentum and impulse	MLT^{-1}	$I_{rel} = l/m_0gt$
Force	MLT^{-2}	$F_{rel} = f/m_0g$
Energy and work	ML^2T^{-2}	$W_{rel} = w/m_0l_0g$
Frequency	T^{-1}	$f_{rel} = f/(g/l)^{1/2}$
Moment	ML^2T^2	$M_{rel} = M/m_0l_0g$
Power	ML^2T^{-3}	$P_{rel} = P/m_0 l^{3/2}g^{1/2}$

circumstances, it may be mathematically more convenient to use force instead of mass as the fundamental unit, consequently $M = f/LT^2$ (Bullimore and Donelan, 2008).

Why all the fuss about dimensions? This is because an analysis of the dimensions responsible for the behavior of a system (aptly termed "dimensional analysis") can provide remarkable insight into the mechanics of the system, even if specific details of the function are not yet known (or understood). Why is this? Simply because physical principles do not depend on arbitrary units of measure, but will be rooted in the fundamental units of the phenomena being investigated. Ignoring the fundamental units is tantamount to ignoring the physical principle of interest. However, analysis of the fundamental units involved can lead to insight into what principles are actually at play. As an example, consider a simple pendulum (the following discussion is based on Barenblatt, 1996).

We are all familiar with the motion of a simple pendulum but, if we did not understand the physical principles behind this motion, how could they be figured out? It is a simple matter to observe that each pendulum has a "natural", or consistent, period of swing regardless of amplitude (within limits). As it is such a simple system, it might be reasonable to assume that the period of oscillation would depend on such things as the length of the supporting cord (no cord, no motion), the size of the swinging mass (no mass, no motion) and the acceleration of gravity (no gravity, no motion). Each of these variables will represent either a fundamental unit, length (L), mass (M), period (T), or a unit derived from fundamental units, acceleration, LT^{-2}.

Imagine what would occur if the "measure" of one of these units were altered; where the length "measure" was decreased by a certain value, then the "measured" length (i.e., the number representing the physical length) would have to increase. The same would occur for a distortion of the "measures" of M or T and, likewise, for any measure derived from them. Consider the ratio L/g. The value of this ratio depends on the size of the units of length and acceleration, where acceleration is dependent on L/T^2. If the value of the length L is increased by a certain proportion, and time is increased by another, then the ratio changes by $L/(L/T^2) = T^2$.

We can use this result to interpret the period of the swinging pendulum by recalling that the period of the pendulum is constant for different amplitudes. Since the period remains constant for different amplitudes, we can assume that there is a factor that changes with period, such that the ratio remains constant when the "measures" are changed – that is, it is independent of the measure (a physical reality, rather than a mathematical accident). We recognize that the units of period are T, so the factor must also vary as T. Since L/g varies as T^2, we can surmise that $(L/g)^{1/2}$ would vary appropriately. As a result, the constant (k) can be defined as $k = P/(L/g)^{1/2}$ or $P = k(L/g)^{1/2}$ (in this case P represents the period of swing). The relationship between the variables, $(L/g)^{1/2}$, is a fundamental descriptor of the behavior of a simple pendulum. The value of the constant k (which is 2π) would have to be determined experimentally (quite a straightforward process) to yield the complete relation, $P = 2\pi (L/g)^{1/2}$.

So, through the use of dimensional analysis, a fundamental feature of the mechanical behavior of the system is identified, even without originally understanding the phenomena responsible for the dynamic behavior of the system.

Dimensional analysis is a tool that should not be neglected. As succinctly stated by Duncan (1953), *"Dimensional analysis is to be regarded as a special and convenient technique for finding the quantitative conditions for similarity of behavior together with the consequences of this similarity."* (p. 3). In the example above, it was the similarity of motion of simple pendula and its consequences that were identified. Similarity in geometry and its comparison over a size range has been discussed in earlier sections of this chapter. Motion can be analyzed in a similar manner to simple geometry, where a special case of consistent motion across a size range is defined as "dynamic similarity" (see Chapter 9).

A key feature of the use of dimensions is recognizing the importance of accounting for the dimensions of a relationship. Fourier recognized that physical laws represent relationships between variables that are independent of the units used to measure them. That is, to form a true equality between physical variables, the dimensions have to also be equivalent. From this, it should be evident that there is a substantial advantage in the search for and, more importantly, the interpretation of the relationship between complex variables, if they are evaluated within the context of their dimension (or converted to a dimensionless form).

By identifying the fundamental dimensions and the composition of derived variables, it is possible to remove dimensional inequalities in the evaluation of the relationship between variables in mechanical comparisons. We have seen a similar strategy in identifying fundamental dimensions in comparing shape in geometrical analysis, where the only complexity comes from the possibility that some systems may be fractal, and so defy this simple comparison. In the case of dynamics, where motion is also important, it is necessary to expand the dimensions considered. A list of variable dimension, and some physically relevant normalizations of mechanical quantities relevant to mammalian locomotion mechanics are listed in Table 8.2.

8.9.3 Froude number: A dimensionless example

The Froude number is an important and commonly used dimensionless number, and is a good example of a functionally derived dimensionless parameter. Understanding why it is useful, how it should be used, and what its limitations are, will likely help in appreciating all dimensionless parameters (also see a discussion of the utilization of the Froude number in studies of mammalian gait in Chapter 9).

Several forms of the Froude number exist. All were originally developed to assess the behavior of ships and design models (Froude, 1876; Vaughan and O'Malley, 2005), but parallels with physical determinants of walking suggest that it should also be applicable in this context (Thompson, 1917; Alexander and Jayes, 1983). In all forms, the Froude number is a dimensionless representation of the relationship between inertia ($I = Mv$) and gravitational forces (Mg).

In walking, the supporting limb functions as a strut that diverts the mass of the animal over the contact in the manner of an inverted pendulum. So, the inertial component of this action is the centrifugal force of the motion, Mv^2/L. The gravitational component is simply the organism's weight, Mg. The ratio of inertial to gravitational forces, then, becomes $(Mv^2/L)/Mg$, which reduces to V^2/gL (Alexander, 1976). The Froude number is sometimes written as $Fr = (v/(gL)^2)$ and is referred to as "relative velocity" (Alexander and Jayes, 1983; Gatesy and Biewener, 1990). Note that $v = LT^{-1}$, $g = LT^{-2}$, so all dimensions cancel (making it a dimensionless parameter – the question is, is it indeed a relevant measure of the mechanical function of the system?).

The Froude number is important in the fluid environment because it is related to a critical flow issue: when the Froude number is unity, the speed of the surface wave and the flow of the medium are equivalent, and the flow is considered to be in a "critical state". Such a critical point is not obvious from physical theory when the Froude number is applied to locomotion, though Alexander has argued that the centripetal force lifting the foot as the CoM passes over the stance limb in the biped (when centripetal force equals body weight, Fr = 1) is equivalent to such a critical point (Alexander, 1980, 1992; Srinivasan and Ruina, 2005; Usherwood, 2005; Usherwood et al., 2008). Such a critical point would indicate a key transition, such as a change to a different gait. However, the transition between walking and running in humans occurs spontaneously at a Froude number closer to 0.5.

The transition between the bounce-like run or trot and the faster asymmetric gallop does appear to occur over a restricted range of relative speed (Alexander and Jayes, 1983). For these gaits, however, the Froude number should not be adequate to capture the fundamental dynamics of the system, where deflections of elastic structures violate the assumptions on which the Froude analysis is based. In such cases, a frequency-based similarity criterion, such as the Strouhal number, must also be applied (Alexander, 1989; Bullimore and Donelan, 2008; Delattre et al., 2009).

In the above example, supporting limb length was chosen as L, and this length was directly related to the centripetal force identified as an important functional characteristic of the system in motion. Dimensional parameters, such as the Froude number, are used as a criterion for comparing motions of different systems (as Froude himself did, when comparing the mechanical behavior of ship hulls and scaled models). The assumption is usually that the systems being compared are geometrically similar. In true geometric similarity, it does not really matter what characteristic length is chosen (as long as it is consistent between compared forms), because all lengths should be related. Recognizing, of course, that this is not precisely the case in comparisons between different mammals, it is best to select a representative length, such as limb length, that is most closely associated with the mechanical function being analyzed. However, if metabolic work or power are the limiting factor, other comparisons are necessary (see Chapter 9, and also Pennycuick, 1992).

▌REFERENCES

Albrecht GH (1988). Simple allometry coefficients – fact or fiction. *American Zoologist* **28**, A172.

Albrecht GH, Gevin BR (1987). The simple allometry equation reconsidered: assumptions, problems, and alternative solutions. *American Journal of Physical Anthropology* **72**, 174.

Alexander RMcN (1976). Estimates of speeds of dinosaurs. *Nature* **261**, 1929–1930.

Alexander RMcN (1977). Allometry of the limbs of antelopes (Bovidae). *Journal of Zoology, London* **183**, 125–146.

Alexander RMcN (1980). Optimal techniques for quadrupeds and bipeds. *Journal of Zoology, London* **192**, 97–117.

Alexander RMcN (1989). Optimization and gaits in the locomotion of vertebrates. *Physiological Reviews* **69**, 1199–1227.

Alexander RMcN (1992). A model of bipedal locomotion on compliant legs. *Philosophical Transactions of the Royal Society of London. Series B* **338**, 189–198.

Alexander RMcN, Jayes AS (1983). A dynamic similarity hypothesis for the gaits of quadrupedal mammals. *Journal of Zoology* **201**(1), 135–152.

Alexander RMcN, Jayes AS, Maloiy GMO, Wathuta EM (1979). Allometry of the limb bones of mammals from shrews (Sorex) to elephant (Loxodonta). *Journal of Zoology, London* **189**, 305–314.

Barenblatt GI (1996). *Scaling, self-similarity, and intermediate asymptotics.* Cambridge University Press, Cambridge, UK.

Bertram JEA (1988). *The biomechanics of bending and its implications for terrestrial support.* Ph.D. dissertation, University of Chicago, Chicago, Il.

Bertram JEA (1989). Size-dependent differential scaling in branches: The mechanical design of trees revisited. *Trees: Structure and Function* **4**, 241–253.

Bertram JEA, Biewener AA (1988). Bone curvature: sacrificing strength for load predictability? *Journal of Theoretical Biology* **131**, 75–92.

Bertram JEA, Biewener AA (1990). Differential scaling of the long bones in the terrestrial Carnivora and other mammals. *Journal of Morphology* **204**, 157–169.

Bertram JEA, Biewener AA (1992). Allometry of curvature in the long bones of quadrupedal mammals. *Journal of Zoology, London* **226**, 455–467.

Bertram JEA, Greenberg LS, Miyake T, Hall BK (1997). Paralysis and long bone growth in the chick: growth trajectories of the pelvic limb. *Growth, Development, and Aging* **61**(2), 51–60.

Biewener AA (1989). Scaling body support in mammals: Limb posture and muscle mechanics. *Science* **245**, 45–48.

Biewener AA (2005). Biomechanical consequences of scaling. *Journal of Experimental Biology* 1665–1676.

Biewener AA, Thomason J, Goodship A, Lanyon LE (1983). Bone stress in the horse forelimb during locomotion at different gaits: a comparison of two experimental methods. *Journal of Biomechanics* **16**, 565–576.

Bou J, Casinos A (1985). Scaling of bone mass to body mass in insectivores and rodents. In: Dunker HR and Fleischer G (eds). *Functional Morphology in Vertebrates*, pp. 61–64. Stuttgart: Gustav Fischer Verlag.

Boxenbaum H (1982). Interspecies scaling, allometry, physiological time, and the ground plan of pharmokinetics. *Journal of Pharmacokinetics and Pharmacodynamics* **10**, 201–227.

Brockmann D, Hufnagel L, Geisel T (2006). The scaling laws of human travel. *Nature* **439**, 462–465.

Brody S (1945). *Bioenergetics and growth: with special reference to the efficiency complex in domestic animals.* Oxford, Reinhold, pp. 1023.

Brown JH, West GB (2000). *Scaling in biology.* Oxford University Press, Oxford.

Bullimore SR, Burn JF (2006). Dynamically similar locomotion in horses. *Journal of Experimental Biology* **209**, 455–465.

Bullimore SR, Donelan JM (2008). Criteria for dynamic similarity in bouncing gaits. *Journal of Theoretical Biology* **250**, 339–348.

Carrano MT (2001). Implications of limb bone scaling, curvature and eccentricity in mammals and non-avian dinosaurs. *Journal of Zoology* **254**, 41–55.

Chaui-Berlinck JG (2006). A critical understanding of the fractal model of metabolic scaling. *Journal of Experimental Biology* **209**, 3045–3054.

Cheverud JM (1982). Relationships among ontogenetic, static, and evolutionary allometry. *American Journal of Physical Anthropology* **59**, 139–149.

Christiansen P (1999). Scaling mammalian long bones: small and large mammals compared. *Journal of Zoology* **247**, 333–348.

Currey JD (1984). *The mechanical adaptations of bones.* Princeton University Press, Princeton, pp. 304.

Dawson TH K (1998). Allometric scaling in biology. *Science* **281**, 751a.

da Silva JKL, Garcia GJM, Barbosa LA (2006). Allometric scaling laws of metabolism. *Physics of Life Reviews* **3**, 229–261.

Delattre N, Lafortune MA, Moretto P (2009). Dynamic similarity during human running: about Froude and Strouhal dimensionless numbers. *Journal of Biomechanics* **42**, 312–318.

Dennis SC, Noakes TD (1999). Advantages of a smaller body mass in humans when distance-running in warm, humid conditions. *European Journal of Applied Physiology and Occupational Physiology* **79**, 280–284.

Doube M, Wiktorowicz-Conroy A, Christiansen P, Hutchinson JR, Shefeline S (2009). Three-dimensional geometric analysis of felid limb bone allometry. *PLoS One* **4**, e4742.

Duncan WJ (1953). *Physical similarity and dimensional analysis: An elementary treatise.* Edward Arnold & Co., London. 156 pp.

Economos AC (1982). On the origin of biological similarity. *Journal of Theoretical Biology* **94**, 25–60.

Economos AC (1983). Elastic and/or Geometric similarity in mammalian design? *Journal of Theoretical Biology* **103**, 167–172.

Feldman HA, McMahon TA (1983). The ¾ mass exponent for energy metabolism is not a statistical artifact. *Respiration Physiology* **52**, 149–163.

Fortelius M, Kappelman J (1993). The largest land mammal ever imagined. *Zoological Journal of the Linnean Society* **108**, 85–101.

Froude W (1876). On the comparative resistances of long ships of several types. *Transactions of the Royal Institution of Naval Architects* **17**, 181–188.

Full RJ (1989). Mechanics and energetics of terrestrial locomotion: From bipeds to polypeds. In: Wieser W, Gnaiger E (eds.) *Energy transformations in cells and animals.* Thieme, Stuttgart, pp. 175–182.

Garcia GJM, da Silva JKL (2004). On the scaling of mammalian long bones. *Journal of Experimental Biology* **207**, 1577–1584.

Garland T Jr., Bennett AF, Rezende EL (2005). Phylogenetic approaches to comparative physiology. *Journal of Experimental Biology* **208**, 3015–3035.

Gatesy SM, Biewener AA (1990). Bipedal locomotion: effects of speed, size and limb posture in birds and humans. *Journal of Zoology, London* **224**, 127–147.

Godek SF, Bartolozzi AR, Burkholder R, Sugarman E, Pedozzi C (2008). Sweat rates and fluid turnover in professional football players: a comparison of National Football League linemen and backs. *Journal of Athletic Training.* **43**, 184–189.

Harris CC (1973). Lilliput revisited, or how fed-up was Gulliver? *Chemical Technology* **3**, 600–602.

Heusner AA (1982). Energy metabolism and body size. I. Is the 0.75 mass exponent of Kleiber's equation a statistical artifact? *Respiration Physiology* **48**, 1–12.

Heusner AA (1991). Body mass, maintenance and basal metabolism in dogs. *Journal of Nutrition* **121**, S8–S17.

Hill AV (1950). The dimension of animals and their muscular dynamics. *Science Progress* **38**, 209–230.

Hill JR, Rahimtulla KA (1965). Heat balance and the metabolic rate of new-born babies in relation to environmental temperature; and the effect of age and of weight on basal metabolic rate. *Journal of Physiology* **180**, 239–265.

Hochachka PW, Darveau C-A, Andrews RD, Suarez RK (2003). Allometric cascade: a model for resolving body mass effects on metabolism. *Comparative Biochemistry and Physiology Part A: Molecular & Integrative Physiology* **134**, 675–691.

Hof AL (1996). Scaling gait data to body size. *Gait & Posture* **4**, 222–223.

Huxley JS (1932). *Problems of relative growth.* Dial Press, New York.

Iannacone PM, Khoka M (1996). Fractal geometry in biological systems: an analytic approaCh CRC Press, Boca Raton, Fl. pp. 360.

Iriarte-Diaz J (2002). Differential scaling of locomotor performance in small and large terrestrial mammals. *Journal of Experimental Biology* **205**, 2897–2908.

Jiang C, Pitt RE, Bertram JEA, Aneshanley DJ (1999). Fractal-based image texture analysis of trabecular bone architecture. *Medical & Biological Engineering & Computing* **37**, 413–418.

Kleiber M (1932). Body size and metabolism. *Hilgardia* **6**, 315–353.

Kleiber M (1961). *The fire of life.* New York, Wiley.

Kolokotrones T, Savage V, Deeds EJ, Fontana W (2010). Curvature in metabolic scaling. *Nature* **464**, 753–756.

Kurz H, Sandau K (1998). Allometric scaling in biology. *Science* **281**, 751a.

Lanyon LE, Baggot DG (1976). Mechanical function as an influence on the structure and form of bone. *Journal of Bone and Joint Surgery* **58-B**, 436–443.

Main RP, Biewener AA (2004). Ontogenetic patterns of limb loading, in vivo bone strains and growth in the goat radius. *Journal of Experimental Biology* **207**, 2577–2688.

Main RP, Biewener AA (2006). In vivo bone strain and ontogenetic growth patterns in relation to life-history strategies and performance in two vertebrate taxa: goats and emu. *Physiological and Biochemical Zoology* **79**, 57–72.

Mandelbrot BB (1983). *The fractal geometry of nature.* Freeman, New York.

McMahon TA (1973). Size and shape in biology. *Science* **179**, 1201–1204.

McMahon TA (1975). Allometry and biomechanics: limb bones in adult ungulates. *American Naturalist* **109**, 547–563.

McMahon TA (1977). *Allometry.* McGraw-Hill Yrbk. Sci. Tech pp. 48–57.

McMahon TA (1984). *Muscles, reflexes, and locomotion.* Princeton, Princeton University Press.

McMahon TA, Bonner JT (1983). *On size and life.* Scientific American Books, New York, 255 pp.

McMahon TA, Kronauer RE (1976). Tree structures: Deducing the principle of mechanical design. *Journal of Theoretical Biology* **59**, 443–466.

Medlar S (2002). Comparative tends in shortening velocity and force production in skeletal muscles. *American Journal of Physiology. Regulatory, Integrative and Comparative Physiology* **283**, R368–R378.

Pedley TJ (1977). *Scale effects in animal locomotion.* Academic Press, London, 545 pp.

Pennycuick CJ (1992). *Newton rules biology. A physical approach to biological problems.* Oxford University Press, New York. 111 pp.

Pennycuick CJ, Kline NC (1986). Units of measurement for fractal extent, applied to the coastal distribution of bald eagle nests in the Aleutian Islands, Alaska. *Oecologia* **68**, 254–258.

Peters RH (1983). *The ecological implications of body size.* Cambridge University Press, Cambridge, 324 pp.

Prange HD, Anderson JF, Rahn H (1979). Scaling of skeletal mass to body mass in birds and mammals. *The American Naturalist* **113**, 103–122.

Prothero J (1995). Bone and fat as a function of body weight in adult mammals. *Comparative Biochemistry and Physiology* **111A**, 633–639.

Prothero DR, Sereno PC (1982). Allometry and paleoecology of medial Miocene dwarf rhinoceroses from the Texas gulf coastal plain. *Paleobiology* **8**, 16–30.

Rubin CT, Lanyon LE (1982). Limb mechanics as a function of speed and gait: a study of functional strains in the radius and tibia of horse and dog. *Journal of Experimental Biology* **101**, 187–211.

Savage VM, Deeds EJ, Fontana W (2008). Sizing up allometric scaling theory. *PLoS Computational Biology* **4**, e1000171.

Schmidt-Nielsen K (1984). *Scaling, Why is animal size so important?* Cambridge University Press, Cambridge, 241 pp.

Silva M (1998). Allometric scaling of body length: elastic or geometric similarity in mammalian design. *Journal of Mammalogy* **79**, 20–32.

Srinivasan M, Ruina A (2005). Computer optimization of a minimal biped model discovers walking and running. *Nature* **439**, 72–75.

Swatrz SM, Biewener AA (1992). Shape and scaling. In: Biewener AA (ed). *Biomechanics – structures and systems*. IRL Press, Oxford, pp. 21–43.

Thompson DW (1917). *On growth and form*. Cambridge University Press, 345 pp.

Usherwood JR (2005). Why not walk faster? *Biology Letters* **1**, 338–341.

Usherwood JR, Szymanek KL, Daley MA (2008). Compass gait mechanics account for top walking speeds in ducks and humans. *Journal of Experimental Biology* **211**, 3744–3749.

Vaughan CL, O'Malley MJ (2005). Froude and the contribution of naval architecture to our understanding of bipedal locomotion. *Gait Posture* **21**(3), 350–362.

Wainwright SA, Biggs WD, Currey JD, Gosline JM (1982). *Mechanical design in organisms*. Princeton University Press, Princeton, 423 pp.

West GB, Brown TH (2005). The origin of allometric scaling laws in biology from genomes to ecosystems: towards a quantitative unifying theory of biological structure and organization. *Journal of Experimental Biology* **208**, 1575–1592.

West GB, Brown JH, Enquist BJ (1997). A general model for the origin of allometric scaling laws in biology. *Science* **276**, 122–126.

West GB, Brown JH, Enquis, BJ (1999). The fourth dimension of life: fractal geometry and allometric scaling of organisms. *Science* **284**, 1677–1679.

West LJ, Pierce CM, Thomas WD (1962). Lysergic acid diethylamide: its effects on a male Asiatic elephant. *Science* **138**, 1100–1103.

Wieser W (1984). A distinction must be made between the ontogeny and phylogeny of metabolism in order to understand the mass exponent of energy metabolism. *Respiration Physiology* **55**, 1–9.

CHAPTER NINE

Accounting for the Influence of Animal Size on Biomechanical Variables: Concepts and Considerations

Sharon Bullimore

Cheltenham, UK

9.1 INTRODUCTION

Frequently in biomechanics research it is necessary to compare measurements made on subjects of different sizes. This may be due to natural variability within a population, or because we have chosen to compare two groups that are inherently different in size – such as males and females, adults and juveniles, or two different species. As explained in Chapter 8, size has a profound influence on many biomechanical and physiological parameters. Some effects are obvious; an elephant will take longer strides and exert higher ground reaction forces than a mouse. Other effects are less obvious; in comparison to a mouse, an elephant uses 40 times less energy per gram of body mass to travel the same distance (Langman *et al.*, 1995).

This pervasive influence of size on biology creates a problem: how do we separate the effects of size from the effects of other factors? For instance, in a comparison of speed in male and female athletes, males would probably be, on average, faster than

Understanding Mammalian Locomotion: Concepts and Applications, First Edition.
Edited by John E. A. Bertram.
© 2016 John Wiley & Sons, Inc. Published 2016 by John Wiley & Sons, Inc.

females, but the males would also be taller. This raises the question of how much of this difference is due to size differences, and how much is due to gender differences. To answer questions of this type, we need some way of removing – or accounting for – the effects of size, so that we can detect and identify other effects. In this chapter, I will discuss some general principles of, and common approaches to, accounting for size effects, and outline some potential pitfalls.

9.2   COMMONLY USED APPROACHES TO ACCOUNTING FOR SIZE DIFFERENCES

9.2.1   Dividing by body mass

A common method of attempting to account for size effects is to divide parameters by body mass, to give a "mass-specific" value. This approach will only be successful, however, in cases where the parameter in question is proportional to body mass – that is, when a good fit to the data is provided by the equation $y = am$, where y is the parameter of interest, a is a constant and m is body mass (and will be used to indicate body mass throughout this chapter). An example is the mass of the lungs. Lung mass is directly proportional to body mass and, on average, is about 1.1% of body mass (Stahl, 1967). Therefore, expressing lung mass as a percentage of body mass, and comparing the result to 1.1%, can tell us whether a species has unusually large or small lungs for its size.

There are many examples, however, of physiological parameters that are not proportional to body mass. For example, stride length depends upon leg length (Alexander and Jayes, 1983), and muscle force is proportional to muscle cross-sectional area (Medler, 2002; Powell *et al.*, 1984), and neither leg length nor muscle cross-sectional area is usually proportional to body mass. The relationships between lengths, areas and masses can be illustrated by considering animals of different sizes that are identical in shape (i.e., "geometrically similar"). While this is an unrealistic assumption, it is often a useful approach to analyzing the effects of size differences, because it allows size differences to be considered independently of anatomical differences. Once the consequences of size differences in geometrically similar animals have been determined, these can be compared to what actually happens in real animals.

In these idealized, perfectly geometrically similar animals, all ratios of length measurements will be the same – so that, for example, all animals will have the same ratio of back length to leg length, and of ear length to head length. Additionally, all area measurements will be proportional to the square of any of the length measurements; femur cross-sectional area will be proportional to the square of trunk length, and body surface area will be proportional to the square of leg length. Similarly, all volumes will be proportional to the cube of any of the length measurements. Finally, if we assume that all the animals have the same density (which seems reasonable, as they are made of the same materials), then we can also say that all volume measurements will be directly proportional to body mass. The preceding statements can be expressed by the following relationships:

$$area \propto (length)^2$$
$$volume \propto (length)^3 \qquad\qquad (9.1)$$
$$volume \propto m$$

Where: *area*, *length* and *volume* refer to any area, length or volume measurement that could be defined for these animals and m is body mass.

Equation 9.1 can be rearranged to give the following relationships:

$$length \propto m^{1/3}$$
$$area \propto m^{2/3}$$

(9.2)

which will be used extensively throughout this chapter.

While real animals of different sizes are not, of course, perfectly geometrically similar, the relationships described by Equation 9.2 often turn out to be good approximations. It has been found that, in terrestrial mammals of different sizes, bone lengths are proportional to $m^{0.35}$, and bone diameters are proportional to $m^{0.36}$ (Alexander *et al.*, 1979). Head and body length is proportional to $m^{0.34}$ in sea mammals, and to $m^{0.31}$ in land mammals (Economos, 1983). These exponents are all close to the exponent of $m^{0.33}$ expected for geometrically similar animals.

On the other hand, some groups of animals, such as the Bovidae (sheep, cattle, goats and antelopes), are better approximated by a type of scaling known as "elastic similarity", in which lengths scale in proportion to $m^{1/4}$, diameters scale in proportion to $m^{3/8}$ and areas scale in proportion to $m^{3/4}$ (McMahon, 1975). Whether animals scale in a geometrically or elastically similar manner, however, lengths and areas are not proportional to body mass, so any parameter that is a function of a length or an area also will not be proportional to body mass. This means that the usefulness of dividing by body mass to account for size effects is limited.

The relationships of many physiological parameters to body mass can be described by the equation:

$$y = am^b$$

where the constant b is often referred to as the scaling exponent. This is sometimes called an "allometric equation". If b is not equal to 1, then dividing y by body mass yields a mass-specific parameter (y/m) that is still a function of body mass. For example, because muscle isometric force is proportional to muscle cross-sectional area, it will be proportional to $m^{2/3}$ in perfectly geometrically similar animals of equal density (Equation 9.2). Therefore the isometric force (f) of, say, the quadriceps muscle, would follow the relationship $f = am^{2/3}$. If we divide by mass to obtain the mass specific muscle force, we get:

$$f/m = am^{-1/3}$$

where the negative scaling exponent indicates that mass specific force will decrease as body mass increases. Alexander (1985) showed that the maximum mass-specific forces exerted by animals do, indeed, decrease with body mass as described by this relationship. In this case, therefore, dividing by body mass does not successfully remove the effects of size differences.

The common belief that ants are unusually strong for their size arises from the incorrect assumption that load-carrying capacity will scale in proportion to body mass. In fact, it is more appropriate to consider load-carrying capacity relative to $m^{2/3}$, because it is muscle forces that are ultimately responsible for supporting the load and, in perfectly geometrically similar ants, muscle forces would be proportional to $m^{2/3}$. When load-carrying capacity is compared in this way, ants turn out to be surprisingly weak. An ant of approximately 5 mg can carry about 150 mg, or 30 times its own body mass (Cerda *et al.*, 1998).

From this information, we can calculate the constant a in the allometric equation, assuming that the scaling exponent, b, is 2/3, and taking into account the fact that the ant's muscles are also supporting its own body mass so that, in total, it is supporting 155 mg. This gives us the equation: $c = 0.53m^{2/3}$, where c is maximum load-carrying capacity in kg (including body mass) and m is ant mass in kg. If we then consider an ant the size of a human – say, 90 kg – the above equation predicts that the giant ant could carry about 11 kg (i.e., it could not even support its own body mass). This demonstrates that the approach that is used for comparing individuals of different sizes can radically alter the conclusions drawn from a given set of data. It also illustrates a very general principle of scaling, which is that animals are adapted physiologically and anatomically to their size, and could not function well if they were very different in size. We should not be surprised, then, that we do not see human-sized ants or ant-sized humans.

9.2.2 Dimensionless parameters

Another commonly used approach for comparing animals of different sizes is to use dimensionless parameters. In this context, the dimensions of a quantity refer to the type of units it is measured in. For example, if we denote the dimensions of length as L, then the dimensions of area are L^2 and the dimensions of volume are L^3 (commonly referred to as being "two-dimensional" and "three-dimensional", respectively). When performing dimensional analysis of mechanical systems, it is conventional to define three mechanical quantities as reference quantities, and to define the dimensions of other quantities in terms of these. When force (F), length (L) and time (T) are used as reference quantities, the dimensions of velocity are LT^{-1}, the dimensions of frequency are T^{-1}, the dimensions of mass are $FL^{-1}T^2$ (i.e., mass = force/acceleration, from Newton's Second Law) and the dimensions of energy are FL (e.g., gravitational potential energy is defined as weight (a force), multiplied by height (a length)). Alternatively, if mass (M), L and T are used as reference dimensions, the dimensions of force are MLT^{-2} and the dimensions of energy are ML^2T^{-2}. These concepts are discussed in detail in Isaacson and Isaacson (1975).

A dimensionless (or "nondimensional") parameter is formed by combining other parameters in such a way that the dimensions cancel, so that the resulting parameter does not have any units. Strain and angle are dimensionless, because strain is defined as change in length divided by initial length, and angle is defined as arc length divided by radius, or a constant multiple of this. Percentages are also dimensionless. When forming a dimensionless parameter, a consistent set of units must be used so that, for example, strain can be calculated by dividing meters by meters, or by dividing inches by inches, but not by dividing meters by millimeters or inches.

Various dimensionless parameters have been used to describe animal locomotion, including stride length divided by leg length ("relative stride length"), peak vertical ground reaction force divided by body weight ("relative peak force"), stance time divided by stride time ("duty factor") and the phase relationships of the limbs – the time at which each limb comes to the ground as a percentage of stride time (Alexander, 1976; Alexander and Jayes, 1983; Donelan and Kram, 2000; Hildebrand, 1977).

The Froude number is a dimensionless parameter that has been widely used as a dimensionless speed in locomotion research. It is named after the shipping engineer William Froude, who showed that model ships of different sizes generate the same wave patterns when towed at speeds proportional to the square roots of their hull lengths (Vaughan and O'Malley, 2005). The Froude number can be defined as u^2/gl, where u is speed, g is gravitational acceleration and l is hull length for model ships, or leg length for animals. Some authors have also used a Froude number that is the square root of the one defined above (Alexander, 1984a; McMahon and Cheng, 1990; Vaughan et al., 2003), which has the advantage of being linearly related to speed.

So, the crucial question is: does expressing gait parameters in dimensionless form eliminate the effects of size differences between animals? For example, if we compared a trotting cat to a trotting horse, could we expect that they would be using approximately the same relative stride length, the same relative peak force, and the same duty factor? Having equal values of mechanical dimensionless parameters is known as "dynamic similarity". Therefore, the above question can be rephrased as, "do animals of different sizes move in a dynamically similar manner?" The answer is: under some circumstances and in some parameters. In order to explain this, it is first necessary to discuss the concept of dynamic similarity in more depth.

Dynamic similarity is often defined as follows: if two moving systems are dynamically similar, the motion of one can be made identical to the motion of the other by multiplying all linear dimensions by a constant factor k_L, all time intervals by a constant factor k_T, and all forces by a constant factor k_F (Alexander and Jayes, 1983). While this definition only explicitly mentions scale factors for force, length and time, all other mechanical parameters can, as mentioned above, be defined in terms of force, length and time. Therefore, dynamic similarity is, in fact, complete mechanical similarity. Scale factors for other mechanical quantities can be calculated from the dimensions of that quantity and the scale factors for force, length and time – for example, the scale factor for velocity would be $k_L k_T^{-1}$.

To see why the above definition of dynamic similarity is equivalent to saying that dynamically similar systems have equal values of all mechanical dimensionless parameters, we can take the Froude number as an example, and consider two animals, A and B, which are moving in a dynamically similar manner. Animal A has a leg length of l_A and is travelling at a mean velocity of u_A on a planet with gravitational acceleration g_A. From the above definition of dynamic similarity, we know that animal B has a leg length of $k_L l_A$ and a mean velocity of $k_L k_T^{-1} u_A$, and is subject to a gravitational acceleration of $k_L k_T^{-2} g_A$ (if both animals are on the same planet, gravitational acceleration will be the same, so that $k_L k_T^{-2}$ must equal 1). Animal A has a Froude number of $u_A^2/l_A g_A$, and animal B has a Froude number of $(k_L k_T^{-1} u_A)^2/k_L l_A\, k_L k_T^{-2} g_A$.

A little algebraic manipulation shows that these two Froude numbers are the same, because the scale factors cancel. Because, in all dimensionless parameters, the numerator and denominator must have the same dimensions, the scale factors will always cancel out in this way, so that all dimensionless parameters are equal in (perfectly) dynamically similar animals. It is often convenient to define dynamic similarity in terms of dimensionless parameters, rather than scale factors, because this makes it easier to compare a few parameters in many animals, rather than many parameters in only two animals.

Alexander (1976) and Alexander and Jayes (1983) hypothesized that animals of different sizes will tend to move in a dynamically similar manner when travelling at speeds corresponding to equal Froude numbers. In making this prediction, they did not suggest a direct cause-and-effect relationship between the Froude number and the dynamics of locomotion but, instead, reasoned that dynamic similarity is to be expected, if animals have evolved so as to minimize the energetic cost of transport. Alexander (1989) clarified this further, pointing out that animals do not necessarily have to move in dynamically similar manner at equal Froude number, but that: *"If one animal has optimized power (or any other mechanical quantity), the other animal can optimize the same quantity by running in dynamically similar fashion when travelling with the same Froude number."*

Experimental studies have supported this "dynamic similarity hypothesis", showing that dimensionless locomotor parameters are independent of size when animals of different species move at speeds corresponding to equal Froude number (Alexander, 1976; Alexander and Jayes, 1983; Blickhan and Full, 1993; Farley *et al.*, 1993). The hypothesis is also supported by studies that have compared individuals of different sizes within a species (Alexander, 1984a; Bullimore and Burn, 2006a; Griffin *et al.*, 2004; Zijlstra *et al.*, 1996). It does not apply, however, to comparisons between cursorial and noncursorial animals (Alexander and Jayes, 1983), or between animals that move with different numbers of legs in contact with the ground at one time (Blickhan and Full, 1993).

Perfect dynamic similarity, like perfect geometric similarity, is, of course, a theoretical idealization: even if we could measure every possible dimensionless parameter, we would never find two moving animals in which these were all identical. The studies cited above, which tested the dynamic similarity hypothesis, were restricted to considering a small number of dimensionless parameters that describe the overall characteristics of gait. This raises the question of whether dynamic similarity of these overall gait parameters can be taken to indicate that other parameters, such as joint angular motions, tendon strains and muscle forces, will be dynamically similar.

Unfortunately this is not the case. The reason for this is that tissue properties, such as muscle strength and tendon elastic modulus, do not scale in a dynamically similar manner, but instead are independent of size (Medler, 2002; Pollock and Shadwick, 1994a). Therefore, to achieve dynamic similarity of overall gait patterns, other parameters have to deviate from dynamic similarity in order to compensate for this. In Bullimore and Burn (2004), we referred to this deviation as a "compensatory distortion", and demonstrated that the increase in limb mechanical advantage with animal size that was observed by Biewener (1990) is sufficient to compensate for the size-independence of tendon elastic modulus, thus making it possible for animals of different sizes to move in a dynamically similar manner in terms of overall gait characteristics.

Some authors have attempted to identify dimensionless parameters, in addition to the Froude number, that must be equal for overall gait dynamics to be dynamically similar. This is important for predicting the conditions under which dynamic similarity of locomotion can be expected, and for explaining observed deviations from dynamic similarity. Alexander (1989) argued that, for gaits such as running, where elastic deformations are important, at least one other dimensionless parameter must be equal in order for locomotion to be dynamically similar, and proposed that the Strouhal number should be used. He did not state, however, that these two conditions would be sufficient to guarantee dynamic similarity.

In theory, the dimensionless parameters that must be equal for motion to be dynamically similar can be determined using a method called "dimensional analysis", which is described for a pendulum in Chapter 8, and also in texts such as Isaacson and Isaacson (1975). This process involves identifying the parameters that govern the mechanics of the system in question and forming them into dimensionless parameters (using a theorem known as Buckingham's Pi theorem to define the necessary number of dimensionless parameters). If two systems have equal values of these governing dimensionless parameters, they will move in a dynamically similar manner.

The problem with a complicated system such as a moving animal is that it is difficult to ensure that you have identified all the parameters that govern its mechanics. However, even without doing this, partial conditions for dynamic similarity can be derived. For example, we can be confident that leg length, forward speed and gravitational acceleration all have a substantial influence on the mechanics of locomotion. Therefore, we can deduce that having equal values of the Froude number is a necessary condition for animals of different sizes to move in a dynamically similar manner.

One approach to identifying the parameters that govern the mechanics of locomotion is to use a mathematical model. If the model provides good predictions of locomotor mechanics, we can use dimensional analysis of the model parameters to determine conditions for dynamic similarity. Bullimore and Donelan (2008) used this approach to obtain conditions for dynamic similarity based on the parameters of the planar spring mass model. This model provides good predictions of the mechanics of bouncing gaits, such as running, trotting and hopping (Blickhan, 1989; Blickhan and Full, 1993; Bullimore and Burn, 2007; He *et al.*, 1991; McMahon and Cheng, 1990). Dimensional analysis of the model parameters revealed that animals moving at constant speed must have equal values of three dimensionless parameters, including the Froude number, for overall gait mechanics to be dynamically similar.

An important caveat, when a model is used in this way, is that the dimensionless parameters obtained can only predict dynamic similarity in situations and parameters for which the model provides accurate predictions. For example, the planar spring-mass model does not provide good predictions of horizontal ground reaction forces (Bullimore and Burn, 2006b, 2007), so it cannot supply useful information about the conditions required for dynamic similarity of these forces.

In summary, a number of studies have shown that expressing gait parameters in dimensionless form and comparing animals at equal Froude number can account for the effects of differences in size, both between and within species. It cannot be assumed, however, that this approach will always be successful, because equal Froude number is not sufficient to guarantee dynamic similarity in locomotion, and because dynamic similarity of overall gait parameters does not necessarily imply that other parameters will also be dynamically similar. Therefore, caution is necessary when using dimensionless parameters to account for size effects in locomotion studies.

Ideally, the efficacy of the approach needs to be established for each parameter and species of interest. For example, in order to determine the age at which the juveniles of a species start to use adult gait patterns, one could obtain data from adults of a wide range of different sizes, and juveniles of various different ages, all moving at the same range of Froude numbers. The gait parameters of interest could then be expressed as dimensionless parameters, and plotted against Froude number to confirm that this approach can remove the effects of size differences between adults. The same dimensionless parameters could then be studied in the juveniles, to determine at what age they begin to move in a manner that is dynamically similar to the adults.

Figure 9.1 shows an example of the use of dimensionless parameters to compare stride lengths in horses and dogs. When stride length is plotted against speed, there is considerable scatter, and larger individuals within each species (indicated by larger

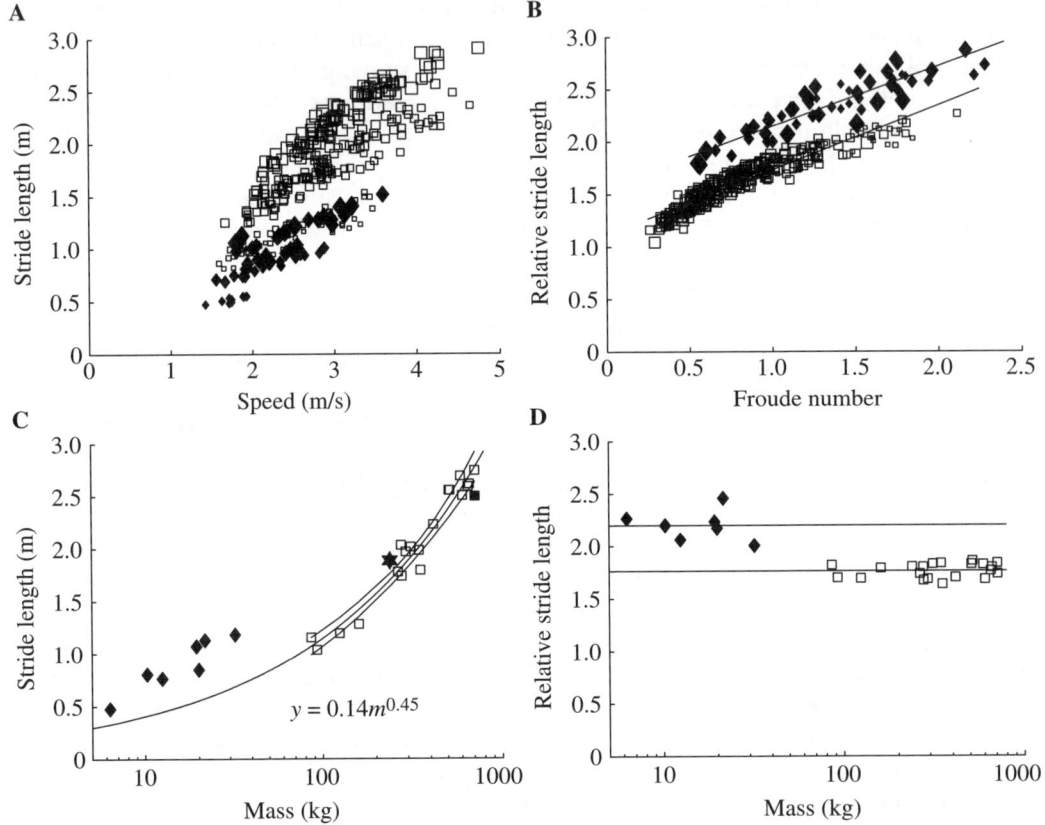

FIGURE 9.1

An example of using dimensionless parameters to compare stride lengths in horses (open squares) and dogs (filled diamonds). The horse data are from Bullimore and Burn (2006a), and the dog data are unpublished data from the same authors. (**A**) Stride length as a function of speed. The size of the symbols represents the body mass of the individual, relative to the mean for that species (although the relationship between marker size and mass is not linear). It can be seen that there is considerable scatter. Horses tend to use longer stride lengths than dogs, and larger individuals within each species tend to use longer strides than smaller animals. There is also quite a bit of overlap, with the smallest horses (which were so-called 'miniature horses') using similar stride lengths to the larger dogs. (**B**) The same data expressed symbols) tend to take longer strides. Plotting relative stride length against Froude number is effective in removing size effects within both species, and clearly demonstrates that dogs use higher relative stride lengths than horses when travelling at the same Froude number.

9.3 EMPIRICAL SCALING RELATIONSHIPS

As mentioned above, the relationship of a physiological parameter to animal size can often be described by an allometric equation of the form $y = ax^b$, where y is the parameter of interest, x is a measure of size, such as leg length, height or body mass, and a and b are constants. Therefore, by fitting an equation of this form to the data, we obtain a description of the effect of animal size on the parameter y. However, we can also get more information than this by considering the position of an individual data point (representing an animal or a species) relative to the fitted line. This allows us to say whether that individual or species has a high or a low value of the parameter in question *for its size*.

An example is shown in Figure 9.1C, where the data for stride length against body mass in horses have been fitted with an allometric equation. For example, we can see that, while the animal designated by a filled square has a longer stride length than the animal designated by a star, the former animal actually has a short stride for its mass, while the latter has a relatively long stride. The information provided by the residuals between the allometric equation and the individual data points has been called the

FIGURE 9.1 (Continued)

using dimensionless parameters as relative stride length against Froude number. The scatter is considerably reduced, collapsing the data onto separate relationships for each species. There is now no effect of size within species and no overlap between the two species. The data for each species have been fitted with a straight line, but it is clear that this is not an ideal fit to the horse data, because the data points fall below the line at both ends. The dog data now lie clearly above the horse data, showing that dogs use a longer relative stride length than horses when trotting at the same Froude number. Parts (**C**) and (**D**) show more clearly how the use of dimensionless parameters has been able to eliminate the effects of size within both species. The stride lengths shown here are for a Froude number of 1 and were obtained by interpolation of the data in (**B**). In (**C**), stride length is plotted against mass for both dogs and horses (with mass on a logarithmic scale). It is clear that stride length increases with size. The dashed line and equation represent an allometric equation that has been fitted to the horse data, and the dotted lines are 95% confidence intervals. The individuals indicated by the star and the filled square are discussed in the text. Part (**D**) shows relative stride length plotted against mass, with the straight lines indicating the mean value for each species. It is clear that using relative stride length (and making the comparison at equal Froude number) has successfully eliminated size effects within both species. The data for individual animals is scattered about the mean values, indicating individual differences due to factors other than size, as well as some experimental error. The greater scatter in the dog data may be due to the greater morphological variability present in this domestic species.

"secondary signal" (Schmidt-Nielsen, 1984). An advantage of this empirical approach is that it does not require any prior assumptions as to the nature of the scaling relationship, nor any knowledge of the physical processes underlying the measured parameters.

Caution must be exercised in interpreting the residuals between the fitted line and the individual data points, however. First, some of the variability around the fitted line will be due to experimental error, rather than true differences, and therefore will not provide useful information. Second, if individual data points differ substantially from the mean value for their size, they can skew the fitted relationship and, consequently, can alter the secondary signals for all the data points. This is particularly likely if these points are at the extremes of the size range, because this will increase their influence on the fitted line. The risk of this can be reduced by having as many data points as possible over the entire size range, so that individual deviations from the general relationship at least partially counterbalance one another (Schmidt-Nielsen, 1984).

9.4 SELECTED BIOMECHANICAL PARAMETERS

It should be apparent from the above discussion that no single method can be universally used to account for size effects, and that the most appropriate approach depends upon the question being addressed and the physical processes that underlie the measured parameters. In this section, I will discuss the best ways of accounting for the effects of size on several widely used biomechanical parameters. At least three of these parameters (muscle force, muscle velocity and jump height) provide examples of measurements for which neither dividing by body mass nor forming dimensionless parameters is usually an appropriate method of accounting for size effects.

9.4.1 Ground reaction force

Based on purely mechanical considerations, we know that the vertical ground reaction force exerted by an animal, when averaged over a stride, must be equal to its body weight. Within this constraint, an animal may apply relatively low forces for a high proportion of the stride cycle (typical of slow gaits), or higher forces for a smaller proportion of the stride cycle (typical of faster gaits). Dividing vertical ground reaction force by body weight, therefore, accounts for weight differences between individuals and allows the influence of other factors, such as speed and gait, to be observed (Figure 9.2).

The above argument does not apply to horizontal ground reaction forces, because these are not involved in supporting body weight. It may still be useful to divide horizontal ground reaction force by body weight to form a dimensionless parameter, if the goal is to determine whether ground reaction force is dynamically similar in animals of different sizes. Alternatively, because force is the product of mass and acceleration (Newton's Second Law), horizontal ground reaction force can be divided by body mass to obtain the horizontal acceleration of the centre of mass. However, because Newton's Second Law applies only to the net force, this approach requires that the horizontal forces exerted by all feet that are in contact with the ground at one time have been summed, and that all other horizontal forces, such as wind resistance, are negligible.

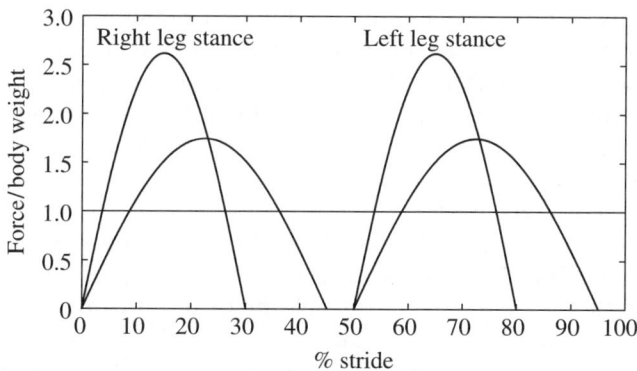

FIGURE 9.2

Schematic representation of the vertical ground reaction forces generated by a running human. Mean vertical ground reaction force must equal body weight (dotted line), but this can be achieved by combining long stance phases and low peak vertical forces (dashed lines), or short stance phases and high peak vertical forces (solid lines). Expressing peak vertical ground reaction force relative to body weight reveals which of these strategies is being used, and allows comparisons to be made between subjects of different body weights.

The above discussion of ground reaction force raises a much more general issue, which is that the most appropriate way of expressing a parameter always depends upon the question being addressed. If we were interested in determining whether animals move in a dynamically similar manner, ground reaction forces divided by body weight would be useful. If we wanted to calculate the appropriate strength and stiffness for a footbridge, then it would be absolute ground reaction forces that were important. If we were interested in the mechanics of foot pads in different species of feline, ground reaction force relative to foot pad cross-sectional area might be the critical factor.

9.4.2 Muscle force

The contractile machinery of muscle is organized into individual contractile units called sarcomeres, which are connected in series to form myofibrils. Each sarcomere is only about 2 µm in length and 1 µm in diameter, but a single muscle cell (or "fiber") contains millions of sarcomeres arranged in series and in parallel. This allows the tiny forces and motions generated by individual sarcomeres to be combined, to produce forces and motions that are significant at the whole organism level. Just as two identical chains joined in series are no stronger than a single chain, but two chains in parallel can support twice the load of a single chain, the force that a muscle fiber can generate depends upon the number of sarcomeres acting in parallel across the fiber, but is not influenced by the number of sarcomeres in series. Therefore, all other things being equal, the maximum force that a muscle fiber can generate is proportional to its cross-sectional area. Dividing muscle fiber force by cross-sectional area (to obtain stress), therefore, allows differences in muscle fiber size to be accounted for, so that the influence of other factors can be detected.

Variations in muscle fiber stress could be due to a number of factors. Probably the most important of these is the proportion of the cell that is taken up by contractile material. Consequently, there is a trade-off in muscle fiber design. If a high proportion of the cell is taken up by myofibrils, then it will be able to exert high stresses but will not have much space for mitochondria, which are important for sustained aerobic contraction, or for sarcoplasmic reticulum, which is important for high frequency contractions (Lindstedt *et al.*, 1998).

Isometric stress can also be influenced by the duty factor of myosin, the motor protein that is responsible for muscle contraction. For example, smooth muscle myosin has a higher duty factor than skeletal muscle myosin, allowing it to develop higher forces for a given amount of protein (Guilford *et al.*, 1997). A third factor which influences muscle stress is resting sarcomere length. While resting sarcomere lengths in vertebrate muscles are fairly constant, at approximately 2 μm, the muscles of crustaceans often have much longer sarcomeres (Taylor, 2000). This allows a greater distance of overlap between the actin and myosin filaments, so that a greater number of myosin molecules can generate force in parallel within a single sarcomere. Taylor (2000) showed that muscle stress increases in proportion to resting sarcomere length, for resting sarcomere lengths between 2–16 μm. The influence of factors such as these can only be detected once differences in muscle fiber size have been accounted for by dividing force by cross-sectional area.

Accounting for size effects in whole muscles is less straightforward than for single fibers, because of the influence of muscle architecture. In parallel-fibered muscles (Figure 9.3A), force can be divided by muscle cross-sectional area, because muscle area is approximately equal to the total cross-sectional area of all the fibers acting in parallel within the muscle. However, in "pennate" (or "pinnate") muscles, where the fibers run at an angle to the long axis of the muscle (Figure 9.3B), the total cross-sectional area of the fibers can greatly exceed the cross-sectional area of the muscle. This allows the muscle to generate a much higher force relative to its volume, compared with a parallel-fibered muscle. Therefore, for pennate muscles, whether of simple unipennate structure (Figure 9.3B) or more complex architecture (Figure 9.3C), muscle force is often divided by the physiological cross-sectional area (PCSA), rather than the physical cross-sectional area.

PCSA can be calculated using the following equation (Sacks and Roy, 1982):

$$\text{PCSA} = \frac{m_{mus} \cdot \cos\theta}{\rho \cdot l_f} \tag{9.3}$$

Where: m_{mus} is muscle mass
θ is mean pennation angle (Figure 9.3B)
l_f is mean fiber length
ρ is muscle density and can be taken as 1.06 g/cm^3 (Mendez and Keys, 1960).
The multiplication by cosθ in Equation 9.3 is used to determine the component of the force that is directed along the long axis of the muscle. The PCSA has been shown to be an excellent predictor of muscle force (Powell *et al.*, 1984).

Some muscles contain muscle fibers that are connected in series; for example, the *rectus abdominis*. When calculating l_f for these muscles, the lengths of the muscle

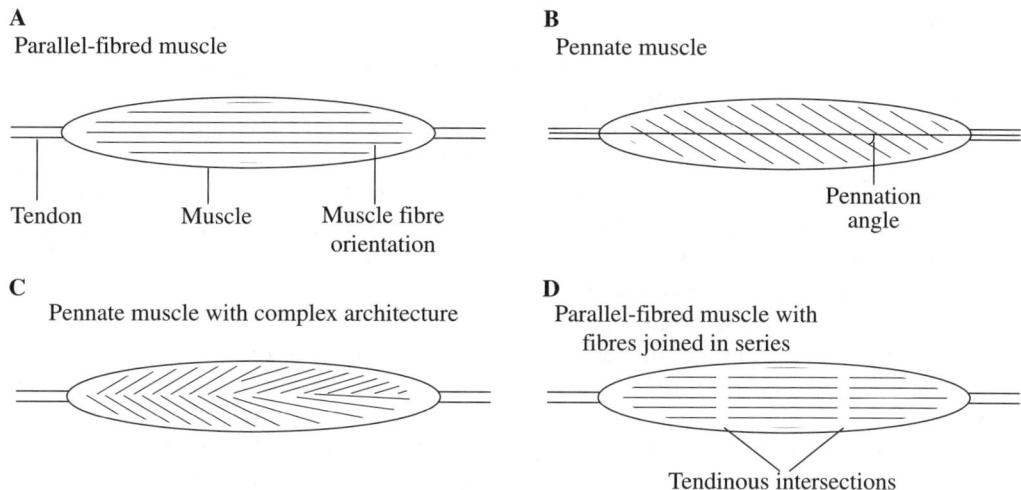

A
Parallel-fibred muscle

Tendon   Muscle   Muscle fibre
orientation

B
Pennate muscle

Pennation
angle

C
Pennate muscle with complex architecture

D
Parallel-fibred muscle with
fibres joined in series

Tendinous intersections

FIGURE 9.3

Schematic illustrating different muscle architectures. (**A**) When the muscle fibers are parallel to the long axis of the muscle, the physiological cross-sectional area (PCSA, Equation 9.3) of the muscle corresponds to its physical cross-sectional area, and muscle fiber length corresponds to muscle length. (**B**) In a pennate muscle, the fibers are orientated at an angle to the long axis of the muscle. PCSA is greater than physical cross-sectional area, and fiber length is less than muscle length. (**C**) In a pennate muscle with complex architecture, the muscle fibers have a distribution of different lengths and pennation angles. PCSA has to be calculated based on estimates of the mean fiber length and pennation angle. (**D**) When muscle fibers are connected in series, their lengths should be summed when calculating the mean fiber length for the determination of PCSA. If this were not done for the muscle illustrated here, the calculated PCSA would be three times too high.

fibers that are in series should be summed, to avoid overestimating the effective PCSA. For example, for the muscle shown in Figure 9.3D, the PCSA as determined without summing the lengths of the muscle fibers that are in series would be three times too high. In addition to the factors listed above for single fibers, the stress exerted by whole muscles can be influenced by the amount of non-contractile connective tissue between the fibers.

As with ground reaction force, the most appropriate method of comparing the forces generated by muscles of different sizes will depend upon the question being addressed. Dividing by PCSA accounts for the effects of differences in architecture, but this may not always be desirable. If the aim is to account purely for differences in muscle size, force should be divided by the physical cross-sectional area of the muscle. In this case, differences in architecture provide one possible explanation for observed differences in stress. Muscle forces can also be divided by $m^{2/3}$ to compare the measured effects of size to what would be expected in geometrically similar animals. Observed differences could then be due to differences in muscle mass relative to body mass, or to differences in muscle architecture, in addition to the other factors listed above.

9.4.3 Muscle velocity

While muscle force depends on the number of sarcomeres acting in parallel, muscle fiber velocity depends upon the number of sarcomeres in series. If one sarcomere shortens at a velocity of v_s nm/s, then 10 sarcomeres joined in series will shorten at $10v_s$ nm/s. It is therefore useful to divide muscle fiber velocity by fiber length, giving velocity in lengths/second (which has dimensions of T^{-1}, because a velocity with dimensions LT^{-1} has been divided by a length). The resulting velocity applies equally to the fiber and to its constituent sarcomeres – if a fiber is shortening at five fiber lengths per second, then its sarcomeres are, on average, shortening at five sarcomere lengths per second. This approach therefore removes the effect of differences in fiber length, and reveals differences in sarcomere velocity. These differences could be due to factors such as temperature, activation level, load or the constituent myosin isoform.

As with muscle force, whole-muscle velocity is also influenced by architecture. In a parallel-fibered muscle (Figure 9.3A,D), velocity can be divided by whole muscle length, which will be proportional to the number of sarcomeres in series (assuming constant sarcomere length as in vertebrate muscles). In a pennate muscle (Figure 9.3B,C), the shortening velocity of the muscle in lengths/s will be slower than the velocities of the constituent muscle fibers, because the direction of fiber shortening is not aligned to the long axis of the muscle. This relationship is further complicated by the fact that pennation angle changes as the muscle shortens. Depending on the question being addressed, it may still be useful to divide muscle velocity by muscle length, while recognizing that the resulting low velocities could reflect the pennate architecture, rather than low intrinsic sarcomere velocities.

9.4.4 Running speed

Two competing factors influence the relationship between running speed and animal size. Larger animals have longer legs, so tend to take longer strides, but smaller animals are able to achieve higher stride frequencies (for reasons that are explained below). Because speed is stride length multiplied by stride frequency, these two factors will tend to counteract one another. Therefore, dividing running speed by leg length, or by overall height in upright species like humans, is rarely useful, because it accounts for the effect of size on stride length but ignores the effect on stride frequency.

Hill (1950) explained why smaller animals are able to achieve higher stride frequencies. He considered animals of different sizes that were perfectly geometrically similar, and of equal density. This allowed him to assume that stride lengths would be proportional to $m^{1/3}$, that muscle cross-sectional areas – and therefore forces – would be proportional to $m^{2/3}$, and the distance through which a muscle can shorten would be proportional to its length, and therefore to $m^{1/3}$ (Equation 9.2). The maximum amount of kinetic energy that a muscle can impart to a limb during protraction ($\Delta KE_{l,max}$) is equal to the maximum amount of mechanical work that the muscle can do ($W_{m,max}$):

$$\Delta KE_{l,max} = W_{m,max} \qquad (9.4)$$

The work that a muscle does is a product of its force ($\propto m^{2/3}$) and the distance through which it shortens ($\propto m^{1/3}$), so that maximum muscle work will be proportional to body mass ($m^{2/3} \cdot m^{1/3} = m$) in geometrically similar animals:

$$W_{m,max} \propto m \tag{9.5}$$

The kinetic energy of a limb is proportional to its mass multiplied by the square of its velocity (v_l). In geometrically similar animals, limb mass will be proportional to body mass, giving:

$$\Delta KE_{l,max} \propto m v_l^2 \tag{9.6}$$

Substituting the right hand sides of Equations 9.5 and 9.6 into Equation 9.4 gives us:

$$m v_l^2 \propto m \Rightarrow v_l^2 \propto 1 \Rightarrow v_l \propto 1 \tag{9.7}$$

This shows that v_l will be independent of size, indicating that geometrically similar animals of different sizes should be able to swing their limbs with the same peak velocity. However, because the distance through which the legs are swung will increase in proportion to $m^{1/3}$, the time taken to swing the leg will also be proportional to $m^{1/3}$. Therefore larger animals take longer to protract their limbs and, thus, have lower maximum stride frequencies.

Hill (1950) concluded that, for idealized, geometrically similar animals, the increase in stride length and the decrease in stride frequency with increasing size would cancel out, so that top speed would be the same in animals of different sizes. In reality, animals are not geometrically similar; some species are adapted for speed, whereas others are adapted to other skills, such as digging. Therefore, we can interpret this as saying that top speed will be independent of animal size, rather than that it will be exactly equal in different species.

Garland (1983) compiled data on the maximum running speeds of 106 mammalian species between 0.016 kg and 6000 kg. He found that, in contrast to Hill's predictions, top speed was dependent upon body mass, with the highest speeds being associated with intermediate body masses of around 120 kg. Within the individual orders Artiodactyla, Carnivora and Rodentia, however, speed was independent of size, suggesting that Hill's predictions apply within groups of animals that are adapted to similar ecological niches and are, therefore, approximately geometrically similar.

In studies of predator-prey interactions, the most appropriate way to compare top speed is usually to consider absolute speed. After all, if a group of lions is chasing a herd of antelope, it is not the speed of an antelope relative to its size that matters, but its speed relative to that of the lions and the other antelope. When comparing the mechanics of locomotion in animals of different sizes, however, it is usually not appropriate to use the same absolute speed, because smaller animals tend to have shorter strides and higher stride frequencies, and these differences are likely to mask differences due to other factors.

Animals of different sizes may even use different gaits at the same speed. For example, at 1.5 m/sec, an average-sized horse would be walking, while a dog would be trotting and a rat would be galloping. Heglund *et al.* (1974) proposed that the speed at which animals transition from a trot to a gallop can be considered a physiologically equivalent speed, and could be used to compare locomotion in animals of different sizes. A difficulty with this approach, however, is that this is a speed at which animals do not usually choose to travel (Hoyt and Taylor, 1981; Pennycuick, 1975).

This problem can be overcome by using speeds that are defined relative to the transition speed – for example, 50% of the trot-gallop transition speed, or midway between the walk-trot and trot-gallop transition speeds. An alternative is to use the preferred speed within a gait. This has the advantage of being more physiological, but the disadvantage that it could be strongly influenced by psychological factors, and by the person who is handling the animal. Perry *et al.* (1988) compared galloping white rats with hopping kangaroo rats and found that, when these two species moved at their preferred speeds, the peak stresses in the ankle extensor muscles were almost identical fractions of maximal isometric stress, despite differences in animal size, gait and ground reaction force. They argued that this supports the notion that preferred speeds represent mechanically equivalent speeds.

Another approach is to compare animals of different sizes moving at speeds corresponding to equal Froude numbers. This method is not necessarily inconsistent with the above approaches, because gait transitions in animals of different sizes typically occur within a narrow range of Froude numbers (Alexander, 1989; Alexander and Jayes, 1983; Gatesy and Biewener, 1991; Griffin *et al.*, 2000; Rubenson *et al.*, 2004). This may also be true of preferred speeds. Heglund and Taylor (1988) found that preferred gallop speed scaled in proportion to $m^{0.18}$ and preferred trot speed scaled in proportion to $m^{0.22}$. At equal Froude number, speed is proportional to the square root of leg length. If we assume leg length is proportional to $m^{0.33}$ (as in geometrically similar animals), then speed at equal Froude number will be proportional to $m^{0.17}$, a scaling exponent close to those measured by Heglund and Taylor (1988). The Froude number has advantages over the above two methods, in that it allows a wide range of different speeds to be compared and is relatively easy to define accurately and objectively.

The reason that the Froude number is particularly useful for comparing animals of different sizes is that, as explained above, it is a necessary condition for dynamically similar locomotion. Animals that are moving in a dynamically similar manner can be thought of as using the same gait patterns or strategies. They swing their limbs through the same angles, spend the same proportion of each stride in contact with the ground, bring their limbs to the ground in the same order and with the same relative timing, and exert ground reaction forces that are the same proportion of body weight. While, as emphasized above, animals moving at equal Froude number will not necessarily move in a dynamically similar manner, they are at least moving at speeds that make this possible. If animals are compared at speeds corresponding to different Froude numbers, they cannot move in a dynamically similar manner, and any differences observed in dimensionless locomotor parameters could be due to the differences in Froude number, rather than to fundamental differences between the animals.

9.4.5 Jump height

"If a flea were as big as a human, it could jump over a house", the old adage tells us. The implicit assumption is that jump height is proportional to animal height, so that a human-sized flea could jump much higher than a normal flea. Is this a reasonable assumption? In the same paper in which he analyzed the relationship of top speed to animal size, Hill (1950) also considered jump height. He concluded that maximum jump height should be independent of animal size, so that a giant flea could not jump any higher than a normal flea.

Scholz *et al.* (2006) argued that the relationship of jump height to animal size depends on how jump height is defined. They divided the jump into a push-off phase and an aerial phase, and pointed out that at least three different definitions of jump height can be used:

(i) the total gain in height of the center of mass during the jump (push-off and aerial phase combined);

(ii) the maximum height above the ground reached by the center of mass;

(iii) the gain in height of the center of mass during the aerial phase.

They argued that only the first of the above, will be independent of animal size. This can be explained by again using the idea of perfectly geometrically similar animals of equal density (Equation 9.2) to separate differences in size from differences in shape. As discussed above for running speed, in perfectly geometrically similar animals with identical muscle properties, the maximum work that a muscle can do will scale in proportion to m. This means that the increase in the energy of the center of mass that can be generated by the extensor muscles during the push-off phase of the jump should also be proportional to m.

Let us consider an animal that starts from rest with an initial center of mass height $h_{initial}$, and performs a vertical jump. During the push-off, its muscles do an amount of work am (where a is a constant), and increase the total kinetic and potential energy of the center of mass by am. During the aerial phase of the jump, the total mechanical energy of the center of mass will not change, but the kinetic energy will gradually be converted to gravitational potential energy until, by the time the center of mass has reached its maximum height, h_{max}, the kinetic energy is zero and the gravitational potential energy has increased by am relative to its value at $h_{initial}$.

This can be written as follows: $mgh_{max} - mgh_{initial} = am$. The m cancels out of this equation, leaving $h_{max} - h_{initial} = a/g$, which tells us that the change in height of the center of mass ($h_{max} - h_{initial}$) should be constant, regardless of the size of the animal. While animals are not perfectly geometrically similar, resulting in variations in jumping performance, this argument implies that, overall, we can expect the total gain in center of mass height to be independent of size.

Scholz *et al.* (2006) pointed out that, if definition (ii) or (iii) is used, jump height will not be independent of animal size. The maximum height that the center of mass reaches above the ground (definition (ii)) will increase with animal size, because initial height, $h_{initial}$, will increase (in proportion to $m^{1/3}$ in geometrically similar animals). Conversely, the gain in height of the center of mass during the aerial phase (definition (iii)) will decrease as animal size increases, because the change in height during the push-off phase will increase with size in proportion to $m^{1/3}$ and, as shown above, total height gain is constant.

This last point has important implications for the ability of an animal to clear an obstacle (which is a fourth possible definition of jump height). This depends on the clearance between the lowest part of the animal (usually the feet) and the ground. If we assume that the animal maintains the limb configuration that it has at take-off throughout the aerial phase, then the maximum height of its feet above the ground will be equal to the increase in height of the center of mass during the aerial phase. This leads to the non-intuitive conclusion that smaller animals would be able to clear larger objects.

Alternatively, we could assume that, while in the air, the animal is able to tuck up its legs to help it to clear the obstacle, gaining a height advantage proportional to its linear dimensions (i.e., $m^{1/3}$). For example, if we assume that, after take-off, the animal is able to tuck up its legs into the configuration they had at the initial center of mass height, $h_{initial}$, then the largest obstacle that could be cleared would be equal to the total change in center of mass height and would therefore, according to the above reasoning for definition (i), be independent of animal size.

Therefore, the relationship of jump height to animal size depends upon which definition is used. The situation is further complicated for very small animals, because air resistance has a disproportionately large effect on them (Bennet-Clark and Alder, 1979). One thing that is clear, however, is that jump height cannot be expected to increase in proportion to animal height, as is often assumed. Therefore, under most circumstances, the best approach to comparing jumping ability in animals of different sizes is to consider absolute jump height. The particular definition of jump height that is chosen will depend upon the question being addressed.

9.4.6 Elastic energy storage

It is widely recognized that tendons reduce the muscular work required for locomotion by storing and releasing strain energy at each stride. This was first demonstrated by Alexander in the 1970s (Alexander, 1974; Alexander and Vernon, 1975). This spring-like action of the tendons is thought to benefit the animal by contributing to the mechanical work of locomotion, thereby increasing efficiency (Alexander, 1984b).

The elastic energy storage capacity of a tendon can be approximated by predicting the maximum force of the associated muscle (from its PCSA), and using tendon dimensions and elastic modulus to estimate the amount of strain energy that would be stored in the tendon if this force were applied to it. This approach has been used to estimate the energy storage capacity of limb extensor tendons of mammalian species of a wide range of different sizes, and has demonstrated that the elastic energy storage capacity per kilogram body mass increases with animal size (Alexander *et al.*, 1981; Pollock and Shadwick, 1994b). This has been taken to indicate that the importance of elastic energy storage in locomotion is greater in larger animals.

This conclusion, however, relies on the assumption that dividing elastic energy storage capacity by body mass is an appropriate way of comparing animals of different sizes. In Bullimore and Burn (2005), we argued that it is more informative to consider the amount of elastic energy storage relative to the mechanical work that is done during locomotion. Although the mechanical work per meter is approximately proportional to body mass (Blickhan and Full, 1993; Heglund *et al.*, 1982), smaller animals use more strides to travel a given distance. Therefore, because elastic energy is stored and released at every stride, this gives them an advantage when it comes to using stored elastic energy to perform mechanical work. We showed that the contribution of elastic energy storage to the mechanical work of locomotion may actually be greater in smaller mammals (Bullimore and Burn, 2005).

9.5 CONCLUSIONS

Accounting for size effects in biomechanical data is a far from trivial problem. No single approach is universally applicable, and each problem needs to be considered individually in the light of the question being addressed and the underlying physical processes. Yet the effort required to determine the most appropriate method for comparing individuals of different sizes is undoubtedly worthwhile, because using an inappropriate method can lead to erroneous conclusions. However, because there is no approach which can be guaranteed to remove all size effects, where possible, groups of individuals that are approximately the same size, or that overlap in size, should be compared.

ACKNOWLEDGEMENTS

I thank Dr. J. Burn for permission to use the data shown in Figure 10.1, Rowan Novinger for providing feedback on a draft of the manuscript, and Dr. John Bertram for inviting me to write this chapter.

▌ REFERENCES

Alexander RM (1974). The mechanics of jumping by a dog (*Canis familiaris*). *Journal of Zoology* **173**, 549–573.

Alexander RM (1976). Estimates of speeds of dinosaurs. *Nature* **261**, 129–130.

Alexander RM (1984a). Stride length and speed for adults, children, and fossil hominids. *American Journal of Physical Anthropology* **63**, 23–27.

Alexander RM (1984b). Elastic energy stores in running vertebrates. *American Zoologist* **24**, 85–94.

Alexander RM (1985). The maximum forces exerted by animals. *Journal of Experimental Biology* **115**, 231–238.

Alexander RM (1989). Optimization and gaits in the locomotion of vertebrates. *Physiological Reviews* **69**, 1199–1227.

Alexander RM, Jayes AS (1983). A dynamic similarity hypothesis for the gaits of quadrupedal mammals. *Journal of Zoology* **201**, 135–152.

Alexander RM, Vernon A (1975). The mechanics of hopping by kangaroos (*Macropodidae*). *Journal of Zoology* **177**, 265–303.

Alexander RM, Jayes AS, Maloiy GMO, Wathuta EM (1979). Allometry of the limb bones of mammals from shrews (*Sorex*) to elephant (*Loxodonta*). *Journal of Zoology* **189**, 315–332.

Alexander RM, Jayes AS, Maloiy GMO, Wathuta EM (1981). Allometry of the leg muscles of mammals. *Journal of Zoology* **194**, 539–552.

Bennet-Clark HC, Alder GM (1979). The effect of air resistance on the jumping performance of insects. *Journal of Experimental Biology* **82**, 105–121.

Biewener AA (1990). Biomechanics of mammalian terrestrial locomotion. *Science* **250**, 1097–1103.

Blickhan R (1989). The spring-mass model for running and hopping. *Journal of Biomechanics* **22**, 1217–1227.

Blickhan R, Full RJ (1993). Similarity in multilegged locomotion: bouncing like a monopode. *Journal of Comparative Physiology A* **173**, 509–517.

Bullimore SR, Burn JF (2004). Distorting limb design for dynamically similar locomotion. *Proceedings of the Royal Society B: Biological Sciences* **271**, 285–289.

Bullimore SR, Burn JF (2005). Scaling of elastic energy storage in mammalian limb tendons: do small mammals really lose out? *Biology Letters* **1**, 57–59.

Bullimore SR, Burn JF (2006a). Dynamically similar locomotion in horses. *Journal of Experimental Biology* **209**, 455–465.

Bullimore SR, Burn JF (2006b). Consequences of forward translation of the point of force application for the mechanics of running. *Journal of Theoretical Biology* **238**, 211–219.

Bullimore SR, Burn JF (2007). Ability of the planar spring-mass model to predict mechanical parameters in running humans. *Journal of Theoretical Biology* **248**, 686–695.

Bullimore SR, Donelan JM (2008). Criteria for dynamic similarity in bouncing gaits. *Journal of Theoretical Biology* **250**, 339–348.

Cerda X, Retana J, Cros S (1998). Prey size reverses the outcome of interference interactions of scavenger ants. *Oikos* **82**, 99–110.

Donelan JM, Kram R (2000). Exploring dynamic similarity in human running using simulated reduced gravity. *Journal of Experimental Biology* **203**, 2405–2415.

Economos AC (1983). Elastic and/or geometric similarity in mammalian design? *Journal of Theoretical Biology* **103**, 167–172.

Farley CT, Glasheen J, McMahon TA (1993). Running springs: speed and animal size. *Journal of Experimental Biology* **185**, 71–86.

Garland T (1983). The relation between maximal running speed and body mass in terrestrial mammals. *Journal of Zoology* **199**, 157–170.

Gatesy SM, Biewener AA (1991). Bipedal locomotion: Effects of speed, size and limb posture in birds and humans. *Journal of Zoology* **224**, 127–147.

Griffin TM, Garcia S, Wickler SJ, Hoyt DF, Kram R (2000). Determinants of the walk-trot transition and preferred walking speeds: insights from intra-specific size comparisons of horses. *American Zoologist* **40**, 1034–1035.

Griffin TM, Kram R, Wickler SJ, Hoyt DF (2004). Biomechanical and energetic determinants of the walk-trot transition in horses. *Journal of Experimental Biology* **207**, 4215–4223.

Guilford WH, Dupuis DE, Kennedy G, Wu J, Patlak JB, Warshaw DM (1997). Smooth muscle and skeletal muscle myosins produce similar unitary forces and displacements in the laser trap. *Biophysical Journal* **72**, 1006–1021.

He JP, Kram R, McMahon TA (1991). Mechanics of running under simulated low gravity. *Journal of Applied Physiology* **71**, 863–870.

Heglund NC, Taylor CR (1988). Speed, stride frequency and energy cost per stride: how do they change with body size and gait? *Journal of Experimental Biology* **138**, 301–318.

Heglund NC, Taylor CR, McMahon TA (1974). Scaling stride frequency and gait to animal size: mice to horses. *Science* **186**, 1112–1113.

Heglund NC, Cavagna GA, Taylor CR (1982). Energetics and mechanics of terrestrial locomotion. 3. Energy changes of the center of mass as a function of speed and body size in birds and mammals. *Journal of Experimental Biology* **97**, 41–56.

Hildebrand M (1977). Analysis of asymmetrical gaits. *Journal of Mammalogy* **58**, 131–156.

Hill AV (1950). The dimensions of animals and their muscular dynamics. *Science Progress* **38**, 209–230.

Hoyt DF, Taylor CR (1981). Gait and energetics of locomotion in horses. *Nature* **292**, 239–240.

Isaacson E, Isaacson M (1975). *Dimensional Methods in Engineering and Physics*. Edward Arnold, London.

Langman VA, Roberts TJ, Black J, Maloiy GM, Heglund NC, Weber JM, Kram R, Taylor CR (1995). Moving cheaply: Energetics of walking in the African elephant. *Journal of Experimental Biology* **198**, 629–632.

Lindstedt SL, McGlothlin T, Percy E, Pifer J (1998). Task-specific design of skeletal muscle: balancing muscle structural composition. *Comparative Biochemistry and Physiology Part B* **120**, 35–40.

McMahon TA (1975). Allometry and biomechanics: Limb bones in adult ungulates. *American Naturalist* **109**, 547–563.

McMahon TA, Cheng GC (1990). The mechanics of running: how does stiffness couple with speed? *Journal of Biomechanics* **23**, 65–78.

Medler S (2002). Comparative trends in shortening velocity and force production in skeletal muscles. *American Journal of Physiology* **283**, R368–R378.

Mendez J, Keys A (1960). Density and composition of mammalian muscle. *Metabolism* **9**, 184–188.

Pennycuick CJ (1975). On the running of the gnu (*Connochaetes taurinus*) and other animals. *Journal of Experimental Biology* **63**, 775–799.

Perry AK, Blickhan R, Biewener AA, Heglund NC, Taylor CR (1988). Preferred speeds in terrestrial vertebrates: are they equivalent? *Journal of Experimental Biology* **137**, 207–219.

Pollock CM, Shadwick RE (1994a). Relationship between body mass and biomechanical properties of limb tendons in adult mammals. *American Journal of Physiology* **266**, R1016–R1021.

Pollock CM, Shadwick RE (1994b). Allometry of muscle, tendon, and elastic energy storage capacity in mammals. *American Journal of Physiology* **266**, R1022–R1031.

Powell PL, Roy RR, Kanim P, Bello MA, Edgerton VR (1984). Predictability of skeletal muscle tension from architectural determinations in guinea pig hindlimbs. *Journal of Applied Physiology* **57**, 1715–1721.

Rubenson J, Heliams DB, Lloyd DG, Fournier PA (2004). Gait selection in the ostrich: Mechanical and metabolic characteristics of walking and running with and without an aerial phase. *Proceedings of the Royal Society B: Biological Sciences* **271**, 1091–1099.

Sacks RD, Roy RR (1982). Architecture of the hind limb muscles of cats: Functional significance. *Journal of Morphology* **173**, 185–195.

Schmidt-Nielsen K (1984). *Scaling: Why is Animal Size so Important?* Cambridge University Press, Cambridge.

Scholz MN, Bobbert MF, Knoek van Soest AJ (2006). Scaling and jumping: Gravity loses grip on small jumpers. *Journal of Theoretical Biology* **240**, 554–561.

Stahl WR (1967). Scaling of respiratory variables in mammals. *Journal of Applied Physiology* **22**, 453–460.

Taylor GM (2000). Maximum force production: Why are crabs so strong? *Proceedings of the Royal Society B: Biological Sciences* **267**, 1475–1480.

Vaughan CL, O'Malley MJ (2005). Froude and the contribution of naval architecture to our understanding of bipedal locomotion. *Gait Posture* **21**, 350–362.

Vaughan CL, Langerak NG, O'Malley MJ (2003). Neuromaturation of human locomotion revealed by non-dimensional scaling. *Experimental Brain Research* **153**, 123–127.

Zijlstra W, Prokop T, Berger W (1996). Adaptability of leg movements during normal treadmill walking and split-belt walking in children. *Gait Posture* **4**, 212–221.

CHAPTER TEN

Locomotion in Small Tetrapods: Size-Based Limitations to "Universal Rules" in Locomotion

Audrone R. Biknevicius[1], Stephen M. Reilly[2] and Elvedin Kljuno[3]

[1] Heritage College of Osteopathic Medicine, Ohio University, USA
[2] Department of Biological Sciences, Ohio University, USA
[3] Russ College of Engineering and Technology, Ohio University, USA

10.1 INTRODUCTION

The quest for unifying principles governing terrestrial locomotion has had a long and fruitful history, from the early musings of Aristotle (Farquharson, 2007) to recent works on the collisional model of locomotion (see other chapters in this book). Yet, these "global principles" may also inadvertently obscure important differences in locomotor dynamics between small- and large-bodied animals. Even gravity – the fundamental force governing most aspects of terrestrial locomotion – has different effects on terrestrial animals, depending on their size. For example, the passive tension of antagonist muscles in the legs of stick insects (*Carausius morosus*) and cockroaches (*Periplaneta americana*) exceed the distracting force of gravity on their lightweight limbs, so that the resting posture of their limbs is unaffected by gravity, and swing phase must be entirely powered by the insects' muscles (Hooper *et al.*, 2009). By

Understanding Mammalian Locomotion: Concepts and Applications, First Edition.
Edited by John E. A. Bertram.

comparison, the heavier limbs of large cursorial mammals will swing pendulum-like with a natural frequency, and muscular effort is used to drive higher or lower swing frequencies (Doke *et al.*, 2005).

That such distinctions in locomotor function exist between terrestrial invertebrates and vertebrates is not surprising, given the great difference in body construction and size, yet size-based differences in the locomotor apparatus and function continue to be observed even across tetrapods (four-legged vertebrates). Indeed, the size range in terrestrial tetrapods is astounding: weighing just 1.8 g, the Etruscan shrew (*Suncus etruscus*) is the smallest mammal, but it is a giant compared with the Brazilian gold frog (*Brachycephalus didactylus*), the smallest tetrapod (0.07 g) while, at the other end of body size extreme, is the African bush elephant (*Loxodonta africana*), the largest extant terrestrial tetrapod, with bulls tipping the scale at 7500 kg (Nowak, 1991). Here, too, size matters. Small mammals can run up inclines (even up vertical surfaces) at about the same speed as they run on level ground (Taylor *et al.*, 1972), a feat unimaginable by mammals of even modest size.

This chapter is not meant to be an exhaustive review of size effects in locomotion but, rather, illustrative of ways in which terrestrial locomotion in smaller tetrapods is not simply a scaled-down version of larger, more cursorially-adapted mammals. Most of the in-depth understanding of locomotor function is derived from this latter group (e.g., humans, horses, dogs), yet establishing a more complete picture of the challenges imposed on less erect and more sprawled tetrapods can yield insight into the evolution of terrestrial locomotion on Earth, as well as future efforts to develop biomimetic technologies (e.g., Spenko *et al.*, 2008).

Before delving into contrasts between small and large tetrapods, it is useful to establish a critical mass that distinguishes these size groups, but this critical mass is not rigidly defined as it varies across locomotor parameters. Depending on the parameter, it may be as low as 0.1 kg (below which stiffness may be a greater criterion for determining limb bone dimensions than safety factor – Biewener, 1990) or as high as 30 kg (above which reductions in relative running speeds become especially pronounced – Iriarte-Diaz, 2002) or even 50 kg (above which skeletal allometry shifts, so that larger mammals have stouter diaphyses relative to long bone length – Christiansen, 1999). Although the distinction between what defines small and large tetrapods varies across studies, approximately 5 kg is a size range for many locomotor parameters where discrete shifts in patterns are observed (Reilly *et al.*, 2007). This, then, is the critical mass adopted for this chapter.

When evaluated at its most comprehensive level – the mass-specific cost of transport (COT – a dimensionless value representing the energetic cost to move a unit mass over a unit distance) – terrestrial locomotion is inherently more costly for small tetrapods, with efficiencies of transport ranging from 0.6% for the small animals to 41.4% for the largest tetrapods (Alexander, 2005; Full, 1989). The long-standing paradigm of decreasing cost of transport with increasing body mass across tetrapods ($M_b^{-0.01}$) (Heglund *et al.*, 1982; Full, 1989) was updated recently, when it was highlighted that terrestrial tetrapods larger than $\approx$ 5 kg displayed this tendency only weakly, whereas body size had a substantially greater effect on locomotor costs in smaller tetrapods (Reilly *et al.*, 2007; Figure 10.1A). Both active (muscle-based) and passive mechanisms that differ between small and large tetrapods are probable causes for the distinct COT patterns.

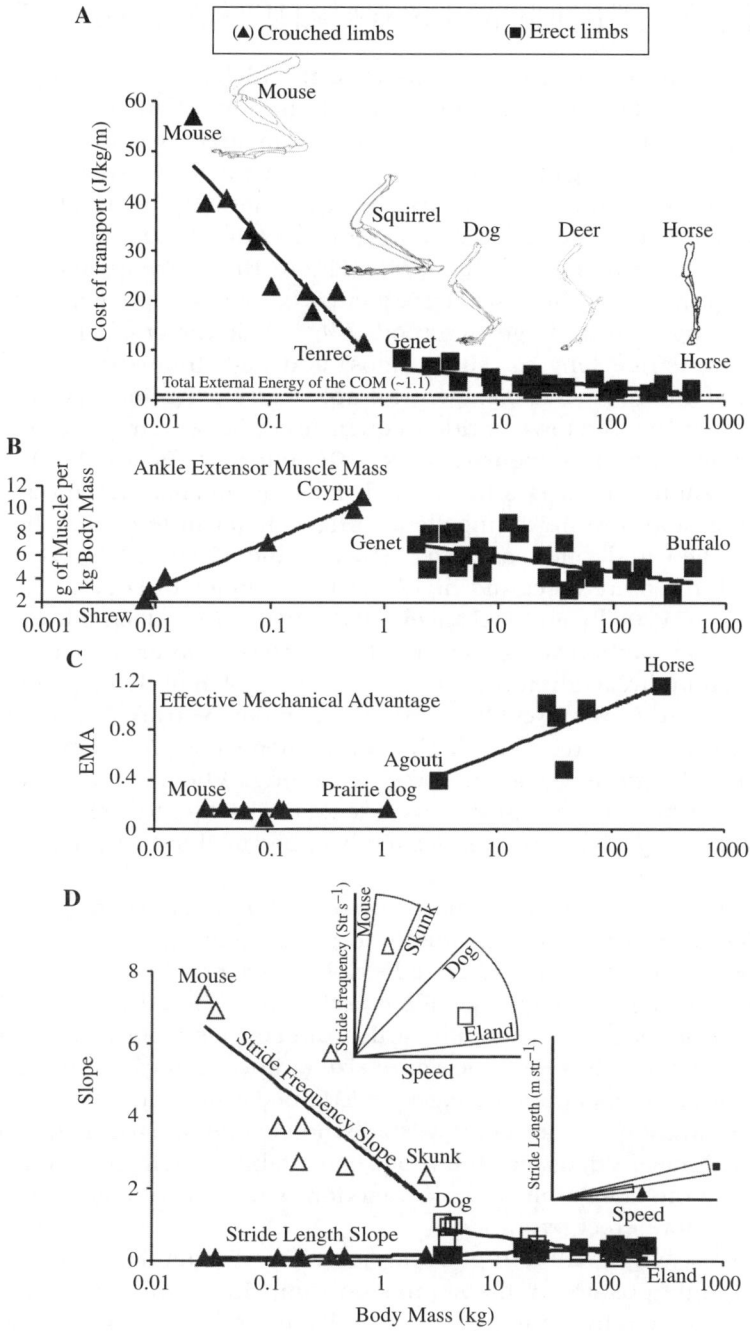

FIGURE 10.1

Mass-related patterns in cost of transport (**A**), muscle mass (**B**), effective mechanical advantage (**C**), and stride frequency (**D**) (Reilly *et al.*, 2007; reproduced with permission by Elsevier).

10.2   ACTIVE MECHANISMS CONTRIBUTING TO THE HIGH COST OF TRANSPORT IN SMALL TETRAPODS

At least three active mechanisms contribute to the high COT in small tetrapods, and each is related to higher levels of muscle activity during different phases of terrestrial locomotion. The first is stride frequency. When compared at equivalent speeds (e.g., preferred speed within a gait or at comparable gait transition speeds), smaller tetrapods move with absolutely higher stride frequencies ($M_b^{-0.15}$; Heglund and Taylor, 1988). Furthermore, the rate of increase in stride frequency with increasing velocity is substantially greater among small tetrapods (Figure 10.1D; Strang and Steudel, 1990).

In order to power these high stride frequencies, small tetrapods have faster rates of muscle shortening than do larger tetrapods ($M_b^{-0.18}$) (Rome *et al.*, 1990). Species that move with short stride lengths (short limbs) and high frequencies are expected to perform a lot of internal mechanical work simply to cycle the limbs (Van Damme *et al.*, 1998). By comparison, the lower stride frequencies of larger tetrapods allow the use of slower, less metabolically expensive muscles (Kram and Taylor, 1990). The second active mechanism that increases the cost of transport in small tetrapods involves the maintenance of limb posture during stance phase. In small tetrapods, limb posture is typically crouched and often sprawled (flexed and abducted limbs, respectively), compared with the more erect and highly adducted limbs of large tetrapods (Gatesy, 1991; Jenkins, 1971; Reilly and DeLancey, 1997; Reilly *et al.*, 2006; Rewcastle, 1980).

Maintenance of limb posture during stance phase requires that joint moments produced by ground reaction forces (F_g) are balanced by muscle moments about those joints ($F_m \times r$, where F_m is muscle force and r is its moment arm). Tetrapods with flexed limb postures experience relatively high joint moments ($F_g \times R$), primarily because the moment arm of the ground reaction force (R) is long. When combined with relatively short moment arms for the antigravity (flexor) muscles, the effective mechanical advantage ($EMA = r/R$) of these muscles tends to be small in tetrapods with habitually flexed limbs.

The more erect limbs of larger, more cursorial tetrapods match shorter lever arms to F_g with longer lever arms to the muscle forces countering the flexor moment of F_g, a configuration yielding greater *EMA*. Indeed, *EMA* scales with $M_b^{0.25}$ over a size range of mice to horses, so that weight-specific muscle forces should scale $M_b^{-0.25}$ (given that peak ground reaction force generally scales directly with body weight) (Biewener, 1989, 2005). Assessing *EMA* over the entire size range of tetrapods masks the fact that nearly all of the size-dependent increases in *EMA* occur in tetrapods larger than ≈ 5 kg, whereas the smallest tetrapods show virtually no change in *EMA* (Figure 10.1C; Reilly *et al.*, 2007). This means that small tetrapods must stabilize their crouched and sprawled postures during the stance phase of locomotion with relatively greater muscle effort than do larger, more erect tetrapods.

One might ask why small tetrapods do not "simply" extend and adduct their limbs towards an upright posture; in theory, this postural shift would increase *EMA*, reduce muscle force and, thereby, limit the loads applied to limb bones. However, mathematical modeling suggests that increases in upright posture may actually increase limb bone stress in species with sprawling postures, not reduce it (Blob, 2001). This is, in part, due to differences in the predominant loading regimes of limb bones: parasagittal limb movements predispose tetrapods with adducted limb postures to high bending loads, whereas sprawling tetrapods (e.g., squamates, crocodilians, turtles) employ long axis rotation of the proximal limb segments and, thus, experience high torsional loads (Blob, 2001; Butcher *et al.*, 2008). It is less clear what the consequences of assuming a more adducted and extended limb posture might be for small and/or generalized mammalian species that move with a lesser degree of humeral and femoral abduction (e.g., echidna, opossum, tree shrew; Jenkins, 1971) than is found in more erect mammals.

More controversial is the third active mechanism contributing to higher costs of transport for small tetrapods, namely, swing phase, or that period between liftoff and touchdown, when the limb is rapidly protracted in preparation for the subsequent support phase. The ability for gravity to initiate swing is, at least theoretically, reduced in small tetrapods if the passive tension of muscles and other connective tissues in their limbs is large, relative to the effect of gravity on their low-weight limbs (Hooper *et al.*, 2009). Under such conditions, the limb muscles of small tetrapods must be active more consistently in order to power and control swing.

Data on the motor control of swing suggest that such a functional difference may exist between small and large tetrapods. Motor neurons to swing muscles fire throughout the entirety of swing phase in small tetrapods (newts, guinea fowl, chicks, mice, cats), whereas motor neuron activity to the iliopsoas of humans and *tensor fasciae latae* of horses is more limited to the beginning of the swing phase (Gatesy, 1999; Hooper *et al.*, 2009; Jacobson and Hollyday, 1982; Prochazka *et al.*, 1989; Székely *et al.*, 1969; Tokuriki and Aoki, 1995). In spite of these differences in motor neuron activity, swing phase in humans (≈60 kg) and guinea fowl (≈1.5 kg) alike represents about one-third of the net energy of locomotion (Doke *et al.*, 2005; Rubenson and Marsh, 2009). Therefore, the cost of swinging a leg alone may not be as distinctive between small and large tetrapods as is suggested by activity patterns of motor neurons.

Together, the absolutely higher rates of cyclically firing and quieting muscles at high stride frequencies, and the relatively higher levels of muscular activity to maintain locomotor posture, confer a disproportionately higher metabolic cost during terrestrial locomotion in small tetrapods. In addition to these active mechanisms, there are also two passive mechanisms that provide significant energy savings during terrestrial locomotion in large tetrapods, but which appear to be vastly less effective, if not ineffective, in small tetrapods.

10.3 LIMITED PASSIVE MECHANISMS FOR REDUCING COST OF TRANSPORT IN SMALL TETRAPODS

The first passive mechanism is associated with inverted pendulum mechanics during walking, whereby external mechanical energy is recoverable, to varying degrees, via a pendulum-like exchange of kinetic energy (E_K) and gravitational potential energy (E_P) of the COM (Cavagna *et al.*, 1977). How these external mechanical energies fluctuate during walking differs across tetrapods, with the greatest distinction occurring in animals with sprawled postures. These tetrapods display substantial fluctuations in the lateral kinetic energy when walking, with magnitudes similar to that of the craniocaudal kinetic energy (Farley and Ko, 1997; Willey *et al.*, 2004).

By comparison, movements of the COM in tetrapods with adducted limbs are largely limited to the sagittal plane (vertically and craniocaudally), so much so that many studies neglect to capture and/or report fluctuations in the lateral kinetic energy, because they contribute triflingly to the fluctuations in the total kinetic energy of the COM (e.g., Cavagna *et al.*, 1977; Rubenson *et al.*, 2004).

Additionally, many tetrapods with sprawled postures have been labeled as "lumbering", which manifests as larger fluctuations in the E_p than E_K during walking (Farley and Ko, 1997; McElroy *et al.*, 2008; Reilly *et al.*, 2006; Willey *et al.*, 2004; Zani *et al.*, 2005). This contrasts with tetrapods with more adducted limbs, for which fluctuations in the total mechanical energy of walking is driven primarily by E_K (Cavanga *et al.*, 1977; Minetti *et al.*, 1999).

It might be surprising, given these differences, that the capacity to recover external mechanical energy via pendular mechanisms during walking is similar across tetrapods, with maximum values commonly reaching or exceeding 50% among quadrupeds, and even as high as 80% in some bipeds, regardless of limb posture (e.g., Cavagna *et al.*, 1977; Farley and Ko, 1997; Griffin *et al.*, 2000, 2004b; Minetti *et al.*, 1999; Reilly *et al.*, 2006; Rubenson *et al.*, 2004; Zani *et al.*, 2005).

That similar *capacities* exist does not necessarily mean that they are expressed similarly across tetrapods, or that these energy savings have a similar effect. First, the expectation that pendulum-based energy recovery would be maximal at moderate walking speeds (near the preferred walking speeds) has only been found in more cursorially-adapted mammals or large terrestrial birds (Cavagna *et al.*, 1977; Griffin *et al.*, 2004a, 2004b; Rubenson *et al.*, 2004). Smaller, more generalized tetrapods show little speed-dependency, with both high and low recoveries occurring at comparable walking speeds (Farley and Ko, 1997; Reilly *et al.*, 2006; Horner and Biknevicius, 2010). This suggests that these animals either are incapable of coordinating movements of the COM to optimize energy recoveries or, more likely, are indifferent to precisely matching COM mechanics with speed and gait (see Section 10.4).

Second, the impact of external mechanical energy savings on the overall COT of walking is very different in small versus large tetrapods. This is because the amount of external mechanical energy needed to move a mass over a given distance varies little with body size ($M_b^{-0.01}$, or approximately 1.1 J kg^{-1} m^{-1}; Full, 1989), but that energy constitutes a small part of the cost of transport in small tetrapods (dashed line in Figure 10.1A), so that even 100% energy recoveries through pendulum-like movements of the COM will convey minimal benefit to small tetrapods (Reilly *et al.*, 2007). It appears that the higher level of muscular effort associated with short effective limb lengths and/or flexed limb postures effectively counteract any energy savings gained by walking with pendulum-like mechanics (Griffin *et al.*, 2000). The penguin provides a telling example: despite enviable pendulum-like recoveries of external mechanical energy (80%), penguins have extremely high COT as they waddle along on their short legs (Griffin *et al.*, 2000; Roberts *et al.*, 1998).

Nor do small tetrapods fair much better with energy savings during running. Most tendons in mammals are built with a high safety factor (ratio of tendon strength to maximum customary stress > 8; Ker *et al.*, 1988), which reflects stiffness in the tendon appropriate for tightly controlling limb segment movement with the associated muscle-tendon units. The distal limbs of large-bodied tetrapods also have tendons with higher elastic moduli, which operate with lower safety factors because they function to store and return elastic strain energy with every step or stride (Alexander, 1988). These same tendons in smaller tetrapods are proportionately thicker and much more stiff (Biewener and Blickhan, 1988; Ettema, 1996), obviating their ability to provide substantial reductions in muscular effort when running.

While many of the examples provided above point to there being "negative" consequences to being small and having flexed limbs, there clearly are advantages to this morphotype. As any school child knows when he has lost his hold on his pet lizard or mouse, many small tetrapods are extremely fast and agile. The same tendons that are too stiff for energy savings via elastic recoil in small tetrapods are ideal for channeling contractile forces of the muscle to the bone for rapid accelerations (Biewener and Blickhan, 1988). Also, a flexed limb posture is associated with greater acceleration and maneuverability than for erect limbs (Biewener, 1983; Hildebrand and Goslow, 2001).

Even the nervous system of small tetrapods appears to be designed for a higher degree of responsiveness (More *et al.*, 2010). While there is more than a 100-fold difference in leg length between shrews and elephants, the axonal diameter in elephants is only about twice that of shrews. Yet maximum axonal conduction velocity is nearly independent of body mass ($M_b^{0.04}$), so that small tetrapods have shorter physiological delays between sensing and responding to stimuli. This suggests that small tetrapods should be able to rely more fully on simple feedback controls to monitor and drive their sensorimotor performance than can large tetrapods.

Collectively, the energetic costs and underlying active and passive mechanisms of terrestrial locomotion in smaller tetrapods cannot universally be summarized as scaled down versions of large tetrapods. The remainder of this chapter highlights two features of terrestrial locomotion – grounded running and intermittent locomotion – with special attention given to how these behaviors differentially affect small and large tetrapods.

10.4   GAIT TRANSITIONS FROM VAULTING TO BOUNCING MECHANICS

Walk-run and walk-trot transitions have become synonymous with shifts between vaulting and bouncing mechanics in bipeds and quadrupeds, respectively (Segers *et al.*, 2007a). However, there is evidence that these transitions are motivated by different mechanical and energetic realities in small and large tetrapods. This section comments first on aspects of walking and running common across tetrapods, and then hones in on some intriguing differences between large and small tetrapods at the boundary between walking and running (for convenience, "walking" and "running" will be used colloquially to reflect moving with vaulting and bouncing mechanics, respectively, and not to imply specific gaits, or footfall patterns).

Walking is typically the slower activity. Walking gaits, such as the striding walk of bipeds, or the lateral-sequence lateral-couplet of quadrupeds, are invariably fully grounded, with at least one limb in contact with the substrate at all times, due to high duty factors (Gambaryan, 1974; Hildebrand, 1976; Howell, 1944). Long periods of multipodal support result in a large base of support for higher stability when moving at slow speeds. While earlier constructions of the inverted pendulum model envisioned the COM vaulting over stiff legs during walking, more contemporary modeling acknowledges that some degree of leg compliancy is necessary, in order to replicate real walking mechanics (Geyer *et al.*, 2006; Seyfarth *et al.*, 2006), and the level of limb compliance may increase at higher walking speeds (Usherwood *et al.*, 2007).

Limb stiffness during walking has not been evaluated in small tetrapods, but it is presumably lower (due to greater habitual limb flexion) than in larger, more cursorial mammals (as it is in trotting; Farley *et al.*, 1993).

By contrast, running *predominately* occurs at faster speeds, with gaits that typically include a ballistic phase when limb pairs and/or all limbs are aerial. These gaits include symmetrical footfall patterns, such as the striding run of bipeds and the running trot of quadrupeds, as well as the asymmetrical footfall patterns such as the gallop (Gambaryan, 1974; Hildebrand, 1976, 1977). The greater compliance of limbs during running results in COM movements that are inconsistent with appreciable recovery of external mechanical energies via pendulum-like mechanics, particularly runs that occur with symmetrical gaits (Cavagna *et al.*, 1977). Rather, the cyclical flexion and then extension of the limbs during each stance phase is conducive to stretching and then recoiling elastic elements in the limbs (Alexander, 1988). Hence, most of the energy recoverable during running is related to spring-mass mechanics.

Detailed studies on the transitions between walking and running are largely limited to large tetrapods. These animals approach the transition velocity and then traverse it quickly (within a stride or two – Hoyt and Taylor, 1981; Minetti *et al.*, 2003). In human bipedalism – the best-studied system – some stride parameters shift discretely at the walk-run transition, as in the inclusion of an aerial phase (Segers *et al.*, 2007b), whereas others display more gradual changes both pre- and post-transition, such as step lengthening (De Smet *et al.*, 2009). Similarly, dogs will transition between a walk and trot within a stride or two (Blaszczyk, 2001). Therefore, steps or strides immediately preceding or following transitions may not be expected to neatly fit the gait and whole-body dynamics profiles of either walking or running.

It is in this transitional zone – between unambiguously defined walks and runs – that "grounded runs" occur. Grounded runs are locomotor chimeras, displaying the in-phase fluctuations of gravitational potential energy and kinetic energy typical of runs (spring-like stance period – Ruina *et al.*, 2005) yet, as the name suggests, lacking an aerial phase, as is found in walks. They are not artifacts of treadmill locomotion (as suggested by Nudds *et al.*, 2011), as grounded runs are well documented across tetrapods, including trackway studies, although they appear to have a more limited distribution among large-bodied species.

Among large quadrupeds, grounded runs occur naturally in elephants (Hutchinson *et al.*, 2003) and certain breeds of domestic horses (e.g., tölting Icelandic horses – Biknevicius *et al.*, 2006). Both species perform grounded runs with footfall patterns that are similar to those used when walking (lateral-sequence singlefoot) – hence, there is no gait transition *per se* (i.e., no shift in footfall pattern), even as the animals shift between vaulting and bouncing mechanics. While elephants employ grounded runs at their fastest speeds, this represents one of many running gaits used by Icelandic horses.

Grounded runs are also a part of the locomotor repertoire in humans, who perform grounded runs when intentionally moving with exaggerated knee flexion (the comedic gait known as "Groucho running"), as well as automatically over compliant substrates or around tight circles (Bertram *et al.*, 2002; McMahon *et al.*, 1987). Non-human primates may also use grounded runs ("amble"; Schmitt *et al.*, 2006; Wallace and Demes, 2008), although the only study to evaluate the whole body mechanics of the amble aligned it with the walk (Robilliard, 2005). Therefore, in these species grounded runs occur as a definitive running gait (elephants, horses) or are situationally employed (humans). Only in ostriches are grounded runs inserted between walks and aerial runs as part of their natural walk-run transition (Rubenson *et al.*, 2004).

While seemingly an oddity among large tetrapods (although this deserves to be more fully explored), grounded runs are astoundingly common among small-bodied tetrapods. They have been reported across the tetrapod phylogeny, from amphibians to tuataras, lizards to birds, and throughout primitive mammals (Biknevicius and Reilly, 2006; Biknevicius *et al.*, 2013; Gatesy and Biewener, 1991; Hancock *et al.*, 2007; McElroy *et al.*, 2008; Nudds *et al.*, 2001; Reilly *et al.*, 2006). For many of these species, grounded runs represent an intermediate stage in the transition between walking and running of small tetrapods (i.e., walk – grounded run – aerial run or trot) while, for some species, grounded runs are the fastest gait attained (e.g., quail; Gatesy and Biewener, 1991; Reilly, 2000).

On theoretical grounds, running with compliant limbs (grounded runs/Groucho running) may provide energetic and mechanical benefits by reducing the energy dissipated by bouncing viscera, and improving robust stability over uneven terrain (Daley and Usherwood, 2010). Presumably these benefits are shared by small and large tetrapods alike – so why, then, are grounded runs rare among large tetrapods, when they represent a normal part of the locomotor repertoire of small tetrapods? The key is probably a combination of motivation and ecological relevance.

While the metabolic consequences of using grounded runs is unknown for elephants and horses, the human experience with grounded runs is of increased effort, associated with muscular exertion used to maintain the flexed knee posture (McMahon *et al.*, 1987; Ortega and Farley, 2005; Skime and Boone, 2011). Additionally, grounded runs may deny large tetrapods the energetic advantages traditionally associated with walking and running:

1. Recouping external mechanical energy via pendulum-like mechanism is lost, since E_P and E_K cycle in-phase of one another during grounded runs.
2. The stretching and recoiling of elastic elements in the limbs associated with running may be limited during grounded runs if the leg springs are not as fully loaded compared with running with aerial phases (as suggested by the shallower vertical oscillations of the COM during grounded runs).

Comparable physiological disincentives for using grounded runs may not exist for small tetrapods that habitually move with crouched or sprawled postures. Indeed, small tetrapods appear to have a more ambivalent approach to grounded runs. Lacking the substantial benefits associated with moving with clear pendulum-like or spring-mass mechanics, these animals more freely interchange COM mechanics and gaits. Indeed, it is common to find vaulting, bouncing and mixed mechanics occurring with similar gaits and at similar speeds (Figure 10.2; Reilly *et al.*, 2006; Biknevicius *et al.*, 2013). In other words, small tetrapods show a much lower fidelity between speed, gaits and COM mechanics than is known for large tetrapods. If small tetrapods experience a similar tendency to reduce limb stiffness at higher walking speed, as is seen in larger tetrapods (Usherwood *et al.*, 2007), then their habitually flexed limbs may unavoidably, and almost imperceptibly, shift from the vaulting mechanics of a walk to the bouncing mechanics of a grounded run.

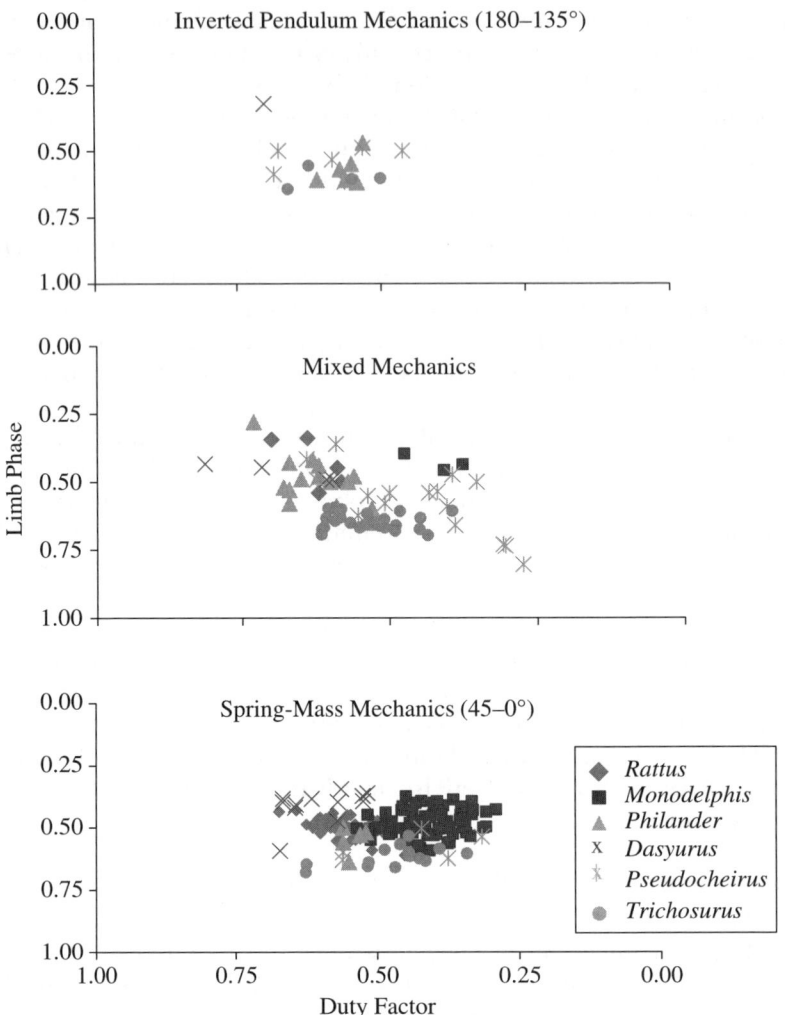

FIGURE 10.2

Trials of (left to right) inverted pendulum or vaulting mechanics (E_p and E_k fluctuating 180–145° out of phase with one another), mixed mechanics and spring-mass or bouncing mechanics (E_p and E_k fluctuating 45–0° in phase with one another), mapped on symmetrical gait plots in small mammals of generalized locomotor morphotype and crouched posture. The different symbols represent five species. All of the bouncing trials with duty factors >0.5 represent grounded runs (modified from Biknevicius *et al.*, 2013).

Revisiting locomotor energetics further illustrates how grounded runs may be more or less energetically inconsequential in small mammals. In most large tetrapods, speed-related changes in COT follow a predictable pattern during walking, summarized by a U-shaped curve (COT is high at both slow and fast walking speeds, and falls to its minimum value at an intermediate speed at, or near, the preferred walking speed), and

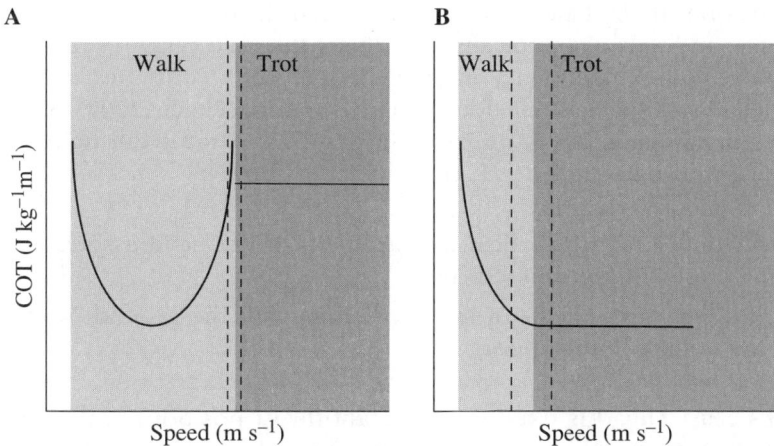

FIGURE 10.3

Relationship between cost of transport (COT), speed and gaits in most large (**A**) and small (**B**) quadrupedal tetrapods (modified from Nudds *et al.*, 2011). Dashed lines in (**B**) indicate the range of speed where grounded runs have been observed in small tetrapods; the same lines in (**A**) reflect where they may occur in large tetrapods.

switching to a running gait at higher speeds confers an energetic advantage (Figure 10.3A; Griffin *et al.*, 2004; Minetti *et al.*, 2003). By contrast, a more monotonic decrease (reversed J-shaped) in COT is found in small tetrapods when they walk, such that there is no reciprocal increase in COT after the minimum value for walking is reached, and the COT of running is near the minimum COT of walking (Figure 10.3B; Fish *et al.*, 2001; Hoyt and Kenagy, 1988; Nudds *et al.*, 2001; Taylor *et al.*, 1970).

Whether this minimum COT represents the preferred walking speed is unknown, because "preferred speeds" is a largely unexplored kinematic concept in small tetrapods, in part because of their proclivity toward locomotor intermittency when moving at slow speeds (Kenagy and Hoyt, 1989). Lizards are the clear exception, as many studies have evaluated locomotor speeds in species that use sit-and-wait versus wide-foraging strategies (e.g., Irschick, 2000), but the relationship to COT is still unclear. If small tetrapods mimic large tetrapods, in preferring to move at speeds at which COT is lowest, then this would correspond to speeds close to the transition between vaulting and bouncing mechanics. With limb compliance hovering between that needed for vaulting and bouncing mechanics, and lacking a clear energetic benefit of choosing one mode over the other, small tetrapods use them interchangeably at these moderate speeds.

A caveat must be acknowledged for head-bobbing birds, in which a non-locomotor benefit has been suggested for grounded runs. Because the avian eye is large relative to the bony orbit that houses it (Martin, 1984), birds rely on head movements, more so than eye movements, to stabilize vision (Haque and Dickman, 2005). Head-bobbing, the fore-aft movement of the head during terrestrial locomotion in many birds, is considered to be an optomotor response, with the hold phase providing detailed vision and motion detection, and the thrust phase enabling depth perception (Davies and Green, 1988; Frost, 1978).

Birds that bob their heads during terrestrial locomotion commonly perform grounded runs, and they also typically coordinate the hold phase of the bobbing cycle (when the head is held stationary in space as the body continues to move forward under the head) with periods of double limb support (Fujita, 2002). Together, these behaviors result in maximally stabilizing the visual apparatus during terrestrial locomotion (Gatesy and Biewener, 1991; Hancock *et al.*, 2007):

1. The birds avoid the large vertical displacement of the head associated with aerial runs by moving with a grounded run.
2. They avoid the jarring effects of limb collisions with the ground by stabilizing the head during double limb support.

Head-bobs cease (head is fixed in a constant thrust position) at the fastest speeds, as the birds sacrifice visual input for the foraging and anti-predatory advantages enabled by moving more quickly with aerial runs (Davies and Green, 1988). That grounded runs may benefit the visual system in some birds does not necessarily mean that these co-evolved, as there are striding birds that do not head bob but still use grounded runs (e.g., ducks – Usherwood *et al.*, 2008).

10.5   THE "UNSTEADINESS" OF MOST TERRESTRIAL LOCOMOTION

Although tetrapods are capable of moving at a range of speeds, many species have a preferred speed within each gait (Blaszczyk, 2001; Griffin *et al.*, 2004a, 2004b; Hoyt and Taylor, 1981; Kenay and Hoyt, 1989; Wickler *et al.*, 2000). Several hypotheses have been forwarded to explain the physiological basis for preferred speeds. While avoiding maximal speeds may be a strategy for moderating musculoskeletal stresses (Perry *et al.*, 1988), energetic economy appears to be a potent factor determining preferred speeds. As noted previously, large tetrapods may minimize the cost of transport by walking with preferred speeds that correspond closely to the speed at which external mechanical energy is best recovered via the pendulum-like exchange of E_P and E_K of the COM (Griffin *et al.*, 2004a, 2004b; Hoyt and Taylor, 1981; Wickler *et al.*, 2000).

The evidence for choosing metabolically economical speed during running is more equivocal, as some studies have recorded U-shaped COT curves for trots and gallops (horses – Hoyt and Taylor, 1981; Wickler *et al.*, 2000), suggesting that the animals are tuning step frequency to the natural frequency of the body's bouncing system (Cavagna *et al.*, 1997), while others have found COTs to remain relatively constant (or decline slightly) with running gaits in large bipeds and quadrupeds alike (donkeys, camels, humans, emu; Maloiy *et al.*, 2009; Margaria, 1938; Watson *et al.*, 2011).

The ecological significance of preferred speeds varies across tetrapods. Because day range scales at $M_b^{0.25}$ (Carbone *et al.*, 2005; Garland, 1983), larger mammals tend to traverse greater distances than do smaller mammals and may, therefore, be expected to utilize their preferred speeds more often. Indeed, ungulates, such as gnu (*Connochaetes taurinus*) and Thomson's gazelle (*Gazella thomsonii*), limit most of their movements to a restricted range of speeds (Pennycuick, 1975), and this may provide a means of moderating the energetic costs of migrations.

Aside from long-distance migrations, however, how often do tetrapods actually move in a consistent manner? Certainly, periods of steady locomotion might be expected for wide-foraging species, such as coyotes (*Canis latrans*), that engage in higher-speed behaviors (trotting and galloping) for 47% of their time budget, while allocating only 7% to slow movements while foraging (Switalski, 2003), or wide-foraging lizards that locomote nearly continuously (e.g., 81% activity time in *Cnemidophorus tigris*) (Anderson, 1993).

At the other extreme are species that move more intermittently. Time budgets of takhi (*Equus ferus przewalskii*) reveal that commuting between favored sites constitute only 13% of their daily activity, compared with 44% of the time walking intermittently while foraging (Boyd and Bandi, 2002). Similarly, golden-mantled ground squirrels (*Spermophilus saturatus*) move slowly through their habitat, foraging for about 27% of their active time, while allocating a mere 4% of time to commuting at higher speeds (Kenagy and Hoyt, 1989). Even lower activity levels have been reported for sit-and-wait lizards (e.g., 1.2% activity time for *Callisurus draconoides* – Anderson and Karasov, 1981).

Thus, while tetrapods may engage in continuous locomotion at preferred speeds for seasonal migration, daily commuting and/or wide-foraging activities, it is also clear, from time budget analyses, that intermittent locomotion is an important and, in some species, dominant locomotor behavior. Intermittent locomotion is characterized as active bouts that are brief (5–60 sec) but frequent and interspersed with pauses (Gleeson and Hancock, 2001). The numerous accelerations and decelerations of intermittent locomotion may be expected to increase the energetic cost, compared with continuous locomotion – yet, when the total cost of activity is determined (both exercise and post-exercise oxygen consumption), the metabolic cost of intermittent locomotion appears similar to that of continuous locomotion (Weinstein and Full, 2000; Edwards and Gleeson, 2001). If there is no energetic loss (or gain) associated with locomotor intermittency, then why do tetrapods do it? There is evidence that moving intermittently can improve foraging success, predator avoidance and overall endurance (Kramer and McLaughlin, 2001).

Given the natural tendency of many tetrapods to use non-steady speed locomotion in the wild, obtaining steady speed events can be a challenge in the laboratory. A study of COM mechanics in lizards noted that only "one out of every 200 trials" qualified as steady speed (Farley and Ko, 1997), and over two-thirds of all runway trials failed to meet the criterion for steady speed in small mammals (Table 10.1; Biknevicius, pers. ob.).

Table 10.1 ■ Occurrence of steady speed locomotion when small mammals move with symmetrical gaits along a level trackway

	Individuals	Body mass (kg)	Symmetrical trials	Steady speed
Dasyurus hallicatus	3	0.81	55	17 (30.9%)
Pseudocheirus peregrinus	5	1.06	88	37 (42.0%)
Trichosurus vulpecula	2	2.03	174	44 (25.3%)
Monodelphis domestica	7	0.10	312	90 (28.8%)
Philander opossum	2	0.83	50	26 (52.0%)
Overall			679	214 (31.5%)

Yet, steady locomotion is a traditional prerequisite for evaluating COM mechanics (Cavagna *et al.*, 1977), primarily because it is presumed that energy-saving mechanisms will be impaired by accelerations and decelerations.

But to what degree do energy recoveries suffer under non-steady speed conditions? In order to approach answering this question, we constructed a computer model of walking, based on a spring-loaded inverted pendulum that included an active element (torque) to assess the effect of acceleration (positive or negative) on energy recovery capacity (Appendix 10A). Separate models were developed for a small-bodied tetrapod with low limb stiffness, and a large-bodied tetrapod with higher limb stiffness. We focused on walking, because it has been noted in nearly every major tetrapod clade (Biknevicius and Reilly, 2006; Cavagna *et al.*, 1977; Farley and Ko, 1997; Griffin *et al.*, 2004b; Hancock *et al.*, 2007; Horner and Biknevicius, 2010; McElroy *et al.*, 2008; Minetti *et al.*, 1999; Reilly *et al.*, 2006).

Our findings are summarized in Figure 10.4. We found that maximum capacity to recover external mechanical energy via pendulum-like mechanics occurs at steady speed (net acceleration = 0), regardless of the initial velocity in both the small tetrapod/low stiffness model and the large tetrapod/high stiffness model. When the models were programmed to accelerate or decelerate from an initial velocity, the overall patterns were again similar between the two models – namely, shifts away from steady speed resulted in reductions in recovery ratios. Thus, the expectation of lower efficiencies in pendulum-like recovery of external mechanical energies during non-steady speed is supported.

On closer inspection, however, distinct responses to accelerations are noted between the two models. In the large tetrapod/high stiffness model, low level accelerations or decelerations have little effect on recovery ratios when initial velocities are slow (i.e., the system was initially tolerant of non-steady speed locomotion – top curves in Figure 10.4), whereas even small accelerations/decelerations in trials with faster initial velocities yield precipitous drops in recovery ratios. The small tetrapod/low stiffness model has the inverse patterns – rapid reductions in recovery ratios when accelerating

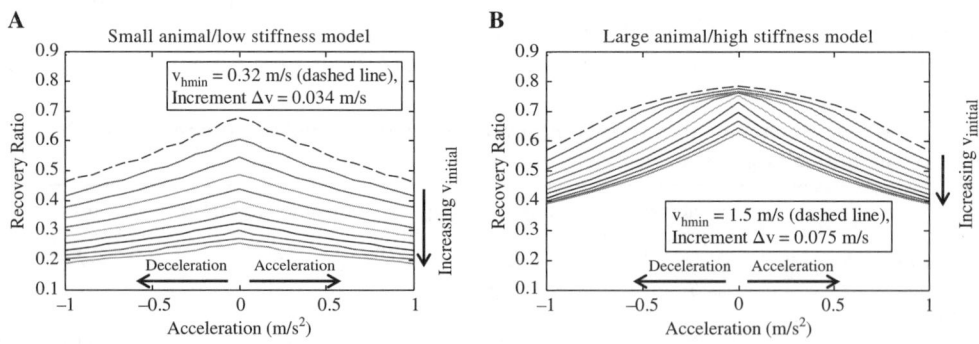

FIGURE 10.4

External mechanical energy recovery via pendulum-like mechanisms during walking based on (**A**) the small tetrapod/low stiffness model and (**B**) the large tetrapod/high stiffness model. Both results are plotted for the same range of accelerations/decelerations for illustrative purposes, although the higher values are impractical.

or decelerating from slow initial velocities, versus more modest reductions at high initial velocities (although these faster trials already have very low recovery ratios).

How do these results inform us about non-steady speed locomotor experiences of small and large tetrapods? We argued earlier in this chapter that the pendulum-like recovery of external mechanical energy during walking has little benefit to small mammals, so any reductions in recovery ratios in these tetrapods can also be viewed as insignificant. It is the pattern observed in large tetrapods that is interpretable energetically and ecologically:

1. The minimal change in the capacity to recover external mechanical energy via the inverted pendulum mechanics during modest accelerations, or deceleration from the slower speeds, may reflect the importance of energy savings in large tetrapods even during intermittent locomotion (e.g., foraging).

2. The more tightly constrained peak in energy recoveries in the faster trials may point to an energetic reason for using preferred speeds, as energy recovery plummets with any deviations from steady speed.

In conclusion, from the mechanics of body weight support to the energetic cost of locomotion, and from experimentally-obtained data to computational models, there is a growing body of evidence that different locomotor advantages and challenges are conferred to tetrapods, depending on their size and correlated limb postures. This does not negate the value of pursuing "global principles" of locomotion but, rather, provides a cautionary note. Just because an animal *can* do something in the lab (e.g., occasionally move with steady speeds, or recover energy by moving with pendulum-like mechanics) does not necessarily mean that it has ecological or economic relevance to that animal in its natural habitat.

APPENDIX – A MODEL OF NON-STEADY SPEED WALKING

By Elvedin Kljuno, Department of Mechanical Engineering, Ohio University Russ College of Engineering, Athens, OH 45701, USA.

10A.1 Spring-mass inverted pendulum model of walking

In its original construct, the inverted pendulum model of walking defaulted to a leg spring of infinite stiffness and no mechanism for input or dissipation of energy (i.e., it was a passive system with the total mechanical energy preserved). This system cannot represent closely a real animal's dynamics, especially if the animal accelerates/decelerates. Hence, we developed a spring-mass inverted pendulum model, as shown in Figure 10A.1(a, b), that consists of:

- a point mass m, which represents total mass of an animal concentrated into the center of mass (COM);

- a spring with stiffness k (generally nonlinear) connecting the COM and the center of pressure (CP) at the contact area between a foot and ground or combined centers of pressure in the case of a multiple support phase;

- a torque at the ground contact point τ as source of energy to accelerate/decelerate walking cycle (τ is zero at steady speed); and
- a dashpot with damping coefficient b.

While the model presented here used a torque to introduce acceleration/deceleration, we acknowledge that varying the spring stiffness would also accomplish the same effect. For example, an average acceleration over the cycle can be achieved (without torque) if the stiffness is continuously changed from a lower value in the first half-cycle, to a higher value within the second half-cycle. Indeed, since the ground force in real walking systems is controlled through the active elements (muscles), then the virtual spring stiffness should be variable, depending on the system variables (angle θ and radial distance r). Nevertheless, the model presented here assumes a constant spring stiffness, and acceleration/deceleration was introduced via the inclusion of a variable torque.

Another reason we included the torque in the model is the fact that the CP point is not equivalent to the ground contact GC point when a non-zero torque is considered. CP is not stationary even during a single support phase, while the GC point in the model is fixed during a single half-cycle. The equivalent CP in the model can be arbitrarily positioned within a certain area using the contact torque – similarly to, for example, the ankle torque in the human body, which can position the CP arbitrarily within the ground-foot contact area. The importance of adding the torque becomes more significant to position the CP for a double or a multiple support phase, when the CP location varies within the convex area bounded by the contact area edges of two or more feet, which is much larger area than just a single foot contact area.

Energy dissipation is modeled by including the dashpot (damper) and an additional friction torque which opposes the angle change θ. The damper dissipates energy through the viscous friction represented by the friction force $F_b = -b\dfrac{dr}{dt}$, where b is the damping coefficient and $\dfrac{dr}{dt}$ is radial velocity.

Relatively high radial force occurs during the collision period. This generates a high radial velocity of the spring and the damper deflection, which causes a proportional damping force. The dashpot force attenuates oscillation within the spring-mass system, generated due to the collision. Since the collision with ground will always generate oscillations, this is the way to account for the energy loss due to the foot/ground collision. The approach deviates from the real-world scenario, since the energy lost due to the collision occurs in a very short portion at the beginning of the walking cycle, while the modeled energy loss occurs over a significant portion of the time period, until the oscillations due to the collision are attenuated. The convenience of this approach is that the analysis is kept relatively simple by avoiding involving a hybrid dynamics analysis (which would include dealing with collision impulses and instantaneous change of velocity) into the model, but still considering the effects of foot-ground collisions – primarily the energy loss and the ground reaction force increase.

One of the accompanying effects of the dashpot inclusion in the model is an effective change of the stiffness. Indeed, the results of the model simulation have shown that the spring-dashpot combination behaves approximately as a spring with higher stiffness if the dashpot is attached in parallel to the spring. This fact complicates the model parameter tuning in order to achieve matching of the real measured data with

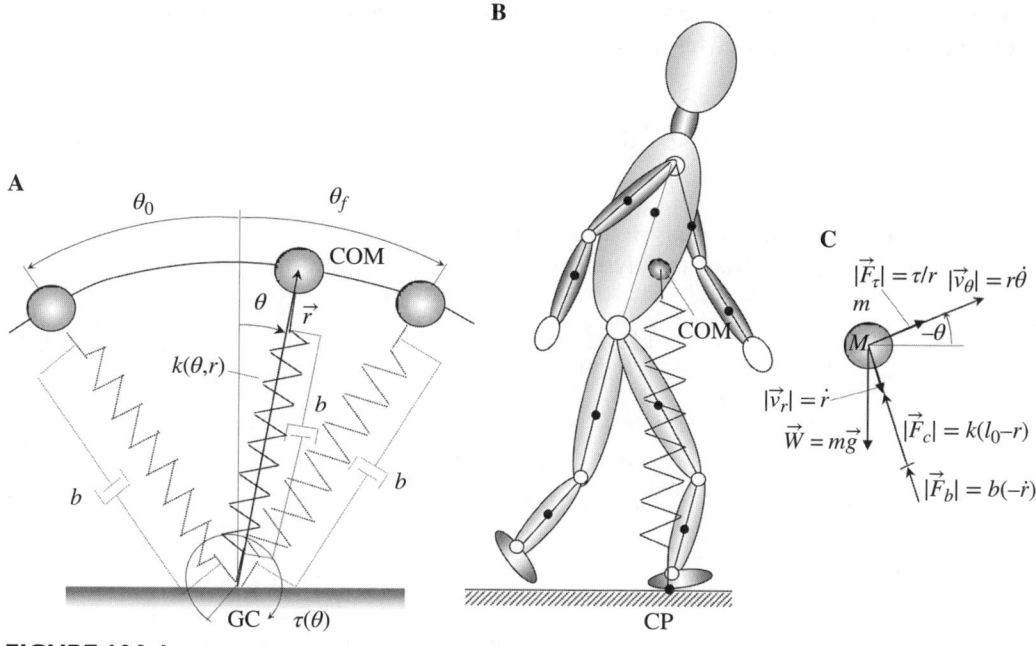

FIGURE 10A.1

The spring-mass inverted pendulum model indicated on a human body architecture (**A**), the spring-mass model (**B**) and the forces acting on the concentrated mass at the COM (**C**).

the results obtained by using the model. Figure 10A.1(c) shows the forces acting on the point mass with respect to polar coordinate system $r - \theta$.

The spring-mass inverted pendulum model has two degrees of freedom: the rotational degree, denoted by θ about GC, and the translational degree in the radial direction, denoted by r. To derive the equations of motion, Lagrange's principle was used.

The kinetic energy using polar coordinates is:

$$E_k = \frac{m}{2}\left(v_r^2 + v_\theta^2\right) = \frac{m}{2}\left(\dot{r}^2 + \left(r\dot{\theta}\right)^2\right) \tag{10A.1}$$

The potential energy consists of the potential energy due to gravity (the ground level is chosen as a datum for the potential energy calculation) and the energy stored in the spring, that is:

$$E_{p-tot} = mgr\cos\left(\theta\right) + \frac{k}{2}\left(l_0 - r\right)^2 \tag{10A.2}$$

Based on Equations 10A.1 and 10A.2, a Lagrange function can be formed:

$$L = E_k - E_{p-tot} = \frac{m}{2}\left(\dot{r}^2 + \left(r\dot{\theta}\right)^2\right) - mgr\cos\left(\theta\right) - \frac{k}{2}\left(l_0 - r\right)^2 \tag{10A.3}$$

Considering the dashpot effect and a rotational viscous friction, the dissipation (Reyleigh) function can be defined as:

$$R = \frac{b}{2}(\dot{r})^2 + \frac{b_\theta}{2}(\dot{\theta})^2 \qquad (10A.4)$$

where b_θ is a rotational viscous friction coefficient.

Using the energy method, the dynamics equations of motion are:

$$\frac{d}{dt}\frac{\partial L}{\partial \dot{\theta}} - \frac{\partial L}{\partial \theta} + \frac{\partial R}{\partial \dot{\theta}} = \tau \qquad (10A.5)$$

$$\frac{d}{dt}\frac{\partial L}{\partial \dot{r}} - \frac{\partial L}{\partial r} + \frac{\partial R}{\partial \dot{r}} = 0 \qquad (10A.6)$$

Where:

$$\frac{\partial L}{\partial r} = mr\dot{\theta}^2 + k(l_0 - r), \frac{\partial L}{\partial \dot{r}} = m\dot{r}, \frac{d}{dt}\frac{\partial L}{\partial \dot{r}} = m\ddot{r},$$

$$\frac{\partial L}{\partial \theta} = mg\sin(\theta), \frac{\partial L}{\partial \dot{\theta}} = mr^2\dot{\theta}, \frac{d}{dt}\frac{\partial L}{\partial \dot{\theta}} = \frac{d}{dt}(mr^2\dot{\theta}) = m(2r\dot{r}\dot{\theta} + r^2\ddot{\theta}),$$

$$\frac{\partial R}{\partial \dot{\theta}} = b_\theta \dot{\theta}, \frac{\partial R}{\partial \dot{r}} = b\dot{r}. \qquad (10A.7)$$

Finally, combining Equations 10A.5, 10A.6 and 10A.7 give:

$$\ddot{r} - r\dot{\theta}^2 + g\cos\theta - \frac{k}{m}(l_0 - r) + \frac{b}{m}\dot{r} = 0 \qquad (10A.8)$$

$$r^2\ddot{\theta} + 2r\dot{r}\dot{\theta} - gr\sin(\theta) = \frac{\tau}{m} - \frac{b_\theta}{m}\dot{\theta} \qquad (10A.9)$$

Equations 10A.8 and 10A.9 represent the equations of motion in radial and perpendicular directions. These equations can be solved numerically or approximately analytically, using certain assumptions (e.g., small angle approximation).

In the context of this analysis, two morphotypes were considered: (a) small tetrapod morphotype with low body mass and compliant limbs, and (b) larger tetrapod morphotype with higher body mass and stiffer limbs (Table 10A.1).

Table 10A.1 ■ Parameters used in constructing a spring-mass inverted pendulum model of small and large tetrapods: _m_, body mass; _k_, spring stiffness (estimated as twice the running stiffness; Figure 3 in Farley et al., 1993); _b_, linear motion damping coefficient; _b_θ, equivalent damping coefficient for rotational motion (about GC)

Model	m (kg)	k (N/m)	b (N/(m/s))	b_θ (Nm/(rad/s))
Small tetrapod/low stiffness	0.5	200	0	0.008
Large tetrapod/high stiffness	30	5000	50	4

10A.2   Recovery ratio calculation

The model represented above was used to calculate the efficiency of energy recovery during steady and non-steady speed locomotion via inverted pendulum mechanics. Traditionally, the recovery ratio R, is defined as (Cavagna et al., 1977):

$$R = \frac{|W_H| + |W_V| - |W_{ext}|}{|W_H| + |W_V|}$$ (10A.10)

where: $|W_H|$ and $|W_V|$ are the integral positive works of the forces in horizontal and vertical directions, respectively; $|W_H| + |W_V|$ is the total energy input (work) over a cycle that would be needed for the motion in case there is no potential/kinetic energy conversion; and $|W_{ext}|$ is an external energy source input (muscles in the case of a biologic walker or motors/actuators in a robotic walker) input (in Equation 10A.10, the symbol $| * |$ denotes that the positive portions of the work are considered; it does not represent the strict mathematical operation of taking the absolute value of the integral work over the cycle. However, the expression is traditionally used in this form).

Equation 10A.10 represents an average recovery ratio for the whole walking cycle. We can define an instantaneous recovery ratio that can be used as a measure of the energy exchange (potential/kinetic) at a particular instant of the cycle, defined in similar way using related powers instead of work.

$$R = \frac{|\dot{W}_H| + |\dot{W}_V| - |\dot{W}_{ext}|}{|\dot{W}_H| + |\dot{W}_V|}.$$ (10A.11)

The traditional recovery ratio definition works very well for steady speed locomotion, where it assumes that the total mechanical energy remains steady at the end of a cycle. This definition must be adjusted for non-steady walks, because there is change in the average kinetic energy throughout the walking cycles obtained through the external work (i.e., the torque modeled into the passive spring-mass mechanism). However, the additional kinetic energy that is obtained through the external work should not represent a measure of the mechanism's inability to recover energy by converting potential energy into kinetic energy and the other way around.

Since use of the traditional recovery ratio expression would include the external work needed for acceleration/deceleration, and necessarily results in a relatively low value, we need to exclude that portion of the external work associated with accelerating/decelerating the system. In this way, we can consider how the conversion between potential and kinetic energies is affected by the acceleration and the corresponding inertial forces. In particular, we can trace the correlation between the energy phases and energy conversion accordingly.

A couple of adjusted definitions have been tested, and the results of the analysis have been used to generate a recommended change in the recovery ratio definition in the cases of accelerating/decelerating walks. The adjusted recovery ratio for non-steady walking cycle with negligible energy losses is:

$$R = \frac{\sum_{i=1}^{n}\Psi\left(\Delta E_{K,V,i}\right)\Delta E_{K,V,i} + \sum_{i=1}^{n}\Psi\left(\Delta E_{P,i}\right)\Delta E_{P,i} + \sum_{i=1}^{n}\Psi\left(\Delta E_{K,F,i}^{*}\right)\Delta E_{K,F,i}^{*} - W_{ext}^{*}}{\sum_{i=1}^{n}\Psi\left(\Delta E_{K,V,i}\right)\Delta E_{K,V,i} + \sum_{i=1}^{n}\Psi\left(\Delta E_{P,i}\right)\Delta E_{P,i} + \sum_{i=1}^{n}\Psi\left(\Delta E_{K,F,i}^{*}\right)\Delta E_{K,F,i}^{*}},$$

$$(10A.12)$$

where: the function $\psi(*) = \frac{1}{2}(1 + \text{sign}(*))$ is the filter function which selects only positive increments in the corresponding energy $*$; $\Delta E_{K,Vi}$ and $\Delta E_{K,Fi}$ are the increments in kinetic energy associated with the vertical motion and forward motion, respectively; ΔE_{Pi} is the increment in potential energy; $W_{ext}^{*} = \sum_{i=1}^{n}\Psi(W_{ext,i})W_{ext,i} - \Delta E_{K,F,cycle}$ is the external work reduced by the quantity of the forward motion kinetic energy change over the walking cycle; and $\Delta E_{K,F,i}^{*}$ is the forward motion kinetic energy increment reduced by the corresponding kinetic energy portion related to the acceleration, which can be expressed as $\frac{\Delta E_{K,F,cycle}}{n}$ in a simplified case of a uniform distribution of the energy change over the cycle.

As long as a constant acceleration $a_{f,av}$ is concerned, the change in the forward motion kinetic energy can be expressed as $\Delta E_{K,F,cycle} = \frac{m}{2}(v_{f,end}^{2} - v_{f,in}^{2}) = \frac{m}{2}((v_{f,in} + a_{f,av}T)^{2} - v_{f,in}^{2})$, where T is the time for completing the cycle.

It is intuitively expected that the recovery ratio for an accelerating walk decreases as the average acceleration of the walking cycle increases. The reason for this is the acceleration-related increase in the external work W_{ext}, which is directly converted into kinetic energy. Figure 10A.2 shows accelerating and decelerating walking cycle potential and kinetic energy plots. It is noticeable that acceleration destroys the symmetry relation, which is directly reflected in a worse recovery ratio.

The potential energy shown in Figure 10A.2 is the gravitational potential energy, which is directly proportional to the COM height. In contrast to the case of the inverted pendulum model, with a rigid link which forces the point mass to follow a circular path, the mass-spring model shows significant influence of the inertial forces on the spring deformation, which reflects the distance change between the COM and the ground CP point. In the first part of the path, kinetic energy decreases at a relatively high rate, while there is no significant change in the potential energy due to the energy accumulation within the spring. However, this energy is mostly dissipated through the radial vibrations and dampers included in the model, representing loss of energy due to collisions with the ground in a real walk.

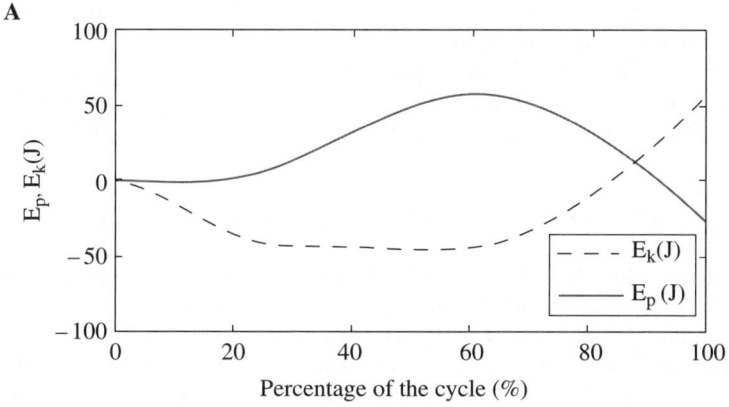

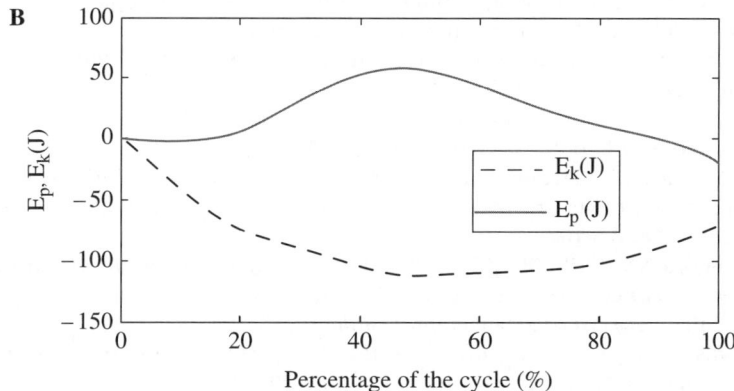

FIGURE 10A.2

Relative potential and kinetic energy plots for: (**A**) an accelerating and (**B**) a decelerating walking cycle (mass = 80 kg; v_{in} = 1.5 m/s; a_{aver} = ± 0.71 m/sec²).

Replacement of the rigid link in an inverted pendulum by a generally nonlinear spring, and adding the dissipation elements (linear and rotational dampers), improved the model results in the sense of narrowing down the difference between the real data and results obtained via the model.

▌REFERENCES

Alexander RM (1988). *Elastic mechanisms in animal movement*. Cambridge University Press, Cambridge.

Alexander RM (2005). Models and the scaling of energy costs for locomotion. *Journal of Experimental Biology* **208**, 1645–1652.

Anderson RA (1993). Analysis of foraging in the lizard, *Cnemidophorus tigris*. In: Wright J, Vitt LJ (Eds). *Biology of Cnemidophorus*, pp. 83–116. University of Oklahoma Press, Norman.

Anderson RA, Karasov WH (1981). Contrasts in energy intake and expenditure in sit–and-wait and widely foraging lizards. *Oecologia* **49**, 67–72.

Bertram JEA, D'Antonio P, Pardo J, Lee DV (2002). Pace length effects in human walking: "groucho" gaits revisited. *Journal of Motor Behavior* **34**, 309–318.

Biewener AA (1983). Allometry of quadrupedal locomotion: the scaling of duty factor, bone curvature and limb orientation to body size. *Journal of Experimental Biology* **105**, 147–171.

Biewener AA (1989). Scaling body support in mammals: limb posture and muscle mechanics. *Science* **245**, 45–48.

Biewener AA (1990). Mechanics of mammalian terrestrial locomotion. *Science* **250**, 1097–1103.

Biewener AA (2005). Biomechanical consequences of scaling. *Journal of Experimental Biology* **208**, 1665–1676.

Biewener AA, Blickhan R (1988). Kangaroo rat locomotion: design for elastic energy storage or acceleration? *Journal of Experimental Biology* **140**, 243–255.

Biknevicius AR, Reilly SM (2006). Correlation of symmetrical gaits and whole body mechanics: debunking myths in locomotor biodynamics. *Journal of Experimental Zoology* **305A**, 923–934.

Biknevicius AR, Mullineaux DR, Clayton HM (2006). Locomotor mechanics of the tölt in Icelandic horses. *American Journal of Veterinary Research* **67**, 1505–1510.

Biknevicius AR, Reilly SM, McElroy EJ, Bennett MB (2013). Symmetrical gaits and center of mass mechanics in small-bodied, primitive mammals. *Zoology* **116**, 67–74.

Blaszczyk JW (2001). Gait transition during unrestrained locomotion in dogs. *Equine Veterinary Journal, Supplement* **33**, 112–115.

Blob, RW (2001). Evolution of hindlimb posture in nonmammalian therapsids: biomechanical tests of paleontological hypotheses. *Paleobiology* **27**, 14–38.

Boyd L, Bandi N (2002). Reintroduction of takhi, *Equus ferus przewalskii*, to Hustai National Park, Mongolia: time budget and synchrony of activity pre- and post-release. *Applied Animal Behaviour Science* **78**, 87–102.

Butcher MT, Espinoza NR, Cirilo SR, Blob RW (2008). *In vivo* strains in the femur of river cooter turtles (*Pseudemys concinna*) during terrestrial locomotion: test of force-platform models of loading mechanics. *Journal of Experimental Biology* **211**, 2397–2407.

Carbone C, Cowlishaw G, Isaac NJB, Rowcliffe JM (2005). How far do animal do? Determinant of day range in mammals. *American Naturalist* **165**, 290–297.

Cavagna GA, Heglund NC, Taylor CR (1977). Mechanical work in terrestrial locomotion: two basic mechanisms for minimizing energy expenditure. *American Journal of Physiology* **233**, 243–261.

Christiansen P (1999). Scaling of mammalian long bones: small and large mammals compared. *Journal of Zoology London* **247**, 333–348.

Daley MA, Usherwood JR (2010). Two explanations for the compliant running paradox: reduced work of bouncing viscera and increased stability in uneven terrain. *Biology Letters* **6**, 418–421.

Davies MNO, Green PR (1988). Head-bobbing during walking, running and flying: relative motion perception in the pigeon. *Journal of Experimental Biology* **138**, 71–91.

De Smet K, Segers V, Lenoir M, De Clercq D (2009). Spatiotemporal characteristics of spontaneous overground walk-to-run transition. *Gait Posture* **29**, 54–58.

Doke J, Donelan JM, Kuo AD (2005). Mechanics and energetics of swinging the human leg. *Journal of Experimental Biology* **208**, 439–445.

Edwards EB, Gleeson TT (2001). Can energetic expenditure be minimized by performing activity intermittently? *Journal of Experimental Biology* **204**, 599–605.

Ettema GJC (1996) Mechanical efficiency and efficiency of storage and release of series elastic energy in skeletal muscle during stretch-shorten cycles. *Journal of Experimental Biology* **199**, 1983–1997.

Farley CT, Ko TC (1997). Mechanics of locomotion in lizards. *Journal of Experimental Biology* **200**, 2177–2188.

Farley CT, Glasheen J, McMahon TA (1993). Running springs: speed and animal size. *Journal of Experimental Biology* **185**, 71–86.

Farquharson ASL (2007). *On the gait of animals by Aristotle.* (http://ebooks.adelaide.edu.au/a/aristotle/gait/complete.html) eBooks@Adelaide, University of Adelaide Library.

Fish FE, Frappell PB, Baudinette RV, Macfarlane PM (2001). Energetic of terrestrial locomotion of the platypus *Onrithorhynchus anatinus. Journal of Experimental Biology* **204**, 797–803.

Frost, BJ (1978). The optokinetic basis of head-bobbing in the pigeon. *Journal of Experimental Biology* **74**, 187–195.

Full RJ (1989). Mechanics and energetics of terrestrial locomotion: from bipeds to polypeds. In: Weisner W, Gnaiger E (Eds). *Energy Transformations in Cells and Animals*, pp. 175–182. Thieme, Stuttgart.

Gambaryan PP (1974). *How animals run: anatomical adaptations.* Wiley, New York.

Garland Jr T (1983). The relation between maximal running speed and body mass in terrestrial mammals. *Journal of Zoology, London* **199**, 157–170.

Gatesy SM (1991). Hind limb movements of the American alligator (*Alligator mississippiensis*) and postural grades. *Journal of Zoology, London* **224**, 577–588.

Gatesy SM (1999). Guineafowl hind limb function. I: Cineradiographic analysis and speed effects. *Journal of Morphology* **240**, 115–125.

Gatesy SM, Biewener AA (1991). Bipedal locomotion: effects of speed, size and limbs posture in birds and humans. *Journal of Zoology* **224**, 127–147.

Geyer H, Seyfarth A, Blickhan R (2006). Compliant leg behaviour explains basic dynamics of walking and running. *Proceedings of the Royal Society B* **273**, 2861–2867.

Gleeson TT, Hancock TV (2001). Modeling the metabolic energetics of brief and intermittent locomotion in lizards and rodents. *American Zoologist* **41**, 211–218.

Griffin TM, Kram R (2000). Penguin waddling is not wasteful. *Nature* **408**, 929.

Griffin TM, Kram R, Wickler SJ, Hoyt DF (2004a). Biomechanical and energetic determinants of the walk–trot transition in horses. *Journal of Experimental Biology* **207**, 4215–4223.

Griffin TM, Main RP, Farley CT (2004b). Biomechanics of quadrupedal walking: how do four-legged animals achieve inverted pendulum-like movements? *Journal of Experimental Biology* **207**, 3545–3558.

Hancock JA, Stevens NJ, Biknevicius AR (2007). Whole-body mechanics and kinematics of terrestrial locomotion in the Elegant-crested Tinamou *Eudromia elegans. Ibis* **146**, 605–614.

Haque A, Dickman JD (2005). Vestibular gaze stabilization: Different behavioral strategies for arboreal and terrestrial avians. *Journal of Neurophysiology* **93**, 1165–1173.

Heglund NC, Taylor CR (1988). Speed, stride frequency and energy cost per stride: how do they change with body size and gait? *Journal of Experimental Biology* **138**, 301–318.

Heglund NC, Cavagna GA, Taylor CR (1982). Energetics and mechanics of terrestrial locomotion, III: energy expenditures of the centre of mass as a function of speed and body size in birds and mammals. *Journal of Experimental Biology* **79**, 41–56.

Hildebrand M (1976). Analysis of tetrapod gaits: general consdierationsd and symmetrical gaits. In: Herman RM, Grillner S, Stein P, Stuart DG (eds). *Neural Control of Locomotion*, Vol. 18, pp. 203–236. Plenum, New York.

Hildebrand M (1977). Analysis of asymmetrical gaits. *Journal of Mammalogy* **58**, 131–155.

Hildebrand M, Goslow TE (2001). *Analysis of vertebrate structure*, 5th edition. Wiley, New York.

Hooper SL, Guschlbauer C, Blumel M, Rosenbaum P, Gruhn M, Akay T, Buschges A (2009). Neural control of unloaded leg posture and of leg swing in stick insect, cockroach, and mouse differs from that in larger animals. *Journal of Neuroscience* **29**, 4109–4119.

Horner AM, Biknevicius AR (2010). A comparison of epigean and subterranean locomotion in the domestic ferret (*Mustela putorius furo*: Mustelidae: Carnivora). *Zoology* **113**, 189–197.

Howell AB (1944). *Speed in animals.* University of Chicago Press, Chicago.

Hoyt DF, Kenagy GJ (1988). Energy costs of walking and running gaits and their aerobic limits in golden-mantled ground squirrels. *Physiol. Zoology* **61**, 34–40.

Hoyt DF, Taylor CR (1981). Gait and the energetics of locomotion in horses. *Nature* **292**, 239–240.

Hutchinson JR, Famini D, Lair R, Kram R (2003). Are fast-moving elephants really running? *Nature* **422**, 493–494.

Iriarte-Díaz J (2002). Differential scaling of locomotor performance in small and large terrestrial mammals. *Journal of Experimental Biology* **205**, 2897–2908.

Irschick DJ (2000). Comparative and behavioral analysis of preferred speed: Anolis lizards as a model system. *Physiological and Biochemical Zoology* **73**, 428–437.

Jacobson RD, Hollyday M (1982). A behavioral and electromyographic study of walking in the chick. *Journal of Neurophysiology* **48**, 238 –256.

Jenkins Jr. FA (1971). Limb posture and locomotion in the Virginia opossum (Didelphis marsupialis) and in other non-cursorial mammals. *Journal of Zoology* **165**, 303–315.

Kenagy GJ, Hoyt DF (1989). Speed and time-energy budget for locomotion in golden-mantled ground squirrels. *Ecology* **70**, 1834–1839.

Ker RF, Alexander RMcN, Bennett MB (1988) Why are mammalian tendons so thick? *Journal of Zoology London* **216**, 309–324.

Kram R, Taylor CR (1990). Energetics of running: a new perspective. *Nature* **346**, 265–267.

Kramer DL, McLaughlin RL (2001). The behavioral ecology of intermittent locomotion. *American Zoologist* **41**, 137–153.

Maloiy GMO, Rugangazai BM, Rowe MF (2009). Energy expenditure during level locomotion in large desert ungulates: the one-humped camel and the domestic donkey. *Journal of Zoology, London* **277**, 248–255.

Margaria R (1938). Sulla fisiologia e specialmente sul consume energetico della marcia e della corsa a varia velocita`ed inclinazione del terreno. *Atti della Accademia Nazionale dei Lincei* **7**, 299–368.

Martin GR (1984). The visual fields of the Tawny Owl, *Strix aluco*. L. *Vision Research* **24**, 1739–1751.

McElroy EJ, Hickey KL, Reilly SM (2008). The correlated evolution of biomechanics, gaits and foraging mode in lizards. *Journal of Experimental Biology* **211**, 1029–1040.

McMahon T A, Valiant G, Frederick EC (1987). Groucho running. *Journal of Applied Physiology* **62**, 2326–2337.

Minetti AE, Ardigò LP, Reinach E, Saibene F. (1999). The relationship between mechanical work and energy expenditure of locomotion in horses. *Journal of Experimental Biology* **202**, 2329–2338.

Minetti AE, Boldrini L, Brusamolin L, Zamparo P, McKee T (2003). A feedback-controlled treadmill (treadmill-on-demand) and the spontaneous speed of walking and running in humans. *Journal of Applied Physiology* **95**, 838–843.

More HL, Hutchinson JR, Collins DF, Weber DJ, Aung SKH, Donelan JM (2010). Scaling of sensorimotor control in terrestrial mammals. *Proceedings of the Royal Society B* **277**, 3563–3568.

Nowak RM (1991). *Walker's Mammals of the World*, 5th edition. The Johns Hopkins University Press, Baltimore.

Nudds RL, Folkow LP, Lees JJ, Stokkan KA and Codd JR (2011). Evidence for energy savings from aerial running in the Svalbard rock ptarmigan (*Lagopus muta hyperborean*). *Proceedings of the Royal Society B: Biological Sciences* [Published online before print February 2, 2011, doi: 10.1098/rspb.2010.2742]

Ortega JD, Farley CT (2005). Minimizing center of mass movement increases metabolic cost of walking. *Journal of Applied Physiology* **99**, 2099–2107.

Pennycuick, CJ (1975). On the running of the gnu (*Connochaetes taurinus*) and other animals. *Journal of Experimental Biology* **63**, 775–799.

Prochazka A, Trend P, Hulliger M, Vincent S (1989) Ensemble proprioceptive activity in the cat step cycle: towards a representative look-up chart. *Progress in Brain Research* **80**, 61–74.

Reilly SM (2000). Locomotion in quail (*Corturnix japonica*): the kinematics of walking and increasing speeds. *Journal of Morphology* **243**(2), 173–185.

Reilly SM, DeLancey MJ (1997). Sprawling locomotion in the lizard *Sceloporus clarkii*: the effects of speed on gait, hindlimb kinematics, and axial bending during walking. *Journal of Zoology, London* **243**, 417–433.

Reilly SM, McElroy EJ, Odum RA, Hornyak VA (2006). Tuataras and salamanders show that walking and running mechanics are ancient features of tetrapod locomotion. *Proceedings of the Royal Society B* **273**, 1563–1568.

Reilly SM, McElroy EJ, Biknevicius AR (2007). Posture, gait and the ecological relevance of locomotor costs and energy-saving mechanisms in tetrapods. *Zoology*, **110**, 271–289.

Rewcastle SC (1980). Form and function in lacertilian knee and mesotarsal joints: a contribution to the analysis of sprawling locomotion. *Journal of Zoology, London* **191**, 147–170.

Roberts TJ, Kram R, Weyand PG, Taylor CR (1998). Energetic of bipedal running. *I. Metabolic cost of generating force. Journal of Experimental Biology* **201**, 2745–2751.

Robilliard J (2005). *Mechanical basis of locomotion with spring-like legs*. PhD thesis, Royal Veterinary College, London, UK.

Rome LC, Sosnicki AA, Goble DO (1990). Maximum velocity of shortening of three fibre types from the horse soleus: Implications for scaling with body size. *Journal of Physiology, London* **431**, 173–185.

Rubenson J, Marsh RL (2009). Mechanical efficiency of limb swing during walking and running in guinea fowl (*Numida meleagris*). *Journal of Applied Physiology* **106**, 1618–1630.

Rubenson J, Heliams DB, Lloyd DG, Fournier PA (2004). Gait selection in the ostrich: mechanical and metabolic characteristics of walking and running with and without an aerial phase. *Proceedings of the Royal Society of London, Series B* **271**, 1091–1099.

Ruina A, Bertram JEA, Srinivasan M. (2005). A collisional model of the energetic cost of support work qualitatively explains leg sequencing in walking and galloping, pseudo-elastic leg behavior in running and the walk-to-run transition. *Journal of Theoretical Biology* **237**, 170–192.

Schmitt D, Cartmill M, Griffin TM, Hanna JB, Lemelin P. (2006). Adaptive value of ambling gaits in primates and other mammals. *Journal of Experimental Biology* **209**, 2042–2049.

Segers V, Aerts P, Lenoir M, De Clercq D (2007a). Dynamics of the body centre of mass during actual acceleration across transition speed. *Journal of Experimental Biology* **210**, 578–585.

Segers V, Lenoir M, Aerts P, De Clercq D (2007b). Kinematic of the transition between walking and running when gradually changing speed. *Gait Posture* **26**, 349–361.

Seyfarth A, Geyer H, Blickhan R, Lipfert S, Rummel J, Minekawa Y, Iida F (2006). Running and walking on compliant legs. *Lecture Notes in Control and Information Sciences* **340**, 383–401.

Skime A, Boone T (2011). Cardiovascular responses during Groucho running. *Journal of Exercise Physiology* online [http://faculty.css.edu/tboone2/asep/JEPonlineApril2011Skime-3.pdf]

Spenko MJ, Haynes GC, Saunders JA, Cutkosky MR, Rizzi AA, Full RJ, Koditschek DE (2008) Biologically inspired coming with a hexapedal robot. *Journal of Field Robotics* **25**, 223–242.

Strang KT, Steudel K (1990). Explaining the scaling of transport costs, the role of stride frequency and stride length. *Journal of Zoology, London* **221**, 343–358.

Switalski TA (2003). Coyote foraging ecology and vigilance in response to gray wolf introduction in Yellowstone National Park. *Can. Journal of Zoology* 985–993.

Székely G, Czéh G, Vörös G (1969). The activity pattern of limb muscles in freely moving normal and deafferented newts. *Experimental Brain Research* **9**, 53 62.

Taylor CR, Schmidt-Nielsen K, Raab JL (1970). Scaling of energetic cost of running to body size in mammals. *American Journal of Physiology* **219**, 1104–1107.

Taylor CR, Caldwell SL, Rowntree VJ (1972). Running up and down hills: Some consequences of size. *Science* **178**, 1096–1097.

Tokuriki M, Aoki O (1995). Electromyographic activity of the hindlimb muscles during the walk, trot and canter. *Equine Veterinary Journal* **18**, 152–155.

Usherwood JR, Williams SB, Wilson AM (2007). Mechanics of dog walking compared with a passive, stiff-limbed, 4-bar linkage model, and their collisional implications. *Journal of Experimental Biology* **210**, 533–540.

Usherwood JR, Szymanek KL, Daley MA (2008). Compass gait mechanics account for top walking speeds in ducks and humans. *Journal of Experimental Biology* **211**, 3744–3749.

Van Damme R, Aerts P, Vanhooydonck B (1998). Variation in morphology, gait characteristics and speed of locomotion in two populations of lizards. *Biological Journal of the Linnean Society* **63**, 409–427.

Wallace IJ, Demes B. (2008). Symmetrical gaits of *Cebus paella*: implications for the functional significance of diagonal sequence gait in primates. *Journal of Human Evolution* **54**, 783–794.

Watson RR, Rubenson J, Coder L, Hoyt DF, Propert MWG, Marsh RL (2011) Gait-specific energetic contributes to economical walking and running in emus and ostriches. *Proceedings of the Royal Society B* **278**, 2040–2046.

Weinstein RB, Full RJ (2000). Intermittent locomotor behavior alters total work. In: Domenici P, Blake RW (eds). *Biomechanics in Animal Behaviour*, pp. 33–48. BIOS Scientific Publishers Ltd., Oxford.

Wickler SJ, Hoyt DF, Cogger EA, Hirschbein MH (2000). Preferred speed and cost of transport: the effect of incline. *Journal of Experimental Biology* **203**, 2195–2200.

Willey JS, Biknevicius, AR, Reilly, SM, Earls, KD (2004). The tale of the tail: limb function and locomotor mechanics in *Alligator mississippiensis*. *Journal of Experimental Biology* **207**, 553–563.

Zani PA, Gottschall JS, Kram R (2005). Giant Galápagos tortoises walk without inverted pendulum mechanical-energy exchange. *Journal of Experimental Biology* **208**, 1489–1494.

CHAPTER ELEVEN

Non-Steady Locomotion

Monica A. Daley

Department of Comparative Biomedical Sciences, Royal Veterinary College, University of London, UK

11.1 INTRODUCTION

Most analyses of locomotion start with the assumption of steady locomotion: that is, movement over completely uniform, level terrain, with a constant velocity and constant total mechanical energy of the body. This simplification facilitates simple mechanical models that identify the basic force and mechanical work requirements of locomotion. A key observation from these studies is that all terrestrial animals follow similar basic movement patterns for steady movement: inverted pendulum dynamics for walking; mass-spring bouncing dynamics during hopping; running and trotting, with both gaits approximated well with a spring-loaded inverted pendulum (SLIP), with appropriate parameters for sweep angle, effective leg stiffness and total mechanical energy (Alexander and Vernon, 1975; Cavagna *et al.*, 1976; McGeer, 1990; McMahon and Cheng, 1990; Farley *et al.*, 1993; Geyer *et al.*, 2006). These movement patterns also emerge from theoretical considerations based only on the minimization of mechanical

Understanding Mammalian Locomotion: Concepts and Applications, First Edition.
Edited by John E. A. Bertram.
© 2016 John Wiley & Sons, Inc. Published 2016 by John Wiley & Sons, Inc.

work, within certain constraints such as force limits for the legs (Ruina *et al.*, 2005). As previous chapters have highlighted, simple models for steady locomotion have provided insights into the ultimate sources of energy cost, limits to maximum speed and the functional differences between gaits.

The aim of this chapter is to extend similar analytical approaches to non-steady locomotion. If we could monitor the activity of animals moving in their natural environment, we would probably find that they spend the majority of their time in non-steady locomotion. Non-steady behaviors include: acceleration and deceleration; gait transitions; jumping; running on slopes; turning; recovering from external disturbances; and negotiating variations in terrain. Certain animal behaviors, such as predator-prey interactions, are inherently non-steady, and such factors as stability, robustness and maneuverability are likely to be at least as important as economy in determining whether an animal eats or gets eaten.

It is important to acknowledge that, once we move into the realm of "non-steady" locomotion, there exists a multitude of behaviors to consider. Any chapter on the topic is necessarily a sampling of highlights, rather than a comprehensive review, and should be treated as such. The purpose of this chapter is to draw attention to some basic physical and geometric principles that can be applied to non-steady locomotion, and to highlight general themes that have emerged from recent work.

Another notable omission of this chapter will be detailed consideration of the interplay of neural control and sensory processing in non-steady locomotion. Although understanding the integration of mechanics and control is particularly important for non-steady behaviors (Marigold and Patla, 2005; Sponberg and Full, 2008; Ting *et al.*, 2009), this is a large topic that deserves its own treatment. Readers are directed to a number of sources for further consideration of this topic (including the works above, as well as Pearson, 2000; Prochazka *et al.*, 2002; Todorov and Jordan, 2002; Pearson *et al.*, 2006; Rossignol *et al.*, 2006).

11.1.1 Why study non-steady locomotion?

Animals must be able to negotiate variable and unpredictable terrain in order to escape predators, catch prey and find mates. Study of non-steady locomotion is not simply a descriptive exercise in exploring an ever-increasing array of animal behavior – it provides insight into how animals have dealt with potential trade-offs between economy and other performance criteria, such as stability, robustness, maneuverability and agility. It is clear that animal form and function does not reflect perfect adaptation for economic steady locomotion; instead, animal morphology and behavior reflects compromises among many different demands. For example, brachiating gibbons could achieve perfectly passive mechanics that would minimize energy cost (Usherwood and Bertram, 2003). However, they instead adopt a strategy that involves higher collision energy loss at handhold transitions, but provides a greater safety factor, making disastrous falls less likely (Usherwood and Bertram, 2003).

Currently, so little is known about the mechanical requirements for non-steady behaviors that trade-offs among different performance criteria are often assumed, with little or no evidence. Such "intuited" relationships may turn out to be quite wrong. For example, a trade-off between maneuverability and stability is often assumed for flying animals; yet, a recent model for yaw turn dynamics suggests that animals specialized for high wing beat frequencies benefit in both maneuverability and stability (Hedrick *et al.*, 2009). However, this specialization is not without trade-offs, because high wing beat frequencies come at the cost of economy. By investigating the performance requirements and movement strategies used by animals for non-steady locomotion, we can begin to develop a clearer picture of the trade-offs and constraints that influenced the evolution of animal morphology and behavior.

Thus, studies of non-steady locomotor behavior serve two main roles:

1. Providing understanding of a broader range of ecologically and evolutionarily important animal behavior.
2. Revealing principles of mechanics and control that cannot be deciphered from studies of steady locomotion alone.

11.2 APPROACHES TO STUDYING NON-STEADY LOCOMOTION

For non-steady behavior, it is important to consider the goal, and possible limiting conditions, of the behavior in question. The simplifying assumptions of steady locomotion do not generally apply. However, reductionist models can still provide insight through analysis of the absolute physical requirements and the potential mechanical and physiological limits that might constrain how the goal behavior is accomplished.

One can consider two broad categories of "non-steady" behavior:

1. Anticipated, goal-directed changes in velocity (speed, direction) or body height.
2. Stabilizing responses to maintain a goal movement in the face of perturbations from it.

The behavioral and mechanical strategies used by animals likely differ between these two broad categories. Goal-directed movements are anticipated and are likely to involve higher-level nervous system control, whereas stabilizing responses are likely to rely to a greater extent on spinal reflex pathways and the intrinsic mechanical response of the system, due to the inherent transmission and excitation-contraction coupling delays of the neuromuscular system. Nonetheless, it should be kept in mind that the distinction in control mechanisms between these categories is not likely to be clear-cut. If an animal knows itself to be in rough, unpredictable terrain, changes in higher-level control strategy might be adopted that facilitate agile and robustly stable locomotion in the current environmental context.

Due to the varied nature of non-steady locomotion, it can be particularly important to address systematically the following questions about any specific behavior of interest. The answers to these questions are not always immediately apparent, but considering the possibilities helps to develop simple models that suggest testable hypotheses and experiments:

- What is the purpose of the movement?
- What are the behavioral and mechanical requirements to achieve it?
- What are the possible mechanical and physiological limits? (Table 11.1)
- What are the possible strategies for achieving the goal within these limits?

11.2.1 Simple mechanical models

One of the main themes of this book is that a great deal can be learned by starting with extremely reductionist mechanical analysis of whole-animal behavior. A simple free-body diagram of the whole body (Figure 11.1) helps to reveal the mechanical absolutes, the morphological factors and alternative possible strategies that may come into play in achieving the mechanical absolutes. Mechanical considerations from simple models form the basis of explicit, testable hypotheses, which then suggest

Table 11.1 ■ Some possible mechanical and physiological limits in locomotion

Power and muscle mass available to deliver it.
Peak forces that can be resisted by the muscles and bones.
Maximum shortening velocity of muscle.
Metabolic power.
Geometry (leg contact conditions, joint ranges of motion, etc.).
Static or dynamic stability.
Friction or grip forces maintaining foot-substrate contact.
Response time due to neural delays.
Frequency/bandwidth at which muscle power can be delivered.
Gearing and other force-velocity trade-offs.

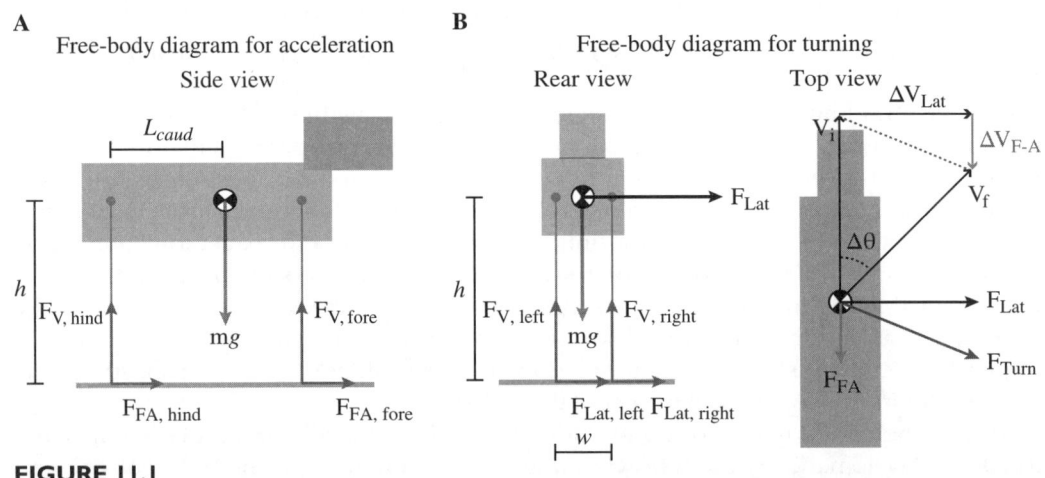

FIGURE 11.1

(**A**) Free body diagram for acceleration, which allows calculation of the maximum acceleration before the body pitches upwards and the forelimbs leave the ground (section 4.1). (**B**) Free body diagrams for turning from two views (section 4.2). The rear view is used in considering roll stability limits in turning and is essentially identical to the acceleration model, but in a different plane. The top view considers the forces required to turn while maintaining the same initial and final velocity, using a simple point-mass model of the body in the yaw plane. Note that the direction designations in the diagram refer to the original velocity heading.

experiments to test the hypotheses. Section 11.5 of this chapter will highlight some themes which have emerged from recent studies that have used simple models to analyze non-steady behaviors.

11.2.2 Research approaches to non-steady locomotion

Studies of non-steady locomotion can take a number of forms, depending on the nature of the question, and sometimes include multiple aspects of the list below:

1. Measurement and analysis of a natural range of behavior, such as acceleration and deceleration (e.g., Lee *et al.*, 1999; Roberts and Scales, 2004; McGowan *et al.*, 2005).

2. Analysis focused on identifying the limits of performance, such as maximum acceleration or maximum speed while turning (Greene and McMahon, 1979; Alexander, 2002; Usherwood and Wilson, 2006; Chang and Kram, 2007; Williams *et al.*, 2009).

3. Manipulation of morphological and physical factors (such as mass distribution, gravity, external force) to test hypotheses about constraints and trade-offs in achieving a particular behavior (Lee *et al.*, 2001; Chang and Kram, 2007).

4. Perturbations (forces, terrain properties, obstacles, etc.) to investigate the mechanical, neuromuscular and behavioral factors in control and stability (Ferris *et al.*, 1998; Jindrich and Full, 2002; Moritz and Farley, 2004; Marigold and Patla, 2005; Daley *et al.*, 2006; Sponberg and Full, 2008; Spence *et al.*, 2010).

As will be clear from some examples, the number of studies on non-steady locomotion has increased rapidly in recent years. Some of the reasons for this include:

1. the clear ecological and evolutionary relevance, as discussed above;

2. the desire to gain "biological inspiration" to improve the agility, stability and robustness of robots; and

3. advances in technology which have made it feasible to study more complex animal behaviors.

Computing and technology advances have made it a trivial matter to measure more complex animal behaviors, using large arrays of force platforms and cameras, or devices attached directly to the animal, such as inertial sensors. However, one should always carefully consider whether added complexity is worth the effort.

This chapter will highlight some relatively simple approaches that lead to predictive models and testable hypotheses for non-steady behaviors. Experimental studies of locomotion are most informative when coupled with mechanical analysis. When combined with simple mechanical models, experiments can test explicit hypotheses suggested by the model. This combined approach provides insight into whether or not our simple models correctly represent the primary mechanical requirements, trade-offs and limits of a particular behavior.

11.3 THEMES FROM RECENT STUDIES OF NON-STEADY LOCOMOTION

11.3.1 Limits to maximal acceleration

Once a simple model has been developed, based on a free-body diagram, and the mechanical requirements of the task have been identified, it is useful to consider the potential physiological and mechanical limits that might influence how these mechanical requirements are met (Table 11.1). Simple models often help identify possible limits. For example, simple models identify a number of possible limits for maximum acceleration:

1. body pitch stability (Figure 11.1), (Gray, 1944; Lee *et al.*, 1999; Alexander, 2002; Aerts *et al.*, 2003; Williams *et al.*, 2009);
2. maximum muscle power (Aerts, 1998; Roberts and Marsh, 2003; Williams *et al.*, 2009);
3. maximum forces (Greene and McMahon, 1979; Alexander, 2002; Usherwood and Wilson, 2006); and
4. friction or grip on the ground (Alexander, 2002).

The pitch stability limit is predicted on the basis of a simple geometric analysis of pitching moments generated about the body CoM when torque-actuated legs generate a net accelerating or decelerating impulse (Lee *et al.*, 1999; Alexander, 2002; Aerts *et al.*, 2003; Williams *et al.*, 2009) (Figure 11.1). When the body is accelerated forwards, a nose-up pitching moment is generated about the body CoM. For forward acceleration, the pitch-avoidance limit relates to body height (h) and the crandio-caudal distance between the hip joint and the body CoM (L_{caud}):

Animals with relatively high bodies and short hip-CoM distances (either due to a short body or a caudally shifted CoM) may be limited in acceleration by pitch stability. To maximize acceleration while avoiding pitch, animals can behaviorally shift the CoM position forward and down or, alternatively, shift foot positioning or weight distribution to move the origin of the ground reaction force vector backward relative to the CoM. Both quadrupeds and bipeds can shift the CoM position by leaning forward or pitching the trunk downward (McGowan *et al.*, 2005; Walter and Carrier, 2009; Williams *et al.*, 2009). Alternatively, they can retract the legs to shift the point of force application backwards (Lee *et al.*, 1999; Roberts and Scales, 2002). Trotting quadrupeds can also redistribute weight support between fore- and hind-limbs to shift the origin of the ground reaction force vector (Lee *et al.*, 1999; Walter and Carrier, 2009).

Interestingly, a pitch stability limit appears to result in spontaneous transition from quadrupedal to bipedal locomotion in some lizards, and bipedal locomotion occurs more frequently among lizards with a caudally shifted CoM position (Aerts *et al.*, 2003; Clemente *et al.*, 2008). In contrast, quadrupeds with relatively short legs relative to body length, long trunks, and cranially shifted CoM positions will be relatively stable in pitch, so acceleration is likely to be limited by available muscle power or other factors.

To maximize acceleration in the face of muscle power limits, animals are likely to maximize the duration over which the muscles can provide power. This can be achieved by increasing the duration of stance and by prolonging the period of muscle contraction through elastic mechanisms. In-series elasticity (tendons) can allow muscles to produce power over a longer period than the external acceleration phase, by storing energy in the spring to be released rapidly, achieving higher average external power output (Roberts and Marsh, 2003). Similar mechanisms are used to maximize power output during jumping (Aerts, 1998; Azizi and Roberts, 2010) – another non-steady locomotor activity that may be limited by the available muscle power.

Similar mechanical and physiological limits apply to deceleration as they do for acceleration; however, the relationships among various factors may differ substantially. The power limit likely differs for energy production and energy dissipation, resulting in a different power limit threshold for deceleration. Additionally, quadrupeds use their hindlimbs to accelerate and their forelimbs to decelerate (Lee *et al.*, 1999, 2004), and both the geometry and the available muscle mass likely differ between the two behaviors.

Similar to acceleration, animals can avoid pitch limits during deceleration by shifting their CoM position up and backward, or by shifting the origin of the ground reaction force vector forward relative to the CoM (Lee *et al.*, 1999; Alexander, 2002; Williams *et al.*, 2009). However, the ability of animals to make such adjustments may differ between acceleration and deceleration, due to anatomical constraints. For example, humans can lean forward, even onto four limbs, to maximize acceleration (as track athletes often do in sprint starts), but cannot lean backward as substantially for deceleration. However, the body could be reoriented to decelerate using lateral forces and lateral body lean. Thus, although similar general principles apply to both acceleration and deceleration, animals likely use different strategies, and experience different ultimate limiting factors, for the two behaviors.

As demonstrated by the example of acceleration mechanics, there are often several possible limits for a given behavior, resulting in a complex interplay among factors. Multiple limits might influence behavior simultaneously, and different limits might dominate, depending on morphology, terrain and behavioral context. For example, the limit to acceleration appears to depend on speed in dogs and horses. Maximum acceleration at low speeds is consistent with a pitch stability limit, but maximum acceleration at high speeds is consistent with a muscle power limit (Williams *et al.*, 2009). Simple models allow systematic consideration of the possible limits, and provide testable hypotheses about which of the potential limits influence behavior and performance in specific circumstances.

11.3.2 Morphological and behavioral factors in turning mechanics

The purpose of a turning maneuver is to change movement direction, and it is likely to be a goal-directed movement that involves anticipation. Simple mechanical models have highlighted two primary tasks for turning:

1. changing velocity heading to the new direction; and
2. rotation of the body about its CoM to face the new velocity heading (Jindrich and Full, 1999; Jindrich *et al.*, 2007; Jindrich and Qiao, 2009).

As discussed below, the behavioral and mechanical strategies used by animals to accomplish these tasks depends on an interplay among a number of factors, including: maximum leg forces (Greene and McMahon, 1979; Greene, 1985; Usherwood and Wilson, 2006; Chang and Kram, 2007; Jindrich and Qiao, 2009); friction or grip of the foot against the ground (Alexander, 2002); body rotational inertia (Lee *et al.*, 2001;

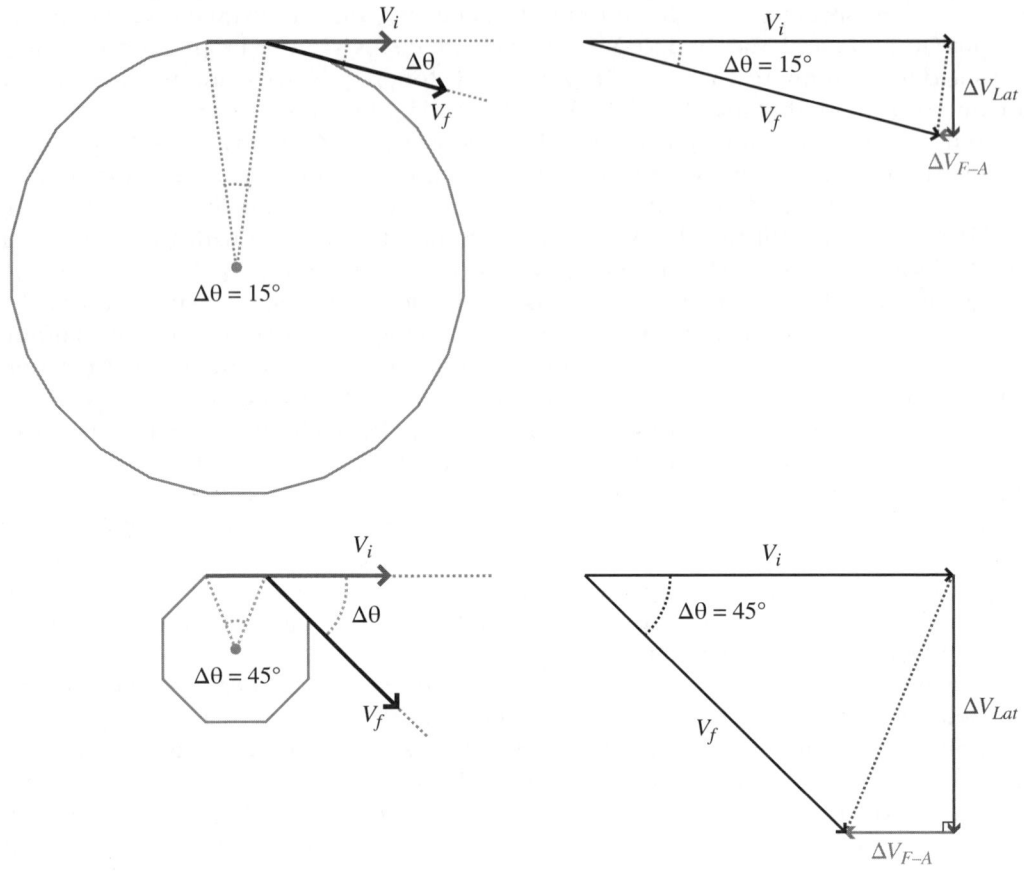

FIGURE 11.2

Running around a bend at a constant speed and step length requires a fixed change in velocity heading per step ($\Delta\theta$). Vector sums (right), can be used to calculate the required changes in velocity in the lateral (ΔV_{Lat}) and fore-aft (ΔV_{F-A}) directions. Turning more sharply (effectively running around a smaller circle) requires a larger $\Delta\theta$ and larger ΔV_{Lat} and ΔV_{F-A}. For small angles, ΔV_{F-A} is much smaller than ΔV_{Lat}, but the ratio of the two reaches unity for a 90° turn. The ground reaction forces required per step can be calculated based on impulse-momentum balance (see section 11.4.2 and Figure 11.3).

Walter, 2003; Jindrich and Qiao, 2009); roll stability limits (Alexander, 2002); and leg geometry (Walter, 2003; Jindrich *et al.*, 2007; Jindrich and Qiao, 2009).

Turning behaviors can be viewed on a continuum between a sudden turn in a single step, to running at a constant speed around a circle (or arc), with each step contributing equally to the turn (bend running). Whether a discrete turning step or a stepping around continuous curve, however, the fundamental mechanical principles are similar.

Because we are considering turns with legs, not wheels, with stance punctuated by swing phases, changes in velocity heading are likely to be discontinuous. From this perspective, bend running is like stepping around a polygon (Figure 11.2).

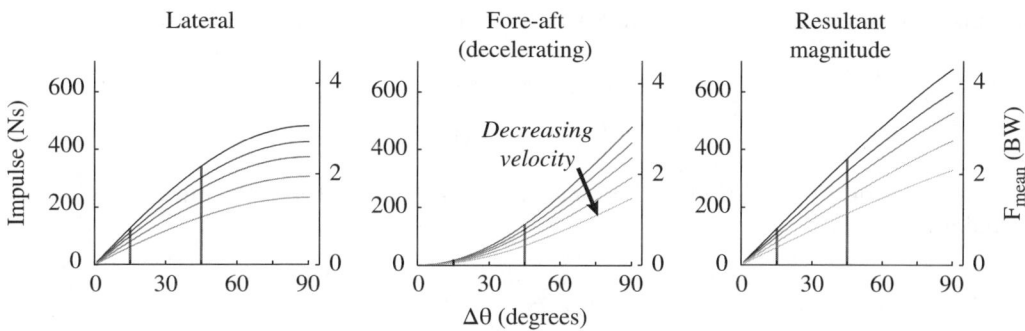

FIGURE 11.3

Ground reaction impulse (left axis) required for each turning step for an 80 kg human running around a bend over a range of velocities (contour lines, velocity decreasing from 6.0 to 2.9 m/s). The mean force during stance (right axis, in units of body weight BW) was calculated by dividing the impulse by the stance period, assuming a fixed stance duration of 0.2 sec. If leg forces are limiting, the runner can use a number of strategies to avoid force limits: 1) slow down (contour lines), 2) divide the heading change into more steps, decreasing theta per step (moving left on the x-axis), or 3) increase stance duration (not shown, but would change right axis scaling). Humans use all three of these strategies when running around very tight bends (Chang and Kram, 2007).

This is particularly true for bends of tight radius. For example, consider a human running around two circles: a large circle of 6 m radius, and a tight circle of 1 m radius. If we assume that the runner maintains a constant step length, it would take about 24 steps to complete the large circle, and only four steps to complete the small circle (approximating a square). The velocity heading per step ($\Delta\theta$) depends on the number of steps taken to complete a full circle (360°).

In the example above, running around the 6 m radius circle would require $\Delta\theta = 15°$, whereas running around the 1 m circle would require $\Delta\theta = 90°$. In fact, runners do not maintain constant step length in bend running, instead using shorter steps in tighter bends (Chang and Kram, 2007). In very tight bends, human runners decrease step length by about half, so that they undergo $\Delta\theta = 46°$ while running around a 1 m radius circle. We will discuss a possible reason for this below.

A change in velocity heading involves acceleration perpendicular to the original velocity heading, so requires a lateral force F_{Lat}. If the total velocity is to remain the same, there must also be net deceleration along the axis of the original heading and, thus, a for-aft force ($-F_{FA}$) (Figure 11.2). The force impulse required in each direction can be calculated based on impulse-momentum balance if $\Delta\theta$ and the initial velocity (V_i) are known:

$$|\Delta V_{Lat}| = |V_i| \sin(\Delta\theta)$$

$$|\Delta V_{FA}| = |V_i|(1 - \cos(\Delta\theta))$$

Impulse-momentum balance applied to each direction with body mass (m) yields:

$$\int F_{Lat} dt = m\Delta V_{Lat}$$

$$\int F_{FA} dt = m\Delta V_{FA}$$

If stance period (T_{st}) is known, the mean ground reaction force during stance can be calculated:

$$\overline{F_{Lat}} = \frac{m\Delta V_{Lat}}{T_{st}}$$

$$\overline{F_{FA}} = \frac{m\Delta V_{FA}}{T_{st}}$$

Similarly, even if the initial and final velocities are different, the impulse and mean stance force required for any turn could be calculated based on vector sums.

If $\Delta\theta$ is small, the decelerating force in the $\overline{F_{FA}}$ direction is small relative to $\overline{F_{Lat}}$, but the ratio $\overline{F_{FA}} / \overline{F_{Lat}}$ approaches unity for $\Delta\theta = 90°$ and exceeds unity for turns greater than 90°. Of course, animals do not necessarily maintain the same velocity from the start to the end of the turn, so the actual leg forces will also depend on the net velocity change.

The ability of an animal to turn sharply could be a critical factor in predator-prey chases. However, if leg forces are limiting, it may be necessary to minimize the mean leg forces required by dividing the turn into multiple steps with a smaller $\Delta\theta$ per step. Running in a large 90° arc requires lower mean leg forces than taking a sharp 90° turn in a single step, because the total impulse required is distributed over multiple steps and a longer total period.

Running around circles with a small radius might be more like a polygon than a circle but, nevertheless, the circular approximation is useful. In circular bend running, each step involves the same average acceleration toward the center of the circle, with a magnitude $a_c = V^2/r$. Similar to the polygon-based analysis above, this observation can be used to calculate the mean horizontal forces required: $\overline{F_C} = mV^2/r$. This approach has provided evidence that maximum leg forces limit top running speeds in humans (Greene and McMahon, 1979; Greene, 1985; Alexander, 2002; Usherwood and Wilson, 2006).

Greene and McMahon, (1979) developed a model to predict decreases in maximum speed associated with bend running of varying path radius, based on this assumption of limited leg force. If the vector sum of the maximum stance mean leg force is assumed to be fixed, then the increase in $\overline{F_C}$ requires a corresponding decrease in the vertical force, $\overline{F_{Vert}}$. Maintaining body weight support with a reduced $\overline{F_{Vert}}$ requires an increase in duty factor. If the distance travelled during stance is constant, and either stride period (Greene and McMahon, 1979) or leg swing period (Usherwood and Wilson, 2006) is fixed, then the increase in duty factor is associated with a decrease in speed. Human runners slow down around bends in a manner consistent with a leg force limit (Greene, 1985; Usherwood and Wilson, 2006).

However, it appears that leg force limits are not the only factor limiting speed in bend running. Green and McMahon's (1979) model overestimates maximum speeds in tight bends. Alexander (2002) suggested that friction limits might become more important for sharper turns, explaining the deviation from Greene's model in tight bends. However, this idea has not been explicitly tested. Additionally, Chang and Kram (2007) found that peak resultant ground reaction forces decrease significantly in very tight bends, and to a greater extent for the inside leg than the outside leg. Interestingly, applying an external centripetal force does not allow humans to run at the same maximum speed as straight running, although it does reduce leg asymmetry (Chang and Kram, 2007). This study suggested that more complex biomechanical factors limit leg forces in very tight bends.

Quadrupeds and hexapods may have an advantage in turning, because they can use differential leg function to execute turns, separating leg force requirements for body weight support and re-direction of velocity heading (Jindrich and Full, 1999; Usherwood and Wilson, 2005; Chang and Kram, 2007). This idea is supported by the observation that racing greyhounds do not change duty factor or slow down when running around bends (Usherwood and Wilson, 2005). Mice executing 90° turns do slow down, but not as much as humans do when running around similar dimensionless bends (Walter, 2003). These observations suggest that the factors influencing turning mechanics and maximum running speeds differ between bipeds, quadrupeds and hexapods.

Similar to pitch stability in acceleration as discussed above, maximum turning rate might also be constrained by a roll stability limit. The lateral forces required to accelerate in the new direction generate a rolling moment about the CoM (Alexander, 2002). The maximum centripetal acceleration $(\overline{a_C} = V^2/r)$ before reaching the roll stability limit depends on the ratio of step width (w) to body CoM height (h) (Figure 11.1) (Alexander, 2002):

$$\overline{a_c} < \frac{gw}{2h}$$

The rolling moment can be minimized by aligning the ground reaction force vector with the CoM, either by shifting the origin of the ground reaction force vector, leaning into the turn to shift the CoM position, or a combination of these. Animals with low step width relative to CoM height must lean relatively more to avoid rolling (Alexander, 2002).

The second requirement of turning is to reorient the body to face the new velocity heading (Jindrich and Full, 1999). The timing of body reorientation does not need to match that of velocity heading change, and tends to vary among animals (Jindrich and Full, 1999; Walter, 2003). Many analyses of turning assume that animals reorient by exerting an external torque to impart angular velocity (Lee *et al.*, 2001; Walter, 2003; Jindrich *et al.*, 2006, 2007; Hutchinson *et al.*, 2007). The torque required to achieve a given angular acceleration is proportional to the moment of inertia (MOI) about the turning axis. Angular velocity is undesirable at the end of the turn, so any yaw rotational energy imparted must be dissipated in some manner. As discussed in further detail below, evidence suggests that turning does, sometimes, involve external torques, so yaw stability may be an important factor in turning behavior.

It is important to note, however, that external torque is not absolutely required to reorient the body. An astronaut in space can reorient without external torques, maintaining zero angular momentum of the body CoM, by moving parts of the body with respect to each other. Likewise, a diver can reorient during the dive and enter the water with zero rotation, and a falling cat can right itself without violating the law of conservation of angular momentum. It is possible that turning animals also do this, for example, by changing velocity heading during stance and reorienting the body during the aerial phase. Whether or not animals actually do this during turning maneuvers remains unknown.

Constraints due to morphology and leg sequencing might make it difficult for some animals to exert the forces required to change heading without exerting a concurrent external yaw torque on the body (Walter, 2003; Jindrich and Qiao, 2009). The external torque can either assist or oppose reorientation of the body in the direction of the turn, depending on the geometry of foot placement relative to the CoM. Animals can adjust foot placement to avoid torques in the undesired direction, but this may not always be possible, due to anatomy or leg sequencing. Animals with larger numbers of legs, such as quadrupeds and hexapods, may have more flexibility in leg positioning with respect to the CoM, allowing them to tune the torques exerted more precisely (Jindrich and Full, 1999).

Animals can also manipulate angular accelerations resulting from the external torque by behaviorally adjusting their MOI about the vertical axis (Walter, 2003). This can be achieved by flexing or extending the trunk, pitching the trunk to a more vertical or horizontal orientation, or moving the limb and head segments relative to the CoM. By adjusting MOI relative to leg forces throughout a turning maneuver, animals could control body rotation and minimize undesired angular accelerations.

Thus, just as for acceleration, turning performance depends on the complex interplay of mechanical, anatomical and behavioral factors. There are a number of different behavioral strategies that animals can use to achieve a desired turning behavior within physiological and mechanical limits. An animal might execute a turn over multiple steps to minimize leg force requirements, maintain stability, and minimize injury risk, if these benefits outweigh the cost of a slower turn. Animals can control external torques through foot placement and leg sequencing, and can control body angular velocity by adjusting moment of inertia.

Some morphological features that increase acceleration limits also increase maximum turning rates; for example, having a low CoM height improves both pitch and roll stability. However, some morphological specializations for one are likely to limit performance in the other. For example, animals with upright, parasagittal limb posture might have differing capacity to exert forces in the fore-aft and medio-lateral directions, compared with those with a sprawling posture. Such differences likely influence the mechanical limits experienced for turning and acceleration, and might necessitate the use of different strategies for these behaviors.

11.4 THE ROLE OF INTRINSIC MECHANICS FOR STABILITY AND ROBUSTNESS OF LOCOMOTION

"Stability" and "robustness" of terrestrial locomotion are challenging topics, in part because it is sometimes difficult to identify with the most relevant task goals and performance variables. Depending on behavioral context, it might be more important to stabilize forward speed, vertical oscillations, rotational motions, or total mechanical energy, or to avoid excessive forces in the legs. Most locomotor behaviors could be achieved using multiple strategies, and animals might choose different strategies, depending on context.

Some factors that might influence the animal's movement strategy include: accuracy of knowledge about the environment and ability to anticipate any upcoming changes in terrain; the movement goals (such as speed, economy, or "smoothness" of gait); and the intrinsic mechanical interaction of the legs and body with the environment.

11.4.1 Some definitions

The concepts of stability and robustness are used in the literature with numerous possible specific meanings. So, before discussing recent research highlights in the area, it may first help to discuss the meaning of terms as they will be used here.

What is stability?

Locomotion typically involves periodic movement patterns (a "periodic trajectory"). In general, dynamic stability is defined on the basis of whether or not, and how quickly, the system returns to its original periodic trajectory following a perturbation. A system is stable (or "asymptotically stable") if it returns to the original periodic trajectory following a perturbation. It is "neutrally" stable if the perturbation remains and neither grows nor decays over time. For example, this is the case if a push forward causes an animal to speed up and continue running at a higher speed. The system is unstable if a perturbation is amplified over time, so that the movement pattern increasingly deviates from the original periodic trajectory (which would eventually lead to catastrophic failure or a fall).

The concepts of static and dynamic stability are related, but different. Static stability refers to whether or not the system returns to its original static position after a perturbation. It can typically be maintained as long as the body CoM remains projected over the base of support created by the feet. This is easier when standing on multiple legs rather than one, and with a wide stance rather than a narrow one.

What is robustness?

In biology, "robust" may often simply refer to an animal with a strong, heavy musculoskeletal build. However, in engineering and system analysis, it generally refers to the ability to maintain performance within an acceptable range in the face of disturbances or uncertainty (such as sensor noise). Robustness is a measure of how large the disturbances can be before the system fails to operate within the acceptable range. We might ask, for example, whether an animal can meet a set of locomotion criteria (e.g., not falling, not breaking, moving forward at an acceptable speed), in the presence of disturbances from the environment, with noisy, incomplete or incorrect sensory information, or changes in the mechanical system itself.

In many cases, robustness may be a more relevant measure than stability for non-steady legged locomotion. Animals do not necessarily need to return to the original periodic trajectory, but do need to avoid certain limits. Robustness of a model can also refer to how well that particular model of the system matches the actual system, despite uncertainty in the model itself or violation of some of the underlying assumptions of the model.

In the analysis of dynamic systems, robustness is sometimes measured on the basis of the size of the "basin of attraction" – a measure of how large a perturbation can be before the system no longer stabilizes to the original periodic trajectory. However, this definition might be too limited for animal movement, because it is not critical that animals return to the original trajectory. Indeed, animals might switch among different behaviors, depending on context, which would require a different dynamic model to approximate and would possibly make the stability analysis of the original model irrelevant.

Practical measures of robustness may need to be defined on a task-specific basis for animal movement, based on the likely limiting factors for the system. At high speeds, for example, animals may operate near peak force limits, so the most relevant robustness measure might be the maximum perturbation before the animal reaches or exceeds force limits. At slower speeds, on the other hand, falls may be more likely, making the most relevant robustness measure the maximum perturbation size before a fall occurs.

What is sensitivity?

The concept of sensitivity is inversely related to robustness. Sensitivity refers to the variation in the output of a system (or model) that results from variation in an input. If a small perturbation in an input (e.g., sensory noise or a terrain change) results in a large change in task-relevant output behavior, the system has high sensitivity and low robustness to the input parameter in question.

In understanding a system through models, it is often very useful to know how sensitive (and robust) the model is to potential error in inputs and parameters. Such a sensitivity analysis can achieve two goals:

1. It can help to identify whether the model is adequate for the scientific question, by evaluating whether the outputs can be predicted reliably, given the uncertainty in any measurements used as inputs.

2. Once a model has been validated against a reference range of data (steady locomotion, for example), variation in inputs can be used to generate quantitative hypotheses about how the animal will respond to specific perturbations.

11.4.2 Measures of sensitivity and robustness

Sensitivity and robustness metrics have recently been suggested as potential alternatives to traditional stability analysis in the evaluation of legged animals and robots. One recent proposal is Gait Sensitivity Norm (Hobbelen and Wisse, 2007). This metric quantifies the sensitivity of gait indicators (such as step time or ground clearance) to terrain disturbances. At least in some legged robot experiments, GSN is a better predictor of whether or not the system will fall, in uneven terrain, than more traditional approaches to stability analysis (Hobbelen and Wisse, 2007).

Another recent suggestion is the calculation of a Normalized Maximum Drop (NMD – see Section 11.4.6), the maximum vertical drop in terrain relative to leg length ($\Delta H_{max}/L_{leg}$), beyond which the leg completely misses a stance event (Daley and Usherwood, 2010). This measure is based on the idea that the most fundamental criterion for animal locomotion is avoidance of an immediate, catastrophic fall. If the runner misses a stance event entirely, the body continues to fall until either the next leg or the body contacts the ground, whichever occurs first. Drops larger than NMD increase the likelihood of a catastrophic fall. If the animal avoids an immediate fall, adjustments can be made to recover in subsequent steps.

The first of the above metrics, Gait Sensitivity Norm, is a continuous quantitative measure that can quantify the system response to any given perturbation. The second metric, NMD, identifies a boundary condition. If a missed stance event occurs, this may require different and more dramatic recovery measures in subsequent steps, thus triggering a switch to a different movement trajectory. However, these are just two recent suggestions among numerous possibilities for evaluating the robustness or sensitivity of a particular system to perturbations. It is not yet clear which measures are most relevant to animal locomotion.

One particular challenge for identifying relevant measures of stability and robustness for animals is that the immediate response to a perturbation is likely to differ substantially from the response in subsequent steps. Many stability and robustness measures used for theoretical models and physical robots assume that the system will maintain the same target periodic trajectory following a perturbation. We do not know if animals "care" whether or not they return to the original periodic trajectory. It may not matter if they speed up or slow down a little, or bounce higher or lower, as long as they do not fall, break their leg, or get eaten immediately. Thus, it is a continuing challenge to identify criteria for stability and robustness that are meaningful for animals.

11.4.3 What do we learn about stability from simple models of running?

The spring-loaded inverted-pendulum model (abbreviated SLIP, aka mass-spring model) describes the forces and energy fluctuations of steady forward walking and running across a diverse spectrum of terrestrial animals (Blickhan, 1989; McMahon and Cheng, 1990; Farley *et al.*, 1993; Schmitt *et al.*, 2002; Geyer *et al.*, 2006). As discussed in previous chapters, this empirical observation highlights fundamental mechanical principles governing the movement strategies used by animals. These simple dynamic models of locomotion are sometimes referred to as "templates" (Full and Koditschek, 1999), because they represent a minimalistic mathematical description of the overall goal task, without the underlying neural and musculoskeletal details. More complex models that consider more detailed neuro-musculo-skeletal mechanisms are referred to as "anchors" in the terminology of Full and Koditschek (1999) (Figure 11.4). Simple models are useful, because they represent a testable hypothesis of the overall strategy used to accomplish a given locomotor task. The predictions of these models can be tested against experimental measures of locomotor dynamics under varied terrain conditions.

Many of the specifics of why animals move the way they do remain poorly understood. Although the body and leg follow relatively simple movement patterns, the underlying joint and muscle-tendon mechanics are much more complex. Individual muscles of the legs – particularly the large proximal muscles at the hip and knee – do not undergo spring-like contraction cycles, and likely produce a substantial amount of positive and negative muscle work to accomplish this pseudo-spring-like leg behavior (Gillis *et al.*, 2005; Biewener and Daley, 2007).

Why do animals choose to follow spring-like whole-body dynamics during running, even though their legs are not actually composed of passive springs? An actual spring-like leg would benefit energetically from this movement pattern by passively storing and releasing energy form the spring, minimizing the amount of active work done

Simple whole-body model

Simplified actuator model

Musculoskeletal model

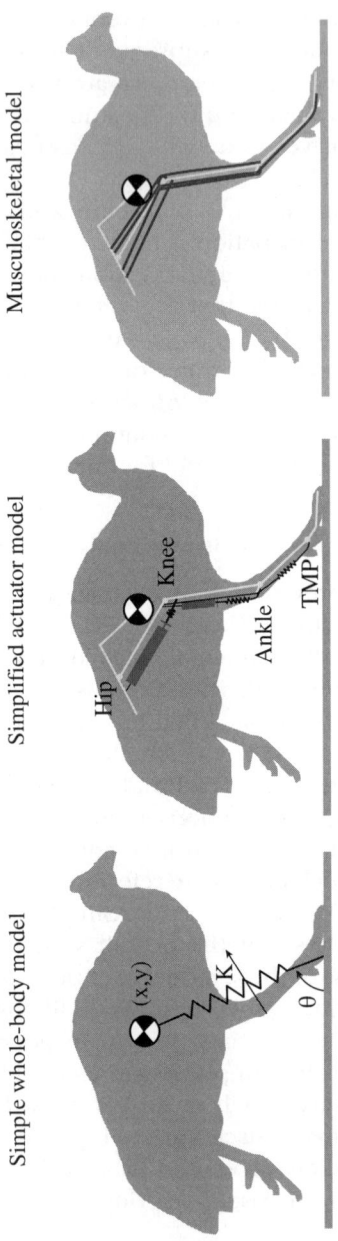

Hip

Knee

Ankle

TMP

(x,y)

K

θ

FIGURE 11.4

Schematic illustrations of different levels of analysis for understanding leg function. The simplest models of leg function require only a few parameters to describe the overall leg dynamics. An example of this is the mass-spring model for running (left), also called the spring-loaded-inverted pendulum (SLIP) (Blickhan, 1989; McMahon and Cheng, 1990; Full and Koditschek, 1999; Full *et al.*, 2002; Geyer *et al.*, 2006), sometimes called a "*template*" model (see section 4.3.3) (Full and Koditschek, 1999). Models of higher complexity ("*anchored*" models, in the terminology of Full and Koditschek) are required to understand the underlying neuromuscular mechanisms. Simplified actuator models can be used to assess the relative contributions of individual joints to the overall actuation and compliance of the leg (middle) (Lee *et al.*, 2008). Some questions may require either detailed musculoskeletal simulations (right) (e.g., (Hutchinson, 2004a, 2004b; Bunderson *et al.*, 2008), direct experimental measures of individual muscles (Biewener and Corning, 2001; Biewener *et al.*, 2004; Daley and Biewener, 2011, in press), or both. In many cases, experimental measures, combined with simple models, can provide insight without the need for more complex models, and we primarily focus on simpler models in this book.

in the system. Although some specialized animals, such as ungulates, kangaroos and wallabies, do have substantial elastic tissue that acts to cycle elastic energy (Alexander and Vernon, 1975; Alexander *et al.*, 1982; Biewener and Baudinette, 1995), most are likely to exhibit more modest amounts of elastic energy cycling (Biewener and Corning, 2001; Daley and Biewener, 2003).

Furthermore, models based on minimizing the mechanical work costs of locomotion predict that animals should run with much stiffer legs than they actually do (Srinivasan and Ruina, 2006). One potential explanation for the relatively compliant spring-like motion of running animals is based on energetic considerations that include energy losses associated with bouncing viscera (Daley and Usherwood, 2010 – and see Chapter 6). However, this idea has not been experimentally tested.

Another possible explanation, which is not mutually exclusive, is that there are stability and neural control benefits to controlling the body around this dynamically simple gait (Full and Koditschek, 1999; Schmitt and Holmes, 2000; Full *et al.*, 2002; Seyfarth *et al.*, 2002, 2003; Koditschek *et al.*, 2004). Mass-spring running dynamics can provide intrinsic stability to perturbations (at least to small perturbations), minimizing the need for rapid neural control to prevent falls. Rather than continuously monitoring feedback and updating neural commands, the system could rely on passive stability to recover from small perturbations, adjusting control only for larger perturbations. This may be particularly important for large animals, which have relatively slower reflex response times, so might have to rely to a greater extent on feed-forward trajectories with intrinsic stability (More *et al.*, 2010).

Analysis of the SLIP model has revealed some simple feed-forward mechanisms that can provide intrinsic stability in uneven terrain. Animals retract the leg backwards relative to the body just before it contacts the ground (Figure 11.5). This leg retraction in late swing phase might serve two roles:

1. "ground speed matching" to minimize the collision forces of the foot against the ground
2. increasing the intrinsic stability of locomotion (Herr *et al.*, 2002; Seyfarth *et al.*, 2003; Blum *et al.*, 2010).

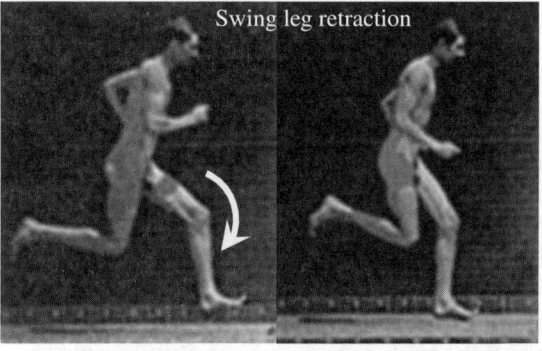

FIGURE 11.5

Swing leg retraction occurs during the last part of the swing phase, just before the leg contacts the ground (stills from plate 62, Muybridge, 1887)).

Leg retraction leads to automatic adjustment of contact angle upon encountering a sudden change in terrain height (Figure 11.6) (Seyfarth *et al.*, 2003; Daley and Biewener, 2006; Grimmer *et al.*, 2008), which improves stability of the SLIP model to terrain perturbations (Seyfarth *et al.*, 2003; Blum *et al.*, 2010). There are other simple

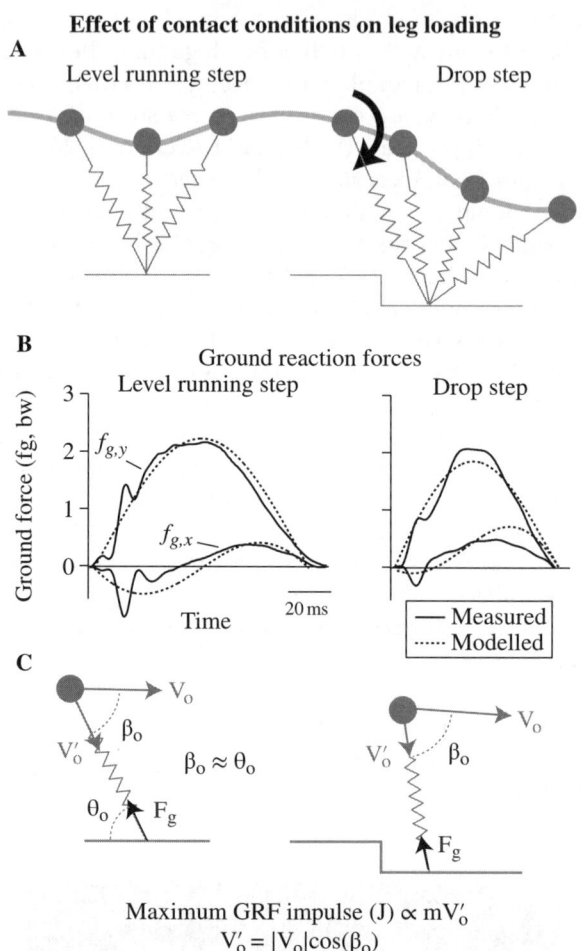

FIGURE 11.6

Leg contact conditions influence leg loading during non-steady locomotion. (**A**) Swing leg retraction leads to automatic adjustment of leg contact angle with changes in terrain height (see sections 4.3.3 and 4.3.4 – example from Daley and Biewener, 2006). (**B**) The SLIP model, with swing retraction, provides reasonably good prediction of the ground reaction forces, during both level running and running over a drop in terrain. (**C**) The maximum impulse of a simple compression leg is proportional to the cosine of the angle between the leg and the body velocity vector. If this angle becomes smaller, leg loading increases and, if this angle becomes larger, loading decreases. A limit is reached when $\beta_o = 90°$, the leg cannot be loaded in compression for simple geometric reasons, and stance phase is missed entirely (Daley and Biewener 2006; Daley and Usherwood 2010).

mechanisms that can improve stability, including feed-forward adjustment of leg length and leg stiffness during late swing (Blum *et al.*, 2010). Adjustments in leg stiffness can also occur through intrinsic mechanical effects when the leg contacts the ground. For example, a sudden change in substrate stiffness can lead to altered joint posture, resulting in changes in mechanical advantage and joint stiffness (Moritz and Farley, 2004).

11.4.4 Limitations to stability analysis of simple models

Any specific dynamic model or template is likely to be valid only over the range of conditions for which the model reasonably matches the true system behavior. For example, the spring-loaded-inverted pendulum (SLIP) model predicts both mechanics and local dynamic stability characteristics of running over uniform terrain, or terrain with small perturbations. However, this analysis is unlikely to predict stability of the real animal for large terrain perturbations. There are two reasons for this:

1. typical stability analyses are based on small local perturbations; and
2. the complex animal musculoskeletal systems are unlikely to approximate the simplified model when operating far outside of the range of data used to fit the model.

Since legged locomotion often involves negotiation of uneven terrain with relatively large perturbations, traditional stability analyses may not be the best way to evaluate the stability of terrestrial animals and legged robots. This is one reason why future work is likely to include analyses of robustness and sensitivity in addition to standard stability analyses (see Section 11.4.2).

It is likely that legged animals switch among locomotion strategies with different target trajectories depending on context. For example, although animals sometimes maintain the same body CoM trajectory in the face of substrate perturbations (Ferris *et al.*, 1998, 1999; Seyfarth *et al.*, 2003; Moritz and Farley, 2005, 2006; Daley and Biewener, 2006; Grimmer *et al.*, 2008), they sometimes stabilize to a different trajectory (Marigold and Patla, 2005; Daley and Biewener, 2006; Daley *et al.*, 2009). The animal can also produce or absorb net energy during the response to a perturbation (Daley and Biewener, 2006; Daley *et al.*, 2007). Such non-conservative energy responses cannot be modeled with a simple mass-spring template.

The specific strategy used by an animal likely depends on a number of factors, including terrain conditions, behavioral context, whether or not a perturbation is anticipated, and the posture and leg loading at the time of ground contact (Marigold and Patla, 2002, 2005; Daley and Biewener, 2006; Daley *et al.*, 2007). Recent theoretical models have begun to incorporate the possibility of non-conservative system behavior, through models that exhibit posture or load dependent leg actuation (Schmitt and Clark, 2009), or a combination of actuation at the hip and damping in the leg (Seipel and Holmes, 2007). These models suggest that a non-conservative energy management strategy might improve robustness to terrain perturbations, although perhaps at a cost to economy. Future work is likely to consider hybrid dynamic models that switch among potential "templates", depending on behavioral and environmental context.

11.4.5 The relationship between ground contact conditions and leg mechanics on uneven terrain

Uneven terrain is an interesting perturbation, because it is ecologically relevant, physically simple, and it relates reasonably well to engineering approaches to system identification and modeling. It is also interestingly complex, because animals could negotiate uneven terrain in quite different ways, and could switch among different strategies and blend them to varying degrees, depending on context. For example, a range of different strategies could be used to negotiate a single obstacle step (Figure 11.7).

The strategy adopted in any given situation will likely depend on morphology, movement speed, terrain environment and behavioral context. Yet, despite these complexities, an emerging theme in recent studies is that ground contact conditions (leg posture and body velocity), have a large effect on the forces and work done by the leg during the stance phase (e.g., Daley and Biewener, 2006, 2011; Daley *et al.*, 2009). This suggests that basic geometrical considerations can be used to identify potential limiting conditions for falls and injury in uneven terrain.

The SLIP model and simple rigid leg models for locomotion assume that the legs resist only compressive loads along the long axis (no tension, shear or torque). Under these assumptions, loading of the leg is related to the angle between the leg and the body velocity vector (angle β, Figure 11.6). This angle, β, is equal to the sum of the leg angle (θ) and the body velocity angle (α): $\beta = \theta + \alpha$. Since the leg can only resist axial compressive loads, the maximum impulse (and maximum change in momentum) during stance is proportional to the magnitude of the velocity directed along the leg axis ($V_{o,||}$), which is a function of β_o:

$$V_{o,||} = |V_o| \cos(\beta_o)$$

In high-speed locomotion, the body velocity vector is near horizontal (forward velocity is high relative to vertical velocity, thus $\alpha \approx 0$). Consequently, β, at the time of ground contact (β_o), is nearly equal to the leg contact angle (θ_o). The larger the angle β_o, the smaller is the fraction of the body velocity that is directed along the long axis of the leg, which reduces the maximum leg impulse during stance.

Animal experiments and simple model simulations reveal that changes in either leg angle or body velocity at contact will lead to an altered ground reaction force impulse during stance (Daley and Biewener, 2006; Schmitt and Clark, 2009). When the animal encounters a change in terrain height, the change in angle β_o depends on the balance between the leg trajectory (leg retraction and/or extension), and the ballistic fall of the body until ground contact. If the leg is retracted very slowly, or not at all, V_o increases and β_o decreases, due to gravitational acceleration and the downward rotation of the body velocity vector during the fall. These two effects would result in an increase in leg loading.

If, on the other hand, the leg is retracted quickly, β_o increases, reducing leg loading. For example, a change in the β_o from 60° to 75° results in reduction in ground reaction force impulse of almost 50%. However, if β_o reaches 90°, the leg is not loaded in compression, and the stance event is missed entirely. This relationship results in an interesting trade-off: high leg retraction velocity during late swing

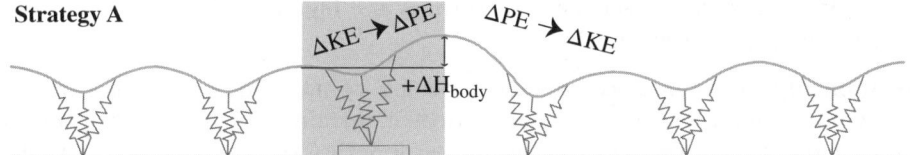

Strategy A

Exchange between kinetic (KE) and potential (PE) energy.

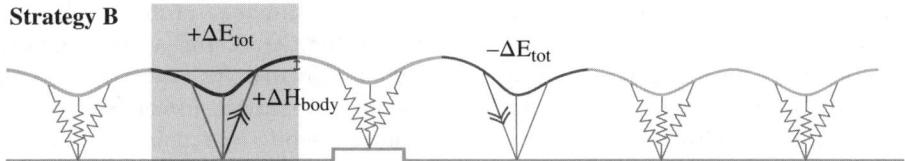

Strategy B

External work (ΔE_{tot}) before and after obstacle.

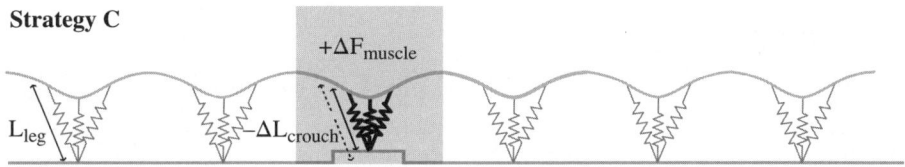

Strategy C

Crouched leg posture ($-\Delta L_{crouch}$); no change in body motion.

FIGURE 11.7

Examples of several different mechanical strategies for negotiating an obstacle in uneven terrain. The specific strategy used by the animal is likely to depend on potential trade-offs between stability, costs due to external work, and costs due to increase muscle forces (and force limits). **Strategy A** is an energy-conservative passive strategy, possible with the SLIP model, in which kinetic energy (KE) and gravitational potential energy (PE) are exchanged while negotiating the drop, resulting in fluctuations in bouncing height and velocity. In **Strategy B**, the runner increases body height through net positive work in anticipation of the step, in order to achieve a steady bounce on the obstacle step, absorbing energy again afterward. In **Strategy C**, the obstacle is accommodated by adopting a more crouched leg posture, avoiding any change in the whole body CoM trajectory. **Strategy A** is passive and energy conservative, but may require more steps for recovery. **Strategy B** involves net external mechanical work, which may incur some cost. **Strategy C** avoids external work and is stable from the perspective of the body CoM trajectory; however, it will require higher muscle forces due to altered gearing with a crouched leg posture.

protects the leg from high forces, but will cause the leg to miss a stance phase for large terrain drops, which might increase the likelihood of catastrophic falls. Low retraction velocity allows the leg to make contact for relatively large drops in terrain, but it requires the leg to resist relatively high forces, which may increase the likelihood of injury.

11.4.6 Compromises among economy, robustness and injury avoidance in uneven terrain

This relationship between body velocity and leg contact angle can also be used to identify a boundary condition as a measure of terrain robustness: the normalized maximum vertical drop (NMD), relative to leg length ($\Delta H_{max}/L_{leg}$), beyond which the leg completely misses a stance event (Figure 11.8) (Daley and Usherwood 2010).

Analysis of NMD for different gaits can help identify potential trade-offs in leg control for economy, robustness and injury avoidance. The estimates of NMD shown here assume constant leg length, constant leg retraction velocity (ω), and approximately zero vertical velocity at the expected time of foot contact. However, the quantity can also be calculated for more complex kinematic trajectories. The simple estimate here demonstrates that running with relatively crouched postures, while increasing the mechanical work of the leg (see Chapter 6), also provides greater robustness to terrain perturbations

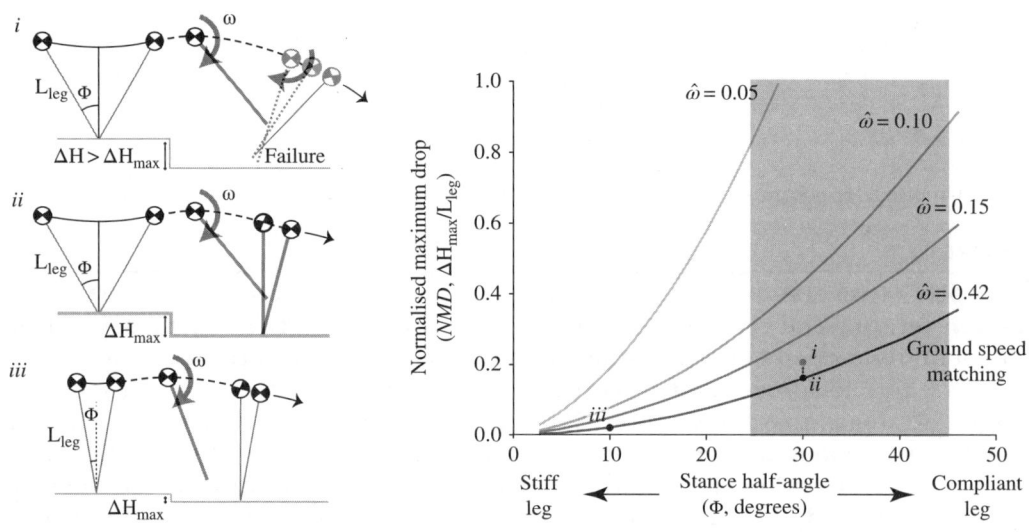

FIGURE 11.8

One example of a possible robustness boundary condition for locomotion over uneven terrain. When the leg is retracted in late swing, as illustrated in Figure 11.6, the leg reaches a steeper leg contact angle upon encountering a drop in terrain. A limit is reached when $\beta_o = 90°$, the leg cannot be loaded in compression for simple geometric reasons, and stance phase is missed entirely (*i*) (Daley and Biewener 2006; Daley and Usherwood 2010). This boundary condition can be estimated based on leg kinematics, and used to predict the Normalized Maximum Drop (NMD), the maximum terrain drop relative to leg length before a stance event is missed (*ii–iii*). For a given running speed, swing period and mean leg retraction velocity ($\bar{\omega}$, shown as dimensionless $\hat{\omega} = \bar{\omega} \big/ 2\pi \left(\sqrt{\left(\frac{2}{3} L_{leg} / g \right)} \right)$), compliant legs (A*ii*, with larger stance half-angle Φ), have higher NMD than stiff legs (A*iii*, with smaller Φ). Higher leg retraction velocities allow ground speed matching and might help minimize foot collision forces. However, they also result in lower NMD.

(Daley and Usherwood, 2010). Analysis of this boundary also highlights a trade-off between "terrain robustness" and "leg force robustness"; high ω allows "ground speed matching" and minimizes excessive forces with terrain variation, but also reduces NMD. Low ω allows higher NMD, but requires the leg to resist higher forces.

This highlights the importance of swing leg trajectory control to achieve an appropriate compromise between stability, economy and injury avoidance in uneven terrain. In addition to selecting among different possible constant leg retraction velocities, as shown in Figure 11.8, animals could also vary ω as a function of time, to achieve a compromise between terrain robustness and injury avoidance. As an example of one potential control strategy, animals could vary ω over time during a drop step to minimize peak force, while exerting sufficient impulse to achieve positive vertical velocity at the end of stance (Daley and Usherwood, 2010, supplementary online material).

High ω early in the fall would minimize peak forces and injury risk. Reducing ω with increasing fall time would ensure that the leg makes contact with the ground and provides enough weight support to achieve positive vertical velocity by the end of stance. However, this does require increasing peak forces for larger drop heights. If leg injury is a more serious risk than missing a single stance event, high ω throughout the fall may be a preferable strategy. On the other hand, if falls must be avoided at all costs and legs are operating at high safety factors, relatively low ω is predicted. Maximum ω is probably also constrained by mechanical and energetic factors, since the required work and force rates increase sharply with leg swing frequency, incurring a greater metabolic cost (Doke *et al.*, 2005).

Compromises among these factors might help explain differences in leg posture and morphology among animals that live in different environments. Such trade-offs could also be used to make predictions about context-dependent changes in locomotor strategies when animals move through different terrain conditions.

11.5 PROXIMAL-DISTAL INTER-JOINT COORDINATION IN NON-STEADY LOCOMOTION

The analyses in the previous examples focused on whole-body mechanical analysis and simple geometric limits in non-steady locomotor behaviors. Of course, the neuromuscular coordination underlying these behaviors is complex. The individual muscles, segments and joints must be coordinated to achieve the desired mechanical tasks of the whole limb and body. To address questions about how musculoskeletal function is coordinated within the limb, different levels of analysis are required (Figure 11.4). The appropriate level of investigation depends on the nature of the question, but good analyses are kept as simple as possible. The examples in this section highlight aspects of muscle and joint coordination within the limb during non-steady tasks.

Due to the complexity of the musculoskeletal system, there are often many different muscle coordination patterns that could be used to achieve a given task (Bernstein, 1967; Bunderson *et al.*, 2008; Ting *et al.*, 2009). In some cases, different solutions may be equivalent at the level of the whole limb, especially for relatively simple tasks. In other cases, however, the specific subset of muscles chosen to coordinate a particular task may have important implications for the stability of the system in response to perturbations (Bunderson *et al.*, 2008). Two activation patterns, using different subsets of muscles, may result in equivalent trajectories for steady locomotion. However, one pattern may result in a more stable response to perturbations than another.

The intrinsic force-velocity and force-length effects on muscle output will vary, depending on muscle-tendon architecture, and these effects could substantially influence the system's response to perturbations. Consequently, if the stability of the two activation patterns were considered, they would not be functionally equivalent, even though they might result in similar steady movement trajectories (Brown and Loeb, 2000). For this reason, perturbation studies can be important for revealing why steady locomotion is coordinated using certain combinations of muscles, rather than others.

It has long been observed that proximal and distal limb segments exhibit different musculoskeletal structure; muscle mass is concentrated proximally, and distal segments tend to be light and elongated, with low muscle mass and substantial elastic connective tissue (e.g., Hildebrand and Goslow, 2001). Additionally, the muscle-tendon architecture differs between proximal and distal segments. Proximal muscles tend to have relatively long fibers and short tendons, while distal muscles tend to have long tendons, shorter fibers and a pinnate fiber arrangement.

This pattern is also reflected in the mechanical function of the joints during locomotion. In steady locomotion, muscles at the proximal joints tend to produce energy, whereas those at the distal joints act as springs and dampers (Biewener and Daley, 2007; Daley et al., 2007; Lee et al., 2008). Concentrating the massive actuators at the proximal joints likely reduces the cost of oscillating the legs. Recent studies also suggest that this mechanical design has implications for stability, agility and robustness in non-steady locomotion.

As discussed in the introduction, non-steady tasks fall into two categories:

1. anticipated, goal directed change in body CoM energy; and
2. stabilizing responses to maintain a goal trajectory in the face of perturbations.

Recent studies suggest that muscles at proximal joints contribute most of the work for anticipated non-steady tasks (such as incline locomotion, acceleration, and jumping), while distal muscles absorb and produce energy to provide stability in response to terrain perturbations. For example, proximal joints appear to contribute the majority of the work for locomotion on inclines and declines (Daley and Biewener, 2003; Biewener et al., 2004; Lee et al., 2008).

When moving uphill or downhill, one might assume that the overall movement goal is to maintain a constant mean velocity, while providing the mechanical work required to increase or decrease potential energy. The total net mechanical work required per step can be calculated on the basis of the step length (L_{step}) and the angle of the slope (φ): $PE = mgL_{step}\sin(\varphi)$ (Daley and Biewener, 2003; Gabaldón et al., 2004). This simple analysis allows the work contribution of individual muscles to be compared to the overall work demand of the task.

In vivo measurements of muscle fascicle and tendon strain allow estimates of the work produced or absorbed by individual muscles. However, these measurements are currently feasible only in distal muscles that have long free tendons (Biewener and Baudinette, 1995; Roberts et al., 1997; Biewener et al., 2004; Gabaldón et al., 2004; McGuigan et al., 2009). Such measurements show that distal muscles do adjust work production for incline and decline locomotion in some animals (Figure 11.9), but they tend to do a small fraction of the total limb work (Daley and Biewener, 2003; Gabaldón et al., 2004; Biewener and Daley, 2007; McGuigan et al., 2009). This implies that

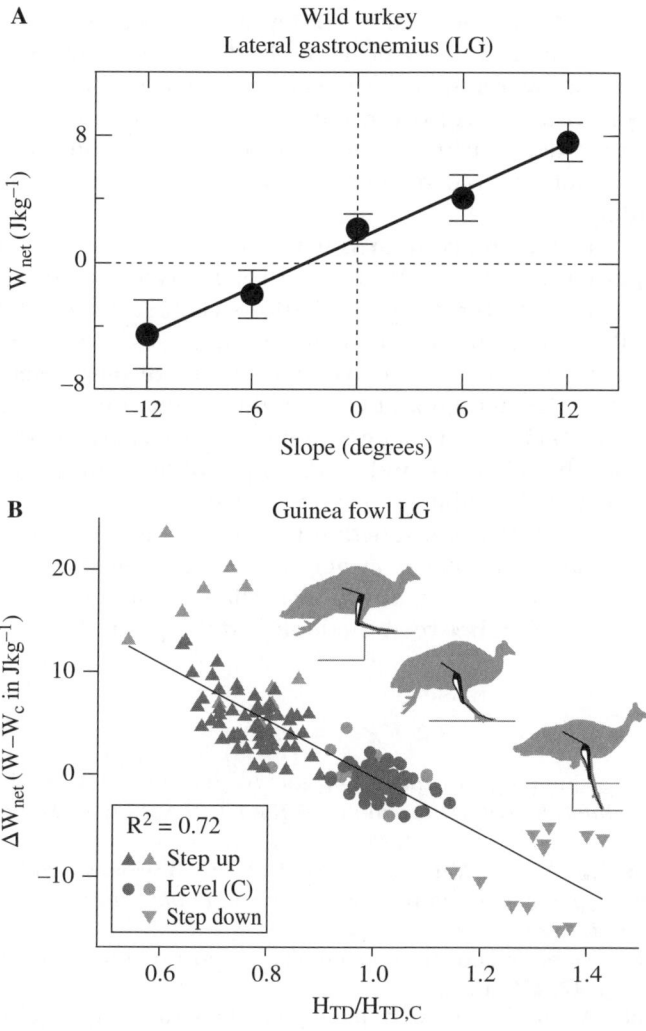

FIGURE 11.9

(**A**) Changes in lateral gastrocnemius (LG) muscle net work output (W_{net}) during incline and decline locomotion in the turkey (from Gabaldón *et al.*, 2004). (**B**) In uneven terrain, the changes in LG muscle net work output (as a change from level control ΔW_{net}), in relation to the hip height at the time of leg touch-down (H_{TD}) (from Daley *et al.*, 2009; Daley and Biewener, 2011). See section 4.4 for further discussion.

muscles at proximal joints contribute the majority of the work demand for these antici-pated non-steady locomotor behaviors, which is consistent with measures of strain and activation in proximal muscles (Gillis and Biewener, 2002; Biewener and Daley, 2007; McGowan *et al.*, 2007).

The distal joints appear to play important roles in stability, providing damping to protect the legs from impact and injury (Wilson *et al.*, 2001; Lee *et al.*, 2008), and exhibiting rapid changes in work output to stabilize the body in negotiating terrain perturbations (Daley *et al.*, 2007, 2009; Daley and Biewener, 2011). The pennate short-fibered muscle architecture typical of distal limb muscles is not suitable for control of joint position over large excursions, because the fascicles can actively contract by only a short distance. However, recent work suggests that this type of architecture allows for load-dependent changes in gearing at the muscle fiber level (Azizi *et al.*, 2008). This can allow rapid switching of muscle mechanical output in response to changes in loading.

Pennate muscle architecture, combined with the low inertia of the distal segments, may explain why some distal muscles exhibit relatively large changes in force and work output in response to terrain perturbations (Figure 11.9) (Daley *et al.*, 2007, 2009; Daley and Biewener, 2011). When the leg contacts the ground with a more crouched or extended leg posture in uneven terrain, the distal extensor muscles operate at very different fascicle length and velocity during contraction (Daley *et al.*, 2009; Daley and Biewener, 2011). These changes in muscle fascicle strain during loading are associated with rapid shifts in force and work output of these muscles with changes in leg posture, which may help stabilize locomotion over uneven terrain.

Consideration of the mechanical function in both steady and non-steady behaviors suggests that proximal joints play a larger role in controlled actuation, and distal joints play a greater role in stabilizing functions, acting as springs, dampers or actuators, depending on the interaction between the limb and the ground.

▌REFERENCES

Aerts P (1998). Vertical jumping in *Galago senegalenis*: the quest for an obligate mechanical power amplifier. *Philosophical Transactions of the Royal Society of London, Series B* **353**, 1607–1620.

Aerts P, Van Damme R, D'Août K, Van Hooydonk B (2003). Bipedalism in lizards: whole-body modelling reveals a possible spandrel. *Philosophical Transactions of the Royal Society of London, Series B* **358**, 1525–1533.

Alexander RM (2002). Stability and manoeuvrability of terrestrial vertebrates. *Integrative and Comparative Biology* **42**, 158–164.

Alexander RM, Vernon A (1975). Mechanics of Hopping by Kangaroos (Macropodidae). *Journal of Zoology London* **177**, 265–303.

Alexander RM, Maloiy GMO, Ker RF, Jayes AS(1982). The Role of Tendon Elasticity in the Locomotion of the Camel (*Camelus dromedarius*). *Journal of Zoology London* **198**, 293–313.

Azizi E, Brainerd EL, Roberts TJ (2008). Variable gearing in pennate muscles. *Proceedings of the National Academy of Sciences* **105**, 1745–1750.

Azizi E, Roberts TJ (2010). Muscle performance during frog jumping: influence of elasticity on muscle operating lengths. *Proceedings of the Royal Society B* **277**, 1523–1530.

Bernstein N (1967). *The Co-ordination and Regulation of Movements*. Oxford, UK, Pergamo.

Biewener AA, Baudinette RV (1995). In vivo muscle force and elastic energy storage during steady-speed hopping of tammar wallabies (*Macropus eugenii*). *Journal of Experimental Biology* **198**, 1829–1841.

Biewener AA, WR Corning (2001). Dynamics of mallard (*Anas platyrhynchos*) gastrocnemius function during swimming versus terrestrial locomotion. *Journal of Experimental Biology* **204**, 1745–1756.

Biewener AA, MA Daley (2007). Unsteady locomotion: integrating muscle function with whole body dynamics and neuromuscular control. *Journal of Experimental Biology* **210**, 2949–2960.

Biewener AA, McGowan C, Card GM, Baudinette RV (2004). Dynamics of leg muscle function in tammar wallabies (*M. eugenii*) during level versus incline hopping. *Journal of Experimental Biology* **207**, 211–223.

Blickhan R (1989). The Spring Mass Model for Running and Hopping. *Journal of Biomechanics* **22**, 1217–1227.

Blum Y, Lipfert SW, Rummel J, Seyfarth A (2010). Swing leg control in human running. *Bioinspiration & Biomimetics* **5**, 026006.

Brown IE, Loeb GE (2000). A reductionist approach to creating and using neuromechanical models.In: Winters JM and Crago PE (eds). *Biomechanics and Neural Control of Posture and Movement*, 148–163. New York, Springer-Verlag.

Bunderson NE, Burkholder TJ, Ting LH (2008). Reduction of neuromuscular redundancy for postural force generation using an intrinsic stability criterion. *Journal of Biomechanics* **41**, 1537–1544.

Cavagna GA, Thys H, Zamboni A (1976). The sources of external work in level walking and running. *Journal of Physiology* **262**, 639–657.

Chang Y-H, Kram R (2007). Limitations to maximum running speed on flat curves. *Journal of Experimental Biology* **210**, 971–982.

Clemente CJ, Withers PC, Thompson G, Lloyd D (2008). Why go bipedal? Locomotion and morphology in Australian agamid lizards. *Journal of Experimental Biology* **211**, 2058–2065.

Daley MA, Biewener AA (2003). Muscle force-length dynamics during level versus incline locomotion: a comparison of in vivo performance of two guinea fowl ankle extensors. *Journal of Experimental Biology* **206**, 2941–2958.

Daley MA, Biewener AA (2006). Running over rough terrain reveals limb control for intrinsic stability. *Proceedings of the National Academy of Sciences* **103**, 15681–15686.

Daley MA, Biewener AA (2011). Leg muscles that mediate stability: Mechanics and control of two distal extensor muscles during obstacle negotiation in the guinea fowl. *Philosophical Transactions of the Royal Society of London, Series B* **366**, 1580–1591.

Daley MA, Usherwood JR (2010). Two explanations for the compliant running paradox: reduced work of bouncing viscera and increased stability in uneven terrain. *Biology Letters* **6**, 418–421.

Daley MA, Usherwood JR, Felix G, Biewener AA (2006). Running over rough terrain: guinea fowl maintain dynamic stability despite a large unexpected change in substrate height. *Journal of Experimental Biology* **209**, 171–187.

Daley MA, Felix G, Biewener AA (2007). Running stability is enhanced by a proximo-distal gradient in joint neuromechanical control. *Journal of Experimental Biology* **210**, 383–394.

Daley MA, Voloshina A, Biewener AA (2009). The role of intrinsic muscle mechanics in the neuro-muscular control of stable running in the guinea fowl. *Journal of Physiology* **587**, 2693–2707.

Doke J, Donelan JM, Kuo AD (2005). Mechanics and energetics of swinging the human leg. *Journal of Experimental Biology* **208**, 439–445.

Farley CT, Glasheen J, McMahon TA (1993). Running springs: speed and animal size. *Journal of Experimental Biology* **185**, 71–86.

Ferris DP, Louie M, Farley CT (1998). Running in the real world: adjusting leg stiffness for different surfaces. *Proceedings of the Royal Society of London B* **265**, 989–994.

Ferris DP, Liang K, Farley CT (1999). Runners adjust leg stiffness for their first step on a new running surface. *Journal of Biomechanics* **32**, 787–794.

Full RJ, Koditschek DE (1999). Templates and Anchors: Neuromechanical hypotheses of legged locomotion on land. *Journal of Experimental Biology* **202**, 3325–3332.

Full RJ, Kubow T, Schmitt J, Holmes P, Koditschek D (2002). Quantifying dynamic stability and maneuverability in legged locomotion. *Integrative and Comparative Biology* **42**, 149–157.

Gabaldón AM, Nelson FE, Roberts TJ (2004). Mechanical function of two ankle extensors in wild turkeys: shifts from energy production to energy absorption during incline versus decline running. *Journal of Experimental Biology* **207**, 2277–2288.

Geyer H, Seyfarth A, Blickhan R (2006). Compliant leg behaviour explains basic dynamics of walking and running. *Proceedings of the Royal Society B* **273**, 2861–2867.

Gillis GB, Biewener AA (2002). Effects of surface grade on proximal hindlimb muscle strain and activation during rat locomotion. *Journal of Applied Physiology* **93**, 1731–1743.

Gillis GB, Flynn JP, McGuigan P, Biewener AA (2005). Patterns of strain and activation in the thigh muscles of goats across gaits during level locomotion. *Journal of Experimental Biology* **208**, 4599–4611.

Gray J (1944). Studies in the mechanics of the tetrapod skeleton. *Journal of Experimental Biology* **20**, 88–116.

Greene PR (1985). Running on flat turns: experiments, theory, and applications. *Journal of Biomechanical Engineering* **107**, 96–103.

Greene PR, McMahon TA (1979). Running in circles. *Physiologist* **22**, S35–S36.

Grimmer S, Ernst M, Günther M, Blickhan R (2008). Running on uneven ground: leg adjustment to vertical steps and self-stability. *Journal of Experimental Biology* **211**, 2989–3000.

Hedrick TL, Cheng B, Deng X (2009). Wingbeat time and the scaling of passive rotational damping in flapping flight. *Science* **324**, 252–255.

Herr HM, Huang GT, McMahon TA (2002). A model of scale effects in mammalian quadrupedal running. *Journal of Experimental Biology* **205**, 959–967.

Hildebrand M, Goslow GE (2001). *Analysis of Vertebrate Structure*. New York, John Wiley & Sons.

Hobbelen DGE, Wisse M (2007). A disturbance rejection measure for limit cycle walkers: The Gait Sensitivity Norm. *IEEE Transactions on Robotics* **23**, 1213–1224.

Hutchinson JR (2004a). Biomechanical modeling and sensitivity analysis of bipedal running ability. I. Extant taxa. *Journal of Morphology* **262**, 421–440.

Hutchinson JR (2004b). Biomechanical modeling and sensitivity analysis of bipedal running ability. II. Extinct taxa. *Journal of Morphology* **262**, 441–461.

Hutchinson JR, Ng-Thow-Hing V, Anderson FC (2007). A 3D interactive method for estimating body segmental parameters in animals: Application to the turning and running performance of *Tyrannosaurus rex*. *Journal of Theoretical Biology* **246**, 660–680.

Jindrich DL, Full RJ (1999). Many-legged maneuverability: Dynamics of turning in hexapods. *Journal of Experimental Biology* **202**, 1603–1623.

Jindrich DL, Full RJ (2002). Dynamic stabilization of rapid hexapedal locomotion. *Journal of Experimental Biology* **205**, 2803–2823.

Jindrich DL, Qiao M (2009). Maneuvers during legged locomotion. *Chaos* **19**, 026105–026105.

Jindrich DL, Besier TF, Lloyd DG (2006). A hypothesis for the function of braking forces during running turns. *Journal of Biomechanics* **39**, 1611–1620.

Jindrich DL, Smith NC, Jespers K, Wilson AM (2007). Mechanics of cutting maneuvers by ostriches (Struthio camelus). *Journal of Experimental Biology* **210**, 1378–1390.

Koditschek DE, Full RJ, Buehler M (2004). Mechanical aspects of legged locomotion control. *Arthropod Structure & Development* **33**, 251–272.

Lee DV, Bertram JEA, Todhunter RJ (1999). Acceleration and balance in trotting dogs. *Journal of Experimental Biology* **202**, 3565–3573.

Lee DV, Walter RM, Deban SM, Carrier DR (2001). Influence of increased rotational inertia on the turning performance of humans. *Journal of Experimental Biology* **204**, 3927–3934.

Lee DV, Stakebake EF, Walter RM, Carrier DR (2004). Effects of mass distribution on the mechanics of level trotting in dogs. *Journal of Experimental Biology* **207**, 1715–1728.

Lee DV, McGuigan MP, Yoo EH, Biewener AA (2008). Compliance, actuation, and work characteristics of the goat foreleg and hindleg during level, uphill, and downhill running. *Journal of Applied Physiology* **104**, 130–141.

Marigold DS, Patla AE (2002). Strategies for dynamic stability during locomotion on a slippery surface: Effects of prior experience and knowledge. *Journal of Neurophysiology* **88**, 339–353.

Marigold DS, Patla AE (2005). Adapting locomotion to different surface compliances: Neuro-muscular responses and changes in movement dynamics. *Journal of Neurophysiology* **94**, 1733–1750.

McGeer T (1990). Passive bipedal running. *Proceedings of the Royal Society of London B* **B240**, 107–134.

McGowan CP, Baudinette RV, Biewener AA (2005). Joint work and power associated with acceleration and deceleration in tammar wallabies (*Macropus eugenii*). *Journal of Experimental Biology* **208**, 41–53.

McGowan CP, Baudinette RV, Biewener AA (2007). Modulation of proximal muscle function during level versus incline hopping in tammar wallabies (*Macropus eugenii*). *Journal of Experimental Biology* **210**, 1255–1265.

McGuigan MP, Yoo E, Lee DV, Biewener AA (2009). Dynamics of goat distal hind limb muscle-tendon function in response to locomotor grade. *Journal of Experimental Biology* **212**, 2092–2104.

McMahon TA, Cheng GC (1990). The mechanics of running: How does stiffness couple with speed? *Journal of Biomechanics* **23**, 65–78.

More HL, Hutchinson JR, Colliins DF, Weber, DJ, Aung SKH, Donelan JM (2010). Scaling of sensorimotor control in terrestrial mammals. *Proceedings of the Royal Society B* **277**, 3563–3568.

Moritz CT, Farley CT (2004). Passive dynamics change leg mechanics for an unexpected surface during human hopping. *Journal of Applied Physiology* **97**, 1313–1322.

Moritz CT, Farley CT (2005). Human hopping on very soft elastic surfaces: implications for muscle pre-stretch and elastic energy storage in locomotion. *Journal of Experimental Biology* **208**, 939–949.

Moritz CT, Farley CT (2006). Human hoppers compensate for simultaneous changes in surface compression and damping. *Journal of Biomechanics* **39**, 1030–1038.

Muybridge E (1887). *Muybridge's complete human and animal locomotion*. Courier Dover Publications.

Pearson K (2000). Motor systems. *Current Opinion in Neurobiology* **10**, 649–654.

Pearson K, Ekeberg O, Büschges A (2006). Assessing sensory function in locomotor systems using neuro-mechanical simulations. *Trends in Neurosciences* **29**, 625–631.

Prochazka A, Gritsenko V, Yakovenko S (2002). Sensory control of locomotion: Reflexes versus higher-level control. *Advances in Experimental Medicine and Biology* **508**, 357–367.

Roberts TJ, Marsh RL (2003). Probing the limits to muscle-powered accelerations: lessons from jumping bullfrogs. *Journal of Experimental Biology* **206**, 2567–2580.

Roberts TJ, Scales JA (2002). Mechanical power output during running accelerations in wild turkeys. *Journal of Experimental Biology* **205**, 1485–1494.

Roberts TJ, Scales JA (2004). Adjusting muscle function to demand: joint work during acceleration in wild turkeys. *Journal of Experimental Biology* **207**, 4165–4174.

Roberts TJ, Marsh RL, Weyand PG, Taylor CR (1997). Muscular force in running turkeys: the economy of minimizing work. *Science* **275**, 1113–1115.

Rossignol S, Dubuc RJ, Gossard J-P (2006). Dynamic sensorimotor interactions in locomotion. *Physiological Reviews* **86**, 89–154.

Ruina A, Bertram JEA, Srinivansan M (2005). A collisional model of the energetic cost of support work qualitatively explains leg sequencing in walking and galloping, pseudo-elastic leg behavior in running and the walk-to-run transition. *Journal of Theoretical Biology* **237**, 170–192.

Schmitt J, Clark J (2009). Modeling posture-dependent leg actuation in sagittal plane locomotion. *Bioinspiration & Biomimetics* **4**, 046005.

Schmitt J, Holmes P (2000). Mechanical models for insect locomotion: dynamics and stability in the horizontal plane I. Theory. *Biological Cybernetics* **83**, 501–515.

Schmitt J, Garcia M, Razo RC, Holmes P, Full RJ (2002). Dynamics and stability of legged locomotion in the horizontal plane: a test case using insects. *Biological Cybernetics* **86**, 343–353.

Seipel J, Holmes P (2007). A simple model for clock-actuated legged locomotion. *Regular and Chaotic Dynamics* **12**, 502–520.

Seyfarth A, Geyer H, Günther M, Blickhan R (2002). A movement criterion for running. *Journal of Biomechanics* **35**, 649–655.

Seyfarth A, Geyer H, Herr H (2003). Swing-leg retraction: a simple control model for stable running. *Journal of Experimental Biology* **206**, 2547–2555.

Spence AJ, Revzen S, Seipel J, Mullens C, Full RJ (2010). Insects running on elastic surfaces. *Journal of Experimental Biology* **213**, 1907–1920.

Sponberg S, Full RJ (2008). Neuromechanical response of musculo-skeletal structures in cockroaches during rapid running on rough terrain. *Journal of Experimental Biology* **211**, 433–446.

Srinivasan M, Ruina A (2006). Computer optimization of a minimal biped model discovers walking and running. *Nature* **439**, 72–75.

Ting LH, van Antwerp KW, Scrivens JE, McKay JL, Welch TDJ, Bingham JT, DeWeerth SP (2009). Neuromechanical tuning of nonlinear postural control dynamics. *Chaos* **19**, 026111–026112.

Todorov E, Jordan MI (2002). Optimal feedback control as a theory of motor coordination. *Nature Neuroscience* **5**, 1226–1235.

Usherwood JR, Bertram JEA (2003). Understanding brachiation: insight from a collisional perspective. *Journal of Experimental Biology* **206**, 1631–1642.

Usherwood JR, Wilson AM (2005). Biomechanics: No force limit on greyhound sprint speed. *Nature* **438**, 753–754.

Usherwood JR, Wilson AM (2006). Accounting for elite indoor 200m sprint results. *Biology Letters* **2**, 47–50.10.

Walter RM (2003). Kinematics of 90° running turns in wild mice. *Journal of Experimental Biology* **206**, 1739–1749.

Walter RM, Carrier DR (2009). Rapid acceleration in dogs: ground forces and body posture dynamics. *Journal of Experimental Biology* **212**, 1930–1939.

Williams SB, Tan H, Usherwood JR, Wilson AM (2009). Pitch then power: limitations to acceleration in quadrupeds. *Biology Letters* **5**, 610–613.

Wilson AM, McGuigan MP, Su A, van den Bogert AJ (2001). Horses damp the spring in their step. *Nature* **414**, 895–899.

CHAPTER TWELVE

The Evolution of Terrestrial Locomotion in Bats: The Bad, the Ugly, and the Good

Daniel K. Riskin[1], John E. A. Bertram[2] and John W. Hermanson[3]

[1] Department of Biology, University of Toronto-Mississauga, CA
[2] Department of Cell Biology and Anatomy, Cumming School of Medicine, University of Calgary, CA
[3] Department of Biomedical Sciences, College of Veterinary Medicine, Cornell University, NY, USA

12.1 BATS ON THE GROUND: LIKE FISH OUT OF WATER?

The birds and the bats are the only two extant vertebrate lineages capable of powered flight, and that ability has awarded both groups extraordinary ecological diversity. However, while many birds have evolved to fill terrestrial niches – some even secondarily losing their ability to fly (most notably the ratites) – the bats have more or less restricted themselves to an aerial existence. There are no flightless bats and, whereas most birds amble quickly on flat ground, almost any bat is slow enough on the ground that it can be caught by hand (Lawrence, 1969).

If there were no associated cost, the ability to walk well would confer a selective advantage to any species of bat. It would enable them to escape danger after an accidental fall and, in many cases, would allow them to exploit a broader range of food resources. However, bats do not occupy the niches of terrestrial mammals, even in the

Understanding Mammalian Locomotion: Concepts and Applications, First Edition.
Edited by John E. A. Bertram.
© 2016 John Wiley & Sons, Inc. Published 2016 by John Wiley & Sons, Inc.

tropics, where the number of bat species is high, and their dietary niches are diverse (Arita and Fenton, 1997). The only real exception to this rule appears to be the short-tailed bat of New Zealand, which spends around 30% of its foraging time on the ground, having evolved such habits in the absence of terrestrial mammals with which to compete (Daniel, 1976). Bats of that species may have the most phylogenetically diverse diet of any bat, eating fruit, nectar, flower fragments, and a broad range of invertebrates (Carter and Riskin, 2006; Lloyd, 2001), which makes the absence of bats from other terrestrial niches all the more conspicuous.

With so much to be gained from foraging terrestrially, why do so few bats take advantage of such opportunities? In this chapter, we survey the diversity of crawling performance among bats, and explore the biomechanical hypotheses that might explain their generally poor terrestrial performance.

12.2 SPECIES-LEVEL VARIATION IN WALKING ABILITY

It is universally true that a flying bat runs the risk of accidentally falling to the ground, and bats of every species are equipped with some strategy for dealing with such an event. Strategies are species-specific; some bats jump immediately into flight, while others shuffle awkwardly to a vertical surface that they can climb to safety, and still others are able to walk away relatively quickly, using stereotyped gaits when they do so. Vaughan (1970) treated these as three distinct categories of performance. To date, very few species have been studied, so the extent to which those groups blend into one another is not known. In this chapter, we call bats belonging to each of Vaughan's three groups the bad crawlers, the ugly crawlers, and the good crawlers, respectively.

The "bad" crawlers appear unable to walk at all on a flat horizontal surface and, instead, always perform flight-initiating jumps. These include the natalids *Natalus stramineus* (Vaughan, 1970) and *N. tumidirostris* (Riskin *et al.*, 2005), several phyllos-tomids, such as *Macrotus californicus* (Vaughan, 1959) and *Leptonycteris* sp. (Dietz, 1973), and at least some rhinolophids, hipposiderids (Lawrence, 1969; Siemers and Ivanova, 2004), and mormoopids (*Mormoops megalophylla*; personal observation). Interestingly, a study by Dietz (1973) demonstrated that at least some bats that are unable to crawl as adults can do so as juveniles, before they are able to fly. That might reflect a biomechanical difference between juveniles and adults that allows juveniles to walk better than adults can, or may result because adults are able to crawl but choose not to since, for them, flying is a much better way to get off the ground.

When grounded, many bats (the "ugly") use a series of asymmetrical movements to push themselves across the substrate. Those motions do not conform to a repeating gait and are, thus, difficult to describe accurately and concisely. The bat's ventral surface faces the floor, and the body moves cranially or craniolaterally. Published accounts of shuffling include those of *Eptesicus serotinus* (Vespertilionidae; Lawrence, 1969), *Artibeus* sp. (Phyllostomidae; Dietz, 1973) and *P. parnellii* (Mormoopidae; Riskin *et al.*, 2005). Although fewer than 5% of bat species have been observed carefully, a great number of those bat species studied so far manoeuvre poorly on the ground (Vaughan, 1970).

The "good" walkers are exemplified by the Common Vampire Bat (Phyllostomidae: *Desmodus rotundus*). This species is probably the most terrestrially agile of all the bats.

It can walk forwards, backwards, and sideways using a walking gait (Altenbach, 1979), run with a bounding gait at speeds greater than 1.1 m/sec (Riskin and Hermanson, 2005), and jump into flight in under 0.03 sec (Schutt *et al.*, 1997). The other two vampire bats (*Diaemus youngi* and *Diphylla ecaudata*) possess similar agility, though they seem more adapted to arboreal climbing than to terrestrial walking (Schutt, 1998). The distantly related Short-tailed Bats of New Zealand (Mystacinidae: *M. tuberculata*) also walk well and leap into flight, though they do not use the running gait seen in *Desmodus* (Riskin *et al.*, 2005).

The short-tailed bats and vampire bats appear to be the most terrestrial bats but, with descriptions lacking for so many species, this statement remains unverified. Several vespertilionids, such as *Antrozous pallidus* and *Nyctalus noctula*, move quickly over ground (Orr, 1954; Lawrence, 1969; Dietz, 1973; Hermanson and O'Shea, 1983), as do a great number of molossid bats (Schutt and Simmons, 2001; Lawrence, 1969; Vaughan, 1959; Dietz, 1973). Anecdotal evidence and morphological analyses also suggest that some of those molossids, such as *Cheiromeles torquatus* and *C. parvidens* of Indonesia and the Philippines, and *Tadarida australis* of Australia, may even be as terrestrially adept as *M. tuberculata* (Schutt and Simmons, 2001; Freeman, 1981; Mills *et al.*, 1996).

The diversity of terrestrial agility in bats varies from species to species, and spans "bad" walkers, "ugly" walkers, and "good" walkers. It seems reasonable that some aspect of morphology should predict terrestrial agility for a species but, until recently, the relationship of form to function in this system was poorly understood.

12.3 HOW DOES ANATOMY INFLUENCE CRAWLING ABILITY?

Many terrestrially agile bats possess anatomical specializations for walking. For example, walking *Cheiromeles* spp. and *M. tuberculata* tuck the tips of their wings into folds or pouches of the wing, probably to prevent these dragging on the ground (Schutt and Simmons, 2001; Dwyer, 1962). Also, whereas the membrane between the legs (uropatagium) of some bats drags on the floor during walking (Lawrence, 1969), bats that walk well lift the uropatagium, either by the tail or by a bony spur of the ankle called the calcar (*Eumops perotis*; Vaughan, 1959; *M. tuberculata*; Dwyer, 1962). In Hairy-legged Vampire Bats (*D. ecaudata*), the calcar is free of the uropatagium, and assists in arboreal locomotion by acting as an opposable sixth digit, giving these animals a means to grip cylindrical branches (Schutt and Altenbach, 1997). A similar role is probably played by the opposable hindlimb digit I (hallux) of *Cheiromeles* spp. (Schutt and Simmons, 2001). It is relatively easy to see how these specializations assist in the terrestrial locomotion of ably walking bats. However, it is far more difficult to explain the anatomical bases of poor crawling in the bad and ugly crawlers.

It would seem reasonable that interspecific variation in terrestrial agility should be reflected by differences in anatomy, since biomechanically relevant differences in bat ecology are often revealed by morphological variation (e.g., Freeman, 1984; Aldridge and Rautenbach, 1987; Norberg and Rayner, 1987; Dumont, 1997). Quadrupedal locomotion has been considered in several studies of chiropteran anatomy (Vaughan, 1959, 1970; Howell and Pylka, 1977; Strickler, 1978; Altenbach, 1979; Schutt and Altenbach, 1997; Schutt and Simmons, 2001), but a clear morphological predictor of walking ability has been elusive.

Several muscles of the forelimb skeleton are larger in species that walk well than in other species, including the *m. pectoralis abdominis, m. subscapularis, m. supraspinatus, m. rhomboideus*, and the *m. triceps brachii* (Strickler, 1978). This suggests that muscle proportions in the chest and shoulders influence terrestrial mobility, but also that the kinds of muscle tissue present may also be of importance. Type II muscle fibers are generally fast and powerful, suitable for powering the downstroke of flying bats. In contrast, the type I fibers are better suited for postural activities or slow, repetitive movements.

Hermanson *et al.* (1993, 1998) found that, whereas the pectoralis muscles of all other bats surveyed possess only type II muscle fibers, those of vampire bats also include type I fibers. Maintaining the upright posture demonstrated by bats that are agile while walking might require slow, fatigue-resistant muscle, and the inability of most bats to hold themselves upright might be caused by the absence of type I fibers. The musculature of other terrestrially agile bats, like *M. tuberculata* and *Cheiromeles* spp. has not been surveyed, but the presence or absence of type I fibers in those species will help clarify the importance of this character.

Hand *et al.* (2009) noted that the morphology of the elbow differed in *M. tuberculata* from the typical chiropteran pattern, and inferred that this reflected their improved ability to jump, compared to other bats. However, there is no evidence that mystacinids are better jumpers than other bats are, and no link between jumping ability and ecology among bats has ever been demonstrated (Gardiner and Nudds, 2011). Indeed, if any trend exists, one would predict that the least agile crawlers should need the ability to jump more than those bats able to crawl to safety, such as *M. tuberculata*. Thus, despite the uniqueness of the mystacinid elbow, the functional relevance of elbow morphology to terrestrial agility is not clear.

Many authors have focused on a possible role for the hindlimbs in terrestrial performance. One explanation for differences in crawling ability is the hindlimb-strength hypothesis of Howell and Pylka (1977). These authors suggested that the bats that walk poorly do so because their slender legs are built for hanging, placing the bone under a predominantly tensile loading regime. They proposed that the femora of bats are too gracile to withstand the compressive forces associated with terrestrial locomotion. As evidence, Howell and Pylka pointed out that the legs of most bats are long and slender, compared with the legs of terrestrial mammals of similar size, but that vampire bats have much thicker hindlimbs, comparable in diameter to those of terrestrial mammals (Figure 12.1).

To test the hindlimb-strength hypothesis, Riskin *et al.* (2005) compared the hindlimb ground-reaction forces of bats that walk well (good walkers) with those of bats that shuffle awkwardly (the ugly). While the hindlimb-strength hypothesis predicted larger forces from the hindlimbs of ably walking bats, it was the poorly walking bats that actually placed the most weight on their legs. Based on these results, and on an engineering model-based re-evaluation of the Howell and Pylka study, the hindlimb-strength hypothesis was not supported (Riskin *et al.*, 2005). Thin legs do not prevent the poorly crawling bats from walking well.

Even though the hindlimb strength hypothesis does not explain the inability of some bats to walk well, hindlimb morphology is important to terrestrial agility, mostly in terms of femur orientation and mobility (Vaughan, 1959, 1970). Bats that do not walk well typically have caudally oriented femora, whereas good crawling bats have

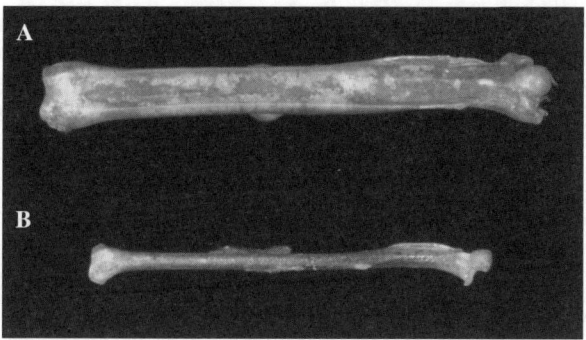

FIGURE 12.1

Femora of a terrestrially agile bat (*Desmodus rotundus*; (**A**) and a bat that is not specialized for terrestrial locomotion (*Rhinolophus clivosus*; (**B**). Bats of these species have
similar body mass ranges, of around 20–30 g. Although it was once thought that poorly
crawling bats were unable to walk because their femora were too fragile to resist the
stresses associated with walking, bats that shuffle awkwardly actually place more
weight on their hindlimbs while crawling than do good walkers. The robustness of
the femur does not prevent bats from walking well. Photograph by M. Brock Fenton.

more laterally-oriented femora. This positions the foot below the crawling body,
providing a sprawling, reptile-like hindlimb posture. Crawling may also be made
possible by specializations of the hip joint that enable the bat to swing its sprawling
legs while walking. The head of the femur is offset from the long axis of the bone to
a greater degree in terrestrial bats than in poorly crawling species (Vaughan, 1970),
and the acetabulum is also wider in terrestrial bats, presumably allowing a greater
range of femoral motion (Dwyer, 1960, 1962).

To facilitate adequate adduction of the hindlimbs during the weight-bearing phase
of terrestrial locomotion, hindlimb adductor muscles must be strong enough to coordinate walking and standing. These adductor muscles (especially the gracilis and the
semitendinosus, which have fused tendons of insertion in bats) tend to be reduced in
mass, compared with those of terrestrial mammals but may serve some of their most
important functions during the downstroke of flight (Vaughan, 1959).

12.4   HINDLIMBS AND THE EVOLUTION OF FLIGHT

Birds fly with their arms and walk with their legs so, despite radical changes in the
forelimbs associated with powered flight, modern birds probably retain the same
terrestrial locomotor kinematics once used by their dinosaur ancestors (Gatesy and
Middleton, 1997). Bats on the other hand, support their wings with all four limbs, and
walk quadrupedally. Thus, the evolution of flight is thought to have imposed biomechanical trade-offs that made bats poor walkers. Just how that occurred is not known,
because fossils clarifying the postcranial anatomy of the mammals that gave rise to
bats have not been uncovered.

Most authors agree that bats evolved flight "trees-down", from an arboreal ancestor (Clark, 1977; Norberg, 1985; Padian, 1987; Simmons and Geisler, 1998; Simmons *et al.*, 2008; Speakman, 2001; Schnitzler *et al.*, 2003), and much of the discussion about how the ability to walk well was lost in that process has been focused on the hindlimbs. The hindlimb of a bat supports the trailing edge of the wing in flight, and it is used to grasp the ceiling during landing (Riskin *et al.*, 2008, 2009). This is different from the condition in most mammals, and the hindlimbs of bats are, therefore, longer and more slender than those of other mammals (Swartz and Middleton, 2007).

The orientation of the hindlimbs is also unusual in bats. During both flight and roosting, the femora of bats extend laterally or caudally and are rotated on their long axes, so that the knees point dorsally, with the plantar surfaces (soles) of the hind feet facing ventrally (Figure 12.2). This is the opposite orientation from the human condition, and it permits the claws to grip when a bat hangs head-down with its chest against a surface. It is also the orientation of the hindlimbs when bats crawl on the ground.

FIGURE 12.2

Roosting posture of Parnell's Mustache Bat (*Pteronotus parnellii*), demonstrating the hindlimb orientation of bats. Note that the knees point laterally and dorsally, whereas the soles of the feet face ventrally. This is typically referred to as "hindlimb reversal", because the toes of bats point in the opposite direction from those of a cat or dog, for example. This prevents bats from using their legs for walking the way most mammals do, and is thought to be one of the main reasons why bats walk poorly. Photograph by M. Brock Fenton.

This rotation, or "hindlimb reversal", prevents bats from positioning their hindlimbs during walking the way most mammals do, and is thought to be one of the main reasons that most bats walk poorly (Vaughan, 1959, 1970).

However, hindlimb reversal is not unique to bats, and is performed by several other mammals that walk perfectly well. Indeed, the ability to point the toes caudally has evolved independently several times among arboreal and climbing mammals, including members of the Multituberculata (Jenkins and Krause, 1983), Marsupiala, Carnivora, Edentata, Primates, Rodentia, and Scandentia (Jenkins and McClearn, 1984). Hindlimb reversal enables animals to hang bat-like from horizontal supports, as evidenced by squirrels (Rodentia), kinkajous (Carnivora) (Jenkins and McClearn, 1984), and by the bats themselves. Importantly though, the hindlimb reversal of non-bat mammals occurs due to specializations of the ankle, especially at the talocrural and subtalar joints, whereas in bats the rotation principally occurs at the hip. The reason for this difference is unknown, but its result is that, while other mammals generally return the hindlimbs to the more typical mammalian orientation when walking on the ground (Jenkins and McClearn, 1984), bats cannot so (Vaughan, 1970).

Hindlimb reversal is common among quadrupedal mammals that maneuver arboreally. It permits a squirrel to descend a tree head-first, while the cat that chased it up there sits helplessly on a branch. This is explained by a free-body diagram of a quadrupedal animal grasping a vertical surface with two limbs at an upper point of contact and two limbs at a lower point of contact (Figure 12.3). Because the gravitational force vector at the animal's center of mass (some distance from the surface) is not aligned with the normal force vector (where the limbs grasp the surface), a torque is created that pulls the animal away from the surface at the upper limbs. To resist falling, an animal must be able to grip the surface at the upper point of contact (Alexander, 2003). From a head-down posture, mammals with cat-like hindlimbs cannot cling to the trunk with their hind claws, because their claws hook upwards, but mammals with reversed hindlimbs can.

Based on the phylogenetically widespread trend toward hindlimb reversal among arboreal mammals, it seems reasonable that the ancestors to bats evolved the reversed hindlimb as they became adapted to an arboreal habitat – before flight. Once hindlimb reversal was achieved, the proto-bat could have adopted the pendulous roosting posture that typifies modern forms. Certainly, a hanging posture would have had important consequences to terrestrial locomotion and, although it has been overlooked in several reviews (Norberg, 1985; Scholey, 1986; Padian, 1987; Arita and Fenton, 1997; Speakman, 2001), the ability to hang by the toes may have also been an important precursor to the evolution of flight. Freed from a role in compressive weight support during roosting, forelimb digits 2 to 5 could become elongated to support the wing membranes (Norberg, 1985).

Of all gliding mammals, only the ancestors to the bats subdivided the main gliding membrane with the bony elements of the hand, and this subdivision has been suggested to have allowed bats to fly by enabling them to perform the differential cambering, tensioning, and folding over of the wing surface necessary for flapping flight (Hill and Smith, 1984; Speakman, 2001). Concomitant with this would be an enlargement of the pectoralis muscle (to nearly 10% of body mass), to power the downstroke phase of the wingbeat and, in some lines, a complex division of labor within the downstroke musculature, to co-opt muscles such as the short head of biceps brachii, and the

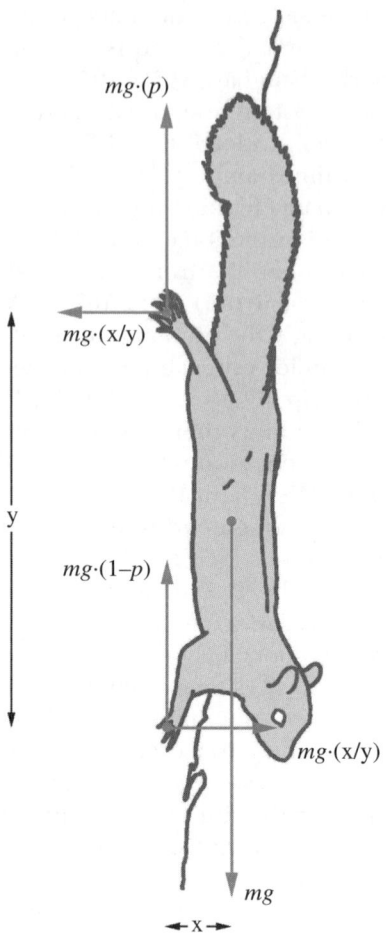

FIGURE 12.3

Hindlimb reversal is a common adaptation to arboreal niches among mammals, suggesting that it evolved in the arboreal ancestor to bats, prior to the evolution of flight. Note that in order to cling head-down to a vertical tree trunk, an animal must pull itself toward the surface with the hindlimbs. The gravitational force, of magnitude mg (where m is mass and g is the gravitational constant), pulls down on the animal's center of mass and is opposed by vertical forces applied at the forelimbs and hindlimbs, where the hindlimbs support any arbitrary proportion (p) of body weight. Because the gravitational and normal forces are separated by some distance (x), a torque acts about the hindlimb of magnitude $mg\cdot(x)$. This is opposed by the forelimbs, which press against the surface with a force of magnitude $mg\cdot(x/y)$, where y is the vertical distance between forelimb and hindlimb attachments. Horizontal forces can only be balanced by the attachment of the hindlimbs to the surface with a force of $mg\cdot(x/y)$ (based on Alexander, 2003).

subscapularis, as powerful contributors to the downstroke. These specializations of the forelimb architecture surely also played a role in the reduction of terrestrial agility, but the biomechanical basis for this is not well understood.

12.5 MOVING A BAT'S BODY ON LAND: THE KINEMATICS OF QUADRUPEDAL LOCOMOTION

The way in which quadrupedal mammals contact the ground with their forelimbs varies among taxa. Bears walk on the palms (carpal bones and phalanges) of their outspread hands, while dogs and cats walk on the distal parts of the fingers (middle and distal phalanges), and ungulates walk on the tips of their fingers (distal phalanges). These are called plantigrade, digitigrade, and unguligrade stances, respectively, and the stance of almost any terrestrial mammal can be assigned to one of these forms (Lovegrove, 2004).

However, because the fingers of bats are so long and slender, none of these stance types are used. Instead, the most terrestrial of bats close the wings and align the proximal digits parallel with the forearms, so that only the palmar surface of the carpus and pollex (wrist and thumb) make contact with the ground. The hindlimbs are not so drastically modified as the forelimbs, and take a more "reptile-like," plantigrade sprawling position. When bats such as *D. rotundus* and *C. torquatus* walk, the femora extend laterally and the soles of the hind feet contact the ground (Altenbach, 1979; W.A. Schutt Jr., personal communication).

An important difference between terrestrially agile bats (the good) and bats that walk awkwardly (the ugly) is that the good crawlers hold the abdomen above the ground at all times, whereas the bodies of other bats collapse to the ground periodically during shuffling (Lawrence, 1969; Riskin *et al.*, 2005). With four limbs making contact with the ground, requirements for stability in bats are no different from those of any other quadrupedal animals (Cartmill *et al.*, 2002). As a result, those bats that walk well use a lateral-sequence symmetrical walking gait, similar to that used by many other tetrapods (Figure 12.4A). The left hind foot moves forward in synchrony with the right forelimb, and vice versa and, as walking bats change speed, they increase their stride frequencies, just as terrestrial mammals do (Riskin and Hermanson, 2005; Riskin *et al.*, 2006).

Many bats possess the ability to hop, or "leap-frog" (Lawrence, 1969), by pressing against the ground with the two forelimbs simultaneously. This results in a brief aerial phase before the bat lands some distance ahead of its original position. Many bats, including vespertilionids, rhinolophids, and miniopterids, use similar kinematics to launch directly into flight with a single jump, with vertical forces of around 3.5–4.5 times body weight and initial speeds of 1.0–1.6 m/sec (Gardiner and Nudds, 2011). This behavior is extremely exaggerated in Common Vampire Bats (*D. rotundus*) which, when jumping, apply a force equivalent to 9.5 body weights, to reach a take-off velocity of 2.38 m/sec in under 0.03 sec (Schutt *et al.*, 1997).

Desmodus rotundus are the only bats known to possess a true running gait. Although peak speeds on treadmill trials only reached 1.1 m/sec (Riskin and Hermanson, 2005), they can probably run as fast as 2.0 m/sec (Altenbach, 1979). The vampire bat running gait is kinematically different from that of any other tetrapod, chiefly because it is forelimb-driven (Figure 12.4B). This gait appears to have evolved

A

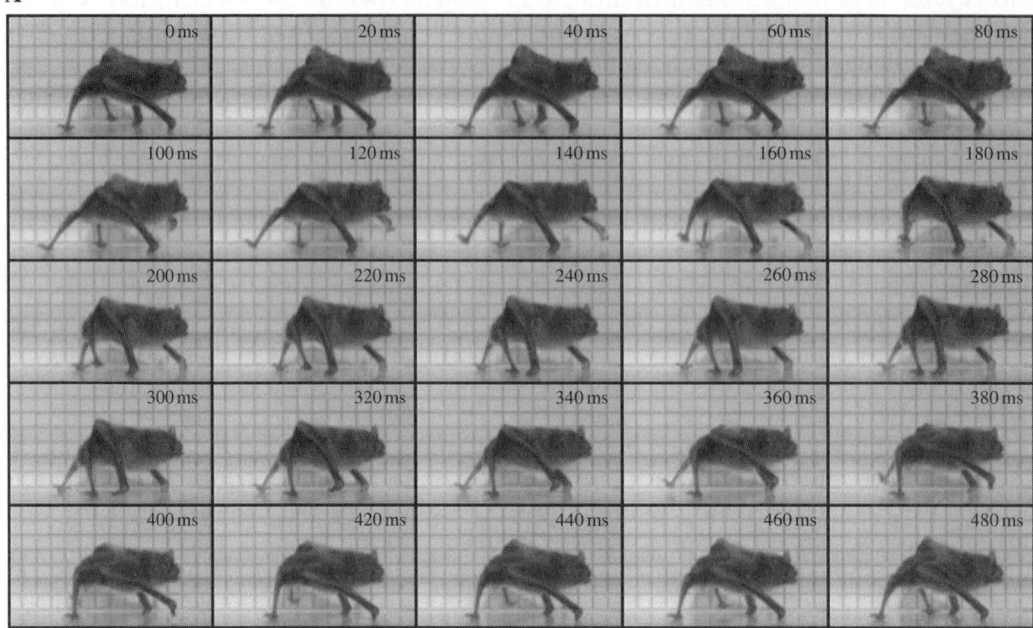

B

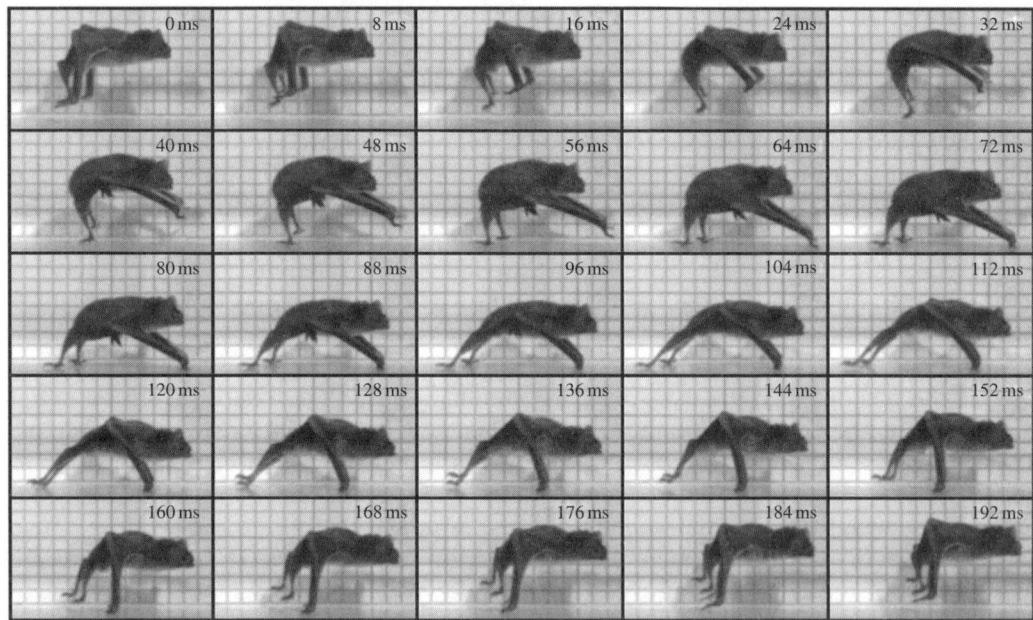

FIGURE 12.4

(**A**) Walking and (**B**) running gaits of Common Vampire Bats (*Desmodus rotundus*). The background is a 1.0 cm² grid. Note that the time scales of these image sequences differ. Some of these images are reprinted from Riskin and Hermanson (2005).

independently within the vampire lineage, probably from the wing-powered jumping behavior mentioned above.

There are very few examples in nature of a tetrapod lineage that has lost the ability to run, only to acquire it afresh some time later (Ahn *et al.*, 2004). The independent evolution of a running gait by *D. rotundus* provides a novel group upon which to test hypotheses regarding cursorial locomotion, enabling researchers to better isolate the roles of biomechanical and phylogenetic constraints on the way in which gaits evolve.

Vampire bats can still fly and have, therefore, not fundamentally altered the forelimbs from a flight structure, or dramatically changed the form of the hind limbs. Rather, terrestrial locomotion is accommodated by using these structures in a manner that allows effective terrestrial locomotion. Also, while the running gait of vampire bats is independently evolved and kinematically distinct from the running gaits of other quadrupeds, its kinetics are strikingly similar to those of other running gaits.

During running, the forelimbs of *D. rotundus* contact the substrate as long struts that reach far in front of the animal's body. This should produce a high angle of incidence with the contact, which will result in substantial momentum loss as the center of mass (COM) is vaulted over the support (Kuo, 2002; Ruina *et al.*, 2005). Next, the COM of the animal is slung relatively low and moves forward on flexed elbows, in a posture that is probably accomplished due to the strength of the downstroke flight muscles. The low vertical position and protrusion of the COM means that the geometric relationship between the COM and substrate contact is similar to that of many other terrestrial forms when the limb contacts the substrate and begins bearing weight. This geometry reduces the momentum loss that is caused by the changing direction of the COM.

While this occurs, the hindlimbs extend caudally. Assuming that they provide power to the COM, hindlimb extension would help to orient the COM velocity vector forward across the weight-bearing forelimbs, reducing the momentum loss. A similar thrust by the ankle extensors in humans can reduce the mechanical cost of walking by 25–75%, depending on details of the timing of the contact and thrust (Kuo, 2002; Ruina *et al.*, 2005), so the hindlimbs of vampire bats may play an important role, despite the obvious dominant role of the forelimbs for bounding.

Following the hindlimb extension that occurs shortly after forelimb contact, the hindlimbs lose contact with the substrate. As the COM passes over the forelimb support, forward and vertical thrust is provided by extending the elbow slightly and by extending the thumb. As in vertical jumping, where action of the thumb accounts for up to 17% of the vertical height of take-off jumps (Schutt *et al.*, 1997), extension of the thumb likely contributes to the forward speed and duration of the flight phase that follows forelimb contact.

As with all terrestrial running gaits that involve a non-contact flight phase, the ballistic portion of the *Desmodus* bounding gait allows forward progression with little energetic investment. However, at the end of the ballistic phase, the COM velocity will have a downward (and forward) direction, as a result of the action of gravity. During the aerial phase, the hindlimbs are brought forward to initiate ground contact. This serves to begin the deflection of the COM path from partially downward to mostly forward, again minimizing the momentum loss that will occur when the forelimbs continue the deflection of the COM. This sequence of contacts parallels the hind limb initiated bound and half-bound gaits seen in some lagomorphs (Bertram and Gutmann, 2009).

12.6   EVOLUTIONARY PRESSURES LEADING TO CAPABLE TERRESTRIAL LOCOMOTION

It would appear that capable terrestrial locomotion in bats has arisen independently more than once, and under different selective pressures. Dwyer (1960) called *M. tuberculata* "the most sure-footed of all bats," and Altenbach (1979) commented that "no other species possess the extreme terrestrial agility (of *D. rotundus*)." Both possess remarkable terrestrial agility, but the phylogenetic relationships of vampires and mystacinids strongly suggests that they evolved their quadrupedal abilities independently (Teeling *et al.*, 2005). We have argued that these lineages evolved their terrestrial abilities under different selective pressures, and that this is reflected in the possession of a running gait in vampire bats, and its absence in short-tailed bats (Riskin *et al.*, 2006).

Vampires possess walking agility as a specialization for their diet – the blood of mammals and birds (Greenhall, 1988). The three vampire species form a monophyletic outgroup (Desmodontinae), and they probably transitioned to blood-feeding from insectivory, not from frugivory (for a review of hypotheses concerning this transition – see Schutt, 1998). Being terrestrially agile is necessary for these animals, in part so that they are able to approach their prey stealthily, by crawling on the ground or along a branch (Greenhall, 1988), but also to avoid injury or predation on the ground, since the neotropical distribution of the vampires overlaps with that of several bat predators.

The New Zealand Short-tailed bats (*Mystacina*) also belong to the Noctilionoidea (Jones *et al.*, 2002) and, like vampire bats, probably descended from an aerial insectivore (Lloyd, 2001). However, the selective pressures that brought about terrestrial agility in mystacinids were almost certainly different from those of blood-feeding vampire bats. *M. tuberculata* is restricted to New Zealand, where it is one of only two endemic non-marine mammals (the other is a distantly related vespertilionid bat, with unexceptional terrestrial abilities, *Chalinolobus tuberculatus*).

In the absence of terrestrial mammals as competitors or predators, *M. tuberculata* adopted a shrew-like niche, foraging in leaf litter for invertebrates (Jones *et al.*, 2003). *M. tuberculata* also feed on fruit and pollen (Daniel, 1976), and take a significant proportion of their prey on the wing, in the manner of typical insectivorous bats (Arkins, 1996). *M. tuberculata* previously shared their terrestrial niche in New Zealand with a sister species, *M. robusta*, that also foraged terrestrially (Hill and Daniel, 1985). However, that species became extinct around 1967, after the introduction of rats to their range.

An alternative hypothesis has been proposed, whereby mystacinids evolved their terrestrial abilities in the predator- and competitor-filled Miocene environment of Australia, before subsequently arriving on the relatively safer island of New Zealand (Hand *et al.*, 2009). This hypothesis is based on the morphological similarity of the elbow in Australian fossils and modern *M. tuberculata*, which has been interpreted to reflect advanced terrestrial agility, especially the ability to perform flight-initiating jumps.

However, flight-initiating jumps are performed by individuals of nearly all bat species, and no biomechanical connection between elbow morphology and walking ability has been demonstrated. Miocene mystacinids might have walked well, just as many bats do, but the extreme terrestrial mobility of *M. tuberculata*, and their exploitation of a terrestrial niche in New Zealand, almost certainly evolved after they arrived in New Zealand. That environment independently brought about flightlessness in several lineages of birds, so we contend that the lack of predators or mammalian competitors in New Zealand was the reason why *M. tuberculata* moved into the terrestrial niches that it now occupies.

It appears that, whereas the vampires evolved terrestrial agility as a means of exploiting a relatively dangerous food resource – at least when feeding on the blood of large mammalian herbivores – the short-tailed bats invaded the ground because there simply were not as many dangers there as were faced by bats on the ground in other ecosystems. New Zealand short-tailed bats walk quickly but, unlike vampire bats, they do not run (Riskin *et al.*, 2006). Also, when *M. tuberculata* have their wingtips tucked away for crawling, they appear unable to launch immediately into flight, but require a few seconds to extend the wings first (DKR, JWH, personal observations). These kinds of restrictions on flight initiation should have important consequences for predator evasion, so it is not surprising that they evolved on an island where predators were absent.

12.7 CONCLUSIONS AND FUTURE WORK

Whereas most birds are able to walk on the ground using their hindlimbs, and fly using their forelimbs, the bats must accomplish both these forms of locomotion using all four limbs simultaneously. Among those bats studied so far, three gross categories of walking ability have been identified (Vaughan, 1970) – bats that do not walk, bats that shuffle awkwardly, and bats that walk well. It is difficult to understand the evolutionary progression that led to capable crawling in these bats, because the basic kinematics are yet to be recorded for almost every species. A broad general survey that puts the abilities of each species into one of Vaughan's three basic categories of terrestrial agility would permit a more detailed analysis of evolutionary patterns. Also, characterization of muscle fiber types in the forelimb and hindlimb muscles of a broader range of species would elucidate the role of muscle fiber types in postural control.

With a range in body sizes spanning nearly three orders of magnitude, and a staggering diversity of physiological and anatomical specializations (Simmons, 2005), the aerial and non-aerial locomotion of bats could provide a convenient model system for the study of biomechanical tradeoffs. However, that would depend on the assumption that most bats crawl poorly as the result of anatomical specializations for flight. Fortunately, this hypothesis is testable. Just as bats possess a broad range of walking abilities, they also demonstrate interspecific differences in flight speed and maneuverability (Norberg and Rayner, 1987).

Whether or not differences in aerial and non-aerial locomotion are correlated begs to be examined. The anatomical requirements of flight and walking could present competing selective pressures on the morphologies of bats, or terrestrial agility might not impose a cost on the ability to fly at all. A trade-off could be quantified through experiments that measure differences in metabolic efficiency during the two modes of locomotion. If the morphological requirements of walking and flight really do conflict, then one would predict that the ably walking bats have lower oxygen consumption rates when walking than do other bats, and that they consume more oxygen in flight or suffer some other cost, such as decreased maneuverability, compared with bats that have not suffered tradeoffs for terrestrial locomotion.

ACKNOWLEDGEMENTS

G.G. Carter, J.M. Ratcliffe, and M.W. Walters provided helpful comments on earlier drafts of this manuscript. M.B. Fenton provided the photographs of *P. parnellii* and of the bat femora. We also thank M.B. Fenton, S. Parsons, W.A. Schutt Jr. and S.M. Swartz for insightful discussions on this topic.

▌ REFERENCES

Ahn AN, Furrow E, and Biewener AA (2004). Walking and running in the red-legged running frog, *Kassina maculata. Journal of Experimental Biology* **207**, 399–410.

Aldridge HDJN, Rautenbach IL (1987). Morphology, echolocation and resource partitioning in insectivorous bats. *Journal of Animal Ecology* **56**, 763–778.

Alexander RMcN (2003). *Principles of Animal Locomotion*. Princeton University Press, Princeton, NJ.

Altenbach JS (1979). Locomotor morphology of the vampire bat, *Desmodus rotundus. Special Publications, American Society of Mammalogists* **6**, 1–137.

Arita HT, Fenton MB (1997). Flight and echolocation in the ecology and evolution of bats. *Trends in Ecology & Evolution* **12**, 53–58.

Arkins AM (1996). *The diet and activity patterns of short-tailed bats (Mystacina tuberculata auporica) on Little Barrier Island*. University of Auckland.

Arkins AM, Winnington AP, Anderson S, Clout MN (1999). Diet and nectarivorous foraging behaviour of the short-tailed bat (*Mystacina tuberculata*). *Journal of Zoology (London)* **247**, 183–187.

Bertram JEA, Gutmann A (2009). Motions of the running horse and cheetah revisited: fundamental mechanics of the transverse and rotary gallop. *Journal of The Royal Society Interface* **6**, 549–559.

Cartmill M, Lemelin P, Schmitt D (2002). Support polygons and symmetrical gaits in mammals. *Zoological Journal of the Linnean Society* **136**, 401–420.

Clark BD (1977). Energetics of hovering flight and the origin of bats. In: Hecht MK, Goody PC and Hecht BM (eds). *Major patterns in vertebrate evolution*. New York: Plenum Press. 423–425.

Daniel MJ (1976). Feeding by the short-tailed bat *Mystacina tuberculata* on fruit and possibly nectar. *New Zealand Journal of Zoology* **3**, 391–398.

Daniel MJ (1979). The New Zealand short-tailed bat *Mystacina tuberculata* a review of present knowledge. *New Zealand Journal of Zoology* **6**, 357–370.

Dietz DL. (1973). Bat walking behavior. *Journal of Mammalogy* **54**, 790–792.

Dumont ER (1997). Cranial shape in fruit, nectar and exudate feeders: implications for interpreting the fossil record. *American Journal of Physical Anthropology* **102**, 187–202.

Dwyer PD (1960). *Studies on New Zealand Chiroptera. Wellington, New Zealand*. Victoria University of Wellington, 166 pp.

Dwyer PD (1962). Studies on the two New Zealand bats. *Zoology Publications from Victoria University of Wellington* **28**, 1–28.

Findley JS, Wilson DE (1974). Observations on the neotropical disk-winged bat, *Thyroptera tricolor* Spix. *Journal of Mammalogy* **55**, 562–571.

Freeman PW (1981). A multivariate study of the family Molossidae (Mammalia, Chiroptera): Morphology, ecology, evolution. *Fieldiana (Zoology)* **7**, 1–173.

Freeman PW (1984). Functional Cranial Analysis of Large Animalivorous Bats (Microchiroptera). *Biological Journal of the Linnean Society* **21**, 387–408.

Gardiner JD, Nudds RL (2011). No apparent ecological trend to the flight-initiating jump performance of five bat species. *Journal of Experimental Biology* **214**, 2182–2188.

Gatesy SM, Middleton KM (1987). Bipedalism, flight, and the evolution of theropod locomotor diversity. *Journal of Vertebrate Paleontology* **17**, 308–329.

Goodwin RE (1970). The ecology of Jamaican bats. *Journal of Mammalogy* **51**, 571–579.

Göpfert MC, Wasserthal LT (1995). Notes on echolocation calls, food and roosting behaviour of the Old World Sucker-footed bat *Myzopoda aurita* (Chiroptera, Myzopodidae). *Zeitschrift für Säugetierkunde* **60**, 1–8.

Greenhall AM (1988). Feeding behavior. In: Greenhall AM and Schmidt U (eds). *Natural history of vampire bats*, pp. 111–131. Boca Raton, CRC Press.

Hand SJ, Weisbecker V, Beck RMD, Archer M, Godthelp H, Tennyson AJD, Worthy TH (2009). Bats that walk: a new evolutionary hypothesis for the terrestrial behaviour of New Zealand's endemic mystacinids. *BMC Evolutionary Biology* **9**, 169–181.

Hermanson JW, O'Shea TJ (1983). *Antrozous pallidus*. *Mammalian Species* **213**, 1–8.

Hermanson JW, Cobb MA, Schutt WA Jr., Muradali F, Ryan JM (1993). Histochemical and myosin composition of vampire bat (Desmodus rotundus) pectoralis muscle targets a unique locomotory niche. *Journal of Morphology* **217**, 347–56.

Hermanson JW, Ryan JM, Cobb MA, Bentley J, Schutt WA Jr (1998). Histochemical and electrophoretic analysis of the primary flight muscle of several phyllostomid bats. *Canadian Journal of Zoology* **76**, 1983–1992.

Hill JE, Daniel MJ (1985). Systematics of the New Zealand short-tailed bat *Mystacina* Gray, 1843 (Chiroptera: Mystacinidae). *Bulletin of the British Museum (Natural History : Zoology)* **48**, 279–300.

Hill JE, Smith JD (1984). *Bats: A natural history*. University of Texas Press, Austin.

Howell DJ, Pylka J (1977). Why bats hang upside down: a biomechanical hypothesis. *Journal of Theoretical Biology* **69**, 625–631.

Jenkins FA, Krause DW (1983). Adaptations for climbing in North American Multituberculates (Mammalia). *Science* **220**, 712–715.

Jenkins FA, McClearn D (1984). Mechanisms of hind foot reversal in climbing mammals. *Journal of Morphology* **182**(2), 197–219.

Johnston DS, Fenton MB (2001). Individual and population-level variability in diets of pallid bats (*Antrozous pallidus*). *Journal of Mammalogy* **82**, 362–373.

Jones G, Webb PI, Sedgeley JA, O'Donnell CFJ (2003). Mysterious *Mystacina*: how the New Zealand short-tailed bat (*Mystacina tuberculata*) locates insect prey. *Journal of Experimental Biology* **206**, 4209–4216.

Jones KE, Purvis A, MacLarnon A, Bininda-Emonds ORP, Simmons NB (2002). A phylogenetic supertree of the bats (Mammalia: Chiroptera). *Biological Reviews* **77**, 223–259.

Kuo AD (2002). Energetics of actively powered locomotion using the simplest walking model. *Journal of Biomechanical Engineering* **124**, 113–120.

Lawrence MJ (1969). Some observations on non-volant locomotion in vespertilionid bats. *Journal of Zoology (London)* **157**, 309–317.

Lindhe Norberg UM (2002). Structure, form, and function of flight in engineering and the living world. *Journal of Morphology* **252**, 52–81.

Lloyd BD (2001). Advances in New Zealand mammalogy 1990–2000: short-tailed bats. *Journal of The Royal Society of New Zealand* **31**, 59–81.

Lovegrove BG (2004). Locomotor mode, maximum running speed, and basal metabolic rate in placental mammals. *Physiological and Biochemical Zoology* **77**, 916–928.

Mills DJ, Norton TW, Parnaby HE, Cunningham RB, Nix HA (1996). Designing surveys for microchiropteran bats in complex forest landscapes – A pilot study from south-east Australia. *Forest Ecology and Management* **85**, 149–161.

Neuweiler G (2000). *The biology of bats*. Oxford University Press, Oxford.

Norberg UM (1985). Evolution of Vertebrate Flight – an Aerodynamic Model for the Transition from Gliding to Active Flight. *American Naturalist* **126**, 303–327.

Norberg UM, Rayner JMV (1987). Ecological Morphology and Flight in Bats (Mammalia, Chiroptera) – Wing Adaptations, Flight Performance, Foraging Strategy and Echolocation. *Philosophical Transactions of the Royal Society of London, Series B* **316**, 337–419.

Orr RT (1954). Natural history of the pallid bat, *Antrozous pallidus* (LeConte). *Proceedings of the California Academy of Sciences* **28**, 165–246.

Padian K (1987). A comparative phylogenetic and functional approach to the origin of vertebrate flight. In: Fenton MB, Racey PA, Rayner JMV (eds). *Recent advances in the study of bats*, pp. 3–22. Cambridge, Cambridge University Press.

Pettigrew JD (1986). Flying primates – megabats have the advanced pathway from the eye to midbrain. *Science* **4743**, 1304–1306.

Ratcliffe JM, Dawson JW (2003). Behavioural flexibility: the little brown bat, *Myotis lucifugus*, and the northern long-eared bat, *M. septentrionalis* both glean and hawk prey. *Animal Behaviour* **66**, 847–856.

Riskin DK, Hermanson JW (2005). Independent evolution of running in vampire bats. *Nature* **434**, 292.

Riskin DK, Bertram JEA, Hermanson JW (2005). Testing the hindlimb-strength hypothesis: Non-aerial locomotion by Chiroptera is not constrained by the dimensions of the femur or tibia. *Journal of Experimental Biology* **208**, 1309–1319.

Riskin DK, Parsons S, Schutt, WA Jr., Carter GC, Hermanson JW (2006). Terrestrial locomotion of the New Zealand Short-tailed Bat *Mystacina tuberculata* and the Common Vampire Bat *Desmodus rotundus*. *Journal of Experimental Biology* **209**, 1725–1736.

Riskin DK, Willis DJ, Iriarte-Diaz J, Hedrick TL, Kostandov M, Chen J, Laidlaw DH, Breuer KS, Swartz SM (2008). Quantifying the complexity of bat wing kinematics. *Journal of Theoretical Biology* **254**, 604–615.

Riskin DK, Bahlman JW, Hubel TY, Ratcliffe JM, Kunz TH, Swartz SM (2009). Bats go head-under-heels: the biomechanics of landing on a ceiling. *Journal of Experimental Biology* **212**, 944–953.

Ruina A, Bertram JEA, Srinivasan M (2005). A collisional model of the energetic cost of support work qualitatively explains leg sequencing in walking and galloping, pseudo-elastic leg behavior in running and the walk-to-run transition. *Journal of Theoretical Biology* **237**, 170–192.

Schnitzler H-U, Moss CF, Denzinger A (2003). From spatial orientation to food acquisition in echolocating bats. *Trends in Ecology & Evolution* **18**, 386–394.

Scholey KD (1986). The evolution of flight in bats. In: Nachtigall W (ed). *Bat flight = Fledermausflug*, pp. 1–12. Stuttgart ; New York, G. Fischer.

Schutt WA Jr. (1998). Chiropteran hindlimb morphology and the origin of blood feeding in bats. In: Kunz TH and Racey PA (eds). *Bat Biology and Conservation*, pp. 157–168. Washington: Smithsonian Institution Press.

Schutt WA Jr., Altenbach JS (1997). A sixth digit in *Diphylla ecaudata*, the hairy legged vampire bat. *Mammalia* **61**, 280–285.

Schutt WA Jr., Simmons NB (2001). Morphological specializations of *Cheiromeles* (naked bulldog bats; Molossidae) and their possible role in quadrupedal locomotion. *Acta Chiropterologica* **3**, 225–235.

Schutt WA Jr., Altenbach JS, Chang YH, Cullinane DM, Hermanson JW, Muradali F, Bertram JEA (1997). The dynamics of flight-initiating jumps in the common vampire bat Desmodus rotundus. *Journal of Experimental Biology* **200**, 3003–3012.

Siemers BM, Ivanova T (2004). Ground gleaning in horseshoe bats: comparative evidence from *Rhinolophus blasii, R. euryale* and *R. mehelyi. Behavioral Ecology and Sociobiology* **56**, 464–471.

Simmons NB (2005). Chiroptera. In: Rose KD and Archibald DJ (eds). *The rise of placental mammals: origins and relationships of the major extant clades*, pp. 159–174. Baltimore, Johns Hopkins University Press.

Simmons NB, Geisler JH (1998). Phylogenetic relationships of Icaronycteris, Archaeonycteris, Hassianycteris, and Palaeochiropteryx to extant bat lineages, with comments on the evolution of echolocation and foraging strategies in Microchiroptera. *Bulletin of the American Museum of Natural History* **235**, 4–182.

Simmons NB, Seymour, KL, Habersetzer, J, Gunnell, GF (2008). Primitive early Eocene bat from Wyoming and the evolution of flight and echolocation. *Nature* **451**, 818–821.

Speakman JR (2001). The evolution of flight and echolocation in bats: another leap in the dark. *Mammal Review* **31**, 111–130.

Strickler TL (1978). *Functional osteology and myology of the shoulder in the Chiroptera.* New York, S. Karger.

Swartz SM, Middleton KS (2007). Biomechanics of the bat limb skeleton: Scaling, material properties and mechanics. *Cells, Tissues, Organs* **187**, 59–84.

Teeling EC, Springer MS, Madsen O, Bates P, O'Brien SJ, Murphy WJ (2005). A molecular phylogeny for bats illuminates biogeography and the fossil record. *Science* **307**, 580–584.

Van den Bussche RA, Hoofer SR (2005). Phylogenetic relationships among recent chiropteran families and the importance of choosing appropriate out-group taxa. *Journal of Mammalogy* **85**, 321–330.

Vaughan TA (1959). Functional morphology of three bats: *Eumops, Myotis, Macrotus. University of Kansas, Museum of Natural History* **12**, 1–153.

Vaughan TA (1970). The skeletal system. In: Wimsatt WA (ed). *Biology of Bats*, pp. 97–138. New York, Academic Press.

Ward S, Möller U, Rayner JMV, Jackson DM, Nachtigall W, Speakman JR (2004). Metabolic power of European starlings *Sturnus vulgaris* during flight in a wind tunnel, estimated from heat transfer modelling, doubly labelled water and mask respirometry. *Journal of Experimental Biology* **207**, 4291–4298.

CHAPTER THIRTEEN

The Fight or Flight Dichotomy: Functional Trade-Off in Specialization for Aggression Versus Locomotion

David R. Carrier

Department of Biology, University of Utah, USA

13.1   INTRODUCTION

Most traits serve multiple functions. When two or more functions impose conflicting demands, optimization is impossible and a trade-off phenotype results (Maynard Smith *et al.*, 1985; Lauder, 1991; Vanhooydonck *et al.*, 2001; Van Damme *et al.*, 2002). Compromises arising from the requirements of locomotion versus those of fighting are an example that may be particularly important in the evolution of phenotypic diversity (Carrier, 2002, 2004; Chase *et al.*, 2002; Pasi and Carrier, 2003; Kemp *et al.*, 2005; Herrel *et al.*, 2009; Helton, 2011; Bro-Jørgensen, 2013; Webster *et al.*, 2014; Morris and Brandt, 2014; Martín-Serra *et al.*, 2014). Both locomotion and physical competition are

Understanding Mammalian Locomotion: Concepts and Applications, First Edition.
Edited by John E. A. Bertram.
© 2016 John Wiley & Sons, Inc. Published 2016 by John Wiley & Sons, Inc.

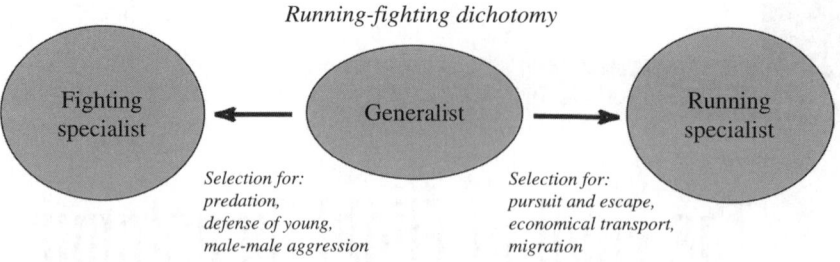

FIGURE 13.1

Illustration of the functional dichotomy between specialization for running versus specialization for fighting. Although most species are locomotor/fighting generalists, being reasonably good at both, strong selection for fighting or locomotor performance will drive species to become fighting or locomotor specialists. Traits that make an individual good at fighting may, in many cases, limit locomotor performance and vice versa.

critical to survival and reproductive fitness in most animal species, but traits that make an individual good at fighting often limit locomotor performance and vice versa (Figure 13.1).

For example, among apes, sexual dimorphism in body size and male-male fighting are most dramatic in gorillas (Nowak & Paradiso, 1983). Large body size allows dominant male gorillas to defend multi-female groups against lone males interested in attracting the females (Fossey, 1983, 1984; Watts, 1989). Large size, however, severely limits the ability of male gorillas to climb and forage in trees (Schaller, 1963). In contrast, both male and female gibbons brachiate with spectacular grace and agility. Gibbons exhibit little or no sexual dimorphism in body size, and mate in monogamous pairs (Nowak & Paradiso, 1983; MacKinnon & MacKinnon, 1984). Both male and female gibbons actively defend their territory, but aggressive encounters rarely involve physical contact (Preuschoft et al., 1984; Mitani, 1987). In this comparison of apes, locomotion is constrained in the fighting specialist, whereas fighting ability is limited in the locomotor specialist.

Functional trade-offs between specialization for running versus specialization for fighting are likely because rapid and economical running is dependent on long, gracile limbs, and muscles that are specialized for the storage and recovery of elastic strain energy (Hildebrand and Goslow, 2001; Taylor, 1994). Fighting ability, on the other hand, appears to be associated with short, stout limbs and muscles specialized for high force production and acceleration.

It is well established that longer limbs reduce the cost of transport in terrestrial species (Kram and Taylor, 1990; Steudel-Numbers and Tilkens, 2004; Pontzer, 2005). Long limbs reduce the number of steps that a running or walking animal must take to cover a given distance (Hildebrand and Goslow, 2001), and reduction of the mass of limb bones and muscles, particularly the distal elements, decreases the energy required to swing the limbs back and forth in each step (Hildebrand and Hurley, 1985; Steudel, 1991). Because the energy required to oscillate the limbs increases dramatically with running speed (Cavagna and Kaneko, 1977; Fedak et al., 1982; Willems et al., 1995; Marsh et al., 2006), selection for low limb mass is expected to be most pronounced in those species that are the fastest runners. Indeed, relatively long and slender limbs have evolved repeatedly in lineages that have become specialized for high speed and economical running (Pough *et al.*, 1999).

Although much less is known about characteristics that enhance fighting ability, strength and agility are generally thought to be important in most types of fighting. During fighting, an individual must produce high forces and mechanical work to accelerate its own body and to manipulate and injure its opponent. Additionally, the direction of force application by limbs and trunk is likely to be much more variable during fighting than during running. Thus, large limb muscles with high mechanical advantage, and stout bones with a circular cross-section, can be expected to enhance fighting performance.

Obviously, species must be competent at many more behaviors than just locomotion and fighting. The musculoskeletal system of mammals also enables the potentially confounding functions of feeding, foraging and prey capture, grooming, climbing, digging, thermoregulation, parental care, pregnancy, birthing, mating, and communication. Nevertheless, fighting and locomotion are important to fitness in many, or most, species of vertebrate, and the case for biomechanical trade-offs between these two functions, as is argued below, is relatively strong. My goal in writing this chapter is to increase awareness among comparative biomechanists and physiologists of the importance of specialization for aggressive behavior, and to illustrate the extent to which specialization for aggression may be incompatible with specialization for locomotor economy and speed.

13.1.1 Why fighting is important

Generally, individuals avoid physical fighting through the development and maintenance of dominance hierarchies and signaling of fighting potential. These signals, however, must convey valid information about fighting potential in order for threat displays to evolve (Maynard Smith and Price 1973; Parker 1974; Enquist and Leimar 1990; Szamado, 2003, 2008). Fighting is defined as physical competition that has potential to seriously injure or kill an opponent.

Actual fighting occurs in situations in which survivorship and reproductive fitness are at stake (Parker, 1974; Hamilton, 1979; Enquist and Leimar, 1983, 1990). Individuals fight:

1. to defend themselves and their young from predators or conspecifics;
2. to defend limiting resources and territory;
3. to compete for access to mates.

Sexual selection has resulted in mating strategies in many species in which males compete for access to reproductively active females through physical fighting, or the threat of physical fighting (Andersson, 1994). Individuals who dominate in these male-male contests have more mating opportunities (Moore, 1990; Francis, 1984; Wolff, 1985; Dewsbury, 1982; Ellis, 1995). In mammals, possibly more than in any other group, fighting performance is central to reproductive fitness. Fatal fighting has been observed in mice (Svare, 1981), musk oxen (5–10% male mortality (Wilkinson and Shank, 1977)), mountain sheep (Geist, 1971), pronghorns (12% of fights between males result in death (Byers, 1998)), cheetahs (Caro and Collins, 1986), lions (Grinnell *et al.*, 1995), hyenas (17% adult mortality (Kruuk, 1972)); 25% of infants die from siblicide (Frank *et al.*, 1991), wolves (as high as 10% mortality of both sexes (Mech *et al.*, 1998)), chimpanzees (as much as 30% male mortality (Wrangham, 1999)), and humans (Beckerman *et al.*, 2009; Chagnon *et al.*, 1979; Daly and Wilson, 1988; Knauft, 1991; Keeley, 1996; Walker, 2001).

Because fighting is central to survival and reproductive success in many species, we expect morphological, physiological and behavioral traits that enhance fighting performance to be retained by natural and sexual selection (Lappin and Husak, 2005; Husak *et al.*, 2006; Briffa and Sneddon, 2007; Lailvaux and Irschick, 2007; Judge and Bonanno, 2008; Bywater *et al.*, 2008). However, if specialization for fighting performance is at odds with specialization for locomotor economy and speed, musculo-skeletal diversity may often reflect adaptive compromises of these two functions, and/or novel traits associated with circumvention of constraint posed by the trade-offs.

13.1.2 Size sexual dimorphism as an indicator of male-male aggression

Comparative physiologists and biomechanists are often able to rank the performance of species. We know, for example, that Blackbuck (*Antelope cervicapra*) can run approximately 25 km/hr faster than Grant's gazelle (*Gazella granti*) (Garland, 1983). With such a species ranking, and knowledge of the anatomical and physiological differences among species, physiologists can develop and test hypotheses of character function and evolution. Performance in fighting, however, is notoriously difficult to quantify (Plavcan and van Schaik, 1997; Plavcan, 1999). Quantification of fighting performance requires extensive periods of field observations, and these observations are often complicated by the subjective interpretations of the observer. Additionally, there is no satisfactory way to equate data from different species because of different fighting behaviors, and because different field workers invariably make the observations, using different subjective behavioral criteria and different methods of observation. These difficulties represent significant obstacles to functional and evolutionary analyses of aggressive behavior.

Given the difficulties associated with quantifying fighting performance, we have adopted body size sexual dimorphism (SSD) as an indicator of the relative level of male-male physical aggression (Carrier *et al.*, 2002; Carrier, 2007). Although other factors may lead to sexual dimorphism, in which males are larger than females (Andersson, 1994; Blanckenhorn, 2005), there is extensive literature documenting the relationship between male-biased SSD and male-male aggression in mammals. Among mammals, species in which males are larger than females tend to have polygynous mating systems, in which males compete physically for reproductive access to females (reviewed in Andersson, 1994). Among primates, there is a positive correlation between SSD and socionomic sex ratio (ratio of adult females to adult males in a breeding group) (Clutton-Brock *et al.*, 1977).

Recent analyses of anthropoid primates show that SSD is strongly associated with both male-male competition levels and operational sex ratio (the ratio of males that are ready to mate to females that are ready to mate) (Plavcan and van Schaik, 1997; Plavcan, 1999, 2004). SSD also increases with level of polygyny among at least three groups of large herbivore: bovids, cervids, and maropods (Clutton-Brock *et al.*, 1980; Jarman, 1983; Alexander *et al.*, 1979). There is also a strong correlation between the degree of polygyny and SSD in pinnipeds (Alexander *et al.*, 1979). Finally, in a seminal game theory study, Parker (1983) showed that, when males compete for mating opportunities, the greater the number of females in the contested group, the higher the evolutionary stable investment in male armament, including investment in larger body size.

There are several factors that prevent SSD from being an ideal indicator of species-level male-male aggression. First, selection for fighting ability in females will result in lower levels of SSD. Predator-prey contests, protection of one's young, female-female competition, and female-male conflict, are all likely to result in selection for fighting ability in females, which is likely to result in the evolution of larger females. Second, the evolution of larger body size due to sexual selection on males could change the natural history of a species, such that the optimal body size of females also increases (Lande, 1980; Chase, *et al.*, 2005). In this case, natural selection would not return females to their original size and, therefore, SSD would not evolve, or would be minimized. Given these limitations, SSD has limited value as an index of levels of male-male aggression in comparisons between two species, but has proven analytical strength to distinguish broad trends in levels of male-male aggression when applied to a large number of closely related species (Carrier *et al.*, 2002; Carrier, 2007).

13.2 TRADE-OFFS IN SPECIALIZATION FOR AGGRESSION VERSUS LOCOMOTION

13.2.1 The evolution of short legs – specialization for aggression?

Among terrestrial tetrapods, long limbs are known to be associated with:

1. economy in both walking and running (Kram and Taylor, 1990; Steudel-Numbers and Tilkens, 2004; Pontzer, 2005);
2. high speed running (Losos, 1990; Pough *et al.*, 1999); and
3. enhanced jumping performance (Losos, 1990; Marsh, 1994; Emerson, 1985).

If these locomotor advantages are real, what, if any, benefit comes from short legs? Short limbs probably improve maneuverability in structurally complex or spatially restrictive habitats, such as thick vegetation or burrows. Smith and Savage (1956) suggested that short limbs increase the mechanical advantage with which animals apply forces to the substrate. Greater leverage is assumed to improve digging performance (Hildebrand and Goslow, 2001) and agility. Short limbs may also be advantageous in aggressive encounters. Increased mechanical advantage increases the forces that can be applied to opponents during grappling, biting and shaking, and striking with the limbs. Additionally, if the center of gravity is lowered, short limbs increase stance stability.

The great apes (i.e., family Hominidae) comprise a group of mammals that is characterized by short legs. Relative hindlimb length decreases with body size in catarrhine primates, and the trend is most dramatic in the hominoids (Jungers, 1984). The relatively short legs of larger primates are thought to represent specialization for climbing.

Short legs improve balance when walking quadrupedally above branches, by lowering the center of mass (Cartmill, 1985; Doran, 1993). Short legs also facilitate climbing broad tree trunks, by allowing the body to be held close to the trunk, thereby lowering the tensile forces required from the forelimbs (Cartmill, 1974; Jungers, 1978; Susman et al., 1985). Short limbs, however, must limit the ability to bridge gaps between possible sites of support when climbing and traveling through the canopy. Indeed, the most arboreal apes, the gibbons, have relatively long hindlimbs for catarrhines of their size (Jungers, 1984). Short limbs, therefore, may facilitate some types of arboreal locomotion, but will limit others.

Physical aggression associated with male-male competition is prevalent in all extant species of great apes (Wrangham and Peterson, 1996; Carrier, 2007) and, as suggested above, is a behavior in which short hindlimbs are expected to improve performance. Although teeth are an important weapon in the Hominoidea, great apes also fight with their forelimbs from a bipedal stance on the ground (Fossey, 1983; Goodall, 1986; Livingstone, 1962; Wrangham and Peterson, 1996; deWaal, 1982, 1986; Kano, 1992; Furuichi, 1997). Characters that improve strength and stability in a bipedal stance should enhance fighting performance.

To test the hypothesis that the evolution of short legs in great apes represents selection for improved fighting performance, I looked at the relationship between limb length and size sexual dimorphism (SSD) and sexual dimorphism in canine height (CSD) in lesser and great apes (Carrier, 2007). Figure 13.2 plots the species values. Analysis of phylogenetic independent contrasts showed that the evolution of hindlimb length in apes is inversely correlated with the evolution of SSD ($R^2 = 0.683$, p-value = 0.0008) and the evolution of CSD ($R^2 = 0.630$, p-value = 0.013). These observations are consistent with the suggestion that selection for fighting performance leads to the evolution of shorter limbs.

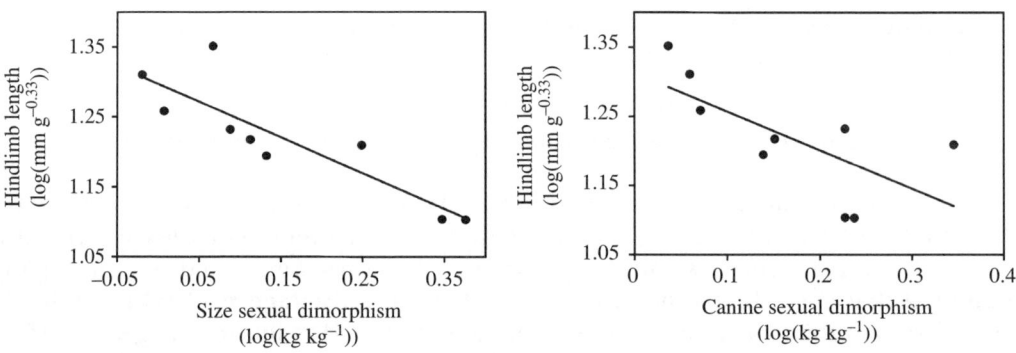

FIGURE 13.2

Relationship between mass-specific hindlimb length and body size sexual dimorphism (SSD) and sexual dimorphism in canine height (CSD) in Hominoidea (i.e., Hylobatidae and Hominidae). SSD is male mass divided by female mass. CSD is male canine height divided by female canine height. Size and canine sexual dimorphism are strong indicators of mating system and levels of male-male competition. Thus, species of ape with the highest levels of male-male competition tend to have relatively short legs. Modified from Carrier (2007).

13.2.2   Muscle architecture of limbs specialized for running versus fighting

We tested a series of hypotheses of functional trade-offs in specialization for fighting versus running in the limb muscles of two breeds of dog, *Canis familiaris* (Figure 13.3; Pasi and Carrier, 2003; Kemp *et al.*, 2005). Although there are well-recognized limitations associated with comparisons of two species or breeds when studying adaptation (Garland and Adolph, 1994), the choice of these breeds ameliorate the problems in substantial ways (see Pasi and Carrier, 2003).

The rotational inertia of oscillating limbs (Cavagna and Kaneko, 1977; Fedak *et al.*, 1982) leads to the expectation that animals specialized for running will have relatively less muscle mass in their distal limbs (Hildebrand and Hurley, 1985; Steudel, 1991). In contrast, the distal limbs of animals specialized for fighting are expected to have large muscles, to allow the production of large forces and high power for agility, balance, and opponent manipulation. Our analysis found that greyhounds had a smaller percentage of appendicular muscle mass in their distal limbs than did pit bulls (Figure 13.4).

Elastic storage and recovery of strain energy in the tendons of distal limb muscles improves the economy of transport during running in many species (Cavagna *et al.*, 1964;

FIGURE 13.3

Comparison of two breeds of dogs – one selected for high-speed, anaerobic stamina (greyhounds), and one selected for dog-dog fighting (pit bulls). On the day this photograph was taken, both dogs weighed 33 kg.

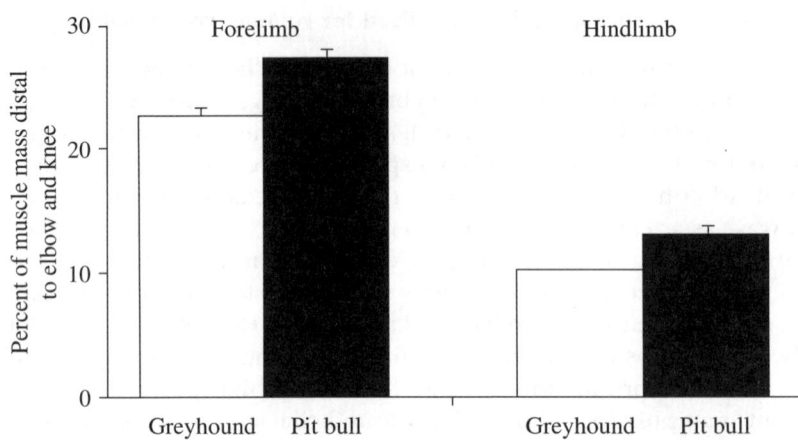

FIGURE 13.4

Percent of limb muscle mass distal to the elbow and knee of greyhounds and pit bulls (Pasi and Carrier, 2003). The greyhounds have a smaller percentage of appendicular muscle mass in their distal limbs than do pit bulls.

Dawson and Taylor, 1973; Alexander, 1984; Roberts *et al.*, 1997; Biewener, 1998). Animals specialized for distance and/or high speed running can be expected to have a high capacity for elastic storage. In contrast, long stretchy tendons in series with extensor muscles would probably be a handicap for any animal attempting to overpower an opponent during physical combat. Therefore, animals specialized for fighting can be predicted to have relatively limited potential for elastic storage.

To assess this prediction in our comparison of greyhounds and pit bulls, we used measures of muscle cross-sectional area to estimate the forces that the muscles could produce, and the length and cross-sectional area of the in-series tendons to estimate the strain energy that these muscle forces could store in the tendons. Our analysis found a two-fold greater potential to store and recover elastic strain energy in the extensor muscle-tendon systems of the ankle joint of greyhounds than in pit bulls (Figure 13.5).

A third prediction is that animals specialized for running have less muscle strength in their forelimbs than in their hindlimbs. This hypothesis emerges from two observations. First, there is a division of labor in the limbs; forelimbs play a greater role in deceleration, whereas hindlimbs play a greater role in acceleration (Cruse, 1976; Cavagna *et al.*, 1977; Jayes and Alexander, 1978; Heglund *et al.*, 1982; Blickhan and Full, 1987; Full *et al.*, 1991).

Second, active skeletal muscle generates much greater force when it is stretched (eccentric contraction) than when it shortens (Katz, 1939). Because the extensor muscles of the limbs must actively stretch to absorb energy during deceleration, but must actively shorten to produce acceleration, less muscle strength is expected in the forelimbs than the hindlimbs of running specialists. In contrast, the extensor muscles of the forelimbs of animals specialized for fighting can be expected to be as large, or larger, than those of the hindlimbs, because forelimb strength is needed to manipulate an opponent and is essential for rapid turning and agility (see Walter, 2003). Our analysis

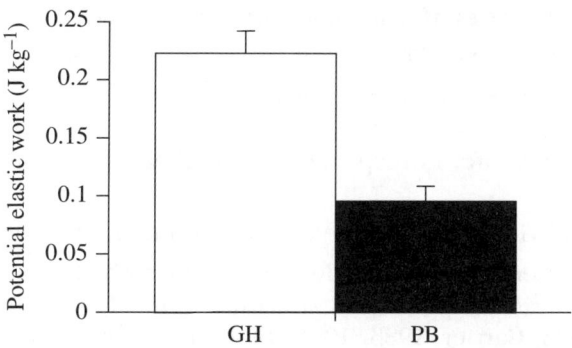

FIGURE 13.5

Comparison of the potential to store and recover elastic strain energy in the extensor muscle-tendon systems of the ankle joints of greyhounds than pit bulls (Pasi and Carrier, 2003).

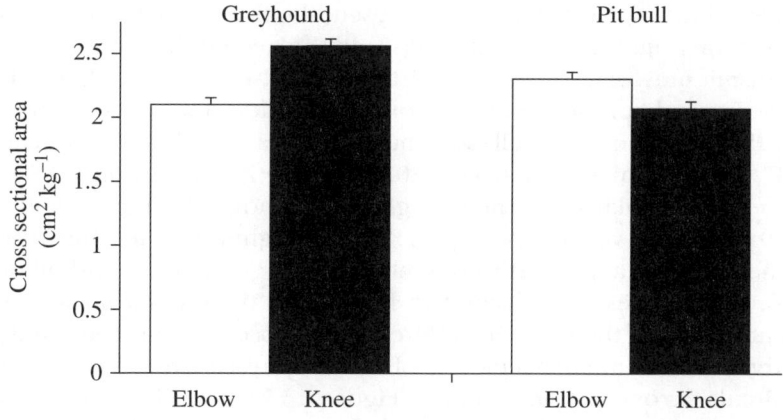

FIGURE 13.6

Average cross-sectional area of the extensor muscles of the knee and elbow of greyhounds and pit bulls (Pasi and Carrier, 2003). Cross-sectional area is a determinant of the force a muscle can produce. Greyhounds have greater cross-sectional area in the extensor muscles of their hindlimbs than their forelimbs. The reverse is true for pit bulls.

revealed that the extensor muscles of the hindlimb of the greyhounds had a greater cross-sectional area than the serially homologous extensor muscles of the forelimb (Figure 13.6). The pit bulls exhibited the opposite pattern.

In summary, our analyses demonstrated that greyhounds differ from pit bulls in having relatively less muscle mass distally in their limbs, a much greater capacity for elastic storage in the extensor muscle-tendon systems of their ankle joints, and relatively weaker muscles in their forelimbs than in their hindlimbs (Pasi and Carrier, 2003). These observations are consistent with the hypothesis that specialization of skeletal muscles for rapid or economical running is often incompatible with specialization for performance in aggressive encounters.

13.2.3 Mechanical properties of limb bones that are specialized for running versus fighting

The mechanical properties of bones can vary:

1. during an organism's life, in response to changing developmental and functional demands;
2. throughout an individual's body, due to varying functional requirements; and
3. among different members of a clade, associated with different life histories and environmental conditions (Currey, 1979; Currey and Pond, 1989; Biewener, 1990; Swartz *et al.*, 1992; Carrier, 1983, 1996; Blob and Biewener, 1999; Heinrich *et al.*, 1999; Blob and LaBarbera, 2001).

Currey (1979) provided a classic illustration of the relationship between bone mechanical properties and function, with a comparison of red deer antler, cow femur, and fin whale tympanic bulla. Of the three types of bone, antler was the least stiff, but absorbed the most energy before it failed, and cow femur resisted the greatest forces in bending. Currey suggested that these differences represent the need of antler to withstand large impact loads during fighting between males, and the need of limb bones to be stiff and strong to transmit muscular forces. This distinction between the mechanical properties of skeletal elements that function as weapons (i.e., antlers and tusks), versus those that function as limb elements, is well documented (Brear *et al.*, 1993; Kitchener, 1991; Currey, 1987, 1989; Blob and LaBarbera, 2001; but see Zioupos *et al.*, 1997).

We wondered if similar differences might exist among the limb bones of animals specialized for running, versus those specialized for fighting. To test this possibility, we compared the mechanical properties of limb bones in greyhounds and pit bulls (Kemp *et al.*, 2005). Limb bones were loaded to fracture in three-point static bending. The proximal limb bones of the pit bulls differed from those of the greyhounds, in having relatively larger second moments of area of mid-diaphyseal cross-sections, and in having more circular cross-sectional shape (Figure 13.7). The limb bones of pit bulls exhibited lower stresses at yield, had lower elastic moduli, and failed at much higher levels of work (Figure 13.8). The results of this analysis suggest that selection for high-speed running is associated with the evolution of relatively stiff limb bones, whereas selection for fighting performance leads to the evolution of limb bones with high resistance to failure.

13.2.4 The function of foot posture: aggression versus locomotor economy

Plantigrade foot posture, in which the heel makes contact with the substrate during a walking or running step, appears to be a derived condition in some lineages of placental mammals. It is found in bears, great apes, and some species of insectivores and rodents. In contrast, most mammalian species, including the fastest and most economical runners, have either digitigrade or unguligrade feet. With these foot postures, the heel is held elevated above the ground, so that the animal walks and runs on the balls of its feet, or on its toes, respectively. Digitigrade and unguligrade postures are thought to improve locomotor economy by increasing step length (Hildebrand and Goslow, 2001), and enhancing storage and recovery of elastic strain energy in the

A

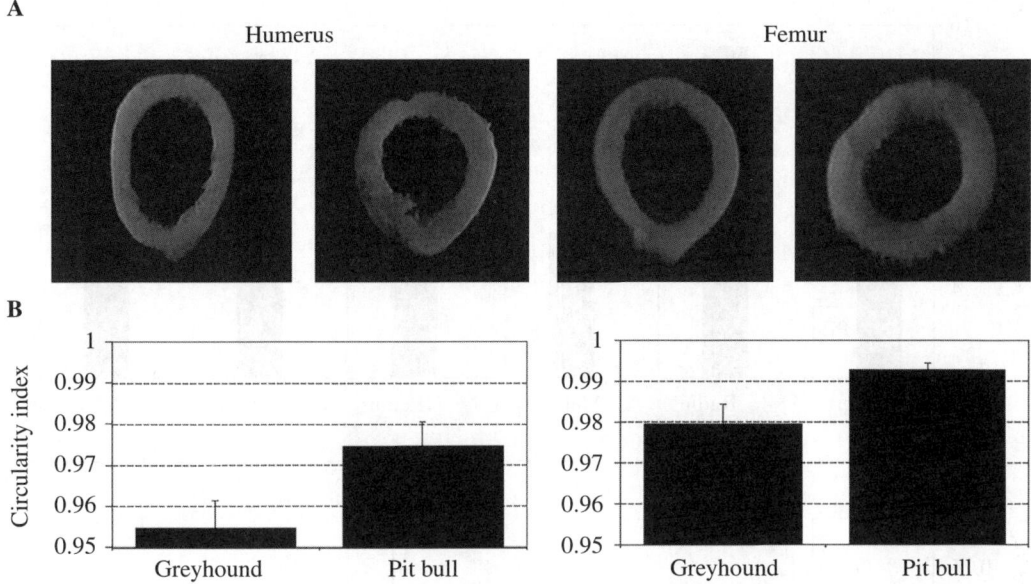

B

FIGURE 13.7

Mid-shaft cross-sectional shape of the humeri and femurs of greyhounds and pit bulls (Kemp *et al.*, 2005). (**A**) Comparisons of representative mid-shaft cross-sections. In both cases, the greyhound is the bone on the left. The anterior-posterior axis is oriented vertically. (**B**) Mean and SEM of circularity index. A circular cross-section has a circularity index of 1.

muscle-tendon systems of the distal limbs. If this is true, why have some groups of mammals evolved plantigrade foot posture?

There are reasons to suspect that plantigrade foot posture improves locomotor economy, particularly during walking. Relative to other species, humans are economical walkers (Steudel-Numbers, 2003; Sockol *et al.*, 2007). Humans also differ from other species that have been studied, in that it costs us substantially less to walk a given distance than to run the same distance (Margaria *et al.*, 1963; Farley and McMahon, 1992; Rubenson *et al.*, 2007). This energetic advantage may be partially due to a reduction in collisional losses of mechanical energy, during the support phase of each step (McGeer, 1990; Ruina *et al.*, 2005), which appears to result from plantigrade feet (Adamczyk *et al.*, 2006). Additionally, humans are known to have greater mechanical advantage at their limb joints during walking than during running (Biewener *et al.*, 2004; Usherwood *et al.*, 2012), and plantigrade foot posture is partially responsible for this difference.

To determine if plantigrade foot posture influences the metabolic cost of locomotion, we investigated the effect of different foot postures on the cost of walking and running in human subjects (Cunningham *et al.*, 2010). We measured oxygen consumption as subjects walked and ran with plantigrade and digitigrade foot postures on a motorized treadmill. Although walking and running on the balls of one's feet are not typical for many of us, they are natural behaviors for humans. Many elite track athletes habitually run on the balls of their feet. Furthermore, during running, unshod individuals avoid landing on their heels and, instead, land on the balls of their feet (Lieberman *et al.*, 2010;

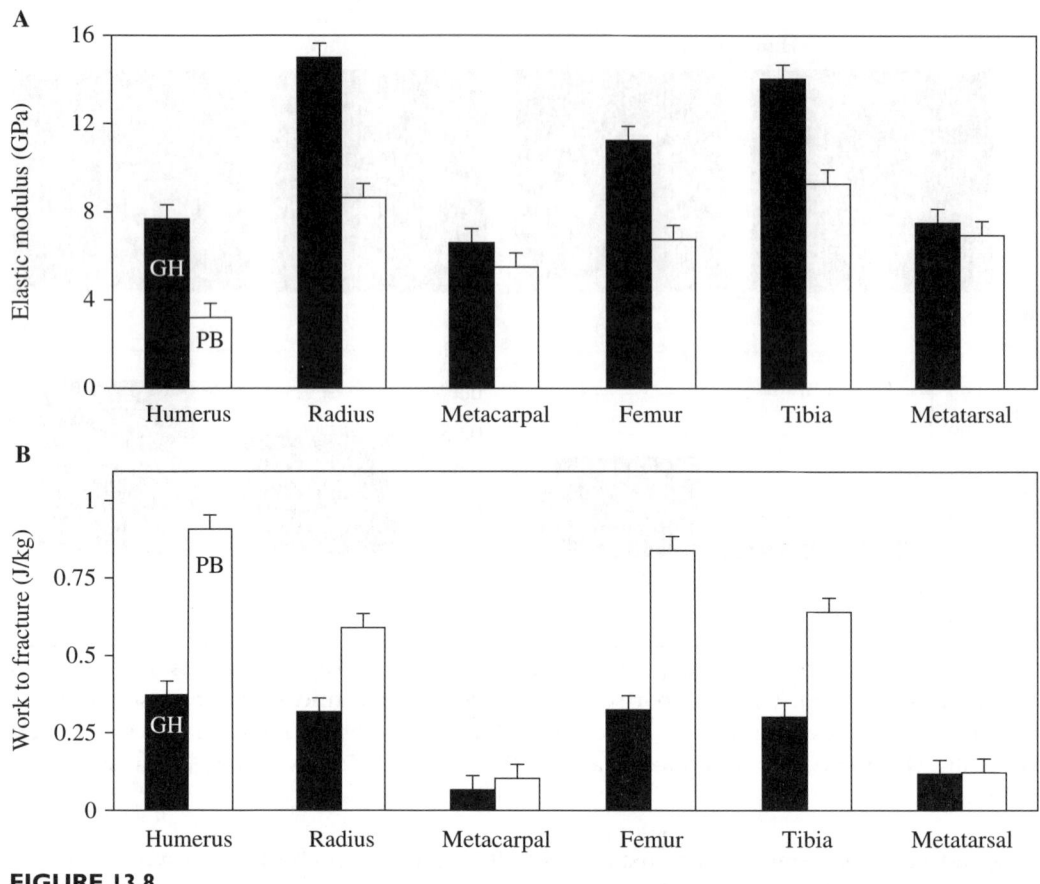

FIGURE 13.8

Average elastic modulus (**A**) and work to fracture (**B**) of the major limb bones of greyhounds and pit bulls (Kemp *et al.*, 2005). Greyhounds have stiffer bone tissue, but the bones of pit bulls are tougher.

Divert *et al.*, 2005). Toe walking is common among young children, and intermittent toe walking is considered normal up to the age of seven years (Kelly *et al.*, 1997; Kalen *et al.*, 1986). It is also common for humans to adopt a "tiptoeing" foot posture when they wish to walk covertly, or with stealth. Thus, the experiments of this study were based on a comparison of natural human movements.

When the subjects walked at their preferred speed with their heels slightly elevated above the surface of the treadmill (i.e., digitigrade) the average oxygen consumed to walk a kilometer (cost of transport, COT) increased by $53 \pm 6\%$ (mean and SEM) above that of walking with plantigrade foot posture (p-value = 0.001; Figure 13.9; Cunningham *et al.*, 2010). In contrast, there was no significant difference in COT when subjects ran with digitigrade versus plantigrade foot posture at four different speeds. These observations suggests that plantigrade posture bestows a biomechanical advantage during walking, but not during running.

In an attempt to discover the mechanical basis for the effect of foot posture on walking COT, we conducted several subsequent experiments. Our analysis of the

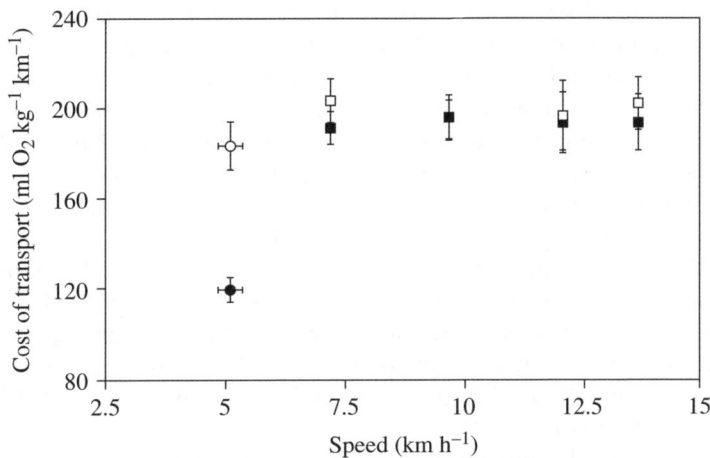

FIGURE 13.9

Cost of transport (mean and SE) versus locomotor speed for human subjects walking (circle symbols) and running (square symbols), with plantigrade (black symbols) and digitigrade (white symbols) foot posture. The horizontal error bars on the circles represent the standard error of the subjects' preferred walking speeds. Modified from Cunningham *et al.* (2010).

activity of the extensor muscles of the ankle, knee, hip and back during walking indicates that the moment and/or the mechanical work done at each of the limb joints increased when the subjects switched from plantigrade to digitigrade posture. Additionally, we found that the greater economy in walking associated with plantigrade foot posture is not related to:

1. an observed increase in stride frequency; or
2. decreases in postural stability.

Instead, the improved economy appears to result from:

1. lower collisional losses;
2. lower ground reaction force moments at the ankle joint; and
3. improved recovery of kinetic and potential energy during each step.

These three energy-saving mechanisms are dependent on heel contact with the substrate at the initiation of support.

For more than 99% of our 2.5 million year history, humans made their living as hunter-gatherers, foraging long distances to acquire edible plants and animals. Each day, modern hunter-gatherers travel an average of 9.5 km (females) and 14.1 km (males) (Marlowe, 2005). Modern hunter-gatherers also have relatively large home ranges – 175 km² (Marlowe, 2005). Thus, it is not surprising that humans are economical walkers, or that we retained a foot posture inherited from our more arboreal ancestors that facilitates economical walking.

But did plantigrade foot posture evolve in great apes to improve the economy of walking? Although this is a possibility (see Gebo, 1996, 1992), great apes are thought to be anatomically specialized for climbing and suspensory locomotion (Ward, 2007), and an analysis of heel contact in a variety of arboreal primates suggests that the plantigrade foot posture did not evolve in a terrestrial habitat (Schmitt and Larson, 2005).

The energetic savings associated with walking with plantigrade posture would likely not have been important to the ancestors of great apes. First, the daily distances traveled by modern apes are not great: orangutans (0.5–0.8 km (Bean, 1999)); chimpanzees (3–10 km (Williams *et al.*, 2002; Bean, 1999)); bonobos (2.4 km (Williams *et al.*, 2002; Bean, 1999)); and gorillas (0.5–2.6 km (Doran and McNeilage, 2001; Bean, 1999)). The average home ranges of great apes are also not large: orangutan, 1.5–30.0 km² (Singleton and van Schaik, 2001; Delgado and van Schaik, 2000; Bean, 1999); chimpanzee, 4–32 km² (Bean, 1999); bonobo, 22–58 km² (Bean, 1999); gorilla, 4–23 km² (Williams *et al.*, 2002; Bean, 1999).

Second, compared to the bipedal gait of humans, the quadrupedal gait of great apes approximately halves the collisional energy losses that occur in a walking stride (Bertram and Gutmann, 2008; Ruina *et al.*, 2005), further reducing the significance of any energetic savings that plantigrade feet provide. Thus, we suspect that the heel-down posture of great apes evolved for reasons other than economical arboreal or terrestrial transport.

Plantigrade feet may improve fighting performance, by allowing the application of larger torques to the ground. Striking and grappling with an opponent applies torques to the body. To avoid spinning of the body, these torques must be resisted through force couples (i.e., free moments) applied to the ground. Free moments can be applied to the ground through two feet, or through one foot that has two points of contact. Thus, the heel and ball/toes of a single plantigrade foot can apply free moments to the ground. Although free moments are poorly studied, we know that they occur when animals walk and run (Holden and Cavanagh, 1991; Li *et al.*, 2001), and when they make sharp turns (Lee *et al.*, 2001; Jindrich *et al.*, 2007; but see Qiao *et al.*, 2014). In general, ground reaction torques are likely to be important to turning agility, and to the application of lateral forces during grappling with and striking an opponent.

In summary, plantigrade foot posture allows the production of larger ground reaction torques, which are likely important for grappling with and striking an opponent during physical combat. Plantigrade posture also enhances locomotor economy during walking, but not during running. Although a heel strike at the beginning of stance has advantages for energetic economy during walking, it probably reduces the potential for storage and recovery of elastic energy at the ankle joint in running gaits. The distal joints – wrist and ankle – are generally believed to be the most important sites for elastic storage in running animals (Alexander, 1984). Thus, plantigrade foot posture appears to be an example in which specialization for aggressive behavior may be compatible with specialization for economical walking, but not compatible with economical running.

13.3 DISCUSSION

These examples suggest that functional trade-offs in specialization for locomotion and fighting broadly impact the design of the musculoskeletal system. Limb length is an obvious example. Although short limbs are likely to increase fighting performance, they reduce locomotor economy. The comparisons of greyhounds and pit bulls suggest that runners differ from fighters in both muscle architecture and bone mechanical properties. Finally, foot posture appears to be an example in which plantigrade posture enhances fighting performance, and is compatible with economical walking. In contrast, digitigrade and unguligrade foot postures likely enhance running economy by increasing stride length and elastic storage at the distal joints, but may reduce fighting ability by limiting the production of ground reaction torques.

Because many, or most, animals must, at some point in their lives, be effective at both locomotion and physical competition, trade-offs intrinsic to these two functions are likely to have influenced the evolutionary trajectories of many species. The comparison of gorillas and gibbons, mentioned above, is an obvious example, and many other likely cases exist. Among felids, for example, cheetahs are exceptionally fast runners, they exhibit little sexual dimorphism in body size (Nowak & Paradiso, 1983) and, although territorial males have been known to kill male intruders (Caro, 1994), cheetahs are recognized as exhibiting relatively little physical aggression toward conspecifics (Eaton, 1974).

In contrast, lions are characterized by intense male-male fighting and sexual dimorphism, in which males are, on average, 35% larger in body mass than females (Nowak & Paradiso, 1983). Consistent with the trade-off hypothesis, lions are not fast runners, and can sprint at only half the maximum speed of cheetahs (Schaller, 1972). Among grouse, Willow and Rock Ptarmigan migrate seasonally (Kaufman, 1996) and are usually monogamous (Johnsgard, 1975). In contrast, Ruffed and Sage Grouse are mainly permanent residents, with little or no seasonal migration (Kaufman, 1996); they are also highly polygynous, with intense male competition that includes physical fighting (Johnsgard, 1975). In this case, the two species with the lower locomotor capacity exhibit greater male-male aggression.

Another noteworthy example in which trade-offs intrinsic to locomotion versus fighting appear to have influenced evolutionary trajectories resulted from the evolution of prey immobilization systems in colubrid snakes (Savitzky, 1980). Basal snakes are thought to have immobilized their prey by constriction (Greene & Burghardt, 1978). In these snakes, the axial musculoskeletal system was used for both locomotion and subduing prey.

Among extant species, powerful constriction appears to preclude rapid locomotion (Ruben, 1977). The evolution of serous poison glands (Duvernoy's gland) in colubrid snakes may have provided a mechanism for uncoupling locomotor and prey capture (Savitzky, 1980). The five genera of colubrids in which serous glands are poorly developed or absent are those that use constriction for prey capture. In contrast, the fastest genera (*Chironius, Coluber, Drymarchon, Drymobius, Masticophis* and *Salvadora*) do not constrict, but have well-developed Duvernoy's glands. Hence, Savitzky (1980) suggests that the dramatic colubrid radiation that occurred during the Miocene may have been a response to the spread of open habitats. Taxa that were characterized by slow locomotion and immobilization of prey through constriction were out-competed by taxa characterized by rapid locomotion and immobilization of prey by envenomation.

Functional trade-offs arising from the fight or flight dichotomy are predicted to be causally associated with the evolution of sexual dimorphism. A recent study has reported sexual dimorphism in the shape of the limb skeleton in gray wolves (*Canus lupus*), a species that was expected to exhibit little or no dimorphism, because of its monogamous mating system and high parental investment by males (Morris and Brandt, 2014). In addition to having broader skulls, male wolves have greater mechanical advantage at their limb joints, and broader limb muscle attachment sites. Morris and Brandt attribute this sexual dimorphism to the different life history roles of male and female wolves. Males play the dominant role in inter-pack aggression, and actively provision the nursing female during the pup-rearing period, often by capturing large prey single-handedly. In contrast, the high energy demands of pregnancy and lactation may place greater selective value on locomotor economy in female wolves.

Functional trade-offs that prevent simultaneous evolution of optimal performance in locomotion and fighting may impact the evolution of life history in ways that are not obvious. For example, although provisioning of offspring by the father (i.e., paternal investment) is relatively rare among vertebrates, it characterizes many species of birds (Lack, 1968; Orians, 1969). The evolution of paternal investment in birds has been attributed to the need to maximize growth rates to avoid high rates of predation on eggs and chicks (Bosque and Bosque, 1995). Survivorship increases substantially once a young bird is able to fly. Thus, selection to reduce the fledging period is thought to have resulted in a suite of avian characteristics, including the highest growth rates among vertebrates (Case, 1978), monogamous mating systems, and paternal investment (Lack, 1968; Orians, 1969).

Nevertheless, another factor that may have contributed to the evolution of monogamous mating systems and paternal provisioning is the loss of fighting ability that resulted from the evolution of flight. Associated with the evolution of flight are the loss of teeth, and modifications of the forelimbs to form wings, making birds much less effective in combat. Thus, birds gave up the ancestral weapons of theropods to improve flight performance. Causally associated with this reduction in lethality may have been a reduction in male-male competition and an associated evolution of male-female pair-bonding and paternal investment.

Given the biomechanical trade-offs that underlie the fight or flight dichotomy, lineages that exhibit high performance in both locomotion and fighting must have evolved characters that circumvent the constraints. Cursorial macropods, such as the red kangaroo, are recognized for their locomotor economy (Dawson and Taylor, 1973) and polygynous mating systems, in which males fight for access to reproductively active females (Jarman, 1983). The evolution of bipedal hopping appears to have allowed the hindlimbs to have evolved effective storage and recovery of elastic strain energy, enhancing locomotor economy, whereas the forelimbs have become specialized for fighting (Jarman, 1989; Warburton *et al.*, 2013).

A second set of examples exists in artiodactyls. Species of both bovids and cervids exhibit very high running speeds (Garland, 1983) and dramatically polygynous mating systems (Clutton-Brock *et al.*, 1977, 1980). Associated with these traits is the evolution of new weapons that are largely independent of the locomotor system – horns in bovids and antlers in cervids. These are weapons that effectively decoupled the evolution of specialization for fighting versus specialization for locomotor performance.

Finally, humans are a remarkable example of a species that appears highly specialized for both aggression and locomotor economy (Carrier, 2004). A strong case can be made that humans rank among the most violent species (Wrangham and Peterson, 1996; Carrier, 2007; Puts, 2010), and are also exceptional endurance runners (Carrier, 1984;

Bramble and Lieberman, 2004; Liebenberg, 2006, 2008). In the case of *Homo*, circumvention of the trade-offs appears to have been associated with the evolution of increased intelligence. The ability of australopiths (i.e., early hominins) to invent tools that could be used as weapons against conspecifics – weapons that were independent of locomotor function – may have allowed early *Homo* to becomes specialized for endurance running (Carrier, 2004).

▌ REFERENCES

Adamczyk PG, Collins SH, Kuo AD (2006). The advantages of a rolling foot in human walking. *Journal of Experimental Biology* **209**, 3953–3963.

Alexander RMcN (1984). Elastic energy stores in running vertebrates. *American Zoologist* **24**, 85–94.

Alexander RD, Hoogland JL, Howard RD, Noonan KM, Sherman PW. (1979). Sexual dimorphisms and breeding systems in pinnipeds, ungulates, primates, and humans. In: Chagnon NA, Irons W (eds). *Evolutionary Biology and Human Social Behavior, An Anthropological Perspective*, pp. 402–435. Duxbury Press, North Scituate.

Andersson M (1994). *Sexual Selection*. Princeton University Press, Princeton, pp. 624.

Beckerman S, Erickson PI, Yost J, Regalado J, Jaramillo L, Sparks C, Iromenga M, Long K (2009). Life histories, blood revenge, and reproductive success among the Waorani of Ecuador. *Proceedings of the National Academy of Sciences* **106**, 8134–8139.

Bean A (1999). Ecology of sex differences in great ape foraging. In: Lee PC (ed). *Comparative primate socioecology*, pp. 339–362, No. 22 Cambridge studies in biological anthropology.

Bertram JEA, Gutmann A (2009). Motions of the running horse and cheetah revisited: fundamental mechanics of the transverse and rotary gallop. *Journal of the Royal Society Interface* **6**, 549–559.

Biewener AA (1990). Biomechanics of mammalian terrestrial locomotion. *Science* **250**, 1097–1103.

Biewener AA (1998). Muscle-tendon stresses and elastic energy storage during locomotion in the horse. *Comparative Biochemistry and Physiology Part B* **120**, 73–87.

Biewener AA, Farley CT, Roberts TJ, Temaner M (2004). Muscular mechanical advantage of human walking and running: implications for energy cost. *Journal of Applied Physiology* **97**, 2266–2274.

Blanckenhorn WU (2005). Behavioral causes and consequences of sexual size dimorphism. *Ethology* **111**, 977–1016.

Blickhan R, Full RJ (1987). Locomotion energetics of the ghost crab. II. Mechanics of the center of mass during walking and running. *Journal of Experimental Biology* **130**, 155–174.

Blob RW, Biewener AA (1999). *In vivo* locomotor strain in the hindlimb bones of *Alligator mississippiensis* and *Iguana iguana*: implications for the evolution of limb bone safety factor and non-sprawling limb posture. *Journal of Experimental Biology* **202**, 1023–1046.

Blob RW, LaBarbera M (2001). Correlates of variation in deer antler stiffness: age, mineral content, intra-antler location, habitat, and phylogeny. *Biological Journal of the Linnean Society* **74**, 113–120.

Bosque C, Bosque MT (1995). Nest predation as a selective factor in the evolution of developmental rates in altricial birds. *American Naturalist* **145**, 234–260.

Bramble DM, Lieberman DE (2004). Endurance running and the evolution of *Homo*. *Nature* **432**, 345–352.

Brear K, Currey JD, Kingsley MCS, Ramsay M (1993). The mechanical design of the tusk of the narwhal (*Monodon monoceros*: Cetacea). *Journal of Zoology, London* **230**, 411–423.

Briffa M, Sneddon LU (2007). Physiological constraints on contest behaviour. *Functional Ecology* **21**, 627–637.

Bro-Jørgensen J. (2013). Evolution of sprint speed in African savannah herbivores in relation to predation. *Evolution* **67**, 3371–3376.

Byers JA (1998). *American Pronghorn: Social adaptations and the ghosts of predators past*, pp. 318. University of Chicago Press, Chicago.

Bywater CL, Angilletta MJ, Wilson RS (2008). Weapon size is a reliable indicator of strength and social dominance in female slender crayfish (*Cherax dispar*). *Functional Ecology* **22**, 311–316.

Caro TM (1994) Cheetahs of the Serengeti Plains. University of Chicago Press, Chicago, pp. 500.

Caro TM, Collins DA (1986). Male cheetahs of the Serengeti. *National geographic Research* **2**, 75–86.

Carrier DR (1983). Postnatal ontogeny of the musculo-skeletal system in the Black-tailed jack rabbit (*Lepus californicus*). *Journal of Zoology, London* **201**, 27–55.

Carrier DR (1984). The energetic paradox of human running and hominid evolution. *Current Anthropology* **25**, 483–495.

Carrier DR (1996). Ontogenetic limits on locomotor performance. *Physiological Zoology* **69**, 467–488.

Carrier DR (2002). Functional tradeoffs in specialization for fighting and running. In: Aerts P, D'Aout K, Herrel A, Van Damme R (eds). *Topics in Functional and Ecological Vertebrate Morphology*, pp. 237–255. Shaker Publishing, Maastricht.

Carrier DR (2004). The running-fighting dichotomy and the evolution of aggression in hominids. In: Meldrum J, Hilton C (eds). *From Biped to Strider: The Emergence of Modern Human Walking, Running, and Resource Transport*, pp. 135–162. Kluwer/Plenum Press, New York.

Carrier DR (2007). The short legs of great apes: evidence for aggressive behavior in australopiths. *Evolution* **61**, 596–605.

Carrier DR, Auriemma J (1992). A developmental constraint on the fledging time of birds. *Biological Journal of the Linnean Society* **47**, 61–77.

Carrier DR, Deban SM, Otterstrom J (2002). The face that sunk the Essex: Potential function of the spermaceti organ in aggression. *Journal of Experimental Biology* **205**, 1755–1763.

Cartmill M (1974). Pads and claws in arboreal locomotion. In FA Jenkins (ed). Primate Locomotion, pp. 45–83. Academic Press, New York.

Cartmill M (1985). Climbing. In: Hildebrand M, Bramble DM, Liem KF, Wake DB (eds). *Functional Vertebrate Morphology*, pp. 73–88. Harvard University Press, Cambridge.

Case TJ (1978). On the evolution and adaptive significance of postnatal growth rates in the terrestrial vertebrates. *Quarterly Review of Biology* **53**, 243–282.

Cavagna GA, Kaneko M (1977). Mechanical work and efficiency in level walking and running. *Journal of Physiology* **268**, 467–481.

Cavagna GA, Saibene FP, Margaria R (1964). Mechanical work in running. *Journal of Applied Physiology* **19**, 249–256.

Cavagna GA, Heglund NC, Taylor CR (1977). Mechanical work in terrestrial locomotion: two basic mechanisms for minimizing energy expenditure. *American Journal of Physiology* **233**, R243–R261.

Chagnon NA, Flinn MV, Melançon T (1979). Sex-ratio variation among the Yanomamö Indians. In: Chagnon N, Irons W (eds). *Evolutionary biology and human social behavior*, pp. 290–320. Duxbury, North Scituate.

Chase K, Carrier DR, Adler FR, Jarvik T, Ostrander EA, Lorentzen TD, Lark KG (2002). Genetic basis for systems of skeletal quantitative traits: principal component analysis of the canid skeleton. *Proceedings of the National Academy of Sciences* **99**, 9930–9935.

Chase K, Carrier DR, Adler FR, Ostrander EA, Lark KG (2005). Interaction between the X chromosome and an autosome regulates size sexual dimorphism in Portuguese Water Dogs. *Genome Research* **15**, 1825–1830.

Clutton-Brock TH, Harvey PH, Rudder B (1977). Sexual dimorphism, socionomic sex ratio and body weight in primates. *Nature* **269**, 797–800.

Clutton-Brock TH, Albon SD, Harvey PH (1980). Antlers, body size and breeding group size in the Cervidae. *Nature* **285**, 565–566.

Cruse H (1976). The function of legs in the free walking stick insect, *Carausius morosus. Journal of Comparative Physiology* **112**, 135–162.

Cunningham CB, Schilling N, Anders C, Carrier DR (2010). The influence of foot posture on the cost of transport in humans. *Journal of Experimental Biology* **213**, 790–797.

Currey JD (1979). Mechanical properties of bone with greatly differing functions. *Journal of Biomechanics* **12**, 313–319.

Currey JD (1987). The evolution of the mechanical properties of amniote bone. *Journal of Biomechanics* **20**, 1035–1044.

Currey, JD (1989). Strain rate dependence of the mechanical properties of reindeer antler and the cumulative damage model of bone fracture. *Journal of Biomechanics* **22**, 469–475.

Currey JD, Pond CM (1989). Mechanical properties of very young bone in the axis deer (*Axis axis*) and humans. *Journal of Zoology (London)* **218**, 59–67.

Daly M, Wilson, M (1988). *Homicide*. Transaction Publishers, New Brunswick.

de Waal FBM (1982). *Chimpanzee politics: power and sex among apes*. Harper and Row, New York, pp. 256.

de Waal FBM (1986). The brutal elimination of a rival among captive male chimpanzees. *Ethology and Sociobiology* **7**, 237–251.

Darwin C (1874). *The Descent of Man, and Selection in Relation to Sex*, 2nd edition. Murray, London, pp. 864.

Dawson TJ, Taylor CR (1973). Energetic cost of locomotion in kangaroos. *Nature* **246**, 313–314.

Dewsbury DA (1982). Dominance rank, copulatory behavior, and differential reproduction. *Quarterly Review of Biology* **57**, 135–159.

Divert C, Mornieux G, Baur H, Mayer F (2005). Mechanical comparison of barefoot and shod running. *International Journal of Sports Medicine* **26**(7), 593–598.

Doran DM (1993). Sex differences in adult chimpanzee positional behavior: the influence of body size on locomotion and posture. *American Journal of Physical Anthropology* **91**, 99–115.

Doran DM, McNeilage A (2001). Subspecific variation in gorilla behavior: the influence of ecological and social factors. In: Robbins, MM, Sicotte, P, Stewart, KJ (eds). *Mountain gorillas: three decades of research at Karisoke*, pp. 123–149. Cambridge, Cambridge University Press.

Eaton RL (1974). *The Cheetah: The Biology, Ecology and Behavior of an Endangered Species*. Van Nostrand Reinhold Co., New York, pp. 192.

Ellis L (1995). Dominance and reproductive success among nonhuman animals: a cross-specific comparison. *Ethology and Sociobiology* **16**, 257–333.

Emerson SB (1985). Jumping and leaping. In: Hildebrand M, Bramble DM, Liem KF, Wake DB (eds). *Functional Vertebrate Morphology*, pp. 58–72. Belknap Press, Cambridge.

Enquist M, Leimar O (1983). Evolution of fighting behaviour: decision rules and assessment of relative strength. *Journal of Theoretical Biology* **102**, 387–410.

Enquist M, Leimar O (1990). The evolution of fatal fighting. *Animal Behaviour* **39**, 1–3.

Farley CT, McMahon TA (1992). Energetics of walking and running: insights from simulated reduced-gravity experiments. *Journal of Applied Physiology* **73**, 2709–2712.

Fedak MA, Heglund NC, Taylor CR (1982). Energetics and mechanics of terrestrial locomotion: II. Kinetic energy changes of the limbs and body as a function of speed and body size in birds and mammals. *Journal of Experimental Biology* **79**, 23–40.

Fossey D (1983). *Gorillas in the mist*. Houghton Mifflin, New York, pp. 400.

Fossey D (1984). Infanticide in mountain gorillas (*Gorilla gorilla beringei*) with comparative notes on chimpanzees. In: Hausfater G, Hardy SB (eds). *Infanticide: Comparative and Evolutionary Perspectives*, pp. 217–236. Hawthorne, New York.

Francis RC (1984). The effects of bidirectional selection for social dominance on agonistic behavior and sex ratios in the paradise fish (*Macropodus opercularis*). *Behaviour* **90**, 25–45.

Frank LG, Glickman SE, Licht P (1991). Fatal sibling aggression, precocial development, and androgens in neonatal spotted hyenas. *Science* **252**, 702–704.

Full RJ, Blickhan R, Ting LH (1991). Leg design in hexapedal runners. *Journal of Experimental Biology* **158**, 369–390.

Furuichi T (1997). Agonistic interactions and matrifocal dominance rank of wild bonobos (*Pan paniscus*) at Wamba. *International Journal of Primatology* **18**, 855–877.

Garland T Jr (1983). The relation between maximal running speed and body mass in terretrail mammals. *Journal of Zoology, London* **199**, 157–170.

Garland T Jr, Adolph SC (1994). Why not to do two-species comparative studies: limitations on inferring adaptation. *Physiological Zoology* **67**, 797–828.

Gebo DL (1992). Plantigrady and foot adaptation in African apes: implications for hominid origins. *American Journal of Physical Anthropology* **89**(1), 29–58.

Gebo DL (1996). Climbing, brachiation, and terrestrial quadrupedalism: historical precursors of hominid bipedalism. *American Journal of Physical Anthropology* **101**, 55–92.

Geist V (1971). *The Mountain Sheep: A Study in Behaviour and Evolution*. Chicago University Press, Chicago, pp. 400.

Goodall J (1986). *The Chimpanzees of Gombe, Patterns of Behavior*. Belknap Press of Harvard University Press, Cambridge, pp. 673.

Greene HW, Burghardt GM (1978). Behavior and phylogeny: constriction in ancient and modern snakes. *Science* **200**, 74–77.

Grinnell J, Packer C, Pusey AE (1995). Cooperation in male lions: kinship, reciprocity, or mutualism? *Animal Behaviour* **49**, 95–105.

Hamilton WD. (1979) Wingless and fighting males in fig wasps and other insects. In: Blum MA, Blum NA (eds). *Sexual selection and reproductive competition in insects*, pp. 167–215. London, Academic Press.

Heglund NC, Cavagna GA, Taylor CR (1982). Energetics and mechanics of terrestrial locomotion. III. Energy changes in the center of mass as a function of speed and body size in birds and mammals. *Journal of Experimental Biology* **79**, 41–56.

Heinrich RE, Ruff CB, Adamczewski JZ (1999). Ontogenetic changes in mineralization and bone geometry in the femur of muskoxen (*Ovibos moschatus*). *Journal of Zoology, London* **247**, 215–223.

Helton WS (2011). Performance constraints in strength events in dogs (*Canis lupus familiaris*). *Behavioural Processes* **86**, 149–151.

Herrel A, Andrade DV, De Carvalho JE, Brito A, Abe A, Navas C (2009). Aggressive behavior and performance in the tegu lizard *Tupinambis merianae*. *Physiological and Biochemical Zoology* **82**, 680–685.

Hildebrand M, Goslow G (2001). *Analysis of vertebrate structure*. John Wiley and Sons, Inc., New York, pp. 660.

Hildebrand M, Hurley JP (1985). Energy of the oscillating legs of a fast-moving cheetah (*Acinonyx jubatus*), pronghorn (*Antilocapra americana*), jackrabbit (*Lepus californicus*) and elephant (*Elephas maximus*). *Journal of Morphology* **184**, 23–32.

Holden JP, Cavanagh PR (1991). The free moment of ground reaction in distance running and its changes with pronation. *Journal of Biomechanics* **24**, 887–897.

Husak JF, Lappin AK, Fox SF, Lemos-Espinal JA (2006). Bite-force performance predicts dominance in male Venerable Collared Lizards (*Crotaphytus antiquus*). *Copeia* **2006**, 301–306.

Jarman PJ (1983). Mating system and sexual dimorphism in large, terrestrial, mammalian herbivores. *Biological Reviews* **58**, 485–520.

Jarman PJ (1989). Sexual dimorphism in Macropodoidea. In: Grigg G, Jarman P, Hume I (eds). *Kangaroos, Wallabies and Rat-kangaroos*, pp. 433–447. Surrey Beatty & Sons Pty Ltd, Chipping Norton.

Jayes AS, Alexander RMcN (1978). Mechanics of locomotion of dogs (*Canis familiaris*) and sheep (*Ovis aries*). *Journal of Zoology (London)* **185**, 289–308.

Jindrich DL, Smith NC, Jespers K, Wilson AM (2007). Mechanics of cutting maneuvers by ostriches (*Struthio camelus*). *Journal of Experimental Biology* **210**, 1378–1390.

Johnsgard PA (1975). *North American Game Birds of Upland and Shoreline*. University of Nebraska Press, Lincoln, pp. 183.

Judge KA, Bonanno VL (2008). Male weaponry in a fighting cricket. *PLoS One* **3**(12), e3980.

Jungers WL (1978). The functional significance of skeletal allometry in *Megaladapis* in comparison to living prosimians. *American Journal of Physical Anthropology* **49**, 303–314.

Jungers WL (1984). Scaling of the hominoid locomotor skeleton with special reference to the lesser apes. In: Preuschoft H, Chivers DJ, Brockelman W, Creel N (eds). *The lesser apes: evolutionary and behavioral biology*, pp. 149–169. Edinburgh University Press, Edinburgh.

Kalen V, Alder N, Bleck EE (1986). Electromyography of idiopathic toe walking. *Journal of Pediatric Orthopaedics* **6**(1), 31–33.

Kano T (1992). *The Last Ape. Pygmy Chimpanzee Behavior and Ecology*. Stanford University Press, Stanford, pp. 276.

Katz B (1939). The relation between force and speed in muscular contraction. *Journal of Physiology* **96**, 45–64.

Kaufman K (1996). *Lives of North American Birds*. Houghton Mifflin Company, Boston, pp. 704.

Keeley LH (1996). *War Before Civilization*. Oxford University Press, New York, pp. 272.

Kelly IP, Jenkinson A, Stephens M, O'Brien T (1997). The kinematic patterns of toe-walkers. *Journal of Pediatric Orthopaedics* **17**(4), 478–480.

Kemp TJ, Bachus KN, Nairn JA, Carrier DR (2005). Functional tradeoffs in the limb bones of dogs selected for running versus fighting. *Journal of Experimental Biology* **208**, 3475–3482.

Kitchener A (1991). The evolution and mechanical design of horns and antlers. In: Rayner JMV, Wooton KJ (eds). *Biomechanics and Evolution*, pp. 229–253. Cambridge University Press, Cambridge.

Knauft BB (1991). Violence and sociality in human evolution. *Current Anthropology* **32**, 391–428.

Kram R, Taylor CR (1990). Energetics of running: a new perspective. *Nature* **346**, 265–267.

Kruuk N (1972). *The Spotted Hyena: A Study of Predation and Social Behavior*. The University of Chicago Press, Chicago, pp. 388.

Lack D (1968). *Ecological Adaptations for Breeding in Birds*. Methuen, London, pp. 422.

Lailvaux SP, Irschick DJ (2007). The evolution of performance-based male fighting ability in Caribbean *Anolis* lizards. *American Naturalist* **170**, 573–586.

Lande R (1980). Sexual dimorphism, sexual selection, and adaptation in polygenic characters. *Evolution* **34**, 292–305.

Lappin AK, Husak JF (2005). Weapon performance, not size, determines mating success and potential reproductive output in the Collared Lizard (*Crotaphytus collaris*) *American Naturalist* **166**, 426–436.

Lauder GV (1991). An evolutionary perspective on the concept of efficiency: how does function evolve? In: Blake RW (ed). *Efficiency and Economy in Animal Physiology*, pp. 169–184. Cambridge University Press, Cambridge.

Lee DV, Walter RM, Deban SM, Carrier DR (2001). Influence of increased rotational inertia on the turning performance of humans. *Journal of Experimental Biology* **204**, 3927–3934.

Li Y, Wang W, Crompton RH, Gunther MM (2001). Free vertical moments and transverse forces in human walking and their role in relation to arm-swing. *Journal of Experimental Biology* **204**, 47–58.

Liebenberg L (2006). Persistence hunting by modern hunter-gatherers. *Current Anthropology* **47**, 1017–1025.

Liebenberg L (2008). The relevance of persistence hunting to human evolution. *Journal of Human Evolution* **55**, 1156–1159.

Livingstone FB (1962). Reconstructing man's Pliocene pongid ancestor. *American Anthropologist* **64**, 301–305.

Losos JB (1990). The Evolution of Form and Function: Morphology and Locomotor Performance in West Indian Anolis Lizards. *Evolution* **44**, 1189–1203.

MacKinnon JR, MacKinnon KS (1984). Territorality, monogamy and song in gibbons and tarsiers. In: Preuschoft H, Chivers DJ, Brockelman WY, Creel N (eds).*The Lesser Apes: Evolutionary and Behavioral Biology*, pp. 291–297. Edinburgh University Press, Edinburgh.

Margaria R, Cerrretelli P, Aghemo P, Sassi G (1963). Energy cost of running. *Journal of Applied Physiology* **18**, 367–370.

Marlowe FW (2005). Hunter-gatherers and human evolution. *Evolutionary Anthropology* **14**(2), 54–67.

Marsh RL (1994) Jumping ability of anuran amphibians. In: Jones JH (ed). *The Advances in Veterinary Science and Comparative Medicine: Comparative Vertebrate Exercise Physiology*, pp. 51–111. Academic Press, New York.

Marsh RL, Ellerby DJ, Henry HT, Rubenson J (2006). The energetic costs of trunk and distal-limb loading during walking and running in guinea fowl *Numida meleagris* I. Organismal metabolism and biomechanics. *Journal of Experimental Biology* **209**, 2050–2063.

Martín-Serra A, Figueirido B, Palmqvist P. (2014). A Three-Dimensional Analysis of Morphological Evolution and Locomotor Performance of the Carnivoran Forelimb. *PLoS One* **9**(1), e85574.

Maynard Smith J, Burian R, Kauffman S, Alberch P, Campbell J, Goodwin B, Lande R, Raup D, Wolpert L (1985). Developmental constraints and evolution. *Quarterly Review of Biology* **60**, 265–287.

Maynard Smith J, Price GR (1973). The logic of animal conflict. *Nature* **246**, 15–18.

McGeer T (1990). Passive dynamic walking. *International Journal of Robotics Research* **9**, 62–82.

Mech LD, Adams LG, Meire TJ, Burch JW, Dale BW (1998). *The Wolves of Denali*. University of Minnesota Press, Minneapolis, pp. 240.

Mitani JC (1987). Territoriality and monogamy among agile gibbons (*Hylobates agilis*). *Behavioral Ecology and Sociobiology* **20**, 265–269.

Moore AJ (1990). The inheritance of social dominance, mating behaviour, and attractiveness to mates in *Nauphoeta cinerea*. *Animal Behaviour* **39**, 288–397.

Morris JS, Brandt EK (2014). Specialization for aggression in sexually dimorphic skeletal morphology in grey wolves (*Canis lupus*). *Journal of Anatomy* **225**, 1–11.

Nowak RM, Paradiso JL (1983). *Walker's Mammals of the World*, Volume 1, 4th Edition. Johns Hopkins University Press, Baltimore.

Orians GH (1969). On the evolution of mating systems in birds and mammals. *American Naturalist* **103**, 589–603.

Parker GA (1974). Assessment strategy and the evolution of fighting behaviour. *Journal of Theoretical Biology* **47**, 223–243.

Parker GA (1983). Arms races in evolution – an ESS to the opponent-independent costs game. *Journal of Theoretical Biology* **101**, 619–648.

Pasi BM, Carrier DR (2003). Functional tradeoffs in the limb muscles of dogs selected for running versus fighting. *Journal of Evolutionary Biology* **16**, 324–332.

Plavcan JM (1999). Mating systems, intrasexual competition and sexual dimorphism in primates. In: Lee PC (ed). *Comparative Primate Socioecology*, pp. 241–269. Cambridge University Press, Cambridge.

Plavcan JM (2004). Sexual selection, measurement of sexual selection, and sexual dimorphism in primates. In: Kappeler PM, van Schaik CP (eds). *Sexual Selection in Primates: New and Comparative Perspectives*, pp. 230–252. Cambridge University Press, Cambridge.

Plavcan JM van Schaik CP (1997). Intrasexual competition and body weight dimorphism in anthropoid primates. *American Journal of Physical Anthropology* **103**, 37–68.

Pontzer H. (2005). A new model predicting locomotor cost from limb length via force production. *Journal of Experimental Biology* **208**, 1513–1524.

Pough FH, Janis CM, Heiser JB (1999). *Vertebrate Life*. Prentice Hall, New Jersey, pp 720.

Preuschoft H, Chivers DJ, Brockelman WY, Creel N (1984). *The Lesser Apes: Evolutionary and Behavioral Biology*. Edinburgh University Press, Edinburgh, pp. 400.

Puts DA (2010). Beauty and the beast: mechanisms of sexual selection in humans. *Evolution and Human Behavior* **31**, 157–175.

Qiao M, Brown B, Jindrich DL (2014). Compensations for increased rotational inertia during human cutting turns. *Journal of Experimental Biology* **217**, 432–443.

Roberts TJ, Marsh RL, Weyand PG, Taylor CR (1997). Muscular force in running turkeys: the economy of minimizing work. *Science* **275**, 1113–1115.

Ruben JA (1977). Morphological correlates of predatory modes in the Coachwhip (*Masticophis flagellum*) and Rosy Boa (*Lichanura roseofusca*). *Herpetologica* **33**, 1–6.

Rubenson J, Heliams DB, Maloney SK, Withers PC, Lloyd DG, Fournier PA (2007). Reappraisal of the comparative cost of human locomotion using gait-specific allometric analyses. *Journal of Experimental Biology* **210**, 3513–3524.

Ruina A, Bertram JEA, Srinivasan M (2005). A collisional model of the energetic cost of support work qualitatively explains leg sequencing in walking and galloping, pseudo-elastic leg behavior in running and the walk-to-run transition. *Journal of Theoretical Biology* **237**, 170–192.

Savitzky AH (1980). The role of venom delivery strategies in snake evolution. *Evolution* **34**, 1194–1204.

Schaller GB (1963). *The Mountain Gorilla: Ecology and Behavior*. University of Chicago Press, Chicago, pp. 450.

Schaller GB (1972). *The Serengeti Lion: A Study of Predator-Prey Relations*. University of Chicago Press, Chicago, pp. 504.

Schmitt D, Larson SG (2005). Heel contact as a function of substrate type and speed in primates. *American Journal of Physical Anthropology* **96**, 39–50.

Singleton I, van Schaik CP (2001). Orangutan home range size and its determinants in a Sumatran swamp forest. *International Journal of Primatology* **22**(6), 877–911.

Smith JM, Savage RJG (1956). Some locomotory adaptations in mammals. *Journal of the Linnean Society (Zoology)* **42**, 603–622.

Sockol MD, Raichlen DA, Pontzer H (2007). Chimpanzee locomotor energetics and the origin of human bipedalism. *Proceedings of the National Academy of Sciences* **104**, 12265–12269.

Steudel K (1991). The work and energetic cost of locomotion: I. The effects of limb mass distribution in quadrupeds. *Journal of Experimental Biology* **154**, 273–286.

Steudel-Numbers KL (2003). The energetic cost of locomotion: humans and primates compared to generalized endotherms. *Journal of Human Evolution* **44**, 255–262.

Steudel-Numbers KL, Tilkens MJ (2004). The effect of lower limb length on the energetic cost of locomotion: implications for fossil hominins. *Journal of Human Evolution* **47**, 95–109.

Susman RL, Stern JT, Jungers WL (1985). Locomotor adaptations in the Hadar Hominids. In: Delson E (ed). *Ancestors: The hard evidence*, pp. 184–192. Alan R. Liss, Inc, New York.

Svare BB (1981). Maternal aggression in mammals. In: Gubernick DJ, Klopfer PH (eds). *Parental Care in Mammals*, pp. 179–210. Plenum, New York.

Swartz SM, Bennett MB, Carrier DR (1992). Wing bone stresses in free flying bats and the evolution of skeletal architecture in flying vertebrates. *Nature* **359**, 726–729.

Szamado S (2003). Threat displays are not handicaps. *Journal of Theoretical Biology* **221**, 327–348.

Szamado, S (2008). How threat displays work: species-specific fighting techniques, weaponry and proximity risk. *Animal Behaviour* **76**, 1455–1463.

Taylor CR (1994). Relating mechanics and energetics during exercise. *Advances in Veterinary Science & Comparative Medicine* **38A**, 181–215.

Usherwood JR, Channon AJ, Myatt JP, Rankin JW, Hubel TY (2012). The human foot and heel–sole–toe walking strategy: a mechanism enabling an inverted pendular gait with low isometric muscle force? *Journal of The Royal Society Interface* **9**, 2396–2402.

Van Damme R, Wilson RS, Vanhooydonck B, Aerts P (2002). Performance constraints in decathletes. *Nature* **415**, 755–756.

Vanhooydonck B, Van Damme R, Aerts P (2001). Speed and stamina trade off in lacertid lizards. *Evolution* **55**, 1040–1048.

Walker PL (2001). A bioarchaelological perspective on the history of violence. *Annual Review of Anthropology* **30**, 573–596.

Walter RM (2003). Kinematics of 90° running turns in wild mice. *Journal of Experimental Biology* **206**, 1739–1749.

Warburton NM, Bateman PW, Fleming PA (2013). Sexual selection on forelimb muscles of western grey kangaroos (Skippy was clearly a female). *Biological Journal of the Linnean Society* **109**, 923–931.

Ward CV (2007). 6 postcranial and locomotor adaptations of hominoids. *Handbook of Paleoanthropology*, pp. 1011–1030.

Watts DP (1989). Infanticide in mountain gorillas: new cases a reconsideration of the evidence. *Ethology* **81**, 1–18.

Webster EL, Hudson PE, Channon SB (2014). Comparative functional anatomy of the epaxial musculature of dogs (*Canis familiaris*) bred for sprinting vs. fighting. *Journal of Anatomy* **225**, 317–327.

Wilkinson GS, Shank CC (1977). Rutting-fight mortality among musk oxen on banks island, Northwest Territories, Canada. *Animal Behaviour* **24**, 756–758.

Willems PA, Cavagna GA, Heglund NC (1995). External, internal and total work in human locomotion. *Journal of Experimental Biology* **198**, 379–393.

Williams JM, Pusey AE, Carlis JV, Farm BP, Goodall J (2002). Female competition and male territorial behaviour influence female chimpanzees' ranging patterns. *Animal Behaviour* **63**(2), 347–360.

Wolff JO (1985). Maternal aggression as a deterrent to infanticide in *Peromyscus leucopus* and *P. maniculatus*. *Animal Behaviour* **33**, 117–123.

Wrangham R (1999). The evolution of coalitionary killing. *Yearbook of Physical Anthropology* **42**, 1–30.

Wrangham R, Peterson D (1996). *Demonic Males: Apes and the Origins of Human Violence*. Houghton Mifflin Co., Boston, pp. 350.

Zioupos P, Currey JD, Casinos A, De Buffrenil V (1997). Mechanical properties of the rostrum of the whale *Mesoplodon densirostris*, a remarkably dense bony tissue. *Journal of Zoology (London)* **241**, 725–737.

CHAPTER FOURTEEN

Design for Prodigious Size without Extreme Body Mass: Dwarf Elephants, Differential Scaling and Implications for Functional Adaptation

John E. A. Bertram

Department of Cell Biology and Anatomy, Cumming School of Medicine, University of Calgary, CA

14.1   INTRODUCTION

Body size is a major determinant of skeletal loading, and large changes in body-size should require skeletal modifications to maintain appropriate functionality. The analysis of such modifications, within the context of the mechanical effects acting at a particular body size, can direct attention to issues that are of functional importance to the organism, or point out aspects of the biology that are or are not influenced by mechanical factors. Dwarf elephant populations from Pleistocene-era Mediterranean islands are particularly informative. They are derived from mainland populations that have robust skeletons adapted for their large body size, the degree of dwarfing was substantial, and high quality skeletal remains are available.

Understanding Mammalian Locomotion: Concepts and Applications, First Edition.
Edited by John E. A. Bertram.
© 2016 John Wiley & Sons, Inc. Published 2016 by John Wiley & Sons, Inc.

Due to differential scaling between small and large mammalian forms, dwarfing that simply reduces the size of individuals of a large body size species, without alteration of structural proportions of the skeleton, should result in a skeleton severely over-built for the mechanical loading regime imposed by the reduced body size. Evaluating the dwarf elephants in this context, it appears that remarkably little structural adaptation occurred in their appendicular skeleton through the dwarfing process.

These results are useful in understanding both the mechanical consequences and biological constraints acting on the appendicular skeleton of mammals. It also demonstrates an apparent insensitivity of the skeleton to substantial mechanical loading changes – results that run counter to generally accepted concepts of functional adaptation of bone, commonly known as "Wolff's Law". Some changes, involving proportionate length of some limb elements and the functional adaptation of cranial components, also suggest some aspects of the functional signals that the dwarf elephant skeleton responds to.

One challenge involved with the investigation of the interaction of the organism with the physical consequences of its environment is identifying the aspects of the interaction that are important to the organism. As complex functional systems that are well integrated with their environment, it is often difficult to identify the factors that are important determinants affecting function and form in mammals. Disconcertingly, it may be a truism that those negative features most critical to the function of an organism may not be readily observable because, if truly critical to function and negative in their effect, their effect will be eliminated or minimized (Srinivasan, 2007).

Normal observation, regardless of how meticulous it might be, may not provide insight if the system has been successful in adapting appropriately to avoid a negative factor. No matter how closely we look, it may be that those features most important in determining an organism's functional form may not be directly observable. A strategy available to approach this problem in a mechanical context is through the use of scaling analyses. Body size has such a powerful effect on mechanics and physiology, that the way a lineage responds over a range of size can point to those features that are forced to adapt to the functional consequences of the size (see Chapters 8 and 9).

The challenge in comparative analysis, including scaling analyses, is the unambiguous identification of the factor(s) responsible for the variations observed. Within the comparative context, it is likely to be impossible to identify all of the factors influencing the measured differences between two systems. However, if the compared species are closely related, and one of the factors has a very large effect on an organ, with a function clearly related to that factor, it is reasonable to apply this approach (with caution) and to infer some relationship between the factor and the observed response.

In this chapter, it is argued that the dwarfing of elephants isolated on Mediterranean islands during the Pleistocene era can be used as a tool to assess the adaptive response of the skeleton to substantial, but natural, changes in mechanical loading. Such size modifications, resulting from isolation events, are referred to as "insular dwarfing", and are well known from a variety of locations throughout the world, having been observed in a variety of species (Van Valen, 1973; Lomolino, 1985; Sondaar, 1977), including even sauropod dinosaurs (Sander *et al.*, 2006).

The potential influence of functional adaptation on skeletal proportions of a species undergoing substantial body size change can only be distinguished if mechanically appropriate bone form at the new body size differs from the proportions present within the original form. Smaller body size species appear to be generally self-similar with regard to skeletal proportions. It appears that, in groups of this body size, changes in limb orientation can mitigate skeletal loading changes that would otherwise arise from increases in body mass (Biewener, 1989, 2005). That is, much of the actual load applied to the limb comes from muscle tension applied across relatively small moment arms acting across joints. As a closed system in which the origin of the muscle is proximal to the joint, this results in compression loading of the element articulating at the joint.

One strategy available for maintaining a consistent overall load level in the supporting skeleton as body size increases is to systematically alter the supporting moment arm ratio, by changing the orientation of the limb as body size increases (limb orientation becomes more upright). For the majority of mammal groups, the differences between isometric scaling (constant shape over a substantial size range) and mechanically relevant skeletal structure are subtle, and any potential differences are overwhelmed by the inevitable variability in bone shape that occurs between distinct species.

The limb reorientation strategy discussed above has a limit, however. Very large terrestrial mammals require different skeletal element proportions in order to allow structural stability at their extreme body mass (Bertram and Biewener, 1990). Thus, very large body size mammals possess skeletal proportions only appropriate for their extreme bulk. Given that large body size groups appear to be derived from smaller body size ancestors (Cope's Rule), skeletal scaling of a large body size clade will be composed of a deflected scaling relation relative to smaller body size groups. This has been referred to as "differential" scaling (Bertram, 1988; Bertram and Biewener, 1990; Christiansen, 1999a, 1999b; Iriarte-Diaz, 2002; and see Chapter 8).

As a consequence, substantial body size reduction in large-bodied mammals without modification of these proportions would produce a measurably over-built skeleton (i.e., isometric (geometric) decrease in size will leave them with proportions apparently inappropriate for their body size). If skeletal form is responsive on an acute level to the new mechanical circumstances determined by the reduced body size (Wolff's Law – Ruff *et al.*, 2006), then skeletal proportions in the dwarf species should approach those of similar smaller body size forms.

Evidence from skeletal form is routinely used to interpret the functional potential and performance limitations of both extant and extinct forms. It is important, therefore, to distinguish between short (functional adaptation within the life span of an individual) and long time scale (genetically determined) modifications, and to be able to recognize what potential the skeleton has to adapt to its mechanical circumstances. Insular dwarfing in Pleistocene Mediterranean elephants presents a strong mechanical effect on the skeleton, which is largely independent of the physiological and pathological consequences that are likely to result from experimental manipulation of activity levels, or artificially contrived changes in skeletal loading. It provides a unique opportunity to distinguish the skeletal response to a substantial change in loading of the skeleton.

14.2 ELEPHANT FORM, MAMMALIAN SCALING AND DWARFING

When comparing elephants to other terrestrial mammals, it is important to distinguish differences due mainly to the extreme size of elephants, from those due to the inherent differences that result from their unique ecology and phylogenetic background (ancestry of the species). However, for elephants, it is difficult to dissociate size, ecology and phylogeny, because these three are intimately associated within the life history of the group (Owen-Smith, 1988; Shoshani, 1998). We can assume that elephants will be subject to the same size-dependent mechanical effects seen in other mammals (McMahon, 1973), whatever the ecologically determined differences that exist between elephants and other terrestrial mammals. Likewise, having factored differences due strictly to the consequences of their prodigious size, we can have some confidence that the remainder of the evident differences between elephants and other terrestrial mammals will be due to inherent "biological" factors. These will exist but, if interpreted in context, such specifics should not interfere with the comparison of form between elephants and other groups.

The elephant skeleton is quite distinctive, with robust limb bones arranged in a columnar "graviportal" fashion (Shoshani, 1998). This is an obvious indication that elephants are designed to carry a large body mass. Less well appreciated is the fact that elephants are particularly long-limbed in comparison to most other mammals, especially in comparison to those with comparably large body mass (Figure 14.1).

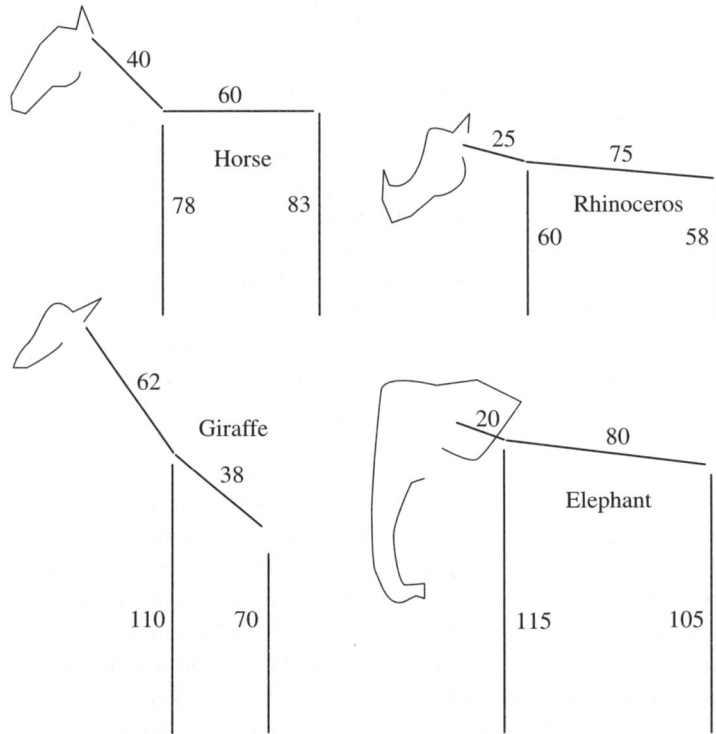

FIGURE 14.1

Relative limb lengths of representative large body size mammals as estimated from lateral photographs. Proportions are represented relative to combined neck and back length, following the style of Hildebrand (1952). Note that the elephant ranks with the horse and giraffe as having particularly long limb proportions. The disproportionate length of the elephant limb is also evident when compared to the rhinoceros, a species that approaches the elephant's body mass.

Both of these factors have an important influence on the interpretation of the skeletal form of elephants, and must be considered when comparing elephant skeletal form to other species.

What are the consequences of size on necessary structural proportions of the support skeleton in terrestrial mammals? This issue has had a long discourse in the literature (Galileo, 1638; Thompson, 1942; Rashevsky, 1948; McMahon, 1973; Alexander *et al.*, 1979; Economos, 1983; Biewener, 1989; Bertram and Biewener, 1990). Such long-standing attention is due both to the fact that size is a critical determinant of structural design requirements (Peters, 1983), and that biological proportions do not appear to fit simple mechanical expectations (Alexander *et al.*, 1979). Recent discussions have suggested that the scaling of small and large animals is *not* equivalent, and that a single scaling relationship should not be expected to cover the entire size range of mammals (Economos, 1983; Bertram and Biewener, 1990; Christiansen, 1999a, 1999b, 2002; Iriarte-Diaz, 2002; and see Chapter 8). There are three essential features of the argument relevant to the current discussion:

a. Isometric scaling (maintaining similar proportions) over a very large size range is structurally impossible, because load carried (largely determined by tissue volume) increases more than support capacity (related to area of support structures).

b. As a consequence of (**a**), proportions must change (scale), but the proportional changes are different for small and large size forms (differential scaling – Bertram, 1988; Bertram and Biewener, 1990).

c. Finally, differences in scaling of limb bones between large and small body size forms described in (**b**) can be explained by a systematic change in limb orientation in small body size species that reduces the effective load applied to the limbs of smaller forms (limbs become more upright as size increases – Biewener, 1991, 2005). At some point, this reorientation reaches a limit, and the largest body size forms must rely on proportions scaled directly to increased volume (load), in order to provide adequate structural support.

The evidence for each of the above will be described in more detail in the discussion that follows. At issue first is how these concepts can be used to evaluate the functional adaptation of the mammalian skeleton and to make relevant comparison between elephants and other terrestrial mammals. The key concept is differential scaling of small and large body-size mammals. This difference means that very large mammals, such as the elephants, will have structural components that are substantially different from, and inappropriate for, a smaller body size. It is argued that the phenomenon of insular dwarfing in large body-size mammals can be used as an opportunity to investigate the influence of dramatic changes in functional loading of the skeleton. The close association between dwarfed island populations, and their larger body size mainland ancestors, make the comparison particularly viable. By comparing the proportions of dwarf forms to their predecessors, it should be possible to determine how the skeleton adapts to changes in loading caused by the dwarfing event.

Insular dwarfing commonly occurs to large body size mammals that become isolated as island populations. Dwarfing can occur rapidly (Lister, 1989) and be quite dramatic (Roth, 1990; Simmons, 1988; Vartanyan *et al.*, 1993). Although evidence of insular dwarfing is cosmopolitan in its distribution, Pleistocene deposits on several Mediterranean islands are particularly useful. The Pleistocene variations in surface level of the Mediterranean Sea resulted in the emergence of several islands (Kurtén, 1968). These islands were eventually colonized by a number of species, but remained relatively depauperate compared to the mainland. Elephants appear to have been particularly adept at gaining access to these islands, possibly due to their well-recognized swimming capabilities (Sanderson, 1896).

Dwarfing events in elephants subsequent to colonization of these islands occurred in parallel at least ten times (Ambrosetti, 1968). Adult dwarfs were as small as 1 m tall at the shoulder (Figure 14.2), giving a mass decrease potentially of an order of magnitude or more (assuming Mass $\propto$ (length)^3). Adult mainland forms are expected to have a shoulder height approaching 3 m and a body mass of at least 3 tonnes, but potentially as great as 5–7 tonnes. The smallest of the dwarf forms are classified as *Elephas falconeri*, included in the same genus as the extant Asian elephant (*Elephas maximus*), although a recent debate regarding ancestry has emerged, based on extracted partial DNA samples (Poulakakis *et al.*, 2002, 2006). The modern Asian elephant is probably somewhat smaller, but in all other ways very similar to the species inhabiting southern Europe in the Pleistocene epoch (*Elephas (Paleoloxodonta) antiquus*), 40,000–20,000 years before the present day. There is evidence that some of these dwarf island populations even persisted into early historic times (Masseti, 2001). Such recent occurrence provides a substantial amount of high-quality skeletal remains, from which scaling analyses can be derived.

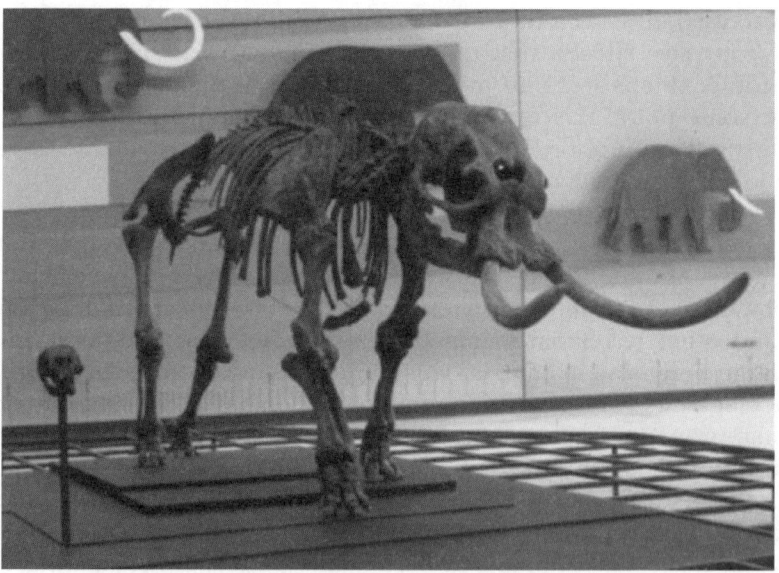

FIGURE 14.2

Elephas falconeri – photo of a mounted specimen at the Senckenberg Museum of Natural History, Frankfurt am Main, Germany. Although this specimen is approximately 1 m at the highest point of its thoracic spine, it is obviously fully mature, as indicated by fully fused epiphyses and adult tusks. Photo courtesy of P. Sereno.

What form should be expected in the dwarf elephant skeleton? If dwarfed forms had a support skeleton shape isometric with their large body size predecessors, bone stress levels would be substantially reduced in the smaller forms. This is because cross-sectional area of the skeleton (the structurally relevant supporting framework) does not decrease in the same proportion as the volume of tissue supported (related to the applied load). The shoulder height of the dwarf elephants was reduced to as much as one-third that of their mainland ancestors (*Elephas maximus* shoulder height 2.5–3 m, Nowak, 1991, while the shoulder height of *Elephas falconeri* has been estimated at 1–1.5 m).

Assuming the dwarf species were isometric with the mainland species from which they arose, and the mass of the mainland species was similar to modern *Elephas*, which have a mass reported to be roughly 2.5–5.5 tonnes, then we would anticipate that the dwarf forms had a mass proportional to between $(1/2.5)^3$ and $(1/3)^3$ tonnes, or roughly 4–6.5% of the mainland species (approximately 100–360 kg). The actual mass of dwarf elephants may have been slightly greater than this, due to the fact that some organs, such as the brain or gut, may not have scaled isometrically (Palombo, 2001). Roth (1987) used the Sicilian dwarf elephant as a case study of body mass estimation based on extant elephant form and weight. Estimated mass ranged from approximately 60–200 kg, depending on assumptions of the model and the skeletal element used as a basis of the estimate.

Limb bones of terrestrial mammals are loaded in complex ways, generally involving a combination of compression and bending, and it is difficult to determine precisely how loading occurred in the extinct dwarf elephant. However, it is possible to estimate the general range of stress reduction that would result from the dwarfing event.

If all of the load borne by the tibia is in pure compression, then an order of magnitude reduction of mass would reduce limb bone stresses to approximately 46% of their original level (i.e., cross-sectional area $\propto$ (length)^2 and load $\propto$ volume $\propto$ (length)^3). At the other extreme, if all load borne by the tibia were in pure bending, then an order of magnitude reduction in mass from a one-third reduction of all linear dimensions would result in stress levels in the dwarf form of approximately 33% of the original large body size form (i.e., relative stress in bending $\propto$ (mass × length)/diameter^3 $\propto$ length).

Given the graviportal nature of limb orientation in the elephant, it is likely that compression dominates in the tibia, but it is unlikely that this element is not also subject to substantial bending loads during locomotion. In the context of current models of functional adaptation (Frost, 2004; Turner and Robling, 2004; Ruimerman *et al.*, 2005), a reduction of this magnitude would elicit a substantial change in the shape or properties of the skeleton of the dwarf form, compared with the original population from which it is derived. To evaluate the functional adaptation of the skeleton of dwarf elephants, limb bone dimensions of dwarf forms can be compared with those of their closest living extant relative. The dimensions can then be compared with scaling relationships for other terrestrial quadrupedal mammals, and shape change (or lack thereof) of skeletal elements, as a result of the dwarfing event, can be evaluated.

The robust nature of the elephant tibia has allowed this bone to be well-represented in fossil deposits of dwarf forms. The tibia is a skeletal element that readily indicates some of the subtler issues of differential scaling between small and large mammals (Bertram and Biewener, 1990). Thus, the tibia represents a unique opportunity to compare scaling of the structural support skeleton in dwarf elephants of the Pleistocene to a range of extant mammalian species. In this analysis, the general structural form of tibiae from a variety of mammalian species, representing a wide range of body size, are compared to the two species of extant elephant and the extinct Pleistocene dwarf form.

14.2.1 Measurements

External dimensions (length and antero-posterior, A-P, diameter) of the complete tibia of 60 Pleistocene dwarf elephants of Sicily were taken from Ambrosetti (1968). Scaling comparisons were made between the tibial dimensions of the dwarf elephants and extant terrestrial carnivorans (100 species – Bertram and Biewener, 1990), and extant ceratomorphs (seven species, rhinoceroses and tapirs – Prothero and Sereno, 1982), from measurements available in the literature. The ceratomorph measurements represent length and medio-lateral diameter, because rhinos have a distinct cnemial crest on the anterior aspect of the tibial diaphysis that obscures the structural form of the supporting shaft. Although not precisely the same measurement in both groups, the length-diameter measurements utilized are intended simply to represent the structural proportions of the supporting bone.

Tibial proportions of extant elephants were taken from six African elephants (*Loxodonta africana*; AMNH specimens 88404, 90176 and 90258 and Smithsonian specimens 163318, 270993 and 588113) and four Asian elephants (*Elephas maximus indicus*, AMNH specimens 39082 and 544452, Smithsonian specimen 266911, and one mounted skeleton on display at the NY State College of Veterinary Medicine, Cornell University).

Due to the influence of exponential terms in the area and volume scaling relationships (described above), it is most common to compare morphological and physiological variables using a power relationship $Y = aX^b$, where Y and X are related variables, and a and b are the equivalence determining coefficient and exponent, respectively. In scaling analyses, dimensions are often plotted on log-log axes, where the slope of the relationship is an estimate of b ($\log Y = \log a + b \log X$). The tibia dimensions of selected groups (as explained in the discussion) were log-transformed and fit with both least-squares and reduced major axis (RMA) linear functions to estimate b, and scaling relationships are compared between groups.

14.2.2 Observations

Within the dwarf elephant population considered, tibial dimensions approach isometry ($b = 0.91$, where geometric similarity predicts $b = 1.0$ for linear dimensions). This value is significantly different from elastic (ES) or static stress (SS) similarity predictions (Table 14.1). Least squares estimates of b are provided simply to allow comparison with previously published literature. In this analysis, it is held that there is no justification for assuming that structural bone length is dependent on bone diameter, as implied by the standard least squares analysis. In fact, there is some logic for the opposite assumption, where the length of the bone determines the functional length of the element, and the diameter is determined as required to provide adequate strength. RMA estimates of b are, therefore, considered most appropriate, and the following report will focus on these values.

Table 14.1 ■ **Results of regression analysis of data provided in Figure 14.3**

	b						Expected		
	LS	RMA	R^2	N	95% CI	St. Err.	GS	ES	SS
Dwarf elephants	0.81	0.909	0.803	60	0.80–1.02	0.053	1.00	0.75	0.50
All elephants	1.29	1.29	0.99	3	0.77–1.82	0.041	1.00	0.75	0.50
Ceratomorphs	0.47	0.48	0.959	7	0.37–0.60	0.044	1.00	0.75	0.50
Carnivorans (large)	0.63	0.73	0.729	14	0.50–0.98	0.111	1.00	0.75	0.50
Carnivorans (small)	0.91	1.02	0.867	104	0.91–1.05	0.036	1.00	0.75	0.50
Carnivorans (all)	0.83	0.87	0.893	118	0.82–0.93	0.027	1.00	0.75	0.50

Table provides the estimated value of b from the allometric relation $Y = aX^b$, where Y is tibia length and X is tibia anterio-posterior diameter. LS: method of least squares (assumes Y dependent on X). RMA: reduced major axis analysis (does not assume dependence between variables). The expected values are the exponent b, as predicted by three common scaling models: GS – geometric similarity (isometry), ES – elastic similarity, SS – static stress similarity. Standard Error (St. Err.) of the slope and the 95% confidence interval (CI) of its estimation are also listed.

Tibial dimensions of dwarf and extant elephants (represented by the means of each species, so that intra- and interspecific variability is not confused) b is estimated at 1.29, but the 95% CI is large (0.77–1.82), due to the small number of distinct species represented. Thus, the available evidence is inconclusive regarding the precise proportions of dwarf and extant elephant tibiae, but it suggests that, with regard to the limb skeleton, the dwarfs may have proportions that are even more robust than their full-size relatives. Previous analyses of the proximal portions of the skeleton suggest that the dwarf forms have similar proportions to extant Asian elephants (Raia *et al.*, 2001), but distal limb bones appear foreshortened (Sondaar, 1977; Roth, 1984, 1990).

Small and large carnivorans have substantially different estimated values of b for tibial dimension. For the small body size group, b = 1.02, not different from the geometric similarity prediction of 1.0, and the large body size group has b = 0.73. Different from both of these groups, the ceratomorphs scaled with b = 0.49. This value was not significantly different from the prediction of static stress (SS) similarity (0.50), a value expected if stress in bending is the limiting factor on body size increases (but note that this predicts much more gracile form if body size decreases).

Tibia diameter of the Pleistocene dwarf elephants is displaced to the right of other mammals with comparable tibia length (Figure 14.3), indicating that the dwarf elephant tibia are substantially more robust than tibiae of comparably sized ceratomorphs or carnivorans.

14.3 INTERPRETATION

Three main observations arise from the scaling of elephant tibial dimensions:

1. Dwarf elephant tibiae scale in almost, but not precisely, an isometric manner. Although the scaling exponent b of 0.91 is not significantly different from the prediction of geometric similarity of 1.0, the value does indicate that some differences

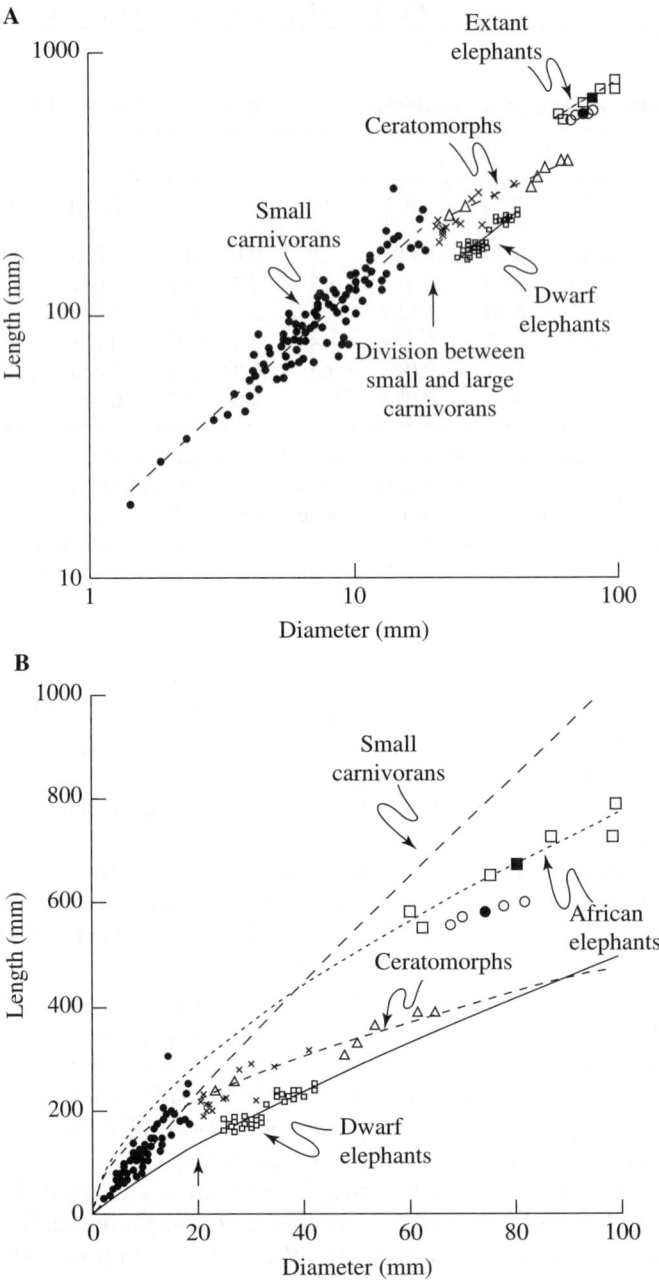

FIGURE 14.3

Scaling relationships for dwarf elephants compared to other terrestrial mammals. (**A**) Sicilian dwarf elephants are indicated by small open squares and show near-isometric scaling, either within the group or when extant elephant tibia are included (*Loxodonta* open squares, *Elephas* larger open circles, averages of each given in closed symbol). Dwarf elephant tibia scaling parallels that of the small carnivorans (small solid circles and long dashed line), but differs substantially from large carnivorans (crosses) or the ceratomorphs (tapirs and rhinoceroses; open triangles). The ceratomorphs scale with a slope approaching 0.5 (short dashed line), which matches the Static Stress similarity in basic shape of the tibia occur within the dwarf population. In Figure 14.3, it can be seen that the range of tibia sizes for dwarf elephants tend to cluster into smaller and larger bone size groups. It is not known whether this indicates a sexual dimorphism, or represents animals at different stages of their adaptation to a dwarfed stature.

2. The tibia of *E. falconeri* is not precisely a scaled-down version of that of the extant elephants, but appears even more robust than occurs in the large body size extant forms. It is well documented that one of the more prominent differences between the dwarf forms and living elephants is the relative shortness of the distal limb elements (Sondaar, 1977; Roth, 1984, 1990).

   Other aspects of the appendicular skeleton of dwarf elephants do appear to be self-similar with extant elephants (Raia *et al.*, 2001). It would appear, then, that the diameter of the dwarf tibiae were maintained during the dwarfing process, while its length relative to some other limb bones was reduced. The femur/tibia length ratio in the full size *E. maximus* ranges between approximately 1.55–1.65, while that of dwarf *E. faconeri* is 1.86. This makes the tibia foreshortened to a greater degree than a geometric reduction in size would require. Similar foreshortening of distal elements, especially the tibia, are seen in other cases of insular dwarfism in proboscidians – for example, the dwarf mammoth *Mammathus exilis* (Roth, 1984).

   Some suggestions have been made regarding the consequences that this might have for locomotion in these mammals, where shorter distal elements are seen as a possible mechanism to alter the lever arm acting at the knee (Raia *et al.*, 2001). However, this is probably a simplistic analysis of the function of the jointed limb in locomotion. Dwarfing, especially under the conditions found on islands, will likely involve both a stunting of development from nutritional deficiency, as well as a genetic size reduction. Nutritional deficiency can alter the growth trajectory of skeletal elements, and this will likely affect those elements that undergo larger proportional growth postnatally, such as the distally positioned tibia (Roth, 1984). However, for dwarfs such as *E. falconeri*, that are the same size, or smaller than, their ancestors at the beginning of post-natal growth, such a shift in growth trajectory cannot be responsible for the majority of dwarfing effects (Roth, 1984).

3. The maintenance of relative tibial diameter with a decrease in relative tibial length means that dwarf elephants had tibia dimensions that are substantially more robust than other mammals of comparable body mass. If the dwarf elephants had activity levels similar to the large body size forms from which they are derived, then it must be concluded that the tibiae of the dwarf elephants are vastly over-built. This is

FIGURE 14.3 (*Continued*)

prediction for maintenance of structural integrity for beams subject to bending loads. (**B**) Scaling relationships as above, but plotted on arithmetic axes. This plot puts the differences between skeletal proportions of dwarf elephants and similarly sized terrestrial mammals in perspective. Curves on this plot are those indicated in A, but each is extrapolated to the axis limit. In each plot, the 20 mm diameter division between "small" and "large" carnivorans is indicated.

indicated in Figure 14.3A; for a given length, the diameter of the dwarf elephant tibia is substantially greater than any other species. In Figure 14.3A, the axes are logarithmic, so that the regression slopes indicate the relative values of the scaling exponent b (from the power function $Y = aX^b$). However, this can give an incorrect visual impression of the differences in dimensions between groups.

In Figure 14.3B, these regressions are shown on arithmetic axes. The *Loxodonta* regression has a lower b than that of the small carnivorans, but the longer relative limb proportions of the elephants are indicated by the extrapolated curve, showing that elephants the size of carnivorans would have longer limbs. This would also be the case for the artiodactyls (McMahon, 1975; Bertram and Biewener, 1990). Thus, the comparison of dwarf elephants to carnivorans can be considered a conservative representation of the difference between normally smaller forms and dwarfed large body forms. The robust nature of the dwarf elephant tibiae suggests that appropriate functional adaptation of this bone does not appear to occur in response to dwarfing and the consequent reduction in skeletal loading.

Evaluated within the context of skeletal proportion changes observed in other groups of mammals, the scaling of the dwarf elephant skeleton provides intriguing evidence regarding the ability of the appendicular skeleton to respond to altered loading conditions. A key feature of skeletal scaling is shown in Figure 14.3: all groups are not well represented by a single relationship.

We have previously shown that, within a monophyletic group having generally similar locomotor function, such as the terrestrial carnivorans, differential skeletal scaling is indicated through a consistent decrease in the scaling exponent for length on diameter of larger body size families (Bertram and Beiwener, 1990). This manifests most obviously as a divergence of the large body size carnivorans (the large cats and bears, indicated by crosses in Figure 14.3) from the trend well established in smaller body size carnivorans.

To illustrate this, the terrestrial carnivorans have been arbitrarily divided at a tibia diameter of 20 mm, allowing species such as the cougar (40–100 kg), snow leopard (25–75 kg), lion (120–250 kg), jaguar (60–120 kg), leopard (30–90 kg), tiger (100–170 kg) and spotted hyena (40–90 kg) to be included in the large size group, and relegating species such as the cheetah (35–70 kg), clouded leopard (16–23 kg), sun bear (25–65 kg), brown and striped hyenas (25–50 kg) to the small body size group. Carnivorans with tibia diameter below 20 mm scale very close to isometry ($b = 1.02$, 1.00 expected), while those above this limit have a significantly lower exponent ($b = 0.74$, 95% CI: 0.50–0.98).

Although the large carnivorans suggest a systematic change in skeletal proportions in larger body size forms, the arbitrary nature of the division of the group invalidates any firm conclusions. How does the skeleton scale in a large body size lineage? For this comparison, the ceratomorphs – tapirs and rhinoceroses – are used. The ceratomorphs are a monophyletic group of moderate to large body size (200–3600 kg), represented by enough extant species to provide a valid indication of the scaling within the group. Ceratomorph tibiae scale remarkably close to the prediction of static stress similarity, where diameter increases in concert with the increases in body mass ($b = 0.49$, 95% CI: 0.37–0.60), static stress similarity predicts $b = 0.50$).

Static stress similarity was proposed by Galileo as the scaling relationship necessary to maintain consistent stress levels in a bending beam (a self-loaded cantilever) as size increases. Although large mammals appear to have vertically oriented limbs, we can assume that bending loads are also important, as has been demonstrated by direct measurement in other animals (Biewener, 1983; Biewener et al., 1983). In the case of the ceratomorphs, skeletal proportions appear to change so that bending stress levels remain a constant function of body mass.

The isometric form of the appendicular skeleton in smaller body size species indicates that they are not subject to the same consequences of size changes that larger body size species are. Even within the smaller body size category, the larger members of the group are not simply scaled-up versions of the smaller forms, as their skeletal dimensions might indicate. Rather, there is a systematic change in limb orientation, which alters the ground reaction force moment arm applied to the limb skeleton. This reduces the relative increase in load applied to the limb skeleton as mass increases, and allows increases in size without requiring changes in limb bone proportion (Biewener, 1989, 2005).

If smaller body size species are able to scale with isometry only because of systematic changes in limb orientation, and mammals larger than approximately 100 kg body mass scale with static stress similarity because they are in a size range beyond the capacity to reorient the limbs (i.e., this strategy has already been used to its fullest by that size), then we can expect that very large body size mammals will have fundamentally different proportions than their smaller body size counterparts. By extension of this, a dwarfing event that simply reduces the size of individuals of a large body size species should result in a population that is over-built for the mechanical loading regime imposed by the smaller body size. This prediction must be tempered, however, by recognition that if the limb skeletons of these animals could functionally adapt (i.e., the shape of the supporting bones within the individual changes to match imposed loads as per Wolff's Law), then the limb bone proportions may not vary substantially from those of other species of body size similar to that of the dwarfed population.

We can test these predictions using the dwarf elephant dimensions in Figure 14.3. The diameter of the tibiae of the dwarf elephants are fully three-fold greater than the average carnivoran tibiae of comparable length. Since cross-sectional area (proportional to diameter^2) or second moment of area (proportional to diameter^4) are more relevant measures of structural capacity to compression or bending loads, this suggests that the tibiae of the dwarf elephants are between 9–80 times more robust than is necessary for their acquired body size!

The above arguments suggest that functional adaptation of limb bones of healthy adult mammals can be far more limited than is commonly expected. On the other hand, we have evidence of changes occurring within the skeleton of dwarf elephants that are, indeed, related to changes in loading environment. For instance, changes in skull form within the dwarf elephant are well documented (Roth, 1992; Palombo, 2001). Some of these relate to allometric changes in relative tooth size, and some may relate to endomorphic changes involved with the mechanisms responsible for the dwarfing process itself.

One major difference between dwarf and normal size elephants is the degree of pneumatization of the posterior aspect of the skull, with this being substantially less in the dwarf forms. It is likely that this is related to the scaling of loads applied by the dorsal cervical musculature; the massive head of normal-sized elephants requires expansion of the posterior skull to accommodate muscle attachment. With a decrease in head volume, a proportionately smaller muscle attachment area is required, so the motivation for skull expansion is removed. Thus, some aspects of skeletal form do appear to adapt to the mechanical circumstances presented by the dwarfing event.

So, there is little evidence from the dwarf elephant tibia that functional adaptation of bone form occurs in spite of large changes in load – yet indications from the skull that it does. There is ample evidence, from a variety of direct experimental manipulations of bone in a variety of species, that indicate that bone does, indeed, respond to its mechanical environment. For instance, changes in load are accommodated over a surprisingly brief period in rats (Schreifer *et al.*, 2005), and these findings are in concert with analyses of the process of load accommodation in human tennis players (Bass *et al.*, 2002), where morphological differences between the playing and nonplaying arms occur soon after the initiation of activity, not after longer-term loading, as previously expected (Jones *et al.*, 1977). How can demonstrated bone mechano-responsiveness be reconciled with the apparent insensitivity of the dwarf elephant limb bones?

One possibility is that we are mistaken in some of the premises of Wolff's Law; namely that external bone form results from a response to mechanical loads. Our modern understanding of genetics and its influence on bone form, unavailable to Wolff, make it common for this aspect of Wolff's Law to be interpreted with substantial "limitation". Studies on recently available strains of mice indicate that genetics determines a large proportion of the responsiveness of the appendicular skeleton (Akhter *et al.*, 1998; Turner and Beamer, 2002). However, it is still understood that the healthy, adult mammalian skeleton responds in a self-regulated manner to changes in the applied load environment (Martin *et al.*, 1998). The dwarf elephant example, however, demonstrates that this is not necessarily the case, even for profound changes in applied load. Bone does have substantial capacity for self-repair. Is it possible that some of the loading procedures employed in experimental manipulation simply invoke aspects of the repair process, but without the requirement of an outright fracture?

Some experimental responses could be direct inflammatory responses invoked by the loading system (Meade *et al.*, 1984). It has long been known that low density periosteal deposition can accompany inflammation or trauma anywhere in the bone – for instance, medullary evacuation (Danckwardt-Lilliestrom *et al.*, 1972; Bab *et al.*, 1985), or irritations can cause bone deposition at a different location (Melcher and Accursi, 1972). Low density woven bone, a standard tissue response in bone injury, has been implicated in the process of bone adaptation and may, itself, be seen as one feature of the adaptive response (Burr *et al.*, 1989).

Techniques have been developed to limit any "pathological" features of exogenous loading (Robling *et al.*, 2001) but, even within these conditions, woven bone and osteophytes are common. Still yet to be resolved is whether woven bone is a part of the normal adaptive process of bone adaptation (Turner *et al.*, 1991), or simply represents one stage in a repair process that may result in increased structural integrity of the bone, but is not necessarily the coordinated response to load, as is generally assumed by Wolff's Law.

Small-scale flaws (microcracks) have also been implicated in functional adaptation of bone (Burr *et al.*, 1985; Martin, 2002; Ruimerman *et al.*, 2005). Again, interpretation of the process of bone adaptation comes down to where the line between fracture, invoking a healing response, and microfracture, presumably stimulating a subtler response, is drawn. Complicating the discussion is the fact that the perspective on functional responsiveness of bone that is the legacy of Wolff's Law generally considers increases and decreases in applied load as simply quantitative variants on the continuum of load circumstances that elicit a response from bone (Carter, 1982). However, as has been witnessed by the poor record of counter-measures to mitigate the consequences of zero gravity during space flight (Lang *et al.*, 2004), direct extrapolation of bone response from laboratory manipulations does not necessarily provide the anticipated results.

Alternatively, the results of the present study could derive from inappropriate interpretation of the form and scaling of dwarf elephants. Dwarf elephant mechanics are developed from simple interpretations of remnant bones. This provides little direct evidence regarding change in overall limb proportions, properties of tissues, or function.

It has been observed that the distal elements of dwarf elephant limbs are proportionately shorter than mainland forms (Sondaar, 1977; Roth, 1990). This has been interpreted as indicating that dwarf elephants may have moved in different ways than their mainland counterparts do (Caloi and Polombo, 1994; Raia *et al.*, 2001). However, the extreme size changes noted would require equally extreme changes in locomotor performance in order to account for the observed lack of bone changes, so it is unlikely that increased activity levels could account for even a small portion of the replacement of skeletal loads lost due to decreased size in these forms.

Could it be that the outer dimensions of the bone do not represent the support capability of the dwarf bones in the same way they do in large forms? Bone scaling in this study has been inferred from outer bone dimensions alone. Although not a large enough sample to definitively indicate diaphyseal proportions, two fractured tibia shafts of dwarf elephants were observed at the British Museum of Natural History (specimen nos. 49213 and 49230). These showed similar proportional cortical thickness to optimization predictions for marrow filled mammalian bones (Currey and Alexander, 1985; Evinger *et al.*, 2005). Diameter/thickness ratios ranged from 4.23–5.69 for specimen 49230 and were in the order of 2.24 for the slightly larger specimen 49213. These are well within the lower range (robust, thick walled) for mammalian long bones (Currey, 2002), indicating that cortical thickness decreases likely do not reduce the load-carrying capacity of the dwarf elephant tibiae.

The dwarfing process itself could involve biological constraints that override mechanical induction of bone form in the insular elephants. This may be similar to some other dwarfing conditions, such as achondroplasia, where skeletal diameter is maintained when length is reduced. A disassociation between bone length and width can also be observed in some normal canine breeds – for example, the Bassett hound, a breed with apparently over-robust limb skeletons for the length of its constituent bones. Similar growth characteristics are seen in psuedoachondroplasia that can be traced directly to control factors in growth plate cells (Breur *et al.*, 1992). It may be that the growth processes of the dwarf elephant long bones are unusual. However, this does not alter the fact that the bones remain functionally overbuilt, in spite of the potential for functional adaptation to take place.

The final explanation for the results, indicating bone response to loading and the dwarf elephant example where this is apparently not seen, may well be somewhere between the options discussed above. The current contribution provides a talking-point designed to stimulate discussion of such issues, in the hopes of encouraging a consideration of alternate points of view. Such alternate perspectives can be stimulated by considering alternative data, such as those documented here, that are not usually available as subject for debate on this issue. The overall goal, of course, is eventually to arrive at a conclusive demonstration of the "rules", both mechanical and biological, that determine the skeletal expression we observe, and to determine the responsiveness of the adult skeleton to its mechanical circumstances. Whatever that solution turns out to be, it must be able to accommodate both observed bone responses occurring in the natural world, such as the insular dwarf elephants, and the results of experimental load manipulations.

ACKNOWLEDGEMENTS

Ms. Hollie Knoll provided measurements of extant elephant tibia dimensions. The assistance of the mammals collection staff at the Field Museum of Natural History, The American Museum of Natural History, and the Smithsonian Institution, are gratefully acknowledged. Special thanks go to the staff of the Paleontology Department at the British Museum of Natural History, London.

▌ REFERENCES

Akhter MN, Cullin DM, Pedersen EA, Kimmel DB, Recker RR (1998). Bone response to *in vivo* mechanical loading in two breeds of mice. *Calcified Tissue International* **63**, 442–449.

Alexander RMcN, Jayes AS, Maloiy GMO, Wathuta EM (1979). Allometry of the limb bones of mammals from shrews (*Sorex*) to elephant (*Loxodonta*). *Journal of Zoology, London* **189**, 305–314.

Ambrosetti P (1968). The Pleistocene dwarf elephants of Spinagallo (Siracusa, south-eastern Sicily). *Geologica Romana* **7**, 277–398.

Bab I, Gazit D, Massarawa A, Sela J (1985). Removal of tibia marrow induces increased formation of bone and cartilage in rat mandibular condyle. *Calcified Tissue International* **37**, 551–555.

Bass SL, Saxon L, Daly RM, Turner CH, Robling AG, Seeman E, Stuckey S (2002). The effect of mechanical loading on the size and shape of bone in pre-, peri-, and postpubertal girls: a study in tennis players. *Journal of Bone and Mineral Research* **17**, 2274–2280.

Bertram JEA (1988). *The biomechanics of bending and its implications for terrestrial support.* Ph.D. dissertation. University of Chicago.

Bertram JEA, Biewener AA (1990). Differential scaling of the long bones in the terrestrial carnivora and other mammals. *Journal of Morphology* **204**, 157–169.

Biewener AA (1983). Locomotor stresses in the limb bones of two small mammals: the ground squirrel and chipmunk. *Journal of Experimental Biology* **103**, 131–154.

Biewener AA (1989). Scaling body support in mammals: limb posture and muscle mechanics. *Science* **245**, 45–47.

Biewener AA (1991). Musculoskeletal design in relation to body size. *Journal of Biomechanics* **24**(Suppl. 1), 19–29.

Biewener AA (2005). Biomechanical consequences of scaling. *Journal of Experimental Biology* **208**, 1665–1676.

Biewener AA, Thomason J, Goodship A, Lanyon LE (1983). Bone stress in the horse forelimb during locomotion at different gaits: a comparison of two experimental methods. *Journal of Biomechanics* **16**, 565–576.

Breur GJ, Farnum CE, Padgett GA, Wilsman NJ (1992). Cellular basis of decreased rate of longitudinal growth of bone in pseudoachondroplastic dogs. *Journal of Bone & Joint Surgery, American Volume* **74**, 516–528.

Burr DB, Martin RB, Schaffler MB, Radin EL (1985). Bone remodeling in response to *in vivo* fatigue microdamage. *Journal of Biomechanics* **18**, 189–200.

Burr DB, Schaffler MB, Yang KH, Lukoschek M, Sivaneri N, Blara JD, Radin EL (1989). Skeletal change in response to altered strain environments: is woven bone a response to elevated strain? *Bone* **10**, 223–233.

Caloi L, Palombo MR (1994). Functional aspects and ecological implications in Pleistocene endemic herbivores of Mediterranean islands. *Historical Biology* **8**, 151–172.

Carter DR (1982). The relationship between *in vivo* stresses and cortical bone remodeling. *Critical Reviews in Bioengineering* **8**, 1–28.

Christiansen P (1999a). Scaling of mammalian long bones: small and large mammals compared. *Journal of Zoology, London* **247**, 333–348.

Christiansen P (1999b). Scaling of the limb long bones to body mass in terrestrial mammals. *Journal of Morphology* **239**, 167–190.

Currey JD (2002). *Bones: structure and mechanics*. Princeton University Press, Princeton, NJ.

Currey JD, Alexander RMcN (1985). The thickness of the walls of tubular bones. *Journal of Zoology, London* **206**, 453–468.

Danckwardt-Lilliestrom G, Grevsten S, Johansson H, Olerud S (1972). Periosteal bone formation on medullary evacuation. *Upsala Journal of Medical Sciences* **77**, 57–61.

Economos AC (1983). Elastic and/or geometric similarity in mammalian design? *Journal of Theoretical Biology* **103**, 167–172.

Évinger S, Suhai B, Bernáth B, Gerico B, Pap I, Horváth G (2005). How does the relative wall thickness of human femora follow the biomechanical optima? An experimental study on mummies. *Journal of Experimental Biology* **208**, 899–905.

Frost HM (2004). A 2003 update of bone physiology and Wolff's Law for clinicians. *Angle Orthodontist* **74**, 3–15.

Galileo Galilei (1638). *Discourse on Mathematical Demonstrations Concerning Two New Sciences*. (Drake S, trans., 1974). Madison, Wisconsin: University of Wisconsin Press, Ltd., pp. 169–170.

Hildebrand M (1952). Body proportions of the Canidae. *American Journal of Anatomy* **99**, 217–256.

Iriarte-Diaz J (2002). Differential scaling of locomotor performance in small and large terrestrial mammals. *Journal of Experimental Biology* **205**, 2897–2908.

Jones HH, Priest JD, Hayes WC, Tichenor CC, Nagel DA (1977). Humeral hypertrophy in response to exercise. *Journal of Bone and Joint Surgery* **59A**, 204–208.

Kurtén B (1968). *Pleistocene mammals of Europe*. Aldine Pub. Co., Chicago.

Lang T, LeBlanc A, Evans H, Lu Y, Genant H, Yu A (2004). Cortical and trabecular bone loss from spine and hip in long-duration space flight. *Journal of Bone and Mineral Research* **19**, 1006–1012.

Lister AM (1989). Rapid dwarfing of red deer on Jersey in the last interglacial. *Nature* **342**, 539–542.

Lomolino MV (1985). Body size of mammals on islands: the island rule reexamined. *American Naturalist* **125**, 31–49.

Martin RB (2002). Is all cortical bone remodeling initiated by microcracks? *Bone* **30**, 8–13.

Martin RB, Burr DB, Sharkey NA (1998). *Skeletal tissue mechanics*. Springer-Verlag, New York, pp. 392.

Masseti M (2001). Did endemic dwarf elephants survive on Mediterranean islands up to proto-historical times? In: Cavarretta G, Gioia P, Mussi M and Palombo MR (eds). *The World of Elephants*, pp. 402–406. Proceedings of the 1st International Congress, Rome. Consiglio Nazionale delle Richerche, Roma.

McMahon TA (1973). Size and shape in biology. *Science* **179**, 1201–1204.

McMahon TA (1975). Allometry and biomechanics: Limb bones in adult ungulates. *American Naturalist* **109**, 547–563.

Meade JB, Cowin SC, Klawitterm JJ, Van Buskirk WC, Skinner HB (1984). Bone remodelling due to continuously applied loads. *Calcified Tissue International* **36**, S25–S30.

Melcher AH, Accursi GE (1972). Transmission of an 'osteogenic message' through intact bone after wounding. *Anatomical Record* **173**, 265–276.

Owen-Smith RN (1988). *Megaherbivores: the influence of very large body size on ecology.* Cambridge University Press, Cambridge. 369 pp.

Palombo MR (2001). Paedomorphic features and allometric growth in the skull of *Elephas falconeri* from Spinagallo (Middle Pleistocene, Sicily). In: Cavarretta, G, Gioia, P, Mussi, M and Palombo, MR (eds). *The World of Elephants*, pp. 492–496. Proceedings of the 1st International Congress, Rome. Consiglio Nazionale delle Richerche, Roma.

Peters RH (1983). *The ecological implications of body size.* Cambridge University Press, Cambridge,UK.

Poulakakis N, Theodorou GE, Zouros E, Mylonas M (2002). Molecular phylogeny of the extinct Pleistocene dwarf elephant *Paleoloxodon antiquus falconeri* from Tilos Island, Dodkanisa, Greece. *Journal of Molecular Evolution* **55**, 364–374.

Poulakakis N, Parmakelis A, Lymberakis P, Mylonas M, Zouros E, Reese DS, Glaberman S, Caccone A (2006). Ancient DNA forces reconsideration of evolutionary history of Mediterranean pygmy elephants. *Biology Letters* **22**, 451–454.

Prothero DR, Sereno PC (1982). Allometry and paleoecology of medial Miocene dwarf rhinoceroces from the Texas Gulf Coastal Plain. *Paleobiology* **8**, 16–30.

Raia P, Barbera C, Conte M (2001). Scaling of proximal limb bones between *Elephas antiquus* and its insular descendant *Elaphas falconeri*. In: Cavarretta G, Gioia P, Mussi M, Palombo MR (eds). *The World of Elephants*, pp. 502–506. Proceedings of the 1st International Congress, Rome. Consiglio Nazionale delle Richerche, Roma.

Rashevsky N (1948). Mathematical Biophysics. University of Chicago Press, Chicago.

Robling AG, Burr DB, Turner CH (2001). Recovery periods restore mechanosensitivity to dynamically loaded bone. *Journal of Experimental Biology* **204**, 3389–3399.

Roth VL (1984). How elephants grow: heterochrony and the calibration of developmental stages in some living and fossil species. *Journal of Vertebrate Paleontology* **4**, 126–145.

Roth VL (1990). Insular dwarf elephants – a case study in body mass estimation and ecological inference. In: Damuth J, MacFadden BJ (eds). *Body size in mammalian paleobiology: Estimation and biological implications*, pp. 151–180. Cambridge University Press, Cambridge,UK.

Roth VL (1992). Inferences from allometry and fossil dwarfing of elephants on islands. *Survey of Evolutionary Biology* **8**, 259–288.

Ruff C, Holt B, Trinkaus E (2006). Who's afraid of the big bad Wolff? "Wolff's law" and bone functional adaptation. *American Journal of Physical Anthropology* **129**, 484–498.

Ruimerman R, Hilbers P, van Rietbergen B, Huiskes R (2005). A theoretical framework for strain-related trabecular bone maintenance and adaptation. *Journal of Biomechanics* **38**, 931–941.

Sander PM, Mateus O, Lacen T, Knötschke N (2006). Bone histology indicates insular dwarfism in a new late Jurassic sauropod dinosaur. *Nature* **441**, 739–741.

Sanderson GP (1896). *Thirteen years among the wild beasts of India.* WH Allen, London (6th edition).

Schriefer JL, Warden S, Saxen L, Robling AG, Turner CH (2005). Cellular accommodation and the response of bone to mechanical loading. *Journal of Biomechanics* **38**, 1838–1845.

Shoshani J (1998). Understanding proboscidian evolution: a formidable task. *Trends in Ecology & Evolution* **13**, 480–487.

Simmons AH (1988). Extinct pygmy hippopotamus and early man on Cyprus. *Nature* **333**, 554–557.

Sondaar PY (1977). Insularity and its effects on mammal evolution. In: Hecht MK, Goody PC, Hecht BM (eds). *Major patterns in vertebrate evolution*, pp. 671–707. Plenum, New York.

Srinivasa M (2007). Bipedal running: "No muscle work and all tendon play" is energetically benefitial even with an energy cost for isometric force production. *SICB*, Phoenix, Jan. 2007.

Thompson DW (1942) *On growth and form*. Cambridge University Press, 1116 pp.

Turner CH, Beamer WG (2002). Is skeletal mechanotransduction under genetic control? *Journal of Musculoskeletal and Neuronal Interactions* **2**, 237–238.

Turner CH, Robling AG (2004). Exercise as an anabolic stimulus for bone. *Current Pharmaceutical Design* **10**, 2629–2641.

Turner CH, Akhter MN, Raab DM, Kimmel DB, Recker RR (1991). A noninvasive, *in vivo* model for studying strain adaptive bone modeling. *Bone* **12**, 73–79.

Van Valen LM (1973). Pattern and balance in nature. *Evolutionary Theory* **1**, 31–49.

Vartanyan SL, Garutt VE, Sher AV (1993). Holocene dwarf mammoths from Wrangle Island in the Siberian arctic. *Nature* **362**, 337–340.

CHAPTER FIFTEEN

Basic Mechanisms of Bipedal Locomotion: Head-Supported Loads and Strategies to Reduce the Cost of Walking

James R. Usherwood[1] and John E. A. Bertram[2]

[1] Royal Veterinary College, University of London, UK
[2] Department of Cell Biology and Anatomy, Cumming School of Medicine, University of Calgary, CA

15.1   INTRODUCTION

Humans employ a specific bipedal gait that has had substantial influence on the survival and proliferation of the species (Marino, 2008; Vaughan and Blaszsyk, 2008). Understanding the mechanics of human locomotion has important consequences for interpreting the adaptive evolution of the species, contending with and rehabilitating dysfunctional movement patterns and maximizing performance and injury prevention (Keefe *et al.*, 2009; Chung and Wang, 2010; Hartigan *et al.*, 2009). In spite of a remarkable amount of research performed on the human bipedal gait annually, and extending over many decades, questions regarding the fundamental dynamics and control of human locomotion still remain. One of these, the apparent cost saving provided by carrying loads on the head, or piled very high over the back, has eluded explicit explanation for some time (Maloiy *et al.*, 1986; Cavagna *et al.*, 2002). Even though carrying

Understanding Mammalian Locomotion: Concepts and Applications, First Edition.
Edited by John E. A. Bertram.
© 2016 John Wiley & Sons, Inc. Published 2016 by John Wiley & Sons, Inc.

loads on the head is widespread in many traditional cultures, the mechanisms through which advantage is gained over carrying via a standard shoulder-supported, torso-level pack are not clear (Cavagna *et al.*, 2002; Bastien *et al.*, 2005b; Minetti *et al.*, 2006).

Although understanding head carrying is only a minor feature of human function, the question may be more important than simply putting the cultural behavior in context. Since the mechanics of its explanation do not appear to fit within standard approaches to human locomotion, it is possible that an explanation that describes the functional advantage of placing loads on the head could also involve fundamental features of human locomotion not addressed by standard approaches. As such, an explanation of the advantage of head loads in locomotion may offer more general insight into the mechanisms important to bipedal walking.

15.2 HEAD-SUPPORTED LOADS IN HUMAN-MEDIATED TRANSPORT

Manually transporting material by carrying it on the head is a common practice in numerous agricultural and traditional cultures. A brief internet search (eHRAF World Cultures – search terms, "headload", "head transport", "head carry") indicated verifiable descriptions of head packing of burdens being utilized by seven cultural groups in Africa, four cultural groups in Asia and at least one cultural group in each of Australia (Chewings, 1936), non-Australian Oceania (Dekeyne, 1966), Central America (Parsons, 1936), and the Caribbean (Kerns, 1984). The practice is notably absent in Europe and much of the Americas, but appears widespread in a large portion of the world.

It has been suggested that this type of load carrying is adopted in areas where the climatic conditions – particularly heat and humidity – do not favor draft animals (Mangin and Arianne Skinner, 1921), and is well suited to areas where the woods are not overly dense (in denser brush, back-supported loads are carried, usually reinforced with a forehead "tumpline" band – Putnam, 1948; Turnbill, 1962).

Large loads are often carried over substantial distances. The women of the Tiv culture of western African, for example, routinely carry head loads of a third to half their body weight for distances of 1–2 km (Bohannan and Bohannan, 1969) and the brick-making cottage industry of India depends on male and female porters routinely carrying head-supported loads of 20–40 kg of bricks to the kilns for firing. Nepalese porters carry loads piled high above their heads, but these are borne on their back with a forehead support. These burdens are 60–200% of body weight (125–150% common – Malville *et al.*, 2001; Bastien *et al.*, 2005a).

The widespread transport of relatively large loads carried on the top of the head (Figure 15.1), suggests that there is an advantage to using this carrying strategy over more standard torso-level packs or arm-carrying. Studies of the metabolic cost of load-carrying (Maloiy *et al.*, 1986) and kinetic energetics (Heglund *et al.*, 1995) of walking by African women, from a variety of tribes, show that up to 20% of the body mass may be carried as a head-supported load, with negligible energetic cost. This contrasts sharply with results for loads applied with backpacks to army recruits (Goldman and Iampietro, 1962; Pandolf *et al.*, 1977), or to the head with men and women inexperienced at carrying such loads (Soule and Goldman, 1969; Maloiy *et al.*, 1986; Heglund *et al.*, 1995), where any added load incurs a measureable cost in metabolic cost that is directly proportional to the added load.

FIGURE 15.1

Photograph of a Luo woman carrying firewood on her head. Such load-carrying is part of normal activities for a variety of cultures worldwide. Photograph courtesy of RF Ker.

The head-load advantage is also contrary to the well-documented direct proportional cost relationship when additional loads are added to running quadrupedal mammals (Taylor *et al.*, 1980). In these cases, an increase in metabolic cost is apparent, even with small loads added to the body. The question of how experienced head-loaded individuals are able to avoid increases in metabolic cost at lower levels of loading, and also enjoy substantially reduced metabolic costs with loads as high as 100% body mass, has not been answered satisfactorily to date (Alexander, 1986; Taylor, 1995).

15.2.1 Can the evidence be depended upon?

What is the existing evidence for a cost advantage for head load carrying? Maloiy *et al.* (1986) measured oxygen consumption of five women of the Luo (head carrying) and Kikuyu (back support with head strap reinforcement) tribes of East Africa with weights of 23–68% of body weight. They found that, at all but the slowest walking speeds, the African women were able to carry these burdens for less metabolic expenditure than untrained individuals from non-African cultures. In further analysis, Heglund *et al.* (1995) found that African women appear to use less mechanical energy to carry head-supported loads. Soule and Goldman (1969) confirmed that head-carrying in individuals untrained in the technique resulted in a proportional increase in cost (a load of 20% body weight resulted in a 20% increase in locomotion cost) but that, in comparison to other means of bearing the load – in the hands or attached to the feet – head-carrying resulted in the least extra cost.

These results are consistent with Balogun *et al.* (1986), who found that carrying load on the head was relatively less costly than other traditional yokes, but also found that cost increased in proportion to load applied, as found in other Western-style carriers. Datta and Ramanathan (1971), however, compared seven modes of load-carrying and found that head-carrying was one of the two least costly modes (insignificantly different from having the weight split between packs on the front and rear of the torso). In this study, the standard backpack had equivalent cost to those of Western trekkers (Bastien *et al.*, 2005b) while head-carrying required 95% of the metabolic cost of the standard backpack. Many of the subjects in Datta and Ramanathan (1971) had prior experience with carrying head loads, although none were routinely engaged in the practice at the time of the study.

Jones *et al.* (1987) questioned the original Maloiy *et al.* (1986) results, and proposed that African women were metabolically more effective at carrying loads than individuals from Western cultures. From their own studies (eight women of the Keneba culture, with body fat ranging between 16–34%), they found that, on a lean body mass basis, African women appeared to be more metabolically cost-effective at walking than individuals from Western cultures. They concluded that the addition of body fat reduced the effectiveness of load-carrying in the African women, to the point where individuals with 40% body fat were as "inefficient" as were untrained head load carriers (Jones *et al.*, 1987, p. 1332).

Substantial work has been conducted on the metabolic cost of load-bearing by Nepalese porters. Although, in this culture, these extreme loads are not carried directly on the head, they are routinely piled high above the head, supported both on the back and shoulders, and stabilized by a "tumpline" across the forehead. Bastien *et al.* (2005a) found that Nepalese porters demonstrated greater economy in carrying loads on level ground than even the African women, and this advantage was larger as loads were increased. Minetti *et al.* (2006) confirmed these results, and noted that this economy of locomotion could account for only a portion of the advantages displayed by these porters when negotiating inclines. They suggested that, as well as economical locomotion, these porters possessed metabolic adaptations that contributed substantially to their load-carrying ability.

Maloiy *et al.* (1986) suggested that African women can carry 20% extra load for no additional cost. However, they also demonstrated that their unloaded cost of locomotion was no different than expected for other humans. This suggests a discontinuity in the relationship between load and transport cost. Although the data could be interpreted this way, it depends on a rectilinear extrapolation of the data to zero load, whereas a curvilinear fit, passing through the metabolic cost point where no additional load is carried, may well reflect reality (as it does for load carrying in Western army recruits (Pandolf *et al.*, 1977), or for the apparent mechanical work performed – Heglund *et al.*, 1995).

15.3 POTENTIAL ENERGY SAVING ADVANTAGES

Although unanswered questions remain, it would appear from the above that carrying loads on the head (or higher, in the case of Nepalese porters) can result in a reduced metabolic cost of load-carrying. Several advantages are possible from either a structural or energetic standpoint:

1. *Passive support of the load through axial and appendicular skeleton.* Load support using the skeleton may not require substantial metabolically active investment to counter the effect of gravity. There is potential for greater load-carrying capacity because the rigid skeleton can support a large portion of the load passively. Muscular investment is required to stabilize joints and balance the load while standing or walking. Both of these features could presumably be accommodated through carrying technique or specifically trained muscle groups. The possibility exists for minimizing metabolic investment through carrying technique, by managing the joints at the singularity point, where the ground reaction force vector passes through the center of rotation of each joint (i.e., where the ground reaction force moment is minimized, and very little muscle moment is required to stabilize the joint). For such a condition, the majority of added load is supported by the (relatively) metabolically inert skeleton. Using relatively passive structures to support added weight could certainly assist in standing weight support, but the motions required of walking will make the required joint alignment extremely difficult to maintain.

   Maloiy *et al.* (1986) report that loads supported by African women during quiet standing had no demonstrable effect on metabolic rate. This might indicate that much of the weight support by head-carrying (Luo women) or head strap-reinforced back support (Kikuyu women) comes from effective use of tissues with a low metabolic response, such as the skeleton. There is ample evidence that the skeleton is highly loaded during carrying.

   Degenerative effects at specific sites, such as the cervical vertebrae, appear to be common in groups that engage in head carrying (Echarri and Forriol, 2002; Joosab *et al.*, 2005; Mahbub *et al.*, 2006). General skeletal robustness appears increased in head load porters in South Africa (Lloyd *et al.*, 2009), but it is not known whether this is due to carrying loads on the head specifically, or is the result of the general loading level of individuals involved in manual labor. However, such evidence does suggest that the skeleton does bear substantial load in this carrying technique, as expected – because the skeleton is, after all, the major compression-resistant component of the body. Likewise, muscle activity and metabolic investment is necessary in all walking (Adamczyk and Kuo, 2009).

2. *Metabolically effective musculature.* Routine carrying of heavy loads, whether on the head or piled high on the back, provides the opportunity for the musculature to adapt to this chore. For those cultures in which this mode of transport is used, most porters are engaged in carrying on a daily basis over an extended number of years, often an entire adult lifetime (Malville *et al.*, 2001). The different fiber types of muscle appear optimized for different activities, and there is evidence that training can alter the proportion of fiber type compliment, by emphasizing the hypertrophy of the type most suitable to the activity engaged in (Simoneau *et al.*, 1985).

There is currently no direct evidence of such training effects that account for a distinct advantage for head load-carrying, but they could be responsible for at least some of the differences noted between indigenous and post-industrialized Western populations. Analysis of Nepalese porters carrying loads up and down hills indicate a substantial functional advantage over Caucasian subjects – even ones who are well acclimated to the altitude and terrain involved (Minetti *et al.*, 2006). Since such carrying activities are an integral part of many traditional cultures, and presumably have been for an extended time period, it may be that muscle fiber distribution favors this type of activity at a population level as well. Physiological differences relevant to manual exertion at altitude have been detected in cardio-respiratory features of the Nepalese populations (Havryk *et al.*, 2002; Kayser *et al.*, 1991; Marconi *et al.*, 2004).

Some indication of a similar muscle-related effect on locomotion economy may exist in tortoises. Analysis of the extremely slow locomotion in these reptiles indicates very little evidence of movement strategies that would reduce the cost of locomotion, yet they have a metabolic cost per distance traveled that is comparable to other land animals (Zani *et al.*, 2005). It is possible that the muscles of these animals are extremely effective at providing locomotion at these slow rates of progress (Griffin, 2006). Unfortunately, more effective muscle function does not explain why trained individuals should not be even more economical than untrained individuals when walking unloaded, or why the advantage exists for carrying loads placed on the head, versus a standard backpack.

3. *Enhanced energy recovery.* It has been long held that exchange (transduction) between available energy forms and, particularly, between potential and kinetic energy, allows for increased energy recovery within the walking stride (Cavagna *et al.*, 1963, 1976). The reduced cost of load-carrying by placing the load on the head has been assumed to result from an increase in the effectiveness of this transduction, likely mediated by the high position of the load augmenting the pendulum-like exchange that occurs in walking (Heglund *et al.*, 1995). Sustained, sophisticated analysis has led to the measurement of only very subtle differences between the apparent exchange between energy forms, however.

Cavagna *et al.* (2002) analyzed the apparent exchange over the course of the stride. They found that the measured exchange between kinetic and potential energy varies substantially over the course of single and double limb contact, ranging from 100%, over brief periods, to zero exchange over others. Their analysis indicated that African women experienced at carrying head loads had altered exchange over the course of the stride, so that some portions of the zero exchange were eliminated. In other areas of the stride, they had somewhat greater exchange than either their inexperienced Western counterparts with a head load, or themselves without a head load. The analysis of relative potential and kinetic energy changes during the stride indicates that exchange is greatest when total energy has zero slope – when changes in PE and KE exactly cancel each other, regardless of how little energy is involved. Apparent exchange can change quickly to zero when either PE or KE have zero slope (no change in magnitude over time), while even the smallest change occurs in the other. In this case, exchange is not possible because changes in one form of energy cannot be transferred to the other (Figure 15.2).

Such subtle differences result in large changes in apparent recovery, even though the manner of movement is relatively consistent, and both kinetic and potential energy continually fluctuate in a cyclic manner. Are such apparently substantive difference in the energetics of walking based on minor differences in exchange? If so, why are these strategies not used to reduce the cost of unloaded walking? Cavagna *et al.* (2002) note that the mechanism responsible for the change in recovery, and the reduction in locomotion cost, has not been identified; "...*the phases of the step mainly affected by loading can now be determined, even though the mechanism of*

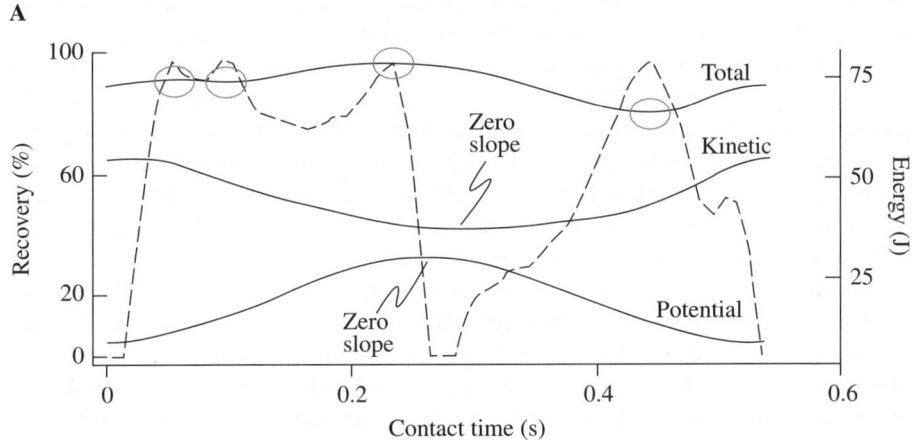

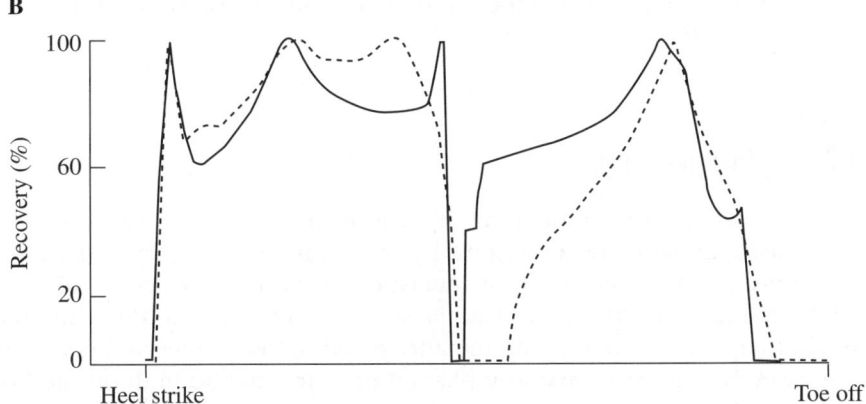

FIGURE 15.2

(A) Plot of the energy fluctuations (kinetic, potential and total – solid lines) over the contact of one limb in walking. Calculated transduction between kinetic and potential energy, labeled recovery (dashed line), varies over the contact. Where total energy has a slope near zero (whether following a rise or fall), the transduction appears high (circles). However, when either kinetic or potential energy have zero slope while the other does not, transduction is negligible (pointers). (B) Apparent recovery over a step for the same experienced head load carrier, without a load (dashed line, or with a head-supported load (solid line). Figure after Cavagna *et al.* (2002).

the observed changes is still unknown." (p. 3421). It is interesting to note that the major changes detected occurred during single stance, when the center of mass (CoM) travels over the support limb. This is a portion of the stride where a substantial amount of positive work is done on the CoM (Donelan *et al.*, 2002) – presumably largely coming from the kinetic energy of the previous step, but also supplemented with some muscular effort (Lee and Farley, 1998).

4. *An alternative approach, adjusting loss rather than recovery.* Why must work be done in carrying the body, loaded or not? Momentum is lost, and work (energy) absorbed, during the transition from one limb to the next (double contact), when the CoM path is deflected from a forward-and-downward to a forward-and-upward trajectory. The positive work added by the muscles in locomotion must equal that lost from all sources during the stride in order to maintain constant average speed (when on a level surface). Thus, identifying the source of loss also explains the requirement for the addition of energy to the stride. The evaluation of sources of loss in walking machines has demonstrated the value of analyzing the cost of the transition (Ruina *et al.*, 2005; Kuo *et al.*, 2005), and has helped with the understanding of human bipedal locomotion (Kuo, 2007; Adamczyk and Kuo, 2009; Bertram and Hasaneini, 2013; and see Chapters 5 and 6).

Could the advantages of carrying head loads (or any load placed high on the body) be explained from this perspective? In the following, we evaluate this through the development of a relatively simple model of bipedal locomotion. Although the model avoids much of the complexity of the human locomotory system, the predictions of the model turn out to be remarkably consistent with observations of head load carriers. Such models, though simple, can be helpful in identifying some of the fundamental features of the cost of locomotion and the strategies available to reduce these costs (Kuo, 2007).

15.4   A SIMPLE ALTERNATIVE MODEL

The observations of apparent reduction in proportional locomotory cost in head-carrying are consistent with the expectations of a simple model that identifies the mechanical energy losses resulting from inelastic collisions associated with walking (McGeer, 1990; Donelan *et al.*, 2002; Garcia *et al.*, 1998; Kuo, 2001). This analysis provides an alternate explanation of the metabolic saving that comes as a consequence of raising the effective center of mass by placing the carried load high on the body.

The analysis indicates that the geometry of the step-to-step transition, where the support of the body and any load is transferred to the next contact limb, can be altered by positioning the load high on the body – in effect providing a longer functional leg length and smaller step angles – and that this can reduce the energetic loss associated with each step, countering some of the cost associated with carrying the extra load. The influence of the geometry of the movement of the body mass (as indicated by the velocity vector of the CoM) and that of the limbs during contact (as indicated by the ground reaction force vectors) is identified as a key feature responsible for the loss of energy during walking, so it directly suggests how this loss might be mitigated by head load-carrying strategies.

The mass of the body M_{body} and load M_{load} may be considered as two point masses of height (during standing) L_{body} and L_{load} respectively (Figure 15.3A). If walking is treated as a pure inverted pendulum with simple compass gait, then the angle below horizontal α made by the velocity vector V of each center of mass, immediately prior to heelstrike, can be found:

$$\alpha = \sin^{-1}\left(\frac{S}{2L}\right). \tag{15.1}$$

Assuming the gait to be symmetrical about the highest or lowest point, the angles relating to each mass, α_{body} and α_{load} (Figure 15.3B), also describe the inclinations above

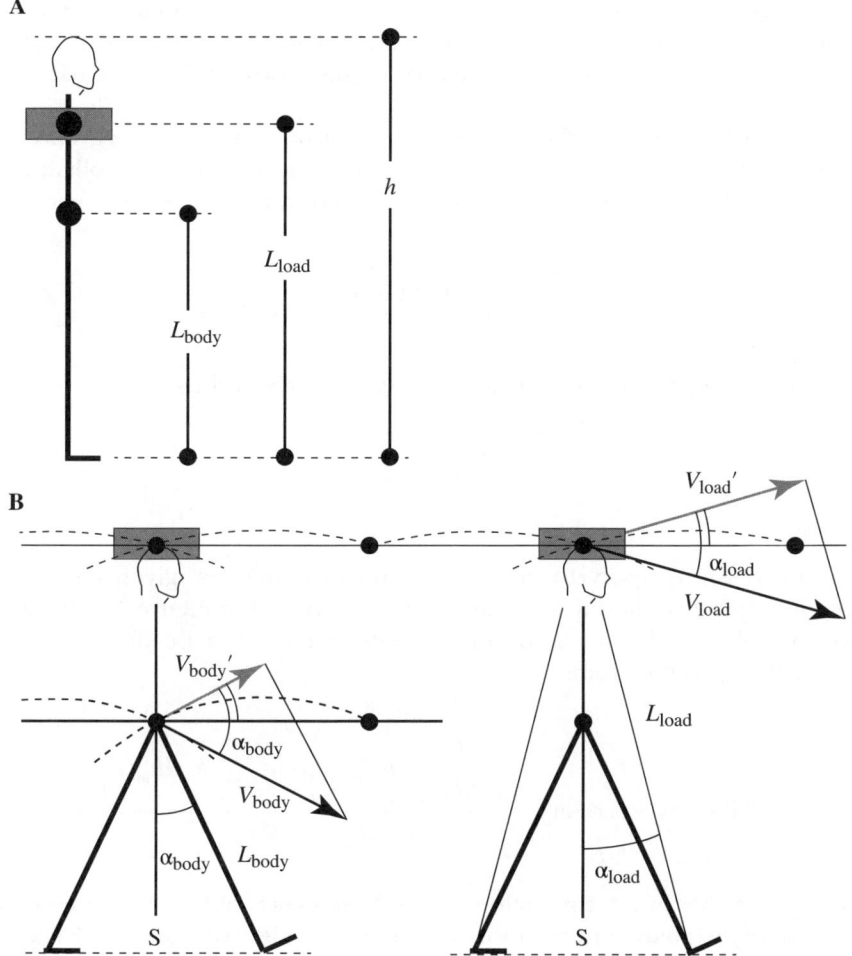

FIGURE 15.3

Parameters (**A**) and construction (**B**) of the geometry of collision losses in bipedal walking while carrying a load.

horizontal for the masses directly after contact. These derivations suppose that the body is stiff at foot contact, and that the paths of each mass can be described as arcs about the contact foot, at least at the moment of contact. It is assumed that reconfiguration within the body (without altering the kinetic or potential energy of the combined center of mass) can be achieved with relatively little cost.

In this scenario, the load mass cannot be maintained directly above the hips (or CoM) without also following an identical path, with radius of leg-length. In this case the load mass would have an equivalent collision geometry (when it is diverted by the limb contact – see Chapters 5 and 6). However, if this "internal" reconfiguration can be realized without prohibitive external or internal power requirements (in a manner analogous to the twisting of a falling cat), an improved, smoother path – and, hence, reduced collision geometry of the load – is perfectly feasible. In effect, having the load well above the hips raises the center of mass of the system, so that the functional leg length is increased, with the center of mass acting as the "functional" hip or center of rotation of a net "virtual" leg. However, to keep track of the collisional implications of a distributed load, the collision geometry of the body and load must be treated separately.

If V is the velocity vector of either mass just prior to foot contact, and V' is the velocity vector immediately after contact, the proportion of the pre-collision velocity remaining in the direction of the new velocity is determined by simple geometry:

$$\frac{|V'|}{|V|} = \cos(2\alpha). \tag{15.2}$$

The ratio of post- to pre- collision kinetic energy KE'/KE is thus:

$$\frac{KE'}{KE} = \left(\frac{|V'|}{|V|}\right)^2. \tag{15.3}$$

The above expressions apply to both body and load masses, given the appropriate "virtual leg lengths" L, denoted with subscripts $_{body}$ and $_{load}$ in Figure 15.3B. Combining the effects of collision for the two components, a term can be derived for "energy recovery" for the whole system:

$$\%\text{Energy Recovery} = \frac{\left(\frac{KE_{body}{}'}{KE_{body}} KE_{body}\right) + \left(\frac{KE_{load}{}'}{KE_{load}} KE_{load}\right)}{KE_{body} + KE_{load}} \tag{15.4}$$

This expression provides a term analogous to those described for the kinetic-potential "pendular" energy exchange in the more conventional view of walking (Cavagna *et al.*, 1976). Unlike the "inverted pendulum" model, collision theory allows predictions to be made for values of energy recovery by accounting for where – and how much – energy is lost, rather than tracking the exchange between available energy forms (Bertram and Hasaneini, 2013). Model values for an individual with parameters intermediate to

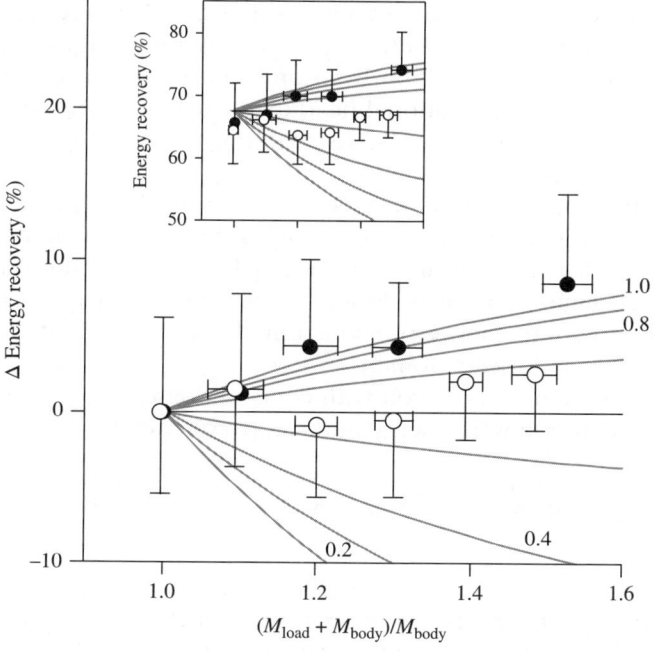

FIGURE 15.4

Recoverable energy as functions of load height and mass. Absolute recoverable energy is shown as an inset; absolute change in percentage recoverable energy from the unloaded state (Δ *Recoverable energy*) as the main panel. Model lines are shown in gray and associated numbers show height of load application as a proportion of body height L_{load}/h. Filled and unfilled circles show values derived from force-plate studies (Heglund *et al.*, 1995) for head-loading in experienced African women and inexperienced subjects respectively. Error bars show one standard deviation.

those described by Maloiy *et al.* (1986): walking velocity = 0.90 ms⁻¹; M_{body} = 53.86 kg; f =1.67 Hz; L_{body}/h = 0.58), are shown in Figure 15.4.

Collision theory, even with the simple geometry described, makes a close prediction for the energy recovery in the unloaded case. Head-loads are predicted to improve energy recovery considerably, and experienced African women (Heglund *et al.*, 1995) appear to utilize a large proportion of the energy recovery made available by the changed geometry of collision-like foot contact. Inexperienced individuals perform poorly, and their energy recovery is no better than that expected for loads added to the center of mass. This may be attributable to losses unrelated to collision; compliance, both due to rotation about the hip and bending of the support leg, would result in negative muscular work, and would reduce the effectiveness of the kinetic-potential energy transfer.

In contrast, experienced African women may be able to utilize the favorable geometry of the collision-like contact, and also avoid negative muscular work, by using a relatively stiff-legged walk: incompliant legs, and little rotation of the body about the hip at foot-contact, enables L_{load} to act as the "virtual leg" applicable for the load, thus improving α, V'/V, and KE'/KE for the load. Although the stride frequency does not appear to change with head loads in these women (Maloiy *et al.*, 1986), some features of stride kinematics do. The transition from one stance limb to the next begins later, and ends earlier, in female occupational head load carriers of the Xhosa tribe of South Africa (Charteris *et al.*, 1989). Such a decrease in transition would indicate an increase in stiffness over the transition process.

A decrease in the transition time results in an increase in single limb stance, which is further increased by a general increase in duty factor, from 0.6 unloaded to 0.65 with a head load (this difference can also be seen in Figure 15.2B, and in Cavagna *et al.* (2002), from which Figure 15.2 was derived). Such a difference in single limb stance may explain the apparent energy transduction differences noted by Cavagna *et al.* (2002) for head loads in African women.

The power requirements associated with collision losses $P_{\text{collision}}$ may be determined given a suitable walking kinetic energy and step frequency f:

$$P_{\text{collision}} = \Delta KE\, f = \left(\left(KE_{\text{body}} - KE_{\text{body}}'\right) + \left(KE_{\text{load}} - KE_{\text{load}}'\right)\right) f \qquad (15.5)$$

$P_{\text{collision}}$ is shown as a function of load height and mass in Figure 15.5. The result for load carried at the top of the head is impressive: reduction in α_{load}, due to a high L_{load}, can potentially reduce the extra energetic cost of carrying the load by 64%. Empirical results derived from previous studies (Maloiy *et al.*, 1986; Jones *et al.*, 1987) show that these saving estimates are excessive for higher loads, if collision losses are considered in isolation as the only energy-saving mechanism. Limitations to the geometry of the simple compass-gait model, leg and body compliance, and variable mechano-chemical efficiencies, are biologically inevitable. However, the principles associated with minimizing collision-loss provide a previously unavailable, and mechanically sound, explanation for what has appeared mysterious for some time. Rather than carrying loads on the head, Nepalese porters bear extremely large loads piled high above their torso. It is possible that their method of transport could also take advantage of center of mass elevation (body plus load). Data from the metabolic cost of load carrying in these porters walking on the level (Bastien *et al.*, 2005b) also fit the model, but at even higher load levels (Figure 15.5).

Numerous cultures who routinely carry head loads master a technique that allows them to efficiently carry these loads, but the benefits seem to elude novice carriers. What is the secret to accomplishing the walking motions that save energy when carrying a head load? We have argued that a technique which alters the geometry of the velocity vector of the CoM of the individual, and the vector of the ground reaction force applied by the lead contact limb, will reduce the energy loss at transition between support limbs. This basic concept is also consistent with gait strategies in quadrupedal mammals (Lee *et al.*, 2011).

But what method is available to accomplish the objective of using the trunk as an extension of the contact limb? During an investigation of optimization strategies for bipedal robotic walking, Wisse *et al.* (2007) demonstrated that a simple "bisecting hip" mechanism, using a spring linkage between the torso and the stance leg of a robot walking bipedally, could increase the energy efficiency of the walking gait. Through the use of this strategy, it was found that raising the CoM with a tall torso, even if it required added mass, decreased the mechanical energy loss incurred by each step.

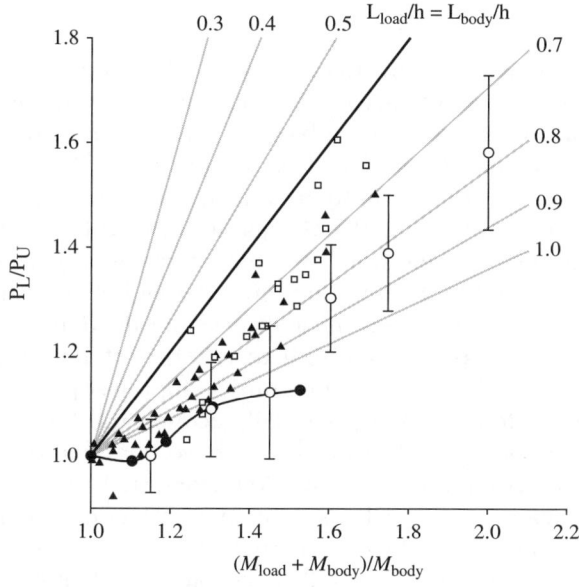

FIGURE 15.5

Proportion of loaded to unloaded power P_L/P_U, as a function of load height and mass. Grey lines relate to modeled $P_{collision}$; associated numbers show height of load application as a proportion of body height L_{load}/h. Bold straight line shows the relationship predicted for loads added at the body center of mass; $P_{collision}$ increases in direct proportion to the load applied. Symbols show data derived from published empirical studies: open squares and filled triangles show metabolic data for two groups of African women carrying loads (Maloiy *et al.*, 1986; Jones *et al.*, 1987; the Jones *et al.* data was converted, assuming that body mass was 129% of the published fat-free mass (their group mean)). Open circles (with standard deviation) are metabolic power values from Nepalese porters (Bastien *et al.,* 2005), and filled circles are mechanical power estimates for African women familiar with head load-carrying from Heglund *et al.* (1995).

This suggests a potential mechanism that could account for the "technique" used by well-trained head-load carriers. It leaves open the possibility that replicating such a mechanism could be achieved on the basis of muscular relationships in human anatomy, resulting in testable hypotheses of changes in activity levels between effective head carrying and normal walking.

We have demonstrated that considering the energy loss associated with the collision-like interactions of the body mass and the stance limb can provide novel insight into energy effective walking strategies. The influence of collision-like interactions on the mechanics of walking may have broad significance; collision costs are reduced with reduced α, associated with elevated center of mass (Figure 15.3), and this may act as a pressure (both behavioral and evolutionary) towards erect posture in animals. Features such as this may have had substantial influence on early hominids, for whom efficiency, presumably, was critical. Consideration of these factors may provide new perspectives on the key innovations that led to the development of the upright, striding walk characteristic of the early human lineage.

15.5 CONCLUSIONS

The mechanism(s) responsible for the load-carrying effectiveness of strategies routinely utilized by numerous cultures worldwide has not yet been established. Several possible mechanisms could contribute to the observed advantages of head load, as in many African and Asian cultures, or high-placed packing as in the Nepalese porters.

Humans naturally alter their gait to minimize energetic expenditure under the circumstances imposed (Bertram, 2005; Gutmann *et al.*, 2006). It is not known why individuals from cultures not conditioned to head-carrying appear unable to adapt to minimizing head load cost. The technique required may be so complex that extensive experience is necessary for the neuromuscular system to identify beneficial options. Alternatively, it may be a case of adapting to a series of limiting features, such as increasing the strength of neck musculature, or developing sophisticated balance capabilities, prior to being able to exploit potentially economical locomotion strategies.

The strategies for reducing load-carrying cost have not been directly tested. Part of the difficulty involves the potential for multiple factors, such as balance technique, and other physiological factors not directly associated with locomotion mechanics (Minetti *et al.*, 2006), to affect the results of experimental analysis. In order to eventually determine the mechanism(s) responsible, it is necessary to identify alternative testable hypotheses capable of explaining the available observations.

▌ REFERENCES

Adamczyk PG, Kuo AD (2009). Redirection of center-of-mass velocity during the step-to-step transition of human walking. *Journal of Experimental Biology* **212**, 2668–2678.

Alexander RMcN (1986). Making headway in Africa. *Nature* **319**, 623–624.

Balogun JA, Robertson RJ, Goss FL, Edwards MA, Cox RC, Metz KF (1986). Metabolic and perceptual responses while carrying external loads on the head and by yoke. *Ergonomics* **29**, 1623–1635.

Bastien GJ, Schepens B, Willems PA, Heglund NC (2005a). Energetics of load carrying in Nepalese Porters. *Science* **308**, 5729.

Bastien GJ, Willems PA, Schepens B, Heglund NC (2005b). Effect of load and speed on the energetic cost of human walking. *European Journal of Applied Physiology* **94**, 76–83.

Bertram JEA (2005). Constrained optimization in human walking: cost minimization and gait plasticity. *Journal of Experimental Biology* **208**, 979–991.

Bertram JEA, Hasaneini SJ (2013). Neglected losses and key costs: Tracking the energetics of walking and running. *Journal of Experimental Biology* **216**, 933–938.

Bohannan P, Bohannan L (1969). Tiv economy. *Science* **164**(3884), 1159–1160.

Cavagana GA, Saibene FP, Margaria R (1963). External work in walking. *Journal of Applied Physiology* **18**, 1–9.

Cavagna GA, Thys H, Zamboni A (1976). The sources of external work in level walking and running. *Journal of Physiology London* **262**, 639–657.

Cavagna GA, Willems PA, Legramandi MA, Heglund NC (2002). Pendular energy transduction within the step in human walking. *Journal of Experimental Biology* **205**, 3413–3422.

Charteris J, Scott PA, Nottrodt JW (1989). Metabolic and kinematic responses of African women headload carriers under controlled conditions of load and speed. *Ergonomics* **32**, 1539–1550.

Chewings C (1936). *Back in the stone age: the natives of central Australia*. Angus & Robertson, Sydney, Australia, pp. 161 (reference appears on pp. 29–30).

Chung MJ, Wang MJ (2010). The change of gait parameters during walking at different percentage of preferred walking speed for healthy adults age 20–60 years. *Gait Posture* **31**, 131–135.

Datta SR, Ramanthan NL (1971). Ergonomic comparison of seven modes of carrying loads on the horizontal plane. *Ergonomics* **14**, 269–278.

Dekeyne RB (1966). Co-operatives of Yega. In: *Orokaiva papers; miscellaneous papers of the Orokaiva of north east Papua*, pp. 53–68. Australian National University, New Guinea Research Unit; Canberra.

Donelan JM, Kram R, Kuo AD (2002). Simultaneous positive and negative work in human walking. *Journal of Biomechanics* **35**, 117–124.

Echarri JJ, Forriol F (2002). Effect of axial load on the cervical spine: a study of Congolese woodbearers. *International Orthopaedics* **26**, 141–144.

Garcia M, Chatterjee A, Ruina A, Coleman M (1998). The simplest walking model: stability, complexity and scaling. *Journal of Biomechanical Engineering* **120**, 281–288.

Goldman RF, Iampietro PF (1962). Energy cost of load carriage. *Journal of Applied Physiology* **17**, 675–676.

Griffin TM (2006). Powering locomotion? It is a loaded question. *Journal of Applied Physiology* **101**, 1273–1274.

Gutmann A, Jacobi B, Butcher MT, Bertram JEA (2006). Constrained optimization in human running. *Journal of Experimental Biology* **209**, 622–632.

Hartigan E, Axe MJ, Snyder-Mackler L (2009). Perturbation training prior to ACL reconstruction improves gait asymmetries in non-copers. *Journal of Orthopaedic Research* **27**, 724–729.

Havryk AP, Gilbert M, Burgess KR (2002). Spirometry values in Himalayan high altitude residents (Sherpas). *Respiratory Physiology & Neurobiology* **132**, 223–232.

Heglund NC, Willems PA, Penta M, Cavagna GA (1995). Energy-saving gait mechanics with head-supported loads. *Nature* **375**, 52–54.

Jones CDR, Jarjou MS, Whitehead RG, Jequier E (1987). Fatness and the energy cost of carrying loads in African women. *Lancet* **XII**, 1331–1332.

Joosab M, Torode M, Prasado Rao PVV (2005). Preliminary findings of the effect of load-carrying on the structural integrity of the cervical spine. *Surgical and Radiologic Anatomy* **16**, 393–398.

Kayser B. Hoppeler H, Claassen H, Cerretelli P (1991). Muscle structure and performance capacity of Himalayan Sherpas. *Journal of Applied Physiology* **70**, 1938–1942.

Keefe DF, Ewert M, Ribarsky W, Chang R (2009). Interactive coordinated multiple-view visualization of biomechanical motion data. *IEEE Transactions on Visualization and Computer Graphics* **15**, 1383–1390.

Kerns V (1984). *Women and the ancestors: Black Carib kinship and ritual*. University of Illinois Press, Urbana, IL., pp. 226 (reference appears on p. 70).

Kuo AD (2001). A simple model of bipedal walking predicts the preferred speed-step length relationship. *Journal of Biomechanical Engineering* **123**, 264–269.

Kuo AD (2007). Six determinants of gait and the inverted pendulum analogy: a dynamic walking perspective. *Human Movement Science* **26**, 617–656.

Kuo AD, Donelan JM, Ruina A (2005). Energetic consequences of walking like an inverted pendulum: step-to-step transitions. *Exercise and Sport Sciences Reviews* **33**, 88–97.

Lee CR, Farley CT (1998). Determinants of the center of mass trajectory in human walking and running. *Journal of Experimental Biology* **201**, 2935–2944.

Lee DV, Bertram JEA, Anttonen J, Ros IG, Harris SL, Biewener AA (2011). The collisional basis of mechanical cost in multi-legged locomotion. *Journal of The Royal Society Interface* **8**, 1480–1486.

Lloyd R, Hind K, Micklesfield LK, Carroll S, Truscott JG, Parr B, Davies S, Cooke C (2009). A pilot study investigation of load-carrying on the head and bone mineral density in premenopausal, black African women. *Journal of Bone and Mineral Metabolism* **28**, 185–190.

Mahbub MH, Laskar MS, Seikh FA, Altaf MH, Inoue M, Yokoyama K, Wakui T, Harada N (2006). Prevalence of cervical spondylosis and musculoskeletal symptoms among coolies of a city of Bangladesh. *Journal of Occupational Health* **48**, 69–73.

Maloiy GMO, Heglund, NC, Prager, LM, Cavagna GA Taylor CR (1986). Energetic costs of carrying loads: have African women discovered an economic way? *Nature* **319**, 668–669.

Malville NJ, Byrnes WC, Lim HA, Basnyat R (2001). Commercial porters of Eastern Nepal: Health status, physical work capacity, and energy expenditure. *American Journal of Human Biology* **13**, 44–56.

Mangin EB, Ariane Skinner E (1921). *Essay on the manners and customs of the Mossi people in the western Sudan*. Augustine Challenul, Paris, pp. 141 (reference appears on p. 76).

Marconi C, Marzorati M, Grassi B, Basnyat B, Colombini A, Kayser B, Cerretelli P (2004). Second generation Tibetan lowlanders acclimatize to high altitude more quickly than Caucasians. *Journal of Physiology* **556**, 661–671.

Marino FE (2008). The evolutionary basis of thermoregulation and exercise performance. *Medicine and Sport Science* **53**, 1–13.

McGeer T (1990). Passive dynamic walking. *International Journal of Robotics Research* **9**, 68–82.

Minetti AE, Formenti F, Ardigo LP (2006). Himalayan porter's specialization: metabolic power, economy, efficiency and skill. *Proceedings of the Royal Society of London B: Biological Sciences* **273**(1602), 2791–2797.

Pandolf KB, Givoni B, Goldman RF (1977). Predicting energy expenditure with loads while standing or walking very slowly. *Journal of Applied Physiology Respirat. Environ. Exer. Physiol.* **43**, 577–581.

Parsons EWC (1936). Milta, town of the souls and other Zapoteco-speaking pueblos of Oaxaca, Mexico. Chicago University Press, Chicago, IL., pp. 590 (reference appears on p. 378).

Putnam P (1948). *The pygmies of the Ituri Forest*. In: Coon CS (ed). A reader in general anthroplogy, pp. 322–334 (reference appears on p. 327). Henry Holt and Co., New York.

Ruina A, Bertram JEA, Srinivasan M (2005). A collisional model of the energetic cost of support work qualitatively explains leg sequencing in walking and galloping, pseudo-elastic leg behavior in running and the walk-to-run transition. *Journal of Theoretical Biology* **237**, 170–92.

Simoneau J-A, Lortie G, Boulay MR, Marcotte M, Thibault M-C, Bouchard C (1985). Human skeletal muscle fiber type alteration with high-intensity intermittent training. *European Journal of Applied Physiology and Occupational Physiology* **54**, 250–253.

Soule RG, Goldman RF (1969). Energy cost of loads carried on the head, hands, or feet. *Journal of Applied Physiology* **27**, 687–690.

Taylor CR (1995). Locomotion. Freeloading women. *Nature* **375**, 17.

Taylor CR, Heglund NC, McMahon TA, Looney TR (1980). Energetic cost of generating muscular force during running: a comparison of large and small animals. *Journal of Experimental Biology* **86**, 9–18.

Turnbill CM (1962). *The forest people*. Simon & Schuster, New York, pp. 295 (reference appears on p. 53).

Vaughan CL, Blaszczyk MB (2008). Dynamic similarity predicts gait parameters for *Homo floresensis* and the Laetoli hominins. *American Journal of Human Biology* **20**, 312–316.

Wisse M, Hobbelen DGE, Schwab AL (2007). Adding an upper body to passive dynamic walking robots by means of a bisecting hip mechanism. *IEEE Transactions on Robotics* **23**, 112–123.

Zani PA, Gottschall JS, Kram R (2005). Giant Galapagos tortoises walk without inverted pendulum mechanical-energy exchange. *Journal of Experimental Biology* **208**, 1489–1494.

CHAPTER SIXTEEN

Would a Horse on the Moon Gallop? Directions Available in Locomotion Research (and How Not to Spend Too Much Time Exploring Blind Alleys)

John E. A. Bertram

*Department of Cell Biology and Anatomy, Cumming School of Medicine,
University of Calgary, CA*

16.1   INTRODUCTION

At the conclusion of a book like this, it might be reasonable to revisit the fundamental question, "What are we trying to accomplish with research into the mechanics of legged locomotion?" As succinctly as I can, I would define our ultimate purpose as: understanding the implications of the form and motions of animals that move across a surface using their limbs. The "implications" referred to here would be performance: speed, stability/maneuverability, endurance, etc., and energetic cost. Understanding these implications, then, would enable the interpretation of the adaptations that allow the variety of terrestrial forms to move within and exploit their environment effectively. That ability to "interpret adaptations" (rather than simply to document

Understanding Mammalian Locomotion: Concepts and Applications, First Edition.
Edited by John E. A. Bertram.
© 2016 John Wiley & Sons, Inc. Published 2016 by John Wiley & Sons, Inc.

them) is a key component of a sophisticated perspective on the evolutionary process, and the role that locomotion has on adapting organisms to their respective survival strategies.

The functional factors listed above (implications) are what need to be understood, but it is important to emphasize that it is the "understanding" that has to be reinforced as the goal. Too often in the history of locomotion research, the identification, description and definition of features has been interpreted as "understanding", when such identification is really only the initial step in the process of placing the form, function and performance in the context of the opportunities and constraints operating in adaptation.

Energy is the currency of life and, as with any prudent investor, animals must spend that currency wisely if bankruptcy is to be avoided. Locomotion is a critical feature of mammalian life but, as a mechanical process, it is a substantial drain on available energy. For example, a migratory ungulate like the wildebeest (*Connochaetes sp.*) is estimated to use approximately 8% of its annual energy intake on locomotion (Pennycuick, 1979). This may seem rather modest, but it should be recognized that this animal invests just over 60% of its total energy on maintaining its basal metabolic processes (keeping all of its organs functioning; calculation based on a basal metabolic rate of 230 W over the course of a year, with a net annual energy intake of 1.2×10^{10} J).

If we assume that the first order of business for an organism is to maintain life, then only 40% of the animal's annual intake of energy is available to spend on processes other than maintaining its metabolism. This means that at least 20% of the animal's *expendable* energy budget goes to locomotion. This will become even greater in times of stress, as in drought, when the net energy input can plummet, and locomotory costs increase as the animal searches more intensely for sparse resources. The less energy invested in locomotion, stride for stride, the more is available for other aspects of survival and reproduction.

These apparently competing factors can be interrelated, where such survival features as avoiding predation and the processes involved with making reproduction possible, like combating competitors (see Chapter 13), may often involve substantial "locomotory" expenditures as well. Locomotion is an activity that is engaged in on a daily basis, so even small savings over the course of each stride can accumulate to represent much larger portions of the daily or annual energy budget. From this perspective, it is possible to see why Alexander (2001) contends that evolution and animal learning are basically "energy-use optimizing processes".

Given the importance of legged locomotion to the survival of terrestrial mammals, and the relatively high energetic investment needed to provide for locomotion, it seems reasonable to expect that numerous important, potentially even crucial, adaptations have taken place in every species (or lineage) where locomotion has an important impact on survival and/or reproductive potential. Up to this time, however, we have been largely left in our investigations and interpretations of the mechanics and energetics of locomotion to work more or less fortuitously, gaining insight into the factors at play almost as much by luck as good management.

Most insights derive from simply observing what animals do (though, admittedly, sometimes quite rigorously, employing high precision techniques). Such "observations" are then compared between different forms, to try to elucidate the "correlations" between form, function and energetics. This can be much like the parable of the group of blind men describing an elephant, each one honestly and accurately describing a portion – with none of the descriptions resembling each other and all basically adding to the confusion (Saxe, 1900). Even though many observations are crafted on the basis of verifiable evidence and presented in good honesty, each contribution to the larger understanding may not end up being of much value. How do we get to "the bigger picture", so that information adds to our understanding, instead of simply providing more confusion?

The answer is to build systematically from the foundation – follow the elephant from the ground up. Given the appropriate resources, of course, a systematic evaluation may start with a valid (complete) description of the elephant, since the elephant is a physical entity. Locomotion, however, is not itself a physical entity but is, instead, a process or "phenomenon", even though it is expressed by physical beings, using their structural components within a material environment. When considering locomotion, where does working "from the ground up" start?

A first step in finding out how to start from a proper foundation is, following the elephant analogy, figuring out "which end is up?" I contend that the field has, for far too long, confused the *mechanism(s)* through which locomotion is produced (and where the most obvious "action" happens, so are most easily observed) with the *phenomenon* of locomotion – where the phenomenon of locomotion is what is actually accomplished. It is the phenomenon of locomotion that ultimately provides the advantages to the animal, largely determines the energetic investments required and defines the basic constraints involved. Effective actualization of the phenomenon of locomotion motivates the development of the specific mechanisms that provide for its achievement. So, what is the "phenomenon" of locomotion that resides as the "root" of the issue?

As it is with all forms of locomotion, whether terrestrial or not, the phenomenon of locomotion remains simply the interaction of the physical organism with its material environment. In the aquatic and aerial environments, it is largely the interaction of the surface of the organism with the dynamics of the surrounding fluid. Motion can be generated through a variety of specific mechanisms, some of which perform better in certain conditions, or provide specific outcomes (endurance versus speed, etc.).

A flying or swimming organism, consequently, represents a "solution" to the problem of physically interacting with its environment. However, there are a variety of "mechanisms" that can be used to accomplish this, and the strategies used in the mechanisms employed affect the opportunities and constraints effective in locomotion. A simple example of this in fish locomotion would be the difference between fish with buoyancy-adjusting swim bladders (most ray-finned fish, Actinopterygii) and those that depend on motion-based lift mechanisms to sustain their position in the water column. In both cases, the phenomenon of locomotion is dominated by the dynamics of moving through the fluid environment while subject to the downward acceleration of gravity. Each of these groups uses a different solution to a portion of this problem, and this alters the possibilities and strategies that exist for each.

Legged locomotion should be considered in the same manner as swimming above – as a "phenomenon" that is accomplished by a range of strategic mechanisms. It is a little less obvious how terrestrial forms "interact with their material environment", but this interaction is the basis of terrestrial locomotion, just as it is for those organisms that function in a fluid-dominated environment (flight or swimming). The movement patterns we observe in legged locomotion, and the mechanisms used to produce them, represent the solution the animal has arrived at to accomplish the phenomenological challenge.

It is necessary to identify the problem(s) the movement patterns solve in order to appropriately interpret what those motions accomplish. Of course, there are many complex problems that come to mind – stability, and its counterpart maneuverability, speed, acceleration, endurance, etc. At the root of these complex demands, however, will be fundamental physical interactions that must be understood as the basis of interpreting the more complex (and potentially subtle) mechanisms implemented to "solve" the higher-level problems.

How, then, should the investigation of terrestrial locomotion be approached? It is easiest to fall into the trap of simply observing, in however sophisticated a manner, because observation is, indeed, relatively (intellectually) easy. It is true that, in certain circumstances (particularly when a field of study has not matured enough), observational studies are valuable (and, in fact, required; see Figure 16.1). However, it is generally only through fortuitous kismet that real insights are stumbled upon through observation alone, especially when a body of observation has already been accumulated. This also holds for "experimental" observations, where circumstances are artificially manipulated so that unusual "observations" can be generated (often dressed up to appear to be an experiment, but not designed to test any real hypothesis). An example of this latter might be providing an organism with a series of obstacles, in order to observe how they negotiate them.

Although new information can be collected by such observations, progress is not usually made simply by asking questions. Rather, progress and insight are gained by proposing answers, then rigorously verifying that the proposed "answers" are correct. Formulating answers requires substantial effort (because if it were easy, someone else would have done it already). It is rare, however, that this "creative" portion of the scientific method is emphasized, and it is almost never discussed in the context of locomotion research (Bertram, 2002).

When the functioning of the system is a mystery, how can proposing "answers" lead to progress? The "answers" I am alluding to are, in fact, an explanation of our understanding of the system of interest. Ultimately, our goal is to understand the physics and physiology of the "system" (not just to publish papers and accumulate an apparent expertise in an area). True experimental science is derived from testing the consequences of a formulated understanding of the system, and this requires that the "answer" should be creatively generated. That is, we formulate an understanding of the world (or that portion of it of current interest), and any understanding has implications or consequences about how it will behave under a given set of circumstances. These "consequences" emerging from a given understanding are just hypotheses, and these can be used to verify that our understanding is not overly flawed.

The hypothesis is a prediction, but it must be derived directly from the ramifications of the understanding of the system in order to be of value (otherwise it is only a "pseudo-hypothesis", and its "test" leads to no verification of a consequence or validation of an understanding). If we believe we understand how a system works (is controlled, limited, depends on, etc.) then, under a specific, defined set of conditions (an experiment), it should be possible to predict what will be the result (this prediction is the hypothesis). If the hypothesis is verified, that tends to validate the understanding. Though an

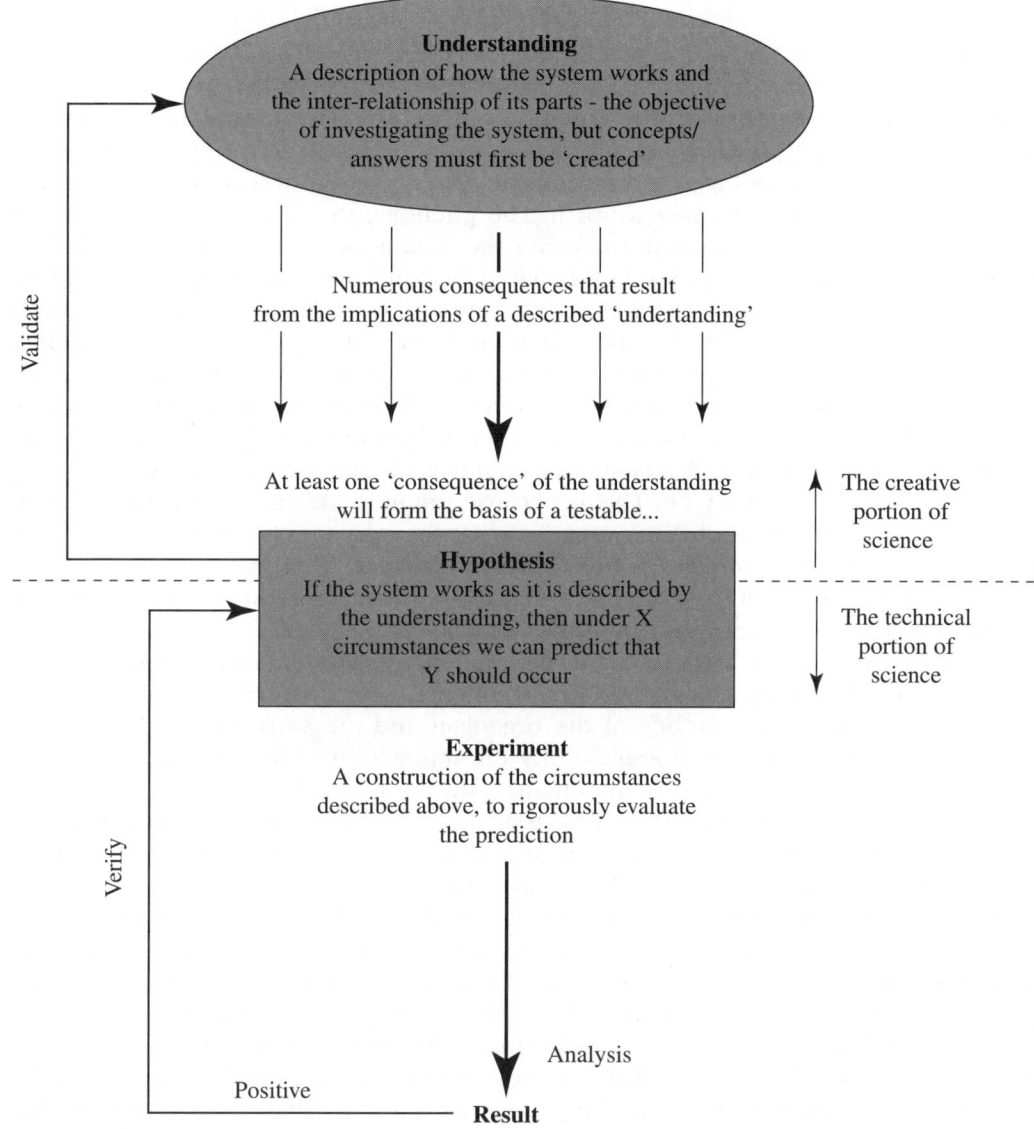

FIGURE 16.1

A diagrammatic representation of the "scientific method" – but one that emphasizes that portion of the process that depends on "creativity". Creativity is not commonly associated with the "hard" sciences, but it is of critical value if true progress is to made. In this scenario, the ultimate goal of scientific investigation is an "understanding" of the system (or portions of the system) of interest. Observation (whether direct observation of organisms in the natural world, or observation of "experiments" interfering with the natural world, then watching the response, or "indirect observations" – those reported by others in the form of the literature) are certainly needed to contribute to the construct of an "understanding", but the development of a *new* understanding depends on "creating" the understanding (by putting the evidence together in new, different – and, sometimes, counter-intuitive – ways).

understanding can never be completely validated (proven), the more different ways it is validated, the less likely it is to be wildly incorrect (Figure 16.1).

How can this approach (i.e., the scientific method) be used to understand locomotion, where so many intricate details make specific "prediction" very complex? As stated above, I advocate the distinction of the "phenomenon of locomotion" from the "mechanisms" that are responsible for it. The phenomenon of locomotion is what is accomplished, while the mechanism(s) are the details of *how* that is accomplished. Unfortunately, given the high level of activity associated with the body parts involved in locomotion, it is very easy to have our attention drawn to the mechanisms (the overt actions) without considering the underlying (and often rather obscure) phenomenon.

It is easy enough to recognize the mechanisms of terrestrial locomotion – this is the subject of substantial rigorous analysis and descriptions of what legged mammals do with their limbs and bodies when in motion. Unfortunately, describing what the limbs do in locomotion, however rigorously, does not usually provide much insight into why those motions occur as they do. This is because the mechanisms are difficult to interpret (understand) outside of the context of what they accomplish – the phenomenon of locomotion. The phenomenon of terrestrial locomotion is progress of the organism through its environment, and involves the consequences of the interaction of the organism with its external (physical) environment.

The phenomenon of locomotion is dominated by different features in different environments. For instance, in fluid environments, it is dominated by drag-and-lift interactions between the surface of the organism and the surrounding fluid. In the terrestrial circumstance, it is dominated by the interaction of masses – the mass that composes the organism and its components, and the mass of the substrate on which the organism moves. Interpreting terrestrial locomotion requires an understanding of the dynamics of solids, just as interpreting aquatic or aerial locomotion requires, fundamentally, an understanding of fluid dynamics.

Details of the phenomenon of locomotion can change as the details of the physical environment change. For instance, moving up a slope is a different locomotion phenomenon than moving on a level surface. This is because moving up a slope requires net positive work to be accomplished, lifting the mass of the organism against the downward acceleration of gravity (steady-speed level locomotion requires active work for several reasons, but the net work on the center of mass is zero). As a result of the requirement of positive net work, the precise mechanisms that operate best in level locomotion will likely not be effective for uphill locomotion. Of course, the organism will be adept at both, so has the ability to (generally rather seamlessly) transition between features of the available mechanisms to suit the locomotion circumstances, even though the details of the internal activity may change. This has been well documented, using the wild turkey as a model (Roberts *et al.*, 2007).

Such changes in operational requirements can be subtle, and the individual may not be aware of them occurring. The example we can most easily identify with is probably the transition in stability of a bicycle. At moderately high speeds, over 4–4.5 m/s (16 km/hr), depending on the particular bicycle, stability is passive; even with a rider, the moving bicycle can reject substantial disturbance without the rider actively participating in the balancing (Meijaard *et al.*, 2007). In contrast, when not moving forward, a bicycle and rider are highly unstable; in order to remain upright on a bicycle at zero forward speed it is necessary for the rider to actively balance, using body positional movements resulting from feedback-driven muscle action.

In between these extremes is a travel speed zone where passive stability can contribute, but is inadequate to stabilize the system fully. As a consequence, the rider must contribute some active stability. So, as experienced bicycle riders slow below 4 m/sec (a relatively fast run, equivalent to a running speed that would give a sub-three hour marathon), they gradually, automatically and largely unconsciously begin supplementing the passive stability of the system with active, muscle-mediated stabilization. In a similar way, we (and other animals) are generally able to transition between different movement challenges without conscious planning. Our understanding of locomotion control would improve considerably through understanding the processes underlying such adaptive response to gradually changing challenges/objectives/opportunities.

In terrestrial locomotion (and other relatively "solid surface" locomotion, such as arboreal), the phenomenon of locomotion is dominated by the interaction of the mass of the organism with the mass of the substrate (see Chapters 5 and 6). There are numerous paths available for the mass to traverse as it moves against gravity and interacts with the substrate. Some of these paths are more favorable than others, for energetic, performance, stability reasons, and so on. The one selected by the organism will depend on the constraints and opportunities acting at that particular moment in time. The center of mass "path" indicates the strategy utilized to have the mass of the organism interact with its supporting surface (which really can be thought of as an interaction of two masses – the organism and the object it is moving on). The "path" is the solid dynamics strategy of locomotion (the phenomenon), while the rest of what we observe – the motion of the limbs, shifts in muscle force and length, and changes in joint position – are the mechanisms that implement that strategy.

If we understood the phenomenon of locomotion, the CoM movement strategies that make for effective locomotion, it would then be possible to generate predictive hypotheses about how the limbs should be used to facilitate these strategies (anticipate the mechanisms that allow the phenomenon). In other words, what we can observe can only be interpreted in the context of what it is trying to accomplish. To a great extent, this issue is largely left out of locomotion as it is currently investigated. However, if we did want to emphasize a search for the "phenomenon" of locomotion, what might this approach look like?

This question brings us to the explanation of the general title of this concluding chapter, "Would a horse on the moon gallop?" This particular question was posed to me by a twelve-year-old science summer camp student. It followed a presentation in which I discussed the influence of gravity on human walking and running (Bertram and Hasaneini, 2013), and explained the mechanics of galloping (Ruina *et al.*, 2005; Bertram and Gutmann, 2009), all on the way to running an activity where I demonstrated that these kids could jump farther, carrying weights in their hands, than they could without (Butcher and Bertram, 2004).

I take it as a testament to the ability of youth to understand complex scientific concepts, if explained in a coherent manner, that my group of adolescents seemed imminently capable of understanding the nuances of what I was telling them about, even to the point of putting the material together in a way that was beyond my ability to answer. One of the students combined two portions of my presentation and asked, "Would a horse on the moon gallop?" I was stumped by a twelve-year-old on her very first introduction to the dynamics of legged locomotion! Yet I purport to be an "expert" in locomotion mechanics, and can indicate that by being able to catalogue how a variety of animals move about in a variety of environments.

What is glaringly missing, however, to achieve the status of a real "expert", is the ability to predict much beyond what my colleagues and I have observed directly. I contend that, at this stage in the history of our field, we should be very much farther along in our understanding of locomotion than we are. It should certainly be possible to understand an important gait such as the gallop, to the point of being able to predict the expected response of a particular species to a range of influences, up to and including those changes in gait strategy expected from the influence of the magnitude of gravitational acceleration. Possibly, on a more practical level, such understanding would also provide insight into changes in strategy that would come from such alternate changes as modification of limb proportion (interpreting morphological adaptation), shifts in performance limits (ultimate speed or endurance), or the advantages and consequences of changes in body size.

As mentioned in the preface of this volume, our limits at this level of understanding, and predictive ability, reside at the conceptual, rather than a technical level. Even if many detailed solutions will require equally detailed analysis of the options available for a system to respond to a given set of conditions (filtered through highly sophisticated modeling), almost all interpretation of such work will depend on a solid conceptual understanding of the factors that primarily influence locomotion strategies. Those are the concepts that will be the foundation of exploring "why" locomotion operates as it does, and seeking, verifying and applying these concepts should be placed at the forefront of our ambitions in locomotion research.

■ REFERENCES

Alexander RMcN (2001). Design by numbers. *Nature* **412**, 591.

Bertram JEA (2002). Hypothesis testing as a laboratory exercise: a simple walking study with a physiological surprise. *Advances in Physiology Education* **26**, 110–119.

Bertram JEA, Gutmann A (2009). The running horse and cheetah revisited: fundamental mechanics of two galloping gaits. *Journal of The Royal Society Interface* **6**, 549–559.

Bertram JEA, Hasaneini SJ (2013). Neglected losses and key costs: Tracking the energetics of walking and running. *Journal of Experimental Biology* **216**, 933–938.

Butcher MT, Bertram JEA (2004). Jump distance increases while carrying hand-held weights: Impulse, history and jump performance in a simple lab exercise. *Journal of Science Education and Technology* **13**, 285–295.

Meijaard JP, Papadopolulos JM, Ruina A, Schwab AL (2007). Linearized dynamics equations for the balance and steer of a bicycle: a benchmark and review. *Proceedings of the Royal Society A* **463**, 1955–1982.

Roberts TJ, Higginson BK, Nelson FE, Gabaldón AM (2007). Muscle strain is modulated more with running slope than speed in wild turkey knee and hip extensors. *Journal of Experimental Biology* **210**, 2510–2517.

Ruina A, Bertram JEA, Srinivasan M (2005). A collisional model of the energetic cost of support work qualitatively explains leg sequencing in walking and galloping, pseudo-elastic leg behavior in running and the walk-to-run transition. *Journal of Theoretical Biology* **237**, 170–192.

Saxe JG (1900). *The poetical works of John Godfrey Saxe.* Cabinet Edition. Houghton, Mifflin and Company. Boston and New York.

Index

Note: Page numbers in *italics* refer to illustrations; those in **bold** refer to tables

Understanding Mammalian Locomotion: Concepts and Applications, First Edition.
Edited by John E. A. Bertram.
© 2016 John Wiley & Sons, Inc. Published 2016 by John Wiley & Sons, Inc.

gestating fetus, metabolic rate, 200, *201*
heat dissipation, 200
heat equilibrium, neonate, 201–202
scaling coefficient, systematic change, 202
surface area-volume relationship, 198
pigmy elephant (*Elephas* (*Paleoloxodon*)
 falconeri), *354*
 dwarfing events, 354
 shoulder height, 355
 tibia, 359
Pinnepedia, 83
pinniped, 329
pitching moment, 95, 165, 282, 288
plantigrade foot posture
 locomotor economy, 335
 placental mammals, 334
pneumatic bulbs, 6
"pogo stick" models, 125
porcupine (*Erithizon dorsatum*), 211
potential energy, 88, 98
power, 68–69
 external mechanical, 99–100
 internal mechanical, 101
 mass specific, 100–101
preferred transition speed, 183
primate, 84, 94
Proboscidia, 83–84
Proboscidians, **196**
pronghorn (*Antilocapra americana*),
 211, 328
pronk, 30, 37–38, 136–137
proximate cause-energetic cost, 154–157
"pseudobipedal" walking machine, 19
pseudo-elastic motion
 cost-effective fast-moving running
 gait, 131
 mechanical spring, 130
 muscle-mediated spring-like motion,
 130–131
 pseudo-elasticity, 131
PTS *see* preferred transition speed
push-off, *121*, 123

quadruped, 31, 37, 95, 101, 111, 124–125,
 132, 133, 135, 158, *159*, 162,
 164, 165
quadrupedal dynamic walking and collisions
 higher speed
 benefits of pacing and trotting, 162
 footfalls in cantering and galloping
 quadrupeds, 162

Kuo's ankle extension, 159
motions of center of mass, *159*
passive, stiff-limbed quadrupedal
 walking., *161*
potential and kinetic energies,
 fluctuations, *160*
reductionist mechanics
 classic free-body diagram, *164*
 cursorial quadruped limbs, 165
 cursorial quadrupeds, 165
 humans power sprinting with
 telescoping legs, 165
 observations, 163
 races, photographs, 163, *163*
 "telescoping", 162
 "wheelie-avoidance" account, 165
 Smith and Berkemeier model, *161*
quadrupedal locomotion *see* quadrupedal
 dynamic walking and collisions
quadrupedal passive dynamics, 20

Rashevsky, N., 7, 15–18, 156, 168, 353
recovery
 energy *see* energy recovery
 ratio calculation
 acceleration/deceleration, 269–270
 definition, 269
 filter function, 270
 gravitational potential energy, 270
 potential energy, 270, *271*
red deer, 334
red kangaroo, 340
reductionist models of walking
 additional biological realism, 148
 ankle extension, 148
 compass-gait, extreme reduction,
 144–147
 "elastic strain energy", 149–150
 foot-foot interchange, compass-gait, 148
 impulses, *149*
 inverted pendulum model, 144
 "kinematic definition", 144
 longest possible step, compass-gait
 model, *147*
 "mechanical definition", 144
 passive, stiff-limbed stance leg, *145*
 prediction of longest walking step length,
 145–146
 SLIP model, 149–150
 springy walking, 148
"regular realizability", concept of, 46